12th International Symposium on Multiscale, Multifunctional and Functionally Graded Materials

(FGM 2012)

AA008857

Journal of Physics: Conference Series Volume 419

Beijing, China
22 October 2012

Editors:

Zhangjian Zhou

ISBN: 978-1-62748-295-0
ISSN: 1742-6588

Printed from e-media with permission by:

Curran Associates, Inc.
57 Morehouse Lane
Red Hook, NY 12571

Some format issues inherent in the e-media version may also appear in this print version.

Printed by Curran Associates, Inc. (2013)

For permission requests, please contact the Institute of Physics at the address below.

Institute of Physics
Dirac House, Temple Back
Bristol BS1 6BE UK

Phone: 44 1 17 929 7481
Fax: 44 1 17 920 0979

techtracking@iop.org

Additional copies of this publication are available from:

Curran Associates, Inc.
57 Morehouse Lane
Red Hook, NY 12571 USA
Phone: 845-758-0400
Fax: 845-758-2634
Email: curran@proceedings.com
Web: www.proceedings.com

TABLE OF CONTENTS

012001 Preparation and Microstructure of Laminated ZrB_2-SiC Ceramics with Porous ZrB_2 Interlayers 1
H Yuan, J Li, Q Shen, L Zhang

012002 Fabrication of Al/Diamond Particles Functionally Graded Materials by Centrifugal Sintered-Casting Method 7
Y Watanabe, M Shibuya, H Sato

012003 Design, Manufacturing and Characterization of Functionally Graded Flextensional Piezoelectric Actuators 11
R C R Amigo, S L Vatanabe, E C N Silva

012004 Functionally Graded Dual-Nanoparticulate-Reinforced Aluminum Matrix Composite Materials 15
H Kwon, G-G Lee, M Leparoux, A Kawasaki

012005 Dynamics of Nonhomogeneous Carbon Steel Plates 19
J-H Kim, L Zhang, R Hebert, J T Wright, A Shukla

012006 Cold Spray and Presureless Sintering of Zirconium Phosphate Bonded Silicon Nitride Ceramics with Porous Gradient Structure 27
F Chen, F Li, Q Shen, L Zhang

012007 Effective Thermoelastic Moduli of FGMs with Temperature-dependent Constituents and Initial Eigenstrains 33
D Boussaa

012008 Precipitation Phases at Different Processes and Heat Treat Ments As Well As Their Effects on the Mechanical Properties of Super-austenitic Stainless Steel 37
H Sun, Z Zhou, M Wang, S Li, L Zhang, L Zou

012009 The Fabrication and Characterization of Polymeric Microcellular Foams with Designed Gradient Density 46
Q Shen, Y Xiong, H Yuan, G Luo, X Liang, L Zhang

012010 Fabrication of Al-W Functionally Graded Impact Material via Vacuum Hot-Pressing Sintering Method 51
P Chen, G Luo, Y Liu, Q Shen, L Zhang

012011 Molecular Dynamic Simulation of Failure of Ettringite 55
W Sun, D Wang, L Wang

012012 Preparation and Properties of ATO Films and Their Effects on the TiO_2/ATO System 61
J A Galaviz-Pérez, F Chen, J R V García, Q Shen, L Zhang

012013 High-temperature Oxidation Behavior of ZrO_2-$ZrSiO_4$/NiCr Composites and Its Applicability for Functionally Graded Materials 65
K Wu, A Rosin, H Park, M Willert-Porada

012014 Microstructure Deformation and Fracture Mechanism of Highly Filled Polymer Composites under Large Tensile Deformation 75
Z J Tao, S D Ping, Z Mei, Z P Cheng

012015 Fracture Analysis of a Viscoelastic Functionally Graded Materials Strip with a Central Crack 82
Z H Wang, L Zhang, L C Guo

012016 The Influencing Mechanism of Modification Layer on the Performance of SiC_3D/Al Multi-function Gradient Composite 86
B Li, Z Ma, Y Wang, F Wang

012017 Preparation and Characterization of PMMA Graded Microporous Foams Via One-step Supercritical Carbon Dioxide Foaming 92
H Yuan, J Li, Y Xiong, G Luo, Q Shen, L Zhang

012018 Fabrication of W/Cu FGM By Aqueous Tape Casting 97
S Liu, Q Shen, G Luo, M Li, L Zhang

012019 Influence of Al_2O_3/YSZ Micro-laminated Coatings on High Temperature Oxidation and Spallation Resistance of MCrAlY Alloys 102
J Yao, Y He, D Wang

012020 Numerical Simulation in Relation to Adiabatic Shearing Behaviors in Titanium Alloy 110
R Gao, Q Fan, F Wang

012021 Microstructural Characterization of the Mg/Cu/Al Diffusion Bonded Joint 117
J Zhang, Q Shen, G Luo, Y Wang, M Li, L Zhang

012022 Study for Double-layered AZO/ATO Transparent Conducting Thin Film 121
M Cao, Y Li, J Yang, Y Chen

012023 Microstructure of Diffusion-Bonded Mg-Ag-Al Multilayer Composite Materials 128
Y Wang, G Luo, J Zhang, Q Shen, L Zhang

012024 Modeling the Dynamic Damage Process of the SiC_{3D}/Al Interpenetrating Phase Composites 132
G Li, Q Fan, Z Wu, X Zhang, Y Wang, F Wang

012025 Mechanical Properties of Tantalum-tungsten Interlayer Between Tungsten Tile and Thimble to Prevent Helium Leak from He-cooled Divertor 138
P Zhang, W Shen, Y Zhou, Q Zhang

012026 The Influence of ALN-Al Gradient Material Gradient Index on Ballistic Performance 145
Y Wang, Q Liu, Y Li, Q Shen

012027 Layered Composite of TiC-TiB_2 to Ti-6Al-4V in Graded Composition by Combustion Synthesis in High-gravity Field 150
X Huang, Z Zhao, L Zhang

012028 Microstructure and Properties of (Ti, Al) N Coatings on WC-Co Cemented Carbide 158
Y Li, Z Liu, J Yang, Y Chen

012029 Fabrication and Mechanical Properties of a 14Cr-ODS Steel 165
H Hu, Z Zhou, L Liao, M Wang, S Li

012030 Preparation and Oxidation of ZrB_2/SiC/$Zr_2Al_4C_5$ Multi-phase Ceramics with Spark Plasma Sintering 170
Q Guo, J Li, Z Ma, Y Nie, Q Shen, L Zhang

012031 Fabrication and Characterization of Multiscale Graded SMAT-MAO Composite Coating Formed on the Surface of 2024 Al Alloy 175
L Wen, Y Wang, Y Jin, D Sun

012032 Fabrication of ATO/Graphene Multi-layered Transparent Conducting Thin Films 183
N Li, F Chen, Q Shen, C Wang, L Zhang

012033 Study on the Preparation and Properties of Colored Iron Oxide Thin Films 188
X Zhao, C Li, Q Liu, Y Duan, J He, S Liu, H Wang, S Liang

012034 Discrete Element Analysis of Sedimentary Body Density of Functionally Graded Materials Derived from Particle Co-Sedimentation 197
J Ye, G Luo, A Yang, Q Shen, L Zhang

012035 Structure and Nanomechanical Properties of $Al_{1-x}Sc_xN$ Thin Films 205
J Yang, Y Li, M Cao, Y Chen

012036 Microstructure and Mechanical Properties of 16 Cr-ODS Ferritic Steel for Advanced Nuclear Energy System 211
S Li, Z Zhou, M Wang, H Hu, L Zou, G Zhang, L Zhang

012037 Effect of Strain Rate on the Plastic Strain Gradient Beneath the Deformed Surface of Iron 218
X Zhang, K Matsuura, M Ohno

012038 Thermal-mechanical Analysis on W/CuCrZr Plasma Facing Component with Functionally Graded Material Interlayer 222
D H Zhu, J L Chen, Z J Zhou, R Yan

012039 Development of Thermal Spraying and Coating Techniques by Using Thixotropic Slurries Including Metals and Ceramics Particles 229
S Kirihara, Y Itakura, S Tasaki

012040 Study on Cd Vacancy in CdZnTe Crystal by Positron Annihilation Technology 233
W Liu, J Min, X Liang, J Zhang, X Sun, L Wang, A Ran, B Ye

012041 Thermal Residual Stresses in SiC Particle Reinforced Aluminium Composites: A Study by the Taylor-based Nonlocal Theory of Plasticity 239
D F Cao, X Liu, H Mei, Z T Lei, L S Liu

012042 Effect of Particle Shapes on Effective Strain Gradient of SiC Particle Reinforced Aluminum Composites 247
X Liu, D F Cao, H Mei, Z T Lei, L S Liu

012043 The Application of Ti/Al_2O_3 Composite in W-Mo-Al System Wave Impedance Graded Flier-plate 254
Z Wang, S Feng, G P Shi, Q Shen, L M Zhang

012044 Numerical Simulation of the Dynamic Performance of the Ceramic Material Affected by Different Strain Rate and Porosity 259
Z Wang, L S Liu, H Mei, X Lai, P C Zhai, D F Cao

012045 Multiscale Numerical Simulation of the Shaped Charge Jet Generated from Tungsten-Copper Powder Liner 268
J Liu, H Cai, F Wang, Q Fan

012046 A Study on Propagation Characteristic of One-dimensional Stress Wave in Functionally Graded Armor Composites 276
S Y Yang, X Liu, D F Cao, H Mei, Z T Lei, L S Liu

012047 Synthesis and Characterization of Dopamine Graft Compound N-methacryloyl 3,4-dihydroxyl-phenylamine 282
X Xiong, S Qu, Y Liu

012048 Functionally Graded Materials using Plasma Spray with Nano Structured Ceramic 289
E L Sioh, A I Y Tok

012049 Analysis of Mechanical Properties of Nanocrystalline Al+α-Al_2O_3 Composites Using Molecular Dynamics Simulation 298
H Mei, L S Liu, X Lai, P C Zhai

012050 Ceramic/Metal Composites with Positive Temperature Dependence of Thermal Conductivity 306
J Li, Q Yu, W Sun, R Zhang, D Ichigozaki, K Wang, J-F Li

012051 Thermal Stress Analysis of W/Cu Functionally Graded Materials by Using Finite Element Method 312
Z Yang, M Liu, C Deng, X Zhang, C Deng

012052 Processing and Structure of a Nitrogen Alloyed Oxide Dispersion Strengthened Austenitic Stainless Steel by Mechanical Alloying 318
Y Xu, Z Zhou

012053 Dielectric Propertiies of the Compositionally Graded $BaZr_xTi_{2-x}O_5$ Thin Films Prepared by Sol-gel Method 324
Y P Gong, D Y Guo, C B Wang, Q Shen, L M Zhang

012054 Numerical Investigation on Anti-penetration Behavior of Ceramic/Metal Target Under Ballistic Impact 328
H Mei, Y C Wang, X Liu, D F Cao, L S Liu

012055 Stress Wave Attenuation in SiC_{3D}/Al Composite 334
C Yuan, Y Wang, G Li, X Zhang, J Gao

012056 The Photocatalytic Activity of Nano-TiO_2 (anatase) Gradient Coating on Tourmaline Beads 340
F Huang, Y Fan, J Sun

012057 Research on Self-propagating High Temperature Synthesis Prepared ZrC-ZrB_2 Composite Ceramic 345
Y Cheng, X Su, G Hou, Y Xing

Author Index

PREFACE

The 12th International Symposium on Multiscale, Multifunctional and Functionally Graded Materials (FGM-2012) was held in Beijing, China, from 22–36 October 2012. This was part of a series of conferences organized every two years endorsed by International Advisory Committee for FGM's, which serves as a forum for scientists, educators, engineers and young students interested in the development of functionally graded materials (FGM). The series continues from the previous international symposium on FGM held in Sendai, Japan (1990), San Francisco, USA (1992), Lausanne, Switzerland (1994), Tsukuba, Japan (1996), Dresden, Germany (1998), Estes Park, USA (2000), Beijing, China (2002), Leuven, Belgium (2004), Hawaii, USA (2006), Sendai, Japan (2008) and Guimaraes, Portugal (2010).

Functionally graded materials are non-uniform materials which are designed with embodied continuous spatial variations in composition and microstructure for the specific purpose of adjusting their thermal, structural, mechanical, biological or functional response to specific application conditions. Such multi-phase materials cover a range of space and time scales, and are best understood by means of a comprehensive multiscale, multiphysics approach. These kinds of materials are presently in the forefront of materials research, receiving worldwide attention. They have a broad range of applications including for example, biomedical, biomechanical, automotive, aerospace, mechanical, civil, nuclear, and naval engineering. New applications are continuously being discovered and developed.

The objective of the FGM-2012 intends to provide opportunities for exchanging ideas and discussing state-of-the-art theories, techniques and applications in the fields of multiscale, multifunctional and FGM, through invited lectures, oral and poster presentations.

FGM-2012 was organized and hosted by University of Science and Technology Beijing, China, together with Tsing-hua University and Wuhan University of Technology, and was supported by National Natural Science Foundation of China. On behalf of the organizing committee of FGM-2012, I express my great appreciation to their support of the symposium.

Nearly 100 scholars and students from Japan, Brazil, Germany, Russia, United States of America, Portugal, France, Switzerland, Turkey, Singapore, China, and so on, attended FGM-2012, and 57 of the presented papers were collected and selected for publication. The subjects of these papers cover simulation and characterization, advanced fabrication technology, novel application of FGM and layer materials. I cordially thank all of the authors and attendees for their support, and my appreciation is also given to the advisory committee, organizing committee, and the conference volunteers for their hard work.

Professor Zhangjian Zhou
Proceedings Editor
Beijing, December 2012

Committees

International Advisory Committee

Professor Glaucio H Paulino, USA
Professor Marek-Jerzy Pindera, USA
Professor Jeong-Ho Kim, USA
Professor Emer Fazil Erdogan, USA
Professor Dr Monika Willert-Porada, Germany
Professor Emer Wolfgang G J Bunk, Germany
Professor Omer Van Der Biest, Belgium
Professor Michael M Gasik, Finland
Professor Evgeny Levashov, Russia
Professor Lianmeng Zhang, China
Professor Qingjie Zhang, China
Professor Wei Pan, China
Professor Chang-Chun Ge, China
Professor Jing-Feng Li, China
Professor Zhangjian Zhou, China
Associate Professor Serkan Dag, Turkey
Professor Fernando A Rochinha, Brazil
Professor Emilio C N Silva, Brazil
Professor Luis August Rocha, Portugal
Dr Sasa Novak, Slovenia
Dr Masayuki Niino, Japan
Professor Akira Kawasaki, Japan
Professor Ichiro Shiota, Japan
Dr Akinaga Kumakawa, Japan
Dr Yoshikazu Shinohara, Japan
Professor Kiyotaka Matsuura, Japan
Professor Yoshinari Miyamoto, Japan
Professor Takashi Goto, Japan
Professor Yoshimi Watanabe, Japan
Professor Kazuhiro Hasezaki, Japan
Professor Soshu Kirihara, Japan
Professor Emer Toshio Hirai, Japan
Mr Choji Endou, Japan
Dr Seiichi Uemura, Japan

Local Organizing Committee

Professor Changchun Ge (Chairman)
Professor Xuanhui Qu
Professor Jingfeng Li
Professor Wei Pan
Professor Lianmeng Zhang; Professor Qingjie Zhang
Professor Qiang Shen
Professor Jianghong Gong
Dr Xianhui Li Dr Min Xia

All papers published in this volume of *Journal of Physics: Conference Series* have been peer reviewed through processes administered by the proceedings Editors. Reviews were conducted by expert referees to the professional and scientific standards expected of a proceedings journal published by IOP Publishing.

Preparation and Microstructure of Laminated ZrB_2-SiC Ceramics with Porous ZrB_2 Interlayers

H Yuan[1,2], J Li[1], Q Shen[1], and L Zhang[1]
[1] State Key Laboratory of Advanced Technology for Materials Synthesis and Processing, Wuhan University of Technology, Wuhan 430070, PR China
[2] Beijing Industrial Technology Research Institute, Beijing 101111, PR China
E-mail: yuan.huiping@whut.edu.cn

Abstract. Laminated ZrB_2–SiC ceramics with porous ZrB_2 interlayers were successfully prepared by Spark plasma Sintering (SPS)/tape casting, laminating and Reactive Spark Plasma Sintering (RSPS). ZrB_2–SiC plates were sintered by SPS. And ZrO_2-B_4C sheets were prepared by tape casting, which will undergo reducing reaction that produces porous ZrB_2 interlayers during SSPS processing. The ZrB_2-SiC plates and ZrO_2-B_4C sheets were laminated alternately and reactive sintered to laminated ceramics with porous ZrB_2 interlayers. The porosity and thickness of the porous layers were controlled by the tape casting processing and sintering pressure. The microstructure of the laminated ZrB_2-SiC ceramics was analysed and discussed in details.

Introduction

Zirconium diboride (ZrB_2) is a member of a family of materials known as ultra-high-temperature ceramics (UHTCs) [1-4]. Several carbides and nitrides of the group IVB and VB transition metals are also considered UHTCs based on melting temperatures in excess of 3000 °C and other properties [1, 5-7]. In addition to high melting temperatures, ZrB_2 has a unique combination of chemical stability, high electrical and thermal conductivities, and resistance to erosion/corrosion that makes them suitable for the extreme chemical and thermal environments associated with hypersonic flight, atmospheric re-entry, and rocket propulsion [8-11].

Thermal conductivity value of single-crystal ZrB_2 is 140 $W \cdot m^{-1} \cdot K^{-1}$ in the basal direction and 100 $W \cdot m^{-1} \cdot K^{-1}$ along the *c*-axis [1, 2]. High thermal conductivity, which leads to low temperature difference between the surface and the bottom of ceramic components, is important for applications such as thermal insulation [12]. In addition to thermal properties, the low fracture toughness directly related to the limited mechanical reliability has long prevented ZrB_2 ceramic from being used in wide structural applications. [13, 14]

Porous materials have advantages, for example, low thermal mass, low thermal conductivity, low density and high specific strength [15-20]. Porous ZrB_2 ceramics are expected to have low density and thermal conductivities. Ronald Loehmen et al. [21] reported that porosity had the greatest effect on the thermal diffusivity and thermal conductivity. In comparison with the ZrB_2-SiC ceramics, the diffusivities of the porous samples are very small. Based on these results, the ceramics designed with low porosity may present low thermal conductivity, and good thermal shock resistance. According to

[1] To whom any correspondence should be addressed.

Published under licence by IOP Publishing Ltd

Chen Yuan *et al.* [22], the pores appear to improve the resistance to thermal shock, because porosity decreases the elastic modulus of the material and hence reduces the driving force for cracking.
The design of ceramic laminate to enhance the fracture toughness of ceramics is considered one of the effective strategies. [13, 14] Previous work has demonstrated the benefits of laminating either strong ceramic layers or fibers with crack deflecting interlayers to produce ceramic components with enhanced resistance in application where there are high thermal loads. [23] In the case of layered ceramics, the low-energy paths for crack propagation in the porous interlayers can be controlled to improve their mechanical properties. [24, 25]
In our former work, Zirconium oxide (ZrO_2) and boron carbide (B_4C) were used to produce ZrB_2 porous ceramic, and the reaction process was investigated in details [26]. And then ZrO_2 and B_4C were added to ZrB_2 raw powders to prepare ZrB_2 porous ceramics by Reactive Spark Plasma Sintering (RSPS). The porosity could be controlled by changing the ratio of synthesized ZrB_2 to raw ZrB_2 powders [27].
In the present work, ZrB_2–SiC plates were sintered by SPS. And ZrO_2-B_4C sheets were prepared by tape casting. Laminated ZrB_2–SiC ceramics with porous ZrB_2 interlayers were successfully prepared by ZrB_2–SiC plates and ZrO_2-B_4C sheets, which will undergo reducing reaction that produces porous ZrB_2 interlayers during SSPS processing. The Microstructure of the laminated ZrB_2-SiC ceramics was investigated in details.

Experimental

The raw materials used in the present study are as follows: commercial monolithic ZrO_2 (monolithic ZrO_2, no stabilizers, particle size ~0.1 μm, Jiaozuo Weina Co., Ltd, Henan, China), B_4C(particle size <10μm, Alfa Aesar Co., Ltd.), ZrB_2 (particle size ~3 μm, Alfa Aesar Co., Ltd.), and SiC (particle size 5.5 μm, Shandong Weifang Huamei Co., Ltd, Shandong, China).

The designed compositions of the laminated ZrB_2–SiC ceramics with porous ZrB_2 interlayers were 80 vol.% ZrB_2 and 20 vol.% SiC. The powder mixtures of ZrB_2 and SiC were grinded using alcohol as the solvent in agate mortar, followed by rotating evaporation to remove the solvent. After evaporation, the obtained mixtures were sintered to ZrB_2-SiC ceramic plates in graphite die under vacuum (~10 Pa) by SPS (Model SPS-1050, Sojitz Machinery Corporation, Japan). The sintering temperature is 1800°C at a heating rate of 100 °C per minute from 600–1800 °C, and a hold time of 3 min at 1800 °C with 30 MPa Pressure, and the heat treatment temperature is 1600 °C with a hold time of 10 min, as seen in Figure 1. The pressure of ~30 MPa at 1800 °C and ~2 MPa at 1600 °C was applied during the sintering processing. Changes in the vacuum level of the chamber and shrinkage of the powder were monitored and recorded during the sintering.

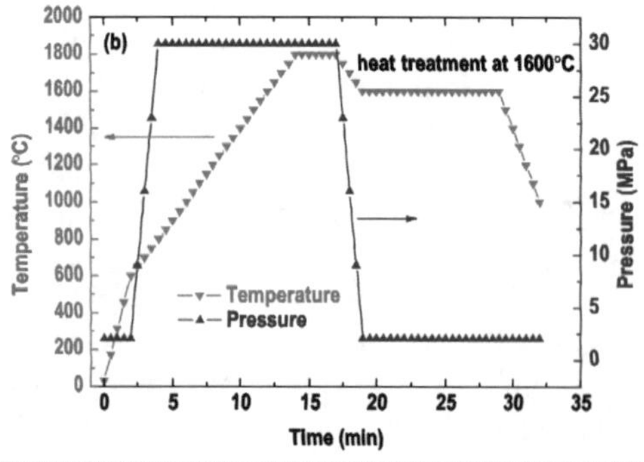

Figure 1. SPS sintering of ZrB_2-SiC ceramic plates

The powder mixtures of ZrO_2 and B_4C (with the mole ratio of 7:5 according to reaction (1)) were ball-mixed for 2 h in a polyethylene bottle using agate balls and ethanol as the grinding media, followed by adding polyvinyl butyral resin (PVB), polyethylene glycol-6000 (PEG-6000), EYPS-3 and acetone as adhesive, plasticizer, disperser and solvent. Then the slurries were ball milled for another10 h. The mass ratio of ZrO_2-B_4C mixture to PVB, PEG-6000, EYPS-3, ethanol and acetone is 10:1:1:0.4:6:6 in slurries. The ZrO_2-B_4C sheets were formed by tape casting.

$$7\ ZrO_2 + 5\ B_4C \rightarrow 7\ ZrB_2 + 3\ B_2O_3\ (l) + 5\ CO\ (g) \quad (1)$$

The ZrB_2-SiC ceramic plates and ZrO_2-B_4C sheets were alternately stacked in a graphite die and sintered to laminated ZrB_2-SiC ceramics with porous ZrB_2 layers by SPS at 1600 °C for 10 min. Different sintering pressure was applied to control the thickness and porosity of the porous ZrB_2 layers. The phase analysis of the ZrB_2-SiC ceramic plates, ZrO_2-B_4C sheet and laminated ZrB_2-SiC ceramics with porous interlayers was carried out using X-Ray Diffraction (XRD, Model Ultima III, Rigaku, Japan). The microstructure features of laminated ZrB_2-SiC ceramics were observed by Scanning Electron Microscope (SEM, Model 4300N, Hitachi, Japan) with simultaneous chemical analysis by energy dispersive spectroscopy (EDS, Hitachi, Japan).

Results and discussion

1.1. Preparation

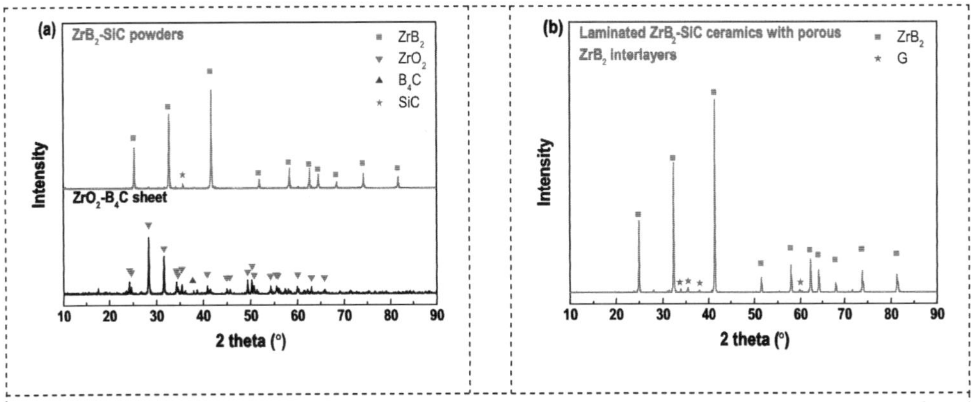

Figure 2. XRD spectra of ZrB_2-SiC powders, ZrO_2-B_4C sheet (a) and the side of the laminated ZrB_2-SiC ceramics with porous ZrB_2 interlayers (b).

The laminated ZrB_2-SiC ceramics were prepared from stacked ZrB_2-SiC plates/ZrO_2-B_4C sheets body. Figure 1 shows the XRD patterns of the ZrB_2-SiC powders, ZrO_2-B_4C and laminated ZrB_2-SiC. According to the XRD results, only the ZrB_2 and SIC phases were detected in the laminated ceramics. The EDS results show that there exists B, Zr, Si and C on the surface that perpendicular to the laminas. It is obvious that the reduction between ZrO_2 and B_4C (reaction (1)) was complicated after the RSPS process at 1600 °C [26].

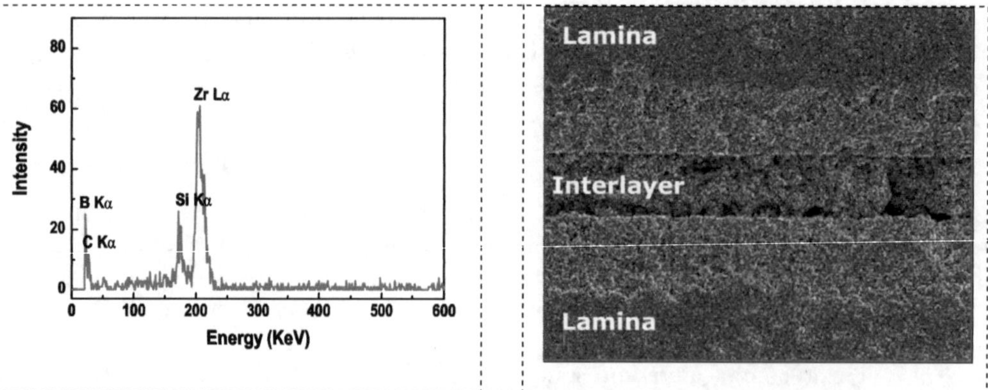

Figure 2. EDS of ZrB_2 porous interlayer and ZrB_2-SiC laminates.

1.2. Microstructure

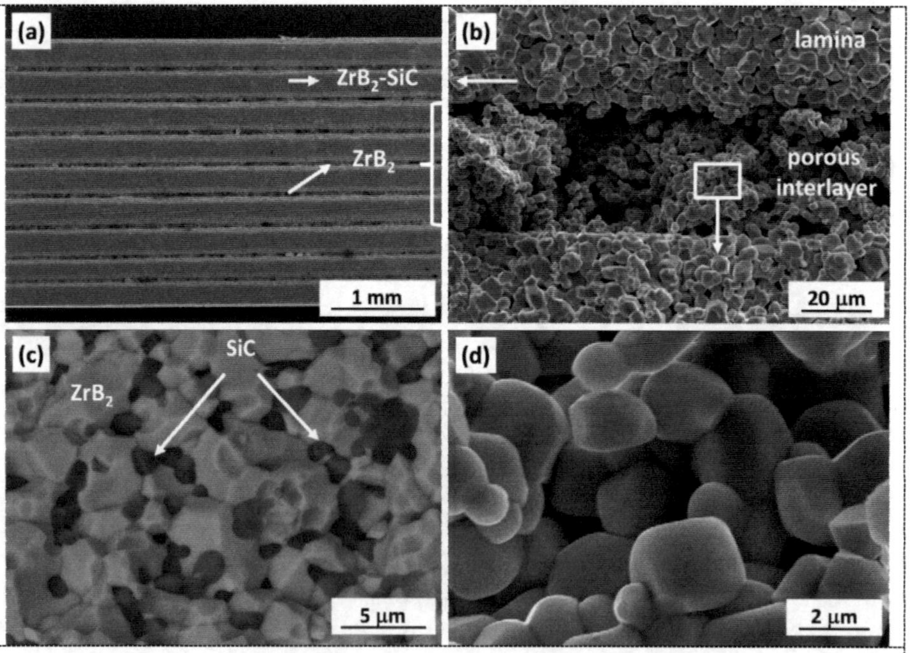

Figure 3. Macro-and microstructure of cross-section of the laminated ZrB_2-SiC ceramic (a-b), ZrB_2-SiC lamina (c) and the ZrB_2 porous interlayer (d).

Figure 3 (a) shows the cross-section of the laminated ZrB_2-SiC ceramic. The uniform, aligned layers in the ceramics are perpendicular to the sintering direction. The alternating interlayers between the ZrB_2-SiC layers and porous ZrB_2 layers are readily observed and the uniform thickness of the different layers is also detected. The porous ZrB_2 layers are bright and thin, while the ZrB_2-SiC layers appear as thick and dark layers separating as the porous layers. It is confirmed by SEM analysis that the thickness of porous layers is about 36 μm.

The microstructure of the ZrB_2-SiC layers is shown in figure 3 (c). The grey phase and black phase is ZrB_2 and SiC, respectively. The SEM image indicates that the SiC grains have been homogeneously

dispersed in the layers. The average thickness and relative density of the ZrB_2-SiC layers is about 245 μm and 92.07%, respectively.
The thickness of ZrO_2-B_4C tape can be controlled by the tape casting process. The average thickness of ZrO_2-B_4C tape used in this research is 192 μm, and the average mass of ZrO_2-B_4C sheets is 28.1 mg. The mass and the porosity of the porous interlayers can be calculated from the mass of the ZrO_2-B_4C sheet according to reaction (1) and the thickness of the porous layers. The porosity of the ZrB_2 interlayers in figure 3 is ~42%, as seen in Figure 3 (b). The grain size of the ZrB_2 in the porous interlayers is about 1~2 μm, figure 3 (d), which is much smaller than that of ZrB_2-SiC laminas, because the ZrB_2 from the reaction between ZrO_2 and B_4C is small than the commercial ZrB_2 particles in the ZrB_2-SiC laminas. This is investigated in our former work [26, 27].
Higher sintering pressure was applied during the RSPS process to adjust the microstructure of the porous interlayers. The thickness and porosity of the porous layers of the samples in figure 4 is ~21 μm and 69%, which is thinner and lower than that of figure 3. The thickness and porosity of the ZrB_2 porous interlayers can be controlled by the pressure during the RSPS process. The thickness of the porous layers decreases with the increase of sintering pressure.

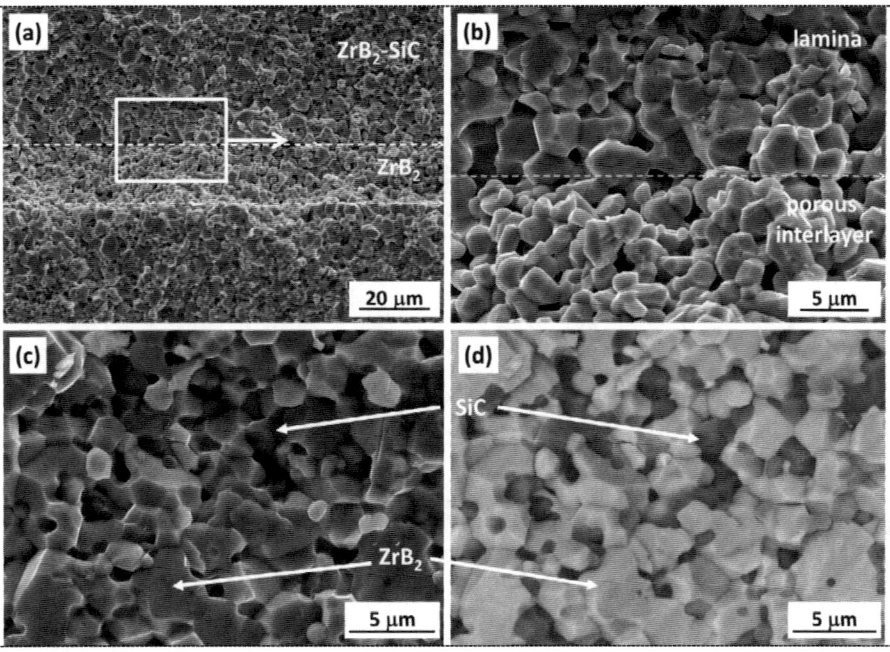

Figure 4. Microstructure of cross-section of the laminated ZrB_2-SiC ceramic (a-b) and ZrB_2-SiC lamina (c-d).

Conclusions

Laminated ZrB_2-SiC ceramics with porous ZrB_2 interlayers were prepared successfully by Reactive Spark Plasma Sintering using the stacking ZrB_2-SiC plates/ZrO_2-B_4C sheets body at 1600 °C 10 minutes holding. The ZrB_2-SiC plates were prepared by Spark Plasma Sintering at 1800 °C, and the ZrO_2-B_4C sheets were prepared by tape casting from slurry, respectively. ZrO_2-B_4C sheets transferred to porous ZrB_2 interlayers in laminated ZrB_2-SiC ceramics, because ZrO_2 react with B_4C to produce ZrB_2 during the RSPS process. The laminated ZrB_2-SiC ceramics with porous ZrB_2 interlayers have uniform thickness of the different layers. The average thickness and relative density of the ZrB_2-SiC layers is about 245 μm and 92.07%, respectively, and the SiC grains have been homogeneously dispersed in the layers. Laminated ZrB_2-SiC ceramics with porosity ZrB_2 interlayers of 42% and 69% relative density were prepared in this research. The thickness and porosity of ZrB_2 interlayers can be controlled by the sintering pressure. As the sintering pressure increase, the thickness of the porous layers decreased and the porosity increased.

Acknowledgments
This work was supported by the Fundamental Research Funds for the Central Universities (2010-ZY-CL-070) and the 111 Project (B13035).

References

[1] Fahrenholtz WG, Hilmas GE, Talmy IG, Zaykoski JA 2007 *J. Am. Ceram. Soc.* **90** 1347
[2] Zimmermann JW, Hilmas GE, Fahrenholtz WG, Dinwiddie RB, Porter WD, Wang H 2008 *J. Am. Ceram. Soc.* **91** 1405
[3] Gupta N, Mukhopadhyay A, Pavani K, Basu B. 2012 *Mater. Sci. Eng.* A **534** 111
[4] Zhang CJ, Kong FT, Xiao SL, Zhao ET, Xu LJ, Chen YY. 2012 *Mater. Sci. Eng.* A **548** 152
[5] Fahrenholtz WG, Hilmas GE, Zhang SC, Zhu S 2008 *J. Am. Ceram. Soc.* **91** 1398
[6] Patel M, Reddy JJ, Bhanu Prasad VV, Subrahmanyam J, Jayaram V 2012 *Mater. Sci. Eng.* A **535** 189
[7] Snyder A, Bo Z, Hodson S, Fisher T, Stanciu L 2012 *Mater. Sci. Eng.* A **538** 98
[8] Van Wie DM, Drewry JR DG, King DE, Hudson CM 2004 **39** 5919
[9] Opeka MM, Talmy IG, Zaykoski JA 2004 *Mater. Sci. Eng.* A **39** 5887
[10] Jackson TA, Eklund DR, Fink AJ 2004 *Mater. Sci. Eng.* A **39** 5905
[11] Tian W, Kan Y, Zhang G, Wang P 2008 *Mater. Sci. Eng.* A **487** 568
[12] Wang Z, Hong C, Zhang X, Sun X, Han J 2009 *Mater. Chem. Phys.* **113** 338
[13] Zhang X, Zhou P, Hu P, Han W 2011 *J. Am. Ceram. Soc.* **31** 2415
[14] Wei C, Zhang X, Hu P, Han W, Li S 2012 *Inter. J. Ref. Metal. Hard. Mater.* **30** 173
[15] Kingery WD, Bowen HK, Uhlmann DR 1976 *Introduaction to Ceramics* (New York: London John Wiley & Sons.)
[16] Loeb AL 1954 *J. Am. Ceram. Soc.* **37** 96
[17] Kingery WD, Quarrie MC 1954 *J. Am. Ceram. Soc.* **37** 67
[18] Francl J, Kingery WD 1954 *J. Am. Ceram. Soc.* **37** 99
[19] Litovsky EY, Shapiro M 1992 *J. Am. Ceram. Soc.* **75** 3425
[20] Litovsky E, Shapiro M, Shavit A 1996 *J. Am. Ceram. Soc.* **79** 1366
[21] Loehman R, Corral E, Dumm HP, Kotula P, Tandon R 2006 *Sandia report SAND 2006-2925*
[22] Chen Y, Vandeperre LJ, Stearn RJ 2008 *J. Mater. Sci* .**43** 4099
[23] Blanks K, Kristoffersson A, Carlstron E, Clegg W 1998 *J. Am. Ceram. Soc.* **18** 1945
[24] Pavese M, Fino P, Ortona A, Badini C 2008 *Ceram. Inter.* **34** 197
[25] Lugovy M, Slyunyayev V, Orlovskaya N, Blugan G, Kuebler J, Lewis M 2005 *Acta Mater.***53** 289
[26] Yuan H, Li J, Shen Q, Zhang L 2012 *Inter. J. Ref. Metal. Hard. Mater.* **34** 3
[27] Yuan H, Li J, Shen Q, Zhang L 2012 *Inter. J. Ref. Metal. Hard. Mater.* **accept**

Journal of Physics: Conference Series **419** (2013) 012002 doi:10.1088/1742-6596/419/1/012002

Fabrication of Al/Diamond Particles Functionally Graded Materials by Centrifugal Sintered-Casting Method

Yoshimi Watanabe[1] **Masafumi Shibuya and Hisashi Sato**
Department of Engineering Physics, Electronics and Mechanics,
Nagoya Institute of Technology, Gokiso-cho, Showa-ku, Nagoya 466-8555, Japan

E-mail: yoshimi@nitech.ac.jp

Abstract. The continuous graded structure of functionally graded materials (FGMs) can be created under a centrifugal force. Centrifugal sintered-casting (CSC) method, proposed by the authors, is one of the fabrication methods of FGM under centrifugal force. This method is a combination of the centrifugal sintering method and centrifugal casting method. In this study, Al/diamond particle FGM was fabricated by the proposed method.

1. Introduction

The continuous graded structure of the functionally graded materials (FGMs) can be created by a centrifugal force. As one of fabrication methods of FGM under the centrifugal force, centrifugal mixed-powder method was developed [1, 2]. However, the mixed powder moves along the flow of the molten matrix metal, when the molten metal is poured into the spinning mold. To overcome this problem, novel FGM fabrication method, centrifugal sintered-casting (CSC) method, is proposed. The schematic illustration of CSC method is shown in **figure 1**. At first, ring-shaped metal-matrix preform including dispersion particles is produced by centrifugal sintering. Powder mixture of dispersion particles and matrix metal particles is inserted into a spinning mold, as shown in **figure 1 (a)**. Subsequently, the powder mixture is sintered under the centrifugal force, as shown in **figure 1 (b)**. Then, the molten matrix metal is poured into the fabricated preform by centrifugal casting, as shown in **figure 1 (c)**. As a result, the molten matrix metal penetrates into the space between the particles by the pressure due to the centrifugal force. At the same time, matrix metal particles are melted by the heat from molten matrix. Finally, an FGM ring with dispersion particles, such as abrasive grains, distributed on its surface can be obtained, as shown in **figure 1 (d)**. In this study, Al/diamond particle FGM has been fabricated by the CSC method.

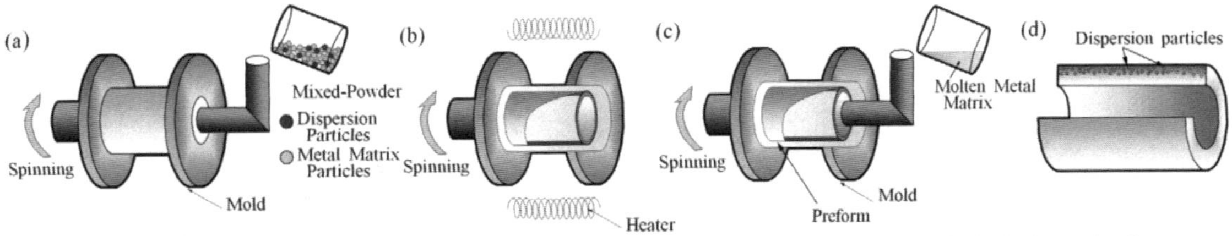

Figure 1. The schematic illustration of the centrifugal sintered-casting (CSC) method.

[1] To whom any correspondence should be addressed.

2. Experimental Procedure

Pure Al particles (purity: 99.99%, diameter: 106 to 180 μm), pure Mg particles (purity: 99.5%, mean diameter: 180 μm), Al-15mass%Si particles (mean diameter: 125 μm) and diamond particles (270 to 325 mesh) were used as primary materials. Three types of mixed powder with pure Al, Al-0.5%Mg and Al-15%Si particles are prepared, and they are named as P1, P2 and P3 samples, respectively. The mixed powder was set in the cylindrical mold having rotation axis with 20 mm in diameter and 30 mm in length, respectively. Then, the mixed powder was sintered in the spinning cylindrical mold under the centrifugal force of 280 *G* at 655 °C, 640 °C and 570 °C for P1, P2 and P3 samples, respectively, for 3 h in vacuum. Here, *G* is the ratio of centrifugal force to gravity (*G* number). The centrifugal sintering conditions of each fabricated preform are listed in **table 1**.

Table 1. Summary of the centrifugal sintering.

Preform sample name		P1	P2	P3
Amount of matrix particles	Al (g)	4.35	4.32	
	Mg (g)		0.02	
	Al-15%Si (g)			4.25
Amount of diamond particles (g)		0.63		
Volume fraction of diamond (vol%)		10		
Sintering temperature (°C)		655	640	570
G number		280		

Pure Al ingot and obtained cylindrical preform were placed in compact vacuum centrifugal casting, as shown in **figure 2**. The Al ingot was melted by heating coil, and then molten Al was poured into the mold with the preform by applied centrifugal force of 34 *G* or 78 *G* at casting temperature of 900 - 1100 °C. **Table 2** summarizes the centrifugal casting conditions with fabricated preforms. For ideal processing, centrifugal direction in centrifugal casting should correspond to the one in centrifugal sintering. However, this centrifugal casting machine was used as a convenient means because of machine performance in this study. The microstructural observations of the fabricated preforms and cast samples were carried out with a scanning electron microscope (SEM). Area fractions of diamond abrasive grains in Al matrix were evaluated from SEM photographs.

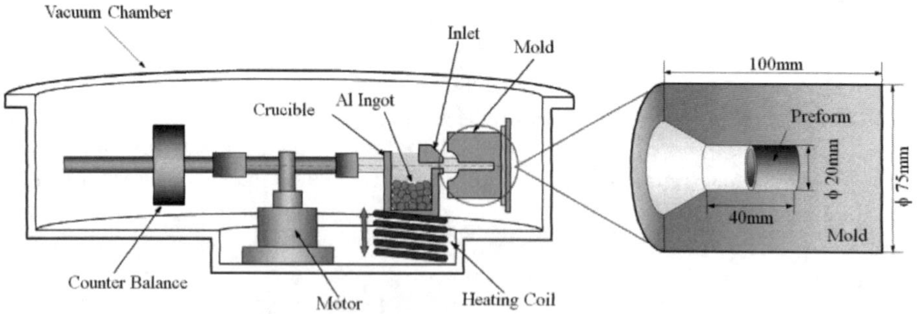

Figure 2. Schematic illustrations of compact vacuum centrifugal machine (left) and mold (right). The shape of products by this machine is rod.

Table 2. Summary of the centrifugal casting.

Cast sample name	C1	C2	C3	C4
Used preform sample	P3			
Amount of Al ingot (g)	35			
Casting temperature (°C)	900	1200	1100	
G number	34			78
Atmosphere	Vacuum		Ar	

3. Results and Discussion

Upper and lower figures of **figure 3 (a)** show the macro photograph and SEM micrograph of P1 sample, respectively. It is seen from these figures, pure Al particles cannot be sintered even though the sintering was carried out at high temperature (5 °C lower than melting point of Al). This is because the Al particles are covered by Al oxide, and higher sintering temperature is necessary for sintering of the Al oxide. To avoid this phenomenon, Mg particles were added into the mixed powder because it has strong reduction ability, and P2 sample was fabricated. The results of P2 sample are shown in **figure 3 (b)**. Although intermetallic compound was formed, good sintering of P2 sample was not found. Alternatively, Al-15%Si alloy particles with lower melting point were used, and macro photograph of P3 perform and its SEM micrograph are shown in **figure 3 (c)**. Ring-shaped Al-15%Si/diamond particles preform has been successfully fabricated using centrifugal sintering method. It is also found that homogeneous distribution of diamond particles was observed in the preform. It is interesting to note here that the microstructure of P3 sample still consists of the spherical-shaped Al particles, similar to those of primary Al-15%Si particles without sintering. However, from magnified SEM photograph, it is found that these spherical-shaped Al has been connected each other. Therefore, it is expected that perform of P3 sample is good for application of CSC method.

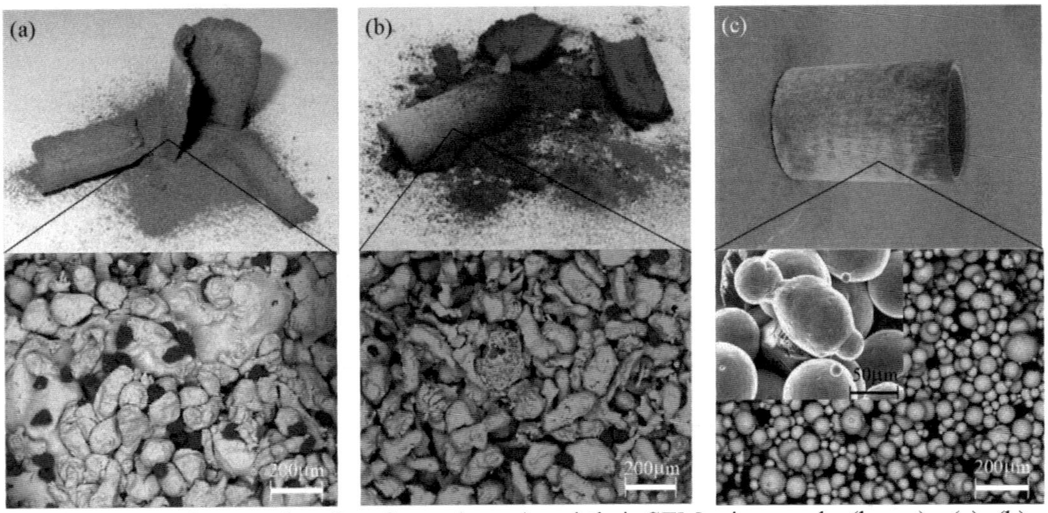

Figure 3. Macro photographs of performs (upper) and their SEM micrographs (lower). (a), (b) and (c) are P1, P2 and P3 samples, respectively.

Since the Al-15%Si/diamond particles preform can be successfully fabricated using centrifugal sintering method, P3 sample will be used for centrifugal casting experiment. **Figures 4 upper (a) to (d)** show the cross sectional views of C1 to C4 samples, respectively. SEM photographs showing microstructures around outer surface of C1 to C4 samples are also shown in **figures 4 lower (a) to (d)**, respectively. Solid samples could not be obtained for the C1 and C2 samples, which were fabricated under vacuum condition with casting temperature of 900 °C or 1200 °C, respectively as shown in **figures 4 upper (a) and (b)**. In our previous study [3], cooling rate distribution of the centrifugal casting using the compact vacuum centrifugal casting machine shown in **figure 2** is estimated by Al-Al_2Cu eutectic lamella structure. It was shown that the casting atmosphere as well as casting temperature and *G* number have affected the cooling rate distribution. Since local cooling rate strongly influences the local microstructure of the cast, centrifugal casting is carried out under Ar gas condition, and the results are shown in **figures 4 (c) and (d)**. It is seen that the solid samples could be fabricated under Ar gas condition. Moreover, as seen in **figure 4 lower (d)**, diamond particles are distributed at only outer surface of C4 sample and volume fraction of diamond particles is 11 vol%, as expected. Therefore, it is concluded that the CSC method is useful to fabricate the rod or ring shaped

FGMs with diamond particles located at outer surface of rod-shaped sample. These rod or ring shaped FGMs dispersed with diamond particles on its outer surface can be used as grading wheels for drilling of carbon fiber reinforced plastic (CFRP) [4].

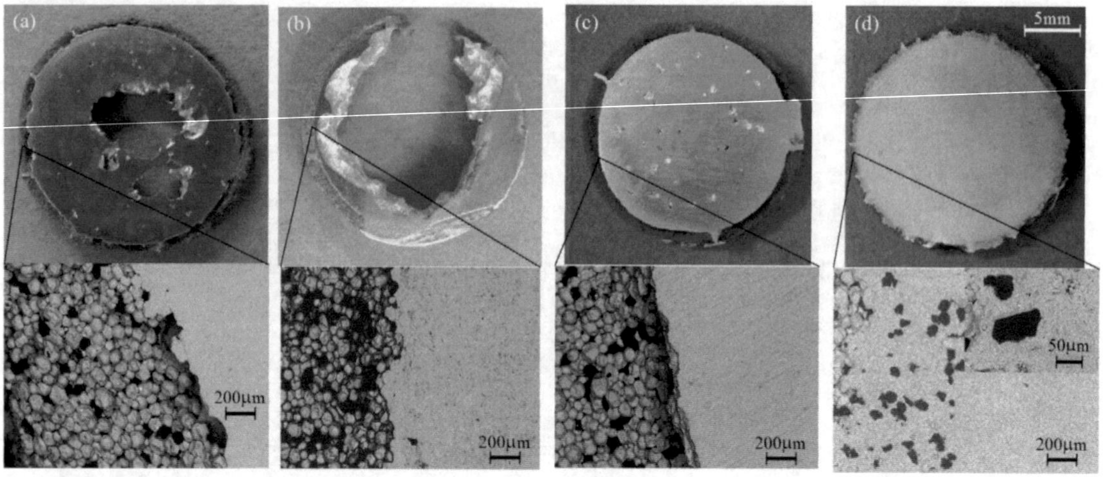

Figure 4. Macro photographs of cast samples (upper) and their SEM micrographs (lower). (a) C1, (b) C2, (c) C3, and (d) C4 samples.

4. Conclusions

In this study, Al/diamond particle FGM has been fabricated by the centrifugal sintered-casting (CSC) method, which is combination of sintering method and casting under the centrifugal force. First, three types of performs are fabricated under the centrifugal force from mixed powder of pure Al, Al-Mg and Al-15%Si alloy particles with diamond particles. Among them, it is found that preform can be obtained from mixed powder of Al-15%Si alloy particles with diamond particles. Using this type of preform, FGM was fabricated by casting under the centrifugal force. It is found that the diamond particles are distributed at only outer surface of casting. In this way, Al/diamond particles FGM can be successfully fabricated by CSC method. This method has advantage of the good control ability of diamond distribution.

Acknowledgments

This research was supported by Regional Innovation Cluster Program (Global Type) "Tokai Region Nanotechnology Manufacturing Cluster" from the Ministry of Education, Culture, Sports, Science and Technology (MEXT) of Japan. One of the authors, YW, acknowledges the financial support from The Light Metal Educational Foundation Inc. of Japan.

References

[1] Watanabe Y, Inaguma Y, Sato H and Miura-Fujiwara E 2009 *Materials* **2** 2510
[2] Watanabe Y and Sato H 2011 Review Fabrication of Functionally Graded Materials under a Centrifugal Force *Nanocomposites with Unique Properties and Applications in Medicine and Industry* ed J Cuppoletti (Rijeka, Shanghai and New York: InTech) chapter 7 pp 133-150
[3] Watanabe Y, Hattori Y and Sato H 2012 unpublished data
[4] Watanabe Y, Miura-Fujiwara E, Sato H, Takekoshi K, Tsuge H, Kaga T, Bando N, Yamagami S, Kurachi K and Yokoyama H 2011 *Inter. J. Mater. Prod. Tech.* **42** 29

Design, Manufacturing and Characterization of Functionally Graded Flextensional Piezoelectric Actuators

R C R Amigo, S L Vatanabe and E C N Silva

Department of Mechatronics and Mechanical Systems Engineering,
Escola Politecnica da Universidade de Sao Paulo,
Av. Prof. Mello Moraes, 2231, 05508-900, Sao Paulo, SP, Brazil.

E-mail: ecnsilva@usp.br

Abstract. Previous works have been shown several advantages in using Functionally Graded Materials (FGMs) for the performance of flextensional devices, such as reduction of stress concentrations and gains in reliability. In this work, the FGM concept is explored in the design of graded devices by using the Topology Optimization Method (TOM), in order to determine optimal topologies and gradations of the coupled structures of piezoactuators. The graded pieces are manufactured by using the Spark Plasma Sintering (SPS) technique and are bonded to piezoelectric ceramics. The graded actuators are then tested by using a modular vibrometer system for measuring output displacements, in order to validate the numerical simulations. The technological path developed here represents the initial step toward the manufacturing of an integral piezoelectric device, constituted by piezoelectric and non-piezoelectric materials without bonding layers.

1. Introduction

Piezoactuators are based on the capability of electromechanical energy conversion presented by some ceramics. To amplify the ceramics displacements, flexible coupling structures are commonly attached to them. This kind of assembly is known as flextensional piezoactuator and it is employed in a wide range of applications, such as ink jet print head [6], compact high-dynamic orientation systems [11] and airplane-wing deformation control [10]. Regarding gains in reliability and performance, flextensional actuators can be designed by considering Functionally Graded Materials (FGMs), which are advanced materials with continuously graded properties due to their spatially graded microstructure. FGMs can be manufactured by using several techniques, among which stands out the Spark Plasma Sintering (SPS), which consists in a powder densification process subjected to pressure and eletromagnetic field, simultaneously [4]. During the last years, the Topology Optimization Method (TOM) has been used as an useful tool for designing flextensional devices [9]. Subsequently, this approach has been extended to the design of FGM piezoelectric actuators [3], by introducing the concept of integral piezoactuators, in which the ceramic-metal interfaces are smoothly graded, eliminating the non-linearities and stress concentrations of bonding layers. Therefore, seeking to expand the role of FGMs in piezoactuators, this work addresses the systematic design and manufacturing of graded flextensional structures.

2. Design Method

2.1. Material Model

The topology optimization method seeks an optimal material distribution inside a fixed domain for given purposes. This is made by defining which points should be solid and which points should be void, assigning pseudo-densities equal to 0 or 1, respectively. However, such problem is ill-posed and the usual way to find a solution is by defining a material model that allows intermediate property values. Although solvable, the relaxed problem produces gray regions, where the presence or absence of material is not clear. To recover the discrete nature of the problem, penalization coefficients are introduced by material models to minimize regions with intermediate densities, as made in the Simple Isotropic Material with Penalization, SIMP [2]. The material model implemented in this work is based on the FGM-SIMP model [7], which has the capability of representing a domain with property changes along certain direction according to a specified law. This model combines the elastic properties of a void material ($\mathbf{c}_{void}$) and two solid isotropic materials ($\mathbf{c}_A$ and $\mathbf{c}_B$) and it estimates the properties of graded materials by the following formulation:

$$\mathbf{c}(\mathbf{x}) = [1 - \rho_1^p(\mathbf{x})]\mathbf{c}_{void} + \rho_1^p(\mathbf{x})\left([1 - \rho_2^q(\mathbf{x})]\mathbf{c}_A + \rho_2^q(\mathbf{x})\mathbf{c}_B\right) \tag{1}$$

where, ρ_1 and ρ_2 are the pseudo-densities regarding topology and gradation, which are penalized by the coefficients p and q, respectively.

2.2. Optimization Problem

The goal of this work is to maximize the output displacement $\mathbf{u}_1$ of a flextensional actuator when the ceramic is submitted to an electric potential φ_1. The Figure 1 shows the load case considered, where Ω_s denotes the design domain and Ω_p denotes the fixed piezoelectric domain. To ensure the minimum structural stiffness, springs are placed at the actuation regions Γ_t. Thus, the formulation of the optimization problem is given by Eq. (2), where Θ_1 is the upper-bound volume constraint and Θ_2 defines the gradation proportions of each material in structure.

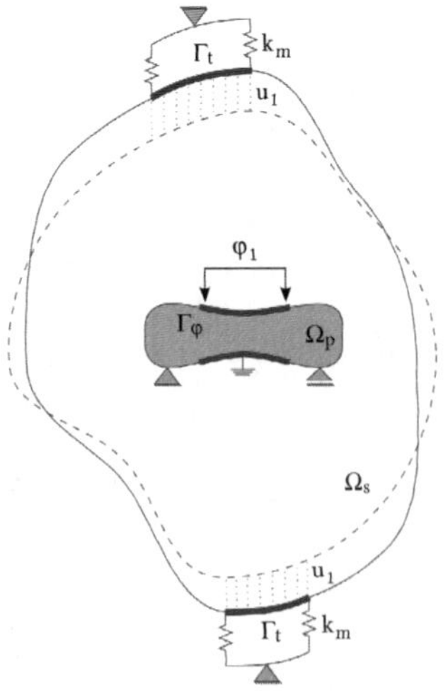

Figure 1. Load case for evaluation of output displacement.

$$\begin{aligned} \text{Maximize}: \quad & L_1 = \int_{\Gamma_t} \mathbf{u}_1(\rho_1, \rho_2)\, d\Gamma \\ \rho_1(\mathbf{x}), \rho_2(\mathbf{x}) \quad & \\ \text{subjected to}: \quad & 0 \le \rho_1, \rho_2 \le 1 \\ & \int_{\Omega_s} \rho_1\, d\Omega_s - \Theta_1 \le 0 \\ & \int_{\Omega_s} \rho_1 \rho_2\, d\Omega_s - \Theta_2 \int_{\Omega_s} \rho_1\, d\Omega_s \le 0 \end{aligned} \tag{2}$$

For the numerical implementation the design domain is meshed using four-node bilinear elements considering plane stress hypothesis and the Sequential Linear Programming (SLP) to solve the optimization problem.

3. Manufacturing and Characterization

By using Spark Plasma Sintering (SPS) process, graded cylinders are sintered. This method has the capability to produce pieces with the desired kind of material gradation, limited by the feasibility of pouring the layers of powder blends as required. Then, this cylinder is machined to achieve pieces with the designed topology, which are bonded to piezoceramics. However, for manufacturing, the gradations must be expressed in terms of mass percentage of each component material, while the property gradation defined by the material model in Eq. (1) is given in terms of stiffness. Thus, in this work, an elastic modulus versus mass percentage curve is approximated by using Hashin-Shtrikman theory [5], in which the moduli of pure Copper and pure Nickel is obtained by characterizing samples sintered in SPS machine by using a non-destructive ultrasonic method [1]. In order to verify the output displacements of prototypes, a *OFV-5000* vibrometer [8] is employed. This equipment operates on the Doppler principle, measuring back-scattered laser light from a vibrating structure.

4. Results

4.1. Numerical

Because of manufacturing constraints, the gradation is restricted to $1mm$-layers, as presented in the results below. The Table 1 presents the displacement values of each actuator case considered.

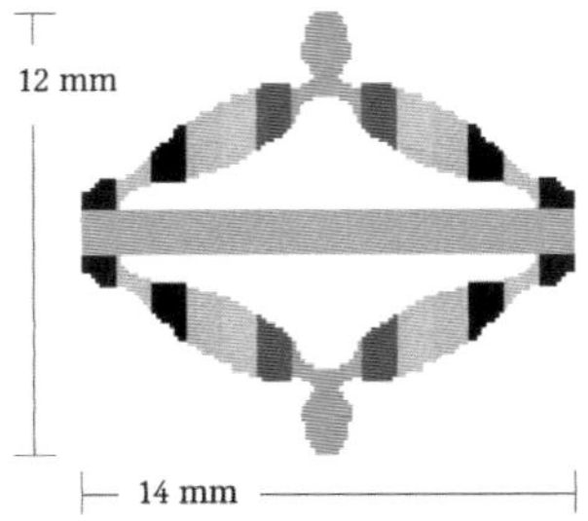

Figure 2. Actuator graded in vertical layers.

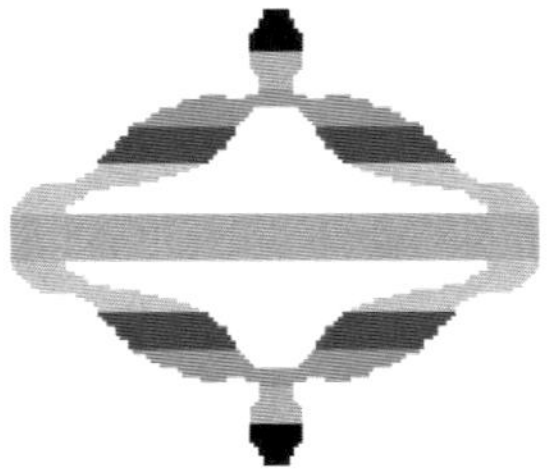

Figure 3. Actuator graded in horizontal layers.

4.2. Experimental

The actuator with vertical gradation layers, constituted by mixture of Cu and Ni, is chosen for manufacturing and it is presented in Figure 4. The output displacement obtained by using the vibrometer, is $0.155\mu m$.

Displacement (μm)	Homogeneous Cu	Homogeneous Ni	Vertical Layers	Horizontal Layers
Numerical	0.1015	0.1149	0.1210	0.1205
Experimental	-	-	0.155	-

Table 1. Output displacements for $20V/4kHz$ excitation.

Figure 4. Prototype of graded actuator.

5. Conclusion

This work demonstrates the contribution of FGM material in optimizing flextensional actuators by the TOM, with slight gains in the studied problem. Therefore, the feasibility of fabricating graded actuators from pieces sintered by SPS is proved. The experimental validation showed up consistency with the numerical results, despite small differences in terms of output displacement values ($0.155\mu m$ in experimental and $0.121\mu m$ in numerical), probably due to the simplifications introduced by the numerical model and imperfections due to bonding process.

References

[1] J C Adamowski, F Buiochi, and R T Higuti. Ultrasonic material characterization using large-aperture pvdf receivers. *Ultrasonics*, 50:110–115, 2010.

[2] M P Bendsoe. Optimal shape design as a material distribution problem. *Structural Optimization*, 1:192–202, 1989.

[3] R C Carbonari, E C N Silva, and G H Paulino. Integral piezoactuator system with optimum placement of functionally graded material - a topology optimization paradigm. *Journal of Intelligent Material Systems and Structures*, 21:1653–1668, 2010.

[4] H Feng, Q Meng, Y Zhou, and D Jia. Spark plasma sintering of functionally graded material in the ti-tib2-b system. *Materials Science and Engineering*, A397:92–97, 2005.

[5] Z Hashin and S Shtrikman. A variational approach to the theory of the elastic behavior of multiphase materials. *Journal of the Mechanics and Physics of Solids*, 11:127–140, 1963.

[6] C G Hyeon and A R M Asipeu. Hybrid type ink jet print head, has ink chamber for storing ink, ben position extended to ink chamber, piezoactuator installed in bottom surface of ink chamber and nozzle formed in chamber bottom. Technical report, Patent Number: KR2010128953-A, 2009.

[7] G H Paulino and E C N Silva. Design of functionally graded structures using topology optimization. *Material Science Forum*, 492-493:435–440, 2005.

[8] Polytec. Ofv-5000 vibrometer controller, 2011.

[9] E C N Silva, S Nishiwaki, and N Kikuchi. Topology optimization design of flextensional actuators. *IEEE Transaction on Ultrasonics Ferroelectrics and Frequency Control*, 47(3):657–671, 2000.

[10] H S Tzou and C I Tseng. Distributed piezoelectric sensor actuator design for dynamic measurement control of distributed parameter-systems: a piezoelectric finite-element approach. *Journal of Sound and Vibration*, 138:17–34, 1990.

[11] T Villgrattner and H Ulbrich. Design and control of a compact high-dynamic camera-orientarion system. *IEEE/ASME Transactions on Mechatronics*, 16,2:221–231, 2011.

Acknowledgements

The authors acknowledge the financial support of CNPq (National Council for Research and Development), under grant 558411/2010-0, and FAPESP (Fundacao de Amparo a Pesquisa do Estado de Sao Paulo), under grant 2011/02387-4.

Functionally Graded Dual-Nanoparticulate-Reinforced Aluminum Matrix Composite Materials

Hansang Kwon[1], Gil-Geun Lee[2], Marc Leparoux[1], Akira Kawasaki[3]

[1]*Empa, Swiss Federal Laboratories for Materials Science and Technology, Advanced Materials Processing, Feuerwerkerstrasse. 39, CH-3602 Thun, Switzerland*

[2]*Department of New Materials System Engineering, Graduate School of Engineering, Pukyong National University, Busan 608-739 Korea*

[3]*Department of Materials Processing, Tohoku University, Aoba-yama 02, Aobaku, 980-8576 Sendai, Japan*

nanocomposites@hotmail.com, hansang.kwon@empa.ch

Abstract. Functionally graded carbon nanotubes (CNT) and nano Silicon carbide (nSiC) reinforced aluminum (Al) matrix composite materials were fully densified by a simple ball milling and hot-pressing processes. The nSiC was used as a physical mixing agent to increase dispersity of the CNT in the Al particles. It was observed that the CNT was better dispersed in the Al particles with a nSiC mixing agent compared to without it used. SEM micrograph showed that the interface of the each layers had very tightly adhesion without any serious pores and micro-cracks. This functionally graded dual-nanoparticulate-reinforced Al matrix composite by powder metallurgical approach could also be applied to comples matrix materials.

Keywords; Carbon nanotubes (CNT), Nano silicon carbide (nSiC), Metal matrix composites (MMCs), Ball milling, Hot-pressing

1. Introduction

Functionally graded materials (FGMs) are compositionally gradient materials which were first proposed in Japan in 1984 [1]. "They have been suggested as a way of fabricating bulk materials with tailored properties because they can achieve the controlled distribution of the desired characteristics within the same bulk material. This FGM can be applied for engineering parts instead of coated materials. However, careful selection of the component materials, particularly the reinforcement, is required to reliably achieve a high performance from the FGM." To evaluate the potential of aluminum light weight composite materials with tailored functionally graded mechanical performances is produced."[1]. "Carbon nanotubes (CNT) are attractive next generation materials due to their unique properties, which lead to high mechanical, electrical, and thermal performance [2-4]. This unique nano order material can not only be utilized on its own in precision industrial fields but can also provide high performance functionality in conjunction with conventional materials. For this reason, many researchers are investigating the fabrication of CNT reinforced metal, ceramic, and polymer matrix composite materials."[5]. However, "The fabrication of CNT-reinforced metal matrix composites appears difficult compared to other matrix materials because of the difficulty of homogeneously dispersing the CNTs into the metal matrix while controlling the interfacial integrity" [6]. It is very important that the achievement of a sufficiently good dispersion, which is a prerequisite to obtaining any benefit from the unique.

In this study, aluminum alloy (Al) will be reinforced with different kinds of particles aiming at tailoring the hardness response of the composite. "Silicon carbide nanoparticles (nSiC) were added as an active solid mixing agent for good homogeneous dispersion of CNTs in the Al powder. SiC is still widely used in the industry because of its excellent temperature tolerance, corrosion resistance, thermal shock resistance, electric conductivity and superior chemical inertness"[6]. Microscale and nanoscale SiC particles as well as carbon nanotubes will be mixed in a high energy mill process with the Al powders to achieve a homogeneous dispersion. The produced blends will then be compacted together in form of multilayer composite materials.

2. Experimental procedure

Multiwalled CNTs (Baytubes C150P, Bayer material science, purity 99.5%, diameter: 20 nm, length: 30 μm) and gas-atomized Al5Mg powder (ECKA Granules, purity 99.5%, mean particle size: 63 μm) were used as the starting materials. The nSiC particles were produced in an inductively coupled plasma (ICP) reactor (the process is described in detail elsewhere [7]), producing an average particle size between 20 and 30 nm. The Al powder and the nanoparticulate materials (5 and 10vol% CNT and/or 1 and 10vol% nSiC) were mixed in a planetary ball mill (Retsch GmbH, PM400) for 20 h under argon atmosphere at 360 rpm, using Ø10 mm balls, a 10:1 ball to powder weight ratio, and 20 wt.% of heptane as a process control agent (PCA). The Al-nSiC-CNT composite powders were assembled in a layered structure inside a 55x55x5 mm die, with compositions ranging from Al5Mg to 1vol%nSiC-5vol%CNT, 10vol%nSiC-10vol%CNT, and 10vol%nSiC, respectively followed by hot-pressing (Walter+bai ag Testing machines, 400kN) consolidating at 500 °C for 10 sec under an uniaxial pressure of 570 MPa. The density of the FGMs were measured according to the Archimedes principle and the micro Vickers hardness (Paar MTH4 microhardness-tester) of the ball milled powders and gradient layers were measured using the loads 0.02 kgf and 20 kgf for at least five measurements per sample. The interlayer microstructure of FGM was observed using optical microscopy (Zeiss Axioplan light microscope), high resolution cold field emission scanning electron microscopy (SEM) (Hitachi, HRCFE-SEM S-4800).

3. Results and Discussion

A cross-section of the plate confirms the high density of the composite, some pores are present in the layer with the highest nSiC and CNT content. Furthermore, the cross-section shows a good cohesion between the layers without cracks either in the layers or at the interfaces (Fig. 1).

Journal of Physics: Conference Series **419** (2013) 012004 doi:10.1088/1742-6596/419/1/012004

Fig. 1. Cross section of the multilayer plate showing the different layers and their interfaces.

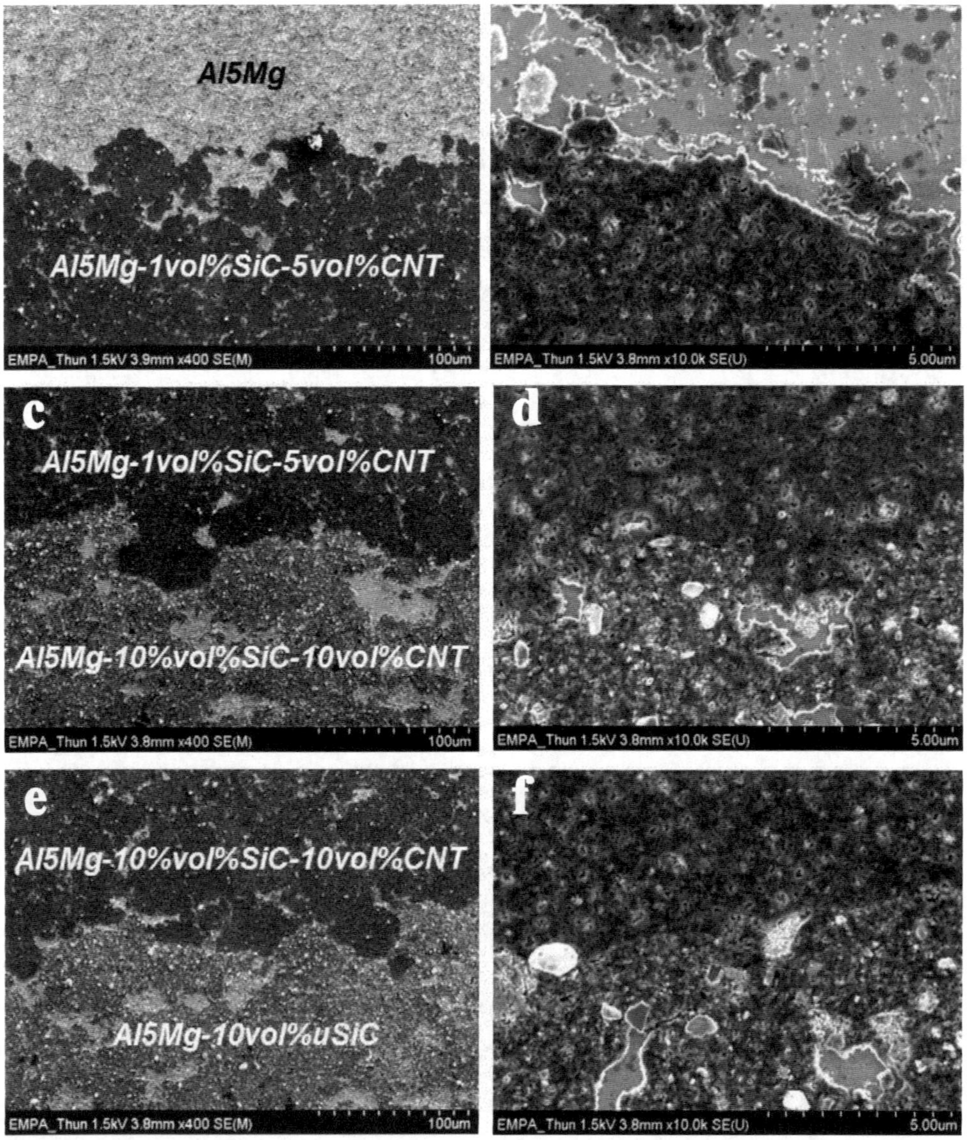

Fig. 2. SEM micrographs of the interfaces between the layers in the composite plate.

Some hardness prints are visible in the different layers in Fig. 1(a). Indeed, hardness measurements were performed in each layer. The results are shown in the table 1 below. Fig. 2 shows SEM micrographs of the interfaces between the functionally graded layers in the composite plate. The interface between Al5Mg and Al5Mg-1vol%nSiC-5vol%CNT(Fig.2a and b), Al5Mg-1vol%nSiC-5vol%CNT and Al5Mg-10vol%nSiC-10vol%CNT (Fig.2c and d), and Al5Mg-10vol%nSiC-10vol%CNT and Al5Mg-10vol%μ-SiC (Fig.2e and f). The multi composition layers were well bonded regardless of the amount of nanoparticles ratio.

Table 1. Hardness measurement of the layers in the final composite material

Samples	HV_{20}	±	$HV_{0.3}$	±
Al5Mg + 10 vol.% μ-SiC (F1200)	121	10	123	3
Al5Mg + 10 vol.% nSiC + 10 vol.% CNT	281	20	269	4
Al5Mg + 1 vol.% nSiC + 5 vol.% CNT	291	10	308	1
Al5Mg	58	1	64	0

4. Conclusions

High energy milling is a possible route to effectively disperse nanoparticulate materials into a metal matrix using a solid mixing. Thus the hardness of an Al5Mg alloy could be increased by a factor of 5-6. Hot pressing of powder blends allows the fabrication of multilayer composite materials. However due to high dilatation stresses during cooling, delamination occurred in the hardest layer. This had then to be "diluted" with pure alloy. One potential improvement of the concept would be to increase the thickness of each layer. This concept if validated is not only interesting for light weight bulletproof vests but can be extended to other metal matrices and to other armor applications like for instance combat aircrafts, tank and battle ship etc.

References

[1] H. Kwon, C.R. Bradbury, M. Leparoux, Advanced Engineering Materials, 13 (2011) 325-329.
[2] S. Iijima, Nature, 354 (1991) 56-58.
[3] M. Endo, T. Koyama, Y. Hishiyama, Jpn. J. Appl. Phys. 15 (1976) 2073-2076.
[4] F.F. Komarov and A.M. Mironov, Phys. Chem. Solid State 5 (2004) 411-429.
[5] H. Kwon and A. Kawasakiin the book "Advances in Composite Materials for Medicine and Nanotechnology" edited by Brahim Attaf, ISBN 978-953-307-235-7, InTech, April 4, 2011
[6] H. Kwon, S. Cho, M. Leparoux, A. Kawasaki, Nanotechnology, 23 (2012) 225704.
[7] Y. Leconte, M. Leparoux, X. Portier, N. Herlin-Boime, Plasma Chem. Plasma Process. 28 (2008) 233-48

Dynamics of Nonhomogeneous Carbon Steel Plates

J-H Kim[1]*, L Zhang[1], R Hebert[2], J T Wright[3], A Shukla[3]
[1]Department of Civil and Environmental Engineering, University of Connecticut, 261 Glenbrook Rd., U-3037, Storrs, CT 06269, USA
[2]Department of Chemical, Materials and Biomolecular Engineering, University of Connecticut,
97N. Eagleville Road, Storrs, CT 06269, USA
[3]Dynamic Photomechanics Laboratory, Department of Mechanical Engineering and Applied Mechanics, University of Rhode Island, Kingston, RI 02881, USA

Email: jhkim@engr.uconn.edu

Abstract. This paper presents finite element modeling procedure for rate-dependent elastic-plastic nonhomogeneous carbon steel plates subjected to dynamic loads. The steel plate consists of two ductile materials. We utilize a modified rule of mixture and linear strain-rate dependence to derive constitutive relations which are then implemented in ABAQUS/Explicit through user subroutine. Validation of the developed procedure is partially made with Steel 1018. Comparative numerical study is done between Steel 1018 and the nonhomogeneous plate. Displacements, stresses and relevant energy terms are calculated and compared for various sets of material configurations.

1. Introduction

A functionally graded material (FGM) is a material concept or a material itself in which the volume fractions of constituent materials vary smoothly resulting in graded macroproperties with a non-uniform microstructure [1, 2]. There has been much research done on the static and dynamic responses of linear-elastic FGMs [3-10]. There has been also some investigations of rate-independent elastic-plastic FGMs. Williamson et al. [11, 12] developed an elastic-plastic model based on a modified rule of mixture so-called Tamura–Tomota–Ozowa (TTO). Giannakopoulos et al. [13] evaluated rate-independent elastic-plastic material properties for a FGM interface layer based on the TTO model and investigated the cyclic thermal response of the layered composite. Gunes et al. [14] studied the elastic-plastic response of a circular FG plate under low velocity impact. There has been a need for the development of rate-dependent elastic-plastic constitutive relations for FG ductile steel plates. Rate-dependent elastic-plastic material models are the key to accurately analyzing FGM subjected to medium and high rate dynamic loading. Li et al. [15, 16] investigated the influence of particle volume fraction, shape, and aspect ratio on the rate-dependent elastic-plastic behavior of a particle-reinforced metal-matrix composite at high strain rates. Li et al. [17] examined the impact response of layered and graded plates.

Herein we develop a rate-dependent elastic-plastic FGM model. The rate-dependent elastic-plastic material properties of the constituent materials are experimentally characterized. The modified rule of mixture is used to calculate material properties. We apply the TTO model to an FGM with two ductile constituent materials. Based on the rate-dependent elastic-plastic FGM model, an ABAQUS/VUMAT subroutine is developed and implemented to simulate the dynamic response of a hypothetic FG plate under dynamic air pressure load.

2. Constitutive Models and Material Characterization

2.1 Strain-rate Dependent Stress-Strain Laws

The flow stress of a material, with no temperature effect, is dependent on the equivalent plastic strain ε^{pl} and the equivalent plastic strain rate $\dot{\varepsilon}^{pl}$

$$\sigma = \sigma(\varepsilon^{pl}, \dot{\varepsilon}^{pl}) \qquad (1)$$

In commonly used rate-dependent plastic material models, such as Johnson-Cook model and Cowper-Symonds model, the strain-rate dependence can be decoupled from the quasi-static stress-strain relation:

$$\sigma = \sigma_{qs}(\varepsilon^{pl}) R(\dot{\varepsilon}^{pl}) \qquad (2)$$

where $\sigma_{qs}(\varepsilon^{pl})$ shows the quasi-static flow stress which is a function of equivalent plastic strain ε^{pl}, and $R(\dot{\varepsilon}^{pl})$ is the strain-rate dependence. There are three models available in ABAQUS/CAE to define strain-rate dependence [18]: yield ratio model, Johnson-Cook strain-rate dependence [19] and Power-law strain-rate dependence [20]. In this work, linear strain-rate dependence is selected due to the nature of experimental data of Steel 1018 (See Figure 1(a)). The linear strain-rate dependence is defined as:

$$R = 1 + C\dot{\varepsilon}^{pl} \qquad (3)$$

where C is a constant representing a linear strain-rate dependence. A bilinear hardening model is widely used in representing the hardening curve due to its descriptive mathematical form. With the bilinear hardening model used, the rate-dependent bilinear hardening model becomes

$$\sigma = (\sigma_y + E_t \varepsilon^{pl})(1 + C\dot{\varepsilon}^{pl}) \qquad (4)$$

where σ_y is the quasi-static yield stress, E_t is the tangent modulus.

2.2 Material Characterization

The Split-Hopkinson Pressure Bar (SHPB) Technique is used to obtain the constitutive relations of steel 1018 for various high strain rates at room temperature. The stress-strain curves of steel 1018 under strain rates (2000/s, 2500/s and 3100/s) obtained by Prof. Arun Shukla's group (See Figure 1(a)). In addition, the quasi-static stress-strain curve is obtained at a rate of 10^{-3}/s in compression (See Figure 1(a)). The strain was determined directly from the instrument, i.e., from the displacement of the cross-head. For the elastic region we use well-known material properties such as Young's modulus=200GPa, Poisson's ratio=0.29. The rate-dependent elastic-plastic material model of steel 1018 is obtained by the curve-fitting method using the experimental data and Eq. (4) as shown in Figure 1(a).

The FG plate is envisioned to consist of Steel 1018 and Steel A36. The stress-strain curves of steel A36 at different strain rates are shown in Figure 1(b) [21]. Steel A36 and Steel 1018 are all low-carbon steel and steel A36 is assumed to share similar strain-hardening and strain-rate dependence. The FG plate considered here has steel A36 at the front, which is more compliant and can absorb more energy by its greater deformation, and steel 1018 at the back which has a larger yield stress. The Young's moduli of these two materials are 200GPa. Linear gradation between the front and back plates is assumed.

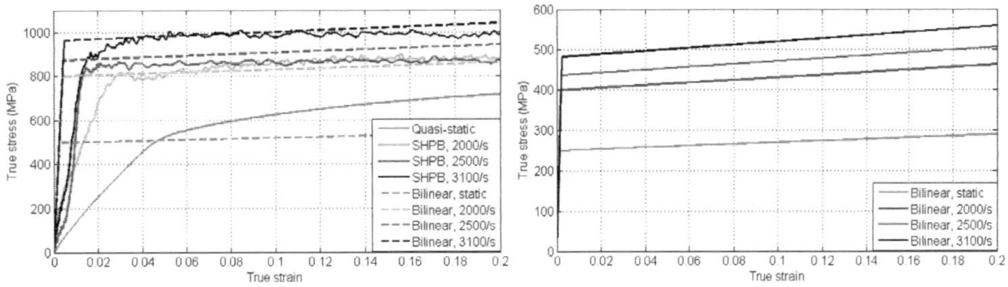

(a) (b)

Figure 1. Stress-strain curves of (a) steel 1018 and (b) A36 [21] at different strain rates

2.3 Strain Rate-dependent FGM model

The rate-dependent FGM model is derived based on the modified rule of mixture [12]. The TTO model relates the uniaxial stress and strain of a two-phase composite to the corresponding uniaxial stresses and strains of two constituent materials by

$$\sigma_c = V_1\sigma_1 + V_2\sigma_2, \varepsilon_c = V_1\varepsilon_1 + V_2\varepsilon_2 \quad (5)$$

where σ_i, ε_i and V_i $(i = 1,2)$ denote the average stresses, average strains and volume fractions of the constituent phases, respectively, and σ_c and ε_c denote the uniaxial stress and strain of the two-phase composite, respectively. The constitutive law of an FGM can be obtained by relating the constitutive curves of the constituent materials through parameter Q at any fixed strain rate, as shown in Figure 2.

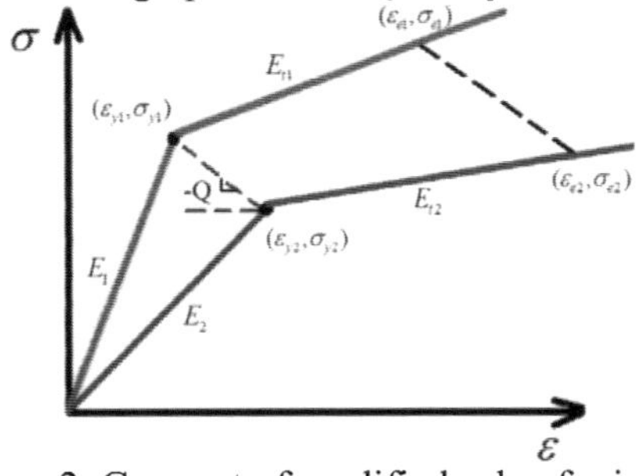

Figure 2. Concept of modified rule of mixture

In this work, the parameter Q value is uniquely defined by the yielding points of ductile constituent materials as:

$$Q = \frac{\sigma_{y1} - \sigma_{y2}}{\varepsilon_{y2} - \varepsilon_{y1}} \quad (6)$$

The effective Young's modulus for a FGM is obtained as [12]

$$E = [V_2E_2\frac{Q+E_1}{Q+E_2} + V_1E_1]/[V_2\frac{Q+E_1}{Q+E_2} + V_1] \quad (7)$$

In the plastic regime, the stress-strain relation is assumed to follow the bilinear hardening curve. For a point in material 2 $(\varepsilon_{e2}, \sigma_{e2})$, its corresponding point in material 1 $(\varepsilon_{e1}, \sigma_{e1})$ can be found by substituting the bilinear hardening model,

$$Q(\varepsilon_{e2} - \varepsilon_{e1}) = \sigma_{e1} - \sigma_{e2} = \sigma_{y1} + E_{t1}(\varepsilon_{e1} - \varepsilon_{y1}) - \sigma_{e2} \quad (8)$$

The corresponding strain in material 1 is

$$\varepsilon_{e1} = \frac{Q\varepsilon_{e2} - \sigma_{y1} + E_{t1}\varepsilon_{y1} + \sigma_{e2}}{Q + E_{t1}}, \sigma_{e1} = \sigma_{y1} + E_{t1}(\varepsilon_{e1} - \varepsilon_{y1}) \quad (9)$$

The corresponsing stress and strain of a FGM are given by

$$\sigma_e = V_1\sigma_{e1} + V_2\sigma_{e2}, \varepsilon_e = V_1\varepsilon_{e1} + V_2\varepsilon_{e2} \quad (10)$$

from which the tangent modulus can be calculated as

$$E_T = \frac{\sigma_e - \sigma_y}{\varepsilon_e - \varepsilon_y} \quad (11)$$

3. Implementation of Rate-dependent FGM model

The VUMAT subroutine in ABAQUS/Explicit and UMAT subroutine in ABAQUS/Standard provide interfaces to define the elastic-plastic material model. Different from the UMAT subroutine, the VUMAT subroutine is called for blocks of material calculation points and there is no need to update the Jacobian matrix in the explicit integration scheme [18]. A subroutine is developed based on the J_2 flow theory with isotropic hardening model. The subroutine is able to read material parameters of constituent materials and calculate material properties of the FGM at each material point. At the beginning of finite element simulation, material constitutive laws of constituent materials are constructed from the VUMAT inputs. In the following steps, volume fractions at material calculation points and the constitutive models of constituent materials are exported to the hardening subroutine VUHARD, and the material properties of FGM at each material calculation point are calculated. In each analysis step, the material is considered to be purely elastic and the trial elastic stresses are calculated at first. Then the von Mises stress can be calculated by trial elastic stresses and used to judge whether the material yields or not. If it doesn't yield, the trial elastic stresses are the true stresses and the analysis continues to the next step. If it yields, the equivalent plastic strain is updated according to the von Mises stress by Newton-Raphson iteration method. Once the equivalent plastic strain is found, the stresses and strains will be updated based on the J_2 flow theory.

4. Shock Tube Test of Steel 1018

The shock tube testing was carried out on a monolithic plate made of steel 1018. The shock tube facility available at the University of Rhode Island (Prof. Arun Shukla) [21-24] is used to generate the dynamic air pressure loads. The shock tube is divided into high-pressure and low-pressure driven sections, which are separated by a destructible diaphragm. The pressure difference between these two sections becomes higher when pressurizing the high-pressure driver section. When the pressure difference reaches a critical value, the diaphragm ruptures and the resulting rapid release of gas forms a one-dimensional shock wave front. When the shock wave reaches the specimen, the dynamic air pressure load is applied to the specimen [21-24]. The reflected dynamic air pressure load generated by the shock tube is measured by the sensor which is mounted 20mm away from the nozzle as shown in Figure 3(a). The measured dynamic air pressure load has an exponential decay behavior shown in Figure 3(b). Due to the lack of experimental data for functionally graded plate, the loading magnitude and history that is obtained for the monolithic plate are assumed to be not affected by material gradation.

The plate is simply supported and located with zero stand-off distance from the nozzle, as shown in Figure 3(a) and 3(c). The size of the plate is 0.203m (8in.) in length, 0.051m (2in.) in width and 0.003m (0.125in.) in thickness. The span between the two supports is around 0.152m (6in.). From the shock tube testing, the time history of the monolithic plate deformation was obtained by a high-speed camera system. During the testing, the rectangular monolithic plate bent to its largest deformation of 14.67mm at 2.5ms (milliseconds), and then oscillated until it reached its stable permanent plastic deformation state.

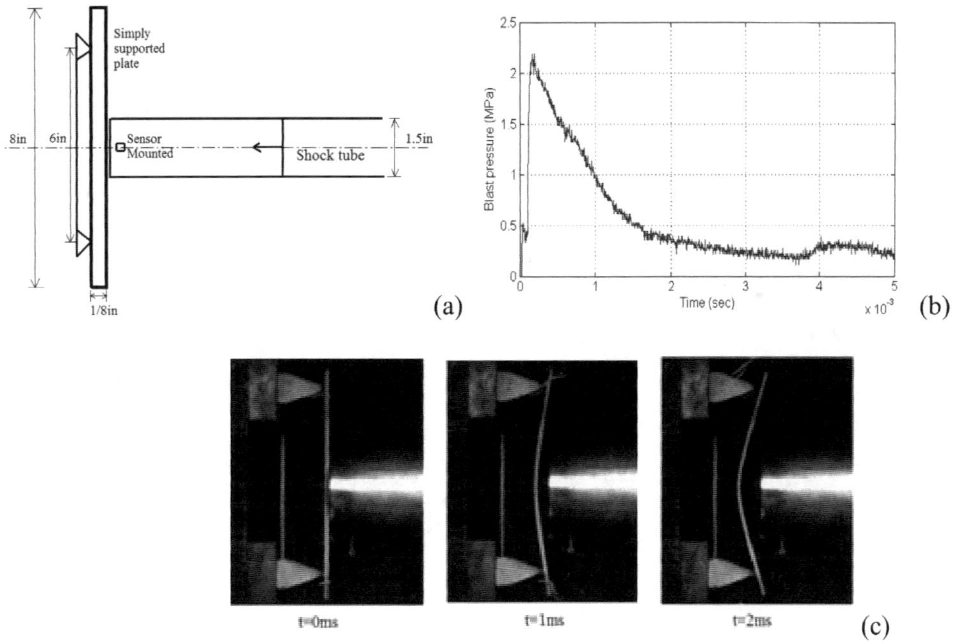

Figure 3. (a) Shock tube testing setup; (b) dynamic air pressure load generated by the shock tube; (c) deformation history

5. Explicit Dynamic Finite Element Analysis

Identical finite element mesh was used for all the simulations as shown in Figure 4(a). Due to model symmetry, a quarter model is built to reduce computational time. Contact with frictionless tangential behavior and hard normal behavior is defined to simulate the simple support between the specimen and the rigid support. The model is meshed using the first-order C3D8I incompatible element, which is the first-order fully integrated C3D8 element enhanced by incompatible modes to improve its bending behavior. The incompatible elements can not only improve the bending analysis but can also reduce the computational time compared to second-order elements. Due to the high rate dynamic air pressure loading during shock tube testing, the explicit algorithm is selected.

It is very challenging to determine the spatial distribution of the pressure loading at each time increment from the shock tube experiment. We observe that when the plate undergoes large deformation, the actual loading area is extended to beyond the muzzle area and the pressure loading at the extended area decays as it moves outward from the shock tube muzzle (see the paper [26] for more information). In this paper, we do not intend to perform fluid-structure analysis but approximate the actual pressure load history. The radius of the loading area is extended from 0.75in (radius of muzzle) to 1.5in to accommodate the extended loading area. The pressure load within the muzzle area is approximately the same as the measured pressure loading. And the pressure load in the extended area is assumed to have a second-order decay behavior. The simulation of Steel 1018 is carried out by the VUMAT subroutine as well as existing bilinear plastic model in ABAQUS/CAE both of which provided identical results. The dynamic behavior of an FG plate under dynamic air pressure loading is also investigated. As a comparative study, simulations of three monolithic plate models made of steel 1018, steel A36 and effective material are also carried out. The effective material property is calculated by considering the averaged material property of steel 1018 and A36.

6. Results and Discussion

6.1 Steel 1018: Test and FEM Results

The displacement history of the center point of Steel 1018 obtained from finite element simulation is compared with shock tube experimental data in Figure 4(b). The maximum deflection from the simulation is 14.33mm at 1.6ms and the maximum deformation from the shock tube test is found to be

14.67mm at 2.5ms. The finite element simulations provide reasonably good comparison with the experiment results.

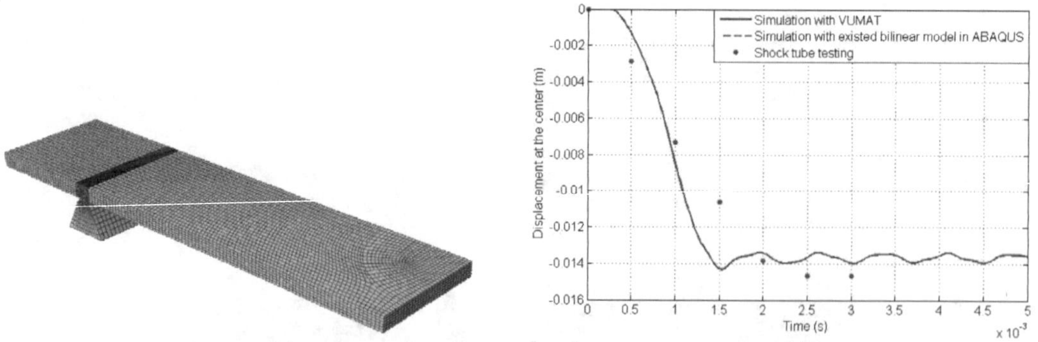

(a) (b)

Figure 4. (a) 3D Finite element mesh for both monolithic and FGM plates (No. of Elements: 24576; No. of Nodes: 29961); (b) Comparison of displacement histories of the center point in Steel 1018

6.2 Functionally Graded Plate: FEM Results

The FGM plate is simulated and compared with Steel 1018, Steel A36 and the effective plate. The displacement histories at the plate center are plotted in Figure 5. As expected, Steel A36 has the largest deformation (maximum 17.71mm), which is due to its lower strain hardening property. Steel 1018 has the smallest deformation (maximum 14.33mm). The displacements of the effective material and the FGM are slightly larger than that of steel 1018, but much smaller than that of steel A36. With the same overall volume fractions of constituent materials, the deformation of the FGM plate is smaller than that of monolithic plate with the effective material property.

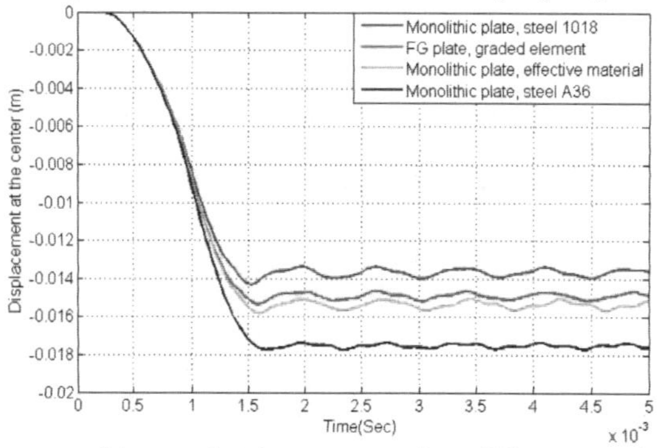

Figure 5. Displacement history for the center point of four material configurations.

For Steel 1018, the energy quantities are shown in Figure 6. The external work is determined by integrating the product of load and the displacement for the entire finite element model. The external work increases until the plate bends to its maximum deformation, and then vibrates while the plate oscillates until it reaches its permanent deflection. The plastic dissipation accumulates starting from 0.6ms and the plastic dissipation increments become smaller as the dynamic air pressure load decays. When the plastic dissipation increases, the kinematic energy decreases correspondingly.

For all four plates, the external work shown in Figure 6(a) increases till the plate bends to its maximum displacement and oscillates little bit when the plate bounces back and forth until its permanent deflection. The external work is largest in Steel A36, and smallest in Steel A36, and in between in the FGM and the effective plate. The external work for the FGM plate is smaller than that for the effective plate. Similar behavior is observed for the internal energy history in Figure 6(b). The kinetic energy of Steel A36 is the largest, while Steel 1018 is the smallest. The plastic dissipation increments decay as the magnitude of dynamic pressure load decays with time. The plastic dissipation of Steel A36 is the largest, while Steel 1018 is the smallest. The plastic dissipation of plate with FGM is smaller than that of a plate with an effective material.

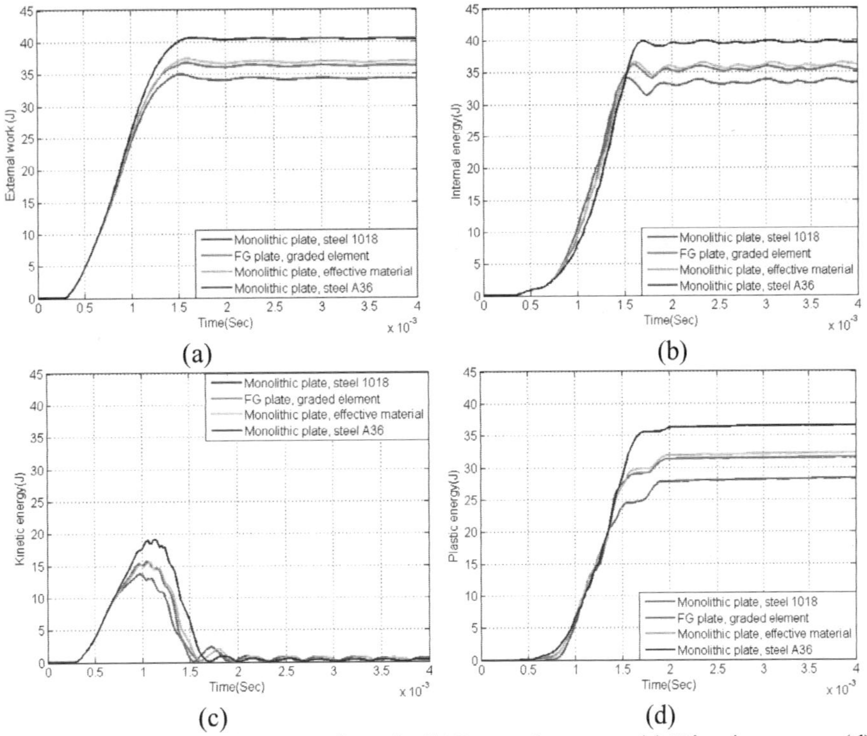

Figure 6. Energy histories: (a) External work; (b) Internal energy; (c) Kinetic energy; (d) Plastic dissipation.

7. Concluding Remarks

This paper addresses rate-dependent constitutive model of FGM. The FGM considered is composed with two ductile constituent materials (steel 1018 and steel A36) with linear gradation along its thickness direction. And, the linear strain-rate dependence and bilinear hardening model are used for the constituent materials. Based on the developed rate-dependent elastic-plastic model for FGM, a VUMAT subroutine is developed and implemented in the finite element simulation on the dynamic response of FGM plate subjected to dynamic air pressure load which is generated by shock tube facility. The displacement history, stress histories at critical locations and energy histories of the FGM are comparatively provided.

Acknowledgement

We gratefully acknowledge the financial support from the U.S. Department of Homeland Security (Award 2008-ST-061-TS0002-02).

References

[1] Hirai T. Functionally gradient materials and nanocomposites. In: Holt JB, Koizumi M, Hirai T, Munir ZA(Eds.), Proceedings of the Second International Symposium on Functionally Gradient Materials, Ceramic Transactions, The American Ceramic Society, Westerville, Ohio, 1993; 34: 11–20.
[2] Ilschner B. Processing-microstructure–property relationships in graded materials. Journal of the Mechanics and Physics of Solids 1996; 44(5): 647–656.
[3] Zenkour AM. A comprehensive analysis of functionally graded sandwich plates: Part I- Deflection and stresses. International Journal of Solids and Structures 2005; 42: 5224-5242.

[4] Qian LF, Batra RC, Chen LM. Static and dynamic deformations of thick functionally graded elastic plates by using higher-order shear and normal deformable plate theory and meshless local Petrov–Galerkin method. Composites: Part B 2004; 35: 685–697.
[5] Yang J, Shen HS. Dynamic response of initially stressed functionally graded rectangular thin plates. Composite Structures 2001; 54: 497-508.
[6] Huang XL, Shen HS. Nonlinear vibration and dynamic response of functionally graded plates in thermal environments. International Journal of Solids and Structures 2004; 41: 2403-2427.
[7] Reddy JN. Analysis of functionally graded plates. International Journal for Numerical Methods in Engineering 2000; 47: 663-684.
[8] Praveen GN, Reddy JN. Nonlinear transient thermoelastic analysis of functionally graded ceramic-metal plates. International Journal of Solids and Structures 1998; 35(33): 4457-4476.
[9] Reddy JN, Chin CD. Thermomechanical analysis of functionally graded cylinders and plates. Journal of Thermal Stresses 1998; 21(6): 593-626.
[10] Chakraborty A, Gopalakrishnan S, Reddy JN. A new beam finite element for the analysis of functionally graded materials. International Journal of Mechanical Sciences 2003; 45: 519-539.
[11] Williamson RL, Rabin BH, Drake JT. Finite element analysis of thermal residual stresses at graded ceramic-metal interfaces. Part I. Model description and geometrical effects. Journal of Applied Physics 1993; 74: 1310-1320.
[12] Tamura I, Tomota Y, Ozawa M. Strength and ductility of Fe-Ni-C alloys composed of austenite and martensite with various strengths. The microstructure and design of alloys: proceedings of the third international conference on the strength of metals and alloys, Cambridge, England 1973; 11: 611-615.
[13] Giannakopoulos AE, Suresh S, Finot M, Olsson M. Elastoplastic analysis of thermal cycling layered materials with compositional gradients. Acta Metal. Mater 1995; 43(4): 1335-1354.
[14] Gunes R, Aydin M, Apalak MK, Reddy JN. The elasto-plastic impact analysis of functionally graded cirvular plates under low-velocities. Composite Structures 2011; 93: 860-869.
[15] Li Y, Ramesh KT. Influence of particle volume fraction, shape and aspect ratio on the behavior of particle-reinforced metal-matrix composites at high rates of strain. Acta Mater 1998; 46(16): 5633-5646.
[16] Li Y, Ramesh KT, Chin ESC. Dynamic characterization of layered and graded structures under impulsive loading. International Journal of Solids and Structures 2001; 38(34-35): 6045-6061.
[17] Li Y, Ramesh KT, Chin ESC. The compressive viscoplastic response of an A359/SiCp metal matrix composite and of the A359 aluminum alloy matrix. International Journal of Solids and Structures 2000; 37: 7547-7562.
[18] ABAQUS 6.9 documentation.
[19] Johnson GR, Cook WH. A constitutive model and data for metals subjected to large strains, high strain rates and high temperatures. In: Proceedings of the 7th International Symposium on Ballistics, The Hague, the Netherlands, 1983, 1–7.
[20] Cowper GR, Symonds PS. Strain hardening and strain rate effect in the impact loading of cantilever beams. Division of Applied Mathematics Report, Brown University, 1957, No.28.
[21] ASTM A36 steel material properties, MatWeb: http://www.matweb.com
[22] Tekalue SA, Bogdanovich AE, Shukla A. Shock loading response of sandwich panels with 3-D woven E-glass composite skins and stitched foam core. Composites Science and Technology 2009; 69(6): 736-753.
[23] Gardner N, Wang E, Kumar P, Shukla A. Blast mitigation in a sandwich composite using graded core and polyurea interlayer. Experimental Mechanics 2011.
[24] Wang E, Shukla A. Analytical and experimental evaluation of energies during shock wave loading. International Journal of Impact Engineering 2010; 37:1188-1196.
[25] Kidane A, Shukla A. Dynamic constitutive behavior of Ti/TiB FGM under thermo-mechanical loading. Journal of Material Science 2008; 43: 2771–2777.
[26] Kumar P, LeBlanc J, Stargel DS, Shukla A. Effect of plate curvature on blast response of aluminum panels. International Journal of Impact Engineering 2012; 46: 74-85.

Cold spray and presureless sintering of zirconium phosphate bonded silicon nitride ceramics with porous gradient structure

Fei Chen[1,2], Feiyu Li[2], Qiang Shen[1*], Lianmeng Zhang[1,2]

[1] *Key Laboratory of Advanced Technology for Specially Functional Materials, Ministry of Education, Wuhan University of Technology, Wuhan 430070, China*

[2] *State Key Laboratory of Advaned Technology for Materials Synthesis and Processing, Wuhan University of Technology, Wuhan 430070, P. R. China*

Email: chenfei81@sohu.com

Abstract. In the present study, silicon nitride (Si_3N_4) electromagnetic wave transparent ceramics with high porosity and porous gradient structure are prepared by cold spray and pressureless sintering technique. Zirconium phosphate solution is used as a binder material instead of the traditional organic materials, in order to prevent the residual carbon which is severe to the dielectric properties of the Si_3N_4 porous ceramics. Firstly, Si_3N_4 ceramic slurries with different phosphorus acid and pore-forming agent contents are prepared. Then the Si_3N_4 slurries are cold sprayed layer by layer to achieve a porous gradient structure, and finally the samples are presurelessly sintered at 1000 °C in a nitrogen atmosphere. The results show that the porosity of the obtained Si_3N_4 ceramics is 20~70 % and the Si_3N_4 ceramics exhibits a good porous graded structure from high to low porosity.

1. Introduction

Functionally graded materials (FGMs) are innovative composite materials which have been intensively studied in recent years. FGMs feature gradual transitions in microstructure and composition that are engineered to meet functional performance requirements that vary with location in a single component and optimize the overall performance of the component [1]. Materials with porous gradient structure are special types of FGMs, and applications of this kind of materials may satisfy particular physical requirements. Silicon nitride (Si_3N_4) ceramics with porous gradient structure are used in numerous applications because of its superior properties, such as high-temperature strength, good oxidation resistance and low thermal expansion coefficient [2-5].

Nowadays, various processing techniques have been developed to prepare ceramic matrix composite with porous gradient structure, such as powder metallurgy, partially sintering, thermal/cold spray, tape casting, in-situ reaction sintering, metal or glass melt infiltration, etc [6-9]. Among these methods, powder metallurgy and partially sintering are considered as the most common methods for the better controlling of each layer's porosity.

However, cracks may appear in the material because of different sintering shrinkage attributing to the different thermal property of each layer. When using tape casting, the porous structure is hard to control because of the excessive organics. For in-situ reaction sintering, some other materials will be introduced during the reaction process. Therefore, from the existing literatures and patents [10-12], thermal/cold spray is the suitable method to prepare Si_3N_4 ceramics with porous gradient structure. S. Thiele [13] has prepared Si_3N_4 ceramic coating on Si_3N_4 substrate using detonation gun spray, high-velocity oxy-fuel (HVOF) thermal spray and plasma spray. However, thermal spray involves the heating of a source material to form molten droplets and tends to produce a splat-like structure, at which temperatures Si_3N_4 have already decomposed [14,15].

Cold spray uses high speed air to accelerate the particles or directly sprays viscosity slurries, avoiding the high temperature effect on coating and substrate material structure [16,17]. P. Holtappels [18] fabricated YSZ $\rightarrow$ LSM $\rightarrow$ LSCo gradient coatings with nine layers on zirconia substrate by cold spray, using $(La_{0.85}Sr_{0.15})_{0.9}MnO_3$ (LSM), $La_{0.84}Sr_{0.16}CoO_3$ (LSCo) and ZrO_2 (YSZ) powders with different contents as the raw materials, adding dispersant and anhydrous ethanol to achieve slurries. This work indicates that cold spray of viscosity slurries combined with heat treatment technology is an effect method to prepare ceramics with gradient structure.

Our previous research [19-21] described a pressureless sintering technique to fabricate α-Si_3N_4 porous ceramics with controlled porosity by using H_3PO_4 as a pore-forming agent and ZrP_2O_7 as a binder. After sintering at 1000~1200 °C, α-Si_3N_4 porous ceramics with controlled porosity of 20~63 %, dielectric properties of 2~5, and bending strengths of more than 100 MPa were obtained. Meantime, Si_3N_4 ceramics with porous gradient structure can be obtained by powder metallurgy and the pressureless sintering. However, it is very difficult to get uniformly mixed power, crack-free material and good adhesion of each layer by using this method.

In this paper, cold spay and pressureless sintering are applied for the preparation of α-Si_3N_4 ceramics with porous gradient structure by using H_3PO_4 as a pore-forming agent and ZrP_2O_7 as a binder. The purpose of this study is to investigate the processing method for preparation of porous α-Si_3N_4 ceramics, and then to explore a proper way to fabricate α-Si_3N_4 ceramics with porous gradient structure.

2. Experimental procedures

2.1. Materials

The raw materials used in the present study were as follows: commercial Si_3N_4 powder whose grain size was about 0.5 μm and content of α phase was more than 93 %, β phase was less than 5 %; commercial ZrO_2 powder whose grain size was around 0.5 μm and the purity was up to 99.9 %; phosphorus acid (H_3PO_4) liquid whose concentration was 85 % and the purity was up to 99.9 %.

2.2. Preparation of ceramics

Si_3N_4 slurries were prepared by mixing Si_3N_4 and ZrO_2 in designed content (the final molar ratio of ZrP_2O_7 and Si_3N_4 is 1:3) at room temperature and then adding excess H_3PO_4 (exceed the stoichiometric ratio of ZrO_2 and H_3PO_4 which formed ZrP_2O_7) to this mixture powder. The different porosity and porous gradient structure was controlled by the different content of excess H_3PO_4. The mixtures have been rotary ball-milled to get homogeneous. Then, the Si_3N_4 slurries were sprayed onto the surface of the substrate layer by layer with compressed air to achieve a porous gradient structure. The preparation of the individual layers and their mutual adhesion was controlled by intermediate pre-heat treatment, which was as follows: the samples were slowly heated to 250 °C and held for 2 h so as to obtain a total formation of ZrP_2O_7 bonded Si_3N_4 porous structure. Finally, the as-obtained Si_3N_4 porous gradient structure were sintered at 1000 °C in a furnace protected by a nitrogen (N_2) atmosphere and held for 2 h to achieve the Si_3N_4 porous ceramic. In this work, excess H_3PO_4 was used as pore-forming agent and ZrP_2O_7 used as binder with no organic matter, which indicates that the porosity will be better controlled. The preparation of Si_3N_4 slurries in a liquid condition indicates the more homogeneous mixture than powder metallurgy.

2.3. Characterization
After sintering, the bulk density of the sintered products is measured according to Archimedes displacement method using distilled water. The porosity is obtained from the relative density that is calculated from the bulk density of the products and the theoretical density of the Si_3N_4 composites. The phase compositions are analyzed by X-ray diffraction (XRD) using a Rigaku-D/Max-IIIA diffractometer with Cu radiation at 40 kV and 50 mA. A scan rate of 10 °/min is used to record the diffraction patterns in 2θ range between 10° and 80°. The microstructure of the fractured surface is observed by Scanning Electron Microscopy (SEM).

3. Results and discussion

3.1 Effect of cold spray parameters on the macrostructure of each layer
Thermal expansion plays a key role when two materials are combined and suffer a temperature change. Prior to preparation, a coefficient of thermal expansion (CTE) match should be considered. For both coating and substrate are using silicon nitride as matrix, the CTE of them are very similar, suggesting a good CTE match between the coating and substrate.

It is very clearly seen from Table 1 that with the increasing of slip solid content from 42 wt.% to 60 wt.% (corresponding to excess H_3PO_4 content decreasing from 95 vol.% to 55 vol.%) and spray time from 8 min to 35 min, coating thickness increases from 56 μm to 82 μm and from 56 μm to 400 μm, respectively. Coating thickness can be well controlled by slip solid content and spray time. Only the thickness of the coating changes, while the microstructure of either the porous coating or dense substrate maintains the same. The thick coating with controlled thickness makes it possible for the preparation of ceramic block materials.

Table 1. Effect of spray parameters on the coating thickness

Slip solid content (wt.%)	42	60	60	60
Excess H_3PO_4 content (vol.%)	95	55	55	55
Spray time (min)	8	8	15	35
Coating thickness (μm)	56	82	163	405

3.2 Phase compositions and porosity of cold sprayed porous Si_3N_4 ceramics

Figure 1 presents the phase compositions of porous Si_3N_4 ceramics with different excess H_3PO_4 content sintered at 1000 °C and the raw Si_3N_4 powder. Owing to the solid state sintering of ZrP_2O_7 binder around the Si_3N_4 grains, sintering of the whole composite is finished at 1000 °C according to the main sintering mechanism. The XRD spectra shows that α-Si_3N_4 is the main matrix phase which maintains the character of Si_3N_4 raw material with just a little amount of β-Si_3N_4 and no apparent phase transformation from α-Si_3N_4 to β-Si_3N_4 is detected. The phases of ZrP_2O_7 and SiP_2O_7 are identified from the XRD spectra. H_3PO_4 has reacted with SiO_2 presenting on the surface of the silicon nitride particles [22]. It is believed that the amount of silicon phosphate is not obvious identified by XRD because the majority of silicon phosphate was amorphous [23].

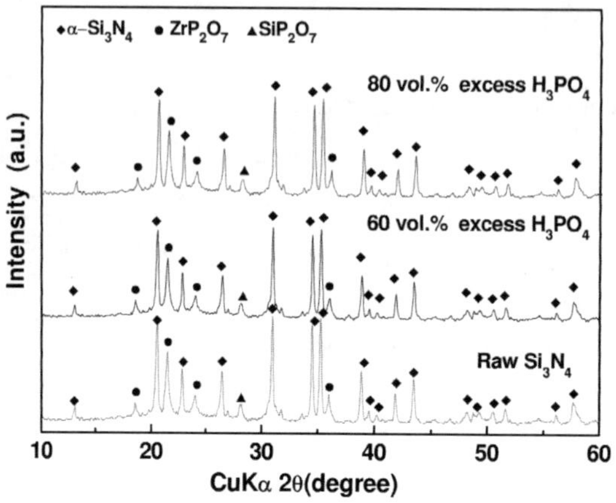

Figure 1. XRD patterns of ZrP_2O_7 bonded Si_3N_4 porous ceramics with different excess H_3PO_4 content sintered at 1000 ℃ (the final molar ratio of ZrP_2O_7 and Si_3N_4 is 1:3)

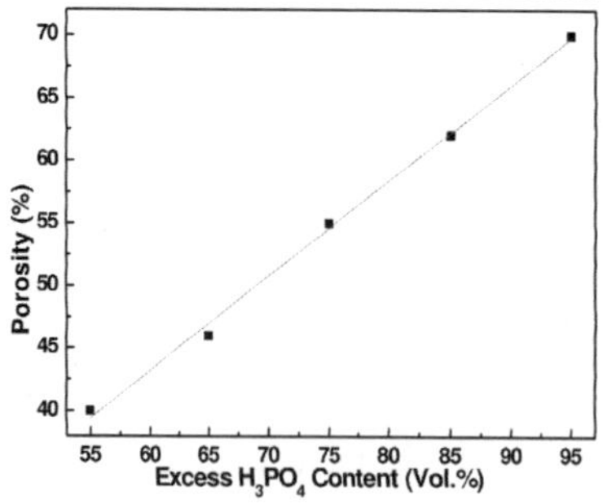

Figure 2. Effect of excess H_3PO_4 content on the porosities of the sintered ZrP_2O_7 bonded Si_3N_4 porous ceramics (the final molar ratio of ZrP_2O_7 and Si_3N_4 is 1:3)

Silicon nitride porous ceramics with various porosity and good processing were prepared by cold spray and pressureless sintering technique. Figure 2 shows the effect of excess H_3PO_4 content on the porosities of the sintered specimens. In general, porosity alters from 40 % to 70 % while the content of excess H_3PO_4 changes from 55 vol% to 95 vol.%. As can be seen clearly, the porosity is increased with increasing content of excess H_3PO_4 and shows almost linear variation with the content of excess H_3PO_4, indicating that the porosity can be well controlled by adjusting the amount of excess H_3PO_4, without changing sintering conditions.

3.3 Microstructure of Si_3N_4 ceramics with porous graded structure

Figure 3 shows the SEM images of fracture surface of the coated specimens. Although their porosities have great or some difference, the interface between the coating and substrate both is well bonded, without microcracks, indicating a good thermal physical match between layers. So the interface between the coating and substrate can be well bonded when their porosities are in the range which is from 20 % to 70 %.

Figure 3. SEM images of fracture surface of the coated specimens (the porosities of coating and substrate are 50 % and 40 %)

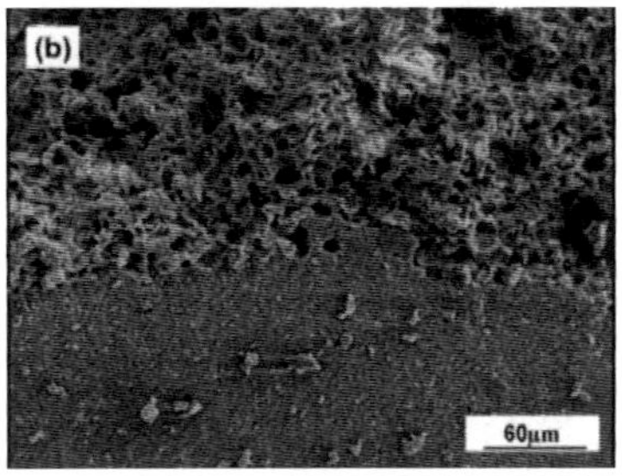

Figure 4. SEM images of fracture surface of the coated specimens (the porosities of coating and substrate are 70 % and 20 %.

For radome material applications, the porous gradient Si_3N_4 ceramics with various prosity were designed. From the top layer to the down layer, the prosity of each layer whose thickness is equal is from 20 % to 70 %, respectively. The pore structure and phase compositions of porous Si_3N_4 structures maintained the characteristic especially sintered at 1000 °C. Moreover, the structure of each layer including thickness and pore structure can be well controlled by cold spray parameters and pressureless sintering technique. So it is by using cold spay and pressureless sintering technique, porous gradient Si_3N_4 ceramics could be obtained, which could be used as radome material.

4. Conclusion

Silicon nitride (Si_3N_4) electromagnetic wave transparent ceramics with high porosity and porous gradient structure are prepared by cold spray and pressureless sintering technique. The results are as follows:

1) The porosity of ZrP_2O_7 bonded α-Si_3N_4 matrix porous ceramics could be well controlled by using H_3PO_4 as pore-forming agents and ZrP_2O_7 as binder. Porosity ranged from 40 % to 70 % while the excess H_3PO_4 pore-forming agent content alters from 55 vol.% to 95 vol.% and ZrP_2O_7 binder content maintains 25 wt.%.
2) Coating thickness could be well controlled by varing the spray parameters, including the slip solid content and spray time. Only the thickness of the coating changes, while the microstructure of either the porous coating or substrate maintains the same.
3) By using cold spray and pressureless sintering technique at 1000℃ in a nitrogen atmosphere, porous gradient Si_3N_4 ceramics can be obtained with the interface between layers clear and well combined.

Acknowledgements

This work is supported by the National Natural Science Fundation of P. R. China under Grant No. 51202171, Research Fund of the Doctoral Program of Higher Education of China (No. 20110143110006) and 111 Project (No. B13035).

References

[1] A. Mortensen and S. Suresh 1995 *Int. Mater. Rev.* **40** 239-265
[2] P. Poper and S. N. Ruddlesden 1961 *Trans. Br. Ceram. Soc.* **60** 603-626
[3] G. Ziegler, J. Heinrich and G. Wotting 1987 *J. Mater. Sci.* **22** 3041-3086
[4] A. J. Pyzik and D. R. Beaman 1993 *J. Am. Ceram. Soc.* **76** 2737-2744
[5] C. J. Lee and D. J. Kim 1999 *J. Am. Ceram. Soc.* **82** 753-756
[6] B. T. Lee, I. C. Kanga and A. K. Gain 2006 *J. Eu. Ceram. Soc.* **26** 3525-3530
[7] K. Nishiyabu, S. Matsuzaki, K. Okubo 2005 *Mater. Sci. Forum* **492-493** 765-770
[8] Y. Kinemuchi, K. Watari, S. Uchimura 2003 *Acta Mater.* **51** 3225-3231
[9] C. H. Chen, K. Takita, S. Honda, H. Awaji 2005 *J. Eu. Ceram. Soc.* **25** 385-391
[10] C. W. Li, D. Raybould and T. E. Strangman. US Patent, 6582779, 2003, 6.
[11] S. Usuba and R. B. Heimann 2006 *J. Thermal Spray Technol.* **15** 356-363
[12] S. Sodeoka, K. Ueno, Y. Hagiwara and S. Kose 1992 *J. Thermal Spray Technol* **1** 153-159
[13] S. Thiele, R. B. Heimann and L. M. Berger 2002 *J. Thermal Spray Technol* **11** 218-225
[14] Rajeev Dhiman, André G. McDonald and Sanjeev Chandra 2007 *Surf. .Coat. Technol.* **201** 7789–7801
[15] Y. Bao, T. Zhang , D.T. Gawne, P. Mason 2009 *J. Eur. Ceram. Soc,* **29** 1521–1528
[16] W. B. Choi, L. Li 2007 *Acta Mater* **55** 857-866
[17] H. Y. Lee, Y. H. Yu. 2003 *J. Thermal Spray Technol.* **13** 184-189
[18] P. Holtappels and C. Bagger 2002 *J. Eu. Ceram. Soc.* **22** 41-48
[19] F. Chen, Q. Shen, F. Q. Yan and L. M. Zhang 2006 *Mater. Sci. Technol.* **22** 915-918
[20] F. Chen, Q. Shen, F. Q. Yan and L. M. Zhang 2007 *J. Am. Ceram. Soc.* **90** 2379-2383
[21] F. Chen, Q. Shen, J. M. Schoenung and L. M. Zhang 2008 *J. Am. Ceram. Soc.* **91** 3173-3180
[22] D. E. Harrison, H. A. Mckinstry and F. A. Hummel 1954 *J. Am. Ceram. Soc.* **37** 277
[23] D. M. Poojary, R. B. Borade, F. L. Campbell and A. Clearfield 1994 *J. Solid State Chem.* **112** 106

FGM 2012 IOP Publishing
Journal of Physics: Conference Series **419** (2013) 012007 doi:10.1088/1742-6596/419/1/012007

Effective thermoelastic moduli of FGMs with temperature-dependent constituents and initial eigenstrains

Djaffar Boussaa
boussaa@lma.cnrs-mrs.fr
LMA, CNRS, UPR 7051, Aix-Marseille Univ, Centrale Marseille, F-13402 Marseille Cedex 20, France

Abstract. We outline a procedure for incorporating the effects on the effective thermoelastic moduli of composites of (i) the temperature dependence of the constituents and (ii) the presence of initial eigenstrains. The procedure uses a unified thermodynamic treatment, which assumes small strains, finite changes in temperature, and initial eigenstrains. The procedure is illustrated by deriving the effective values of the isothermal compliance tensor, the thermal expansion tensor, and the heat capacity per unit reference volume at constant stress. The procedure is of particular interest in the modeling of thermoelastic FGMs operating under thermal conditions where the material properties are liable to depend on temperature.

1. Introduction

Many FGM applications are intended to operate under severe thermomechanical conditions that involve considering both the temperature dependence of the constituents [1, 2] and the residual stresses [3, 4]. The micromechanical models usually used to estimate the effective thermomechanical moduli of FGMs, on the other hand, are based on linear thermoelasticity, where (i) the changes in temperature are assumed to be small, (ii) the material moduli are assumed to be independent of temperature, and (iii) the effects of initial eigenstrains associated with the residual stresses are neglected. A method to take into account the temperature dependence of the constituents on the effectives thermoelastic moduli has recently been proposed by the present author [5]. The present study outlines a procedure to extend this method to include the effects of initial eigenstrains.

More specifically, the purpose of the paper is to illustrate the procedure by developing the effective values of three moduli—the isothermal compliance tensor, the thermal expansion tensor, and the heat capacity per unit reference volume at constant stress—the definitions of which are recalled in Section 2. The thermodynamic potential most convenient to use for that purpose is the Gibbs potential. Its expression in the case of the homogeneous material is developed in Section 3. Its analogue in the case of the heterogeneous material, the macroscopic Gibbs potential, is obtained in Section 4, along with the corresponding localization relations. The effective moduli are obtained from the macroscopic potential in Section 5.

2. The moduli of the material

The moduli of the material whose effective values are developed below are the isothermal compliance tensor $\boldsymbol{M}$, the thermal expansion tensor $\boldsymbol{\alpha}$, and the heat capacity per unit reference

Journal of Physics: Conference Series **419** (2013) 012007 doi:10.1088/1742-6596/419/1/012007

volume at constant stress $C_{\boldsymbol{\sigma}}$. These moduli are defined as follows:

$$\boldsymbol{M} = \left(\frac{\partial \boldsymbol{\epsilon}}{\partial \boldsymbol{\sigma}}\right)_T, \qquad \boldsymbol{\alpha} = \left(\frac{\partial \boldsymbol{\epsilon}}{\partial T}\right)_{\boldsymbol{\varepsilon}}, \qquad C_{\boldsymbol{\sigma}} = T\left(\frac{\partial s}{\partial T}\right)_{\boldsymbol{\sigma}}, \tag{1}$$

where $\boldsymbol{\epsilon}$ is the strain tensor, $\boldsymbol{\sigma}$ is the stress tensor, T is the temperature, and s is the entropy.

The Gibbs potential, denoted by g, is most convenient to implement the thermodynamic approach to obtain the effective values of these moduli. The stress and the temperature are the natural variables for this potential, to which the stress and the entropy are related by

$$\boldsymbol{\epsilon} = -\left(\frac{\partial g}{\partial \boldsymbol{\sigma}}\right)_T, \qquad s = -\left(\frac{\partial g}{\partial T}\right)_{\boldsymbol{\sigma}}. \tag{2}$$

3. The Gibbs potential of a homogeneous material

To obtain the Gibbs potential for a homogeneous material, we start from the following state equation:

$$\boldsymbol{\epsilon} = \boldsymbol{M}(T)\boldsymbol{\sigma} + \boldsymbol{m}_\theta(T) + \boldsymbol{m}_0 \tag{3}$$

where $\boldsymbol{M}$ is assumed to depend on the temperature, $\boldsymbol{m}_\theta$ is the thermal strain, and $\boldsymbol{m}_0$ is the constant initial eigenstrain, the physical origin of which is irrelevant for the present analysis.

From this equation, it is clear that the thermal expansion tensor is independent of $\boldsymbol{m}_0$, as

$$\boldsymbol{\alpha}(\boldsymbol{\sigma}, T) = \frac{\mathrm{d}\boldsymbol{M}}{\mathrm{d}T}\boldsymbol{\sigma} + \frac{\mathrm{d}\boldsymbol{m}_\theta}{\mathrm{d}T}.$$

In view of (3), integrating (2-1) gives

$$g(\boldsymbol{\sigma}, \boldsymbol{T}) = -\frac{1}{2}\boldsymbol{\sigma}\cdot\boldsymbol{M}(T)\boldsymbol{\sigma} - \boldsymbol{m}(T)\cdot\boldsymbol{\sigma} + \phi(T), \tag{4}$$

where $\boldsymbol{m}$ denotes the total eigenstrain, $\boldsymbol{m} = \boldsymbol{m}_\theta + \boldsymbol{m}_0$, and ϕ is a function only of the temperature.

The function ϕ can be related to $C_{\boldsymbol{\sigma}}$ as follows. From (1-3), (2-2) and (4),

$$C_{\boldsymbol{\sigma}}(\boldsymbol{\sigma}, \boldsymbol{T}) = T\left(\frac{1}{2}\boldsymbol{\sigma}\cdot\frac{d^2\boldsymbol{M}}{dT^2}(T)\boldsymbol{\sigma} + \frac{d^2\boldsymbol{m}}{dT^2}(T)\cdot\boldsymbol{\sigma} - \frac{d^2\phi}{dT^2}\right).$$

Writing this equation for $\boldsymbol{\sigma} = \boldsymbol{0}$, dividing both sides by T, and integrating the resulting equation twice with respect to temperature gives

$$\phi(T) = -s_0(T - T_0) - \int_{T_0}^{T}\left(\int_{T_0}^{\xi}\frac{C_{\boldsymbol{\sigma}}(\boldsymbol{0}, \nu)}{\nu}d\nu\right)d\xi + g_0, \tag{5}$$

where s_0 and g_0 are the entropy and the Gibbs energy in the state characterized by $\boldsymbol{\sigma} = \boldsymbol{0}$ and $T = T_0$, respectively.

Substituting (5) into (4) yields

$$g(\boldsymbol{\sigma}, T) = -\frac{1}{2}\boldsymbol{\sigma}\cdot\boldsymbol{M}(T)\,\boldsymbol{\sigma} - \boldsymbol{m}(T)\cdot\boldsymbol{\sigma} - s_0(T - T_0) \quad - \int_{T_0}^{T}\left(\int_{T_0}^{\xi}\frac{C_{\boldsymbol{\sigma}}(\boldsymbol{0}, \nu)}{\nu}d\nu\right)d\xi + g_0. \tag{6}$$

A second state equation is readily obtained from this equation and (2-2) as

$$s = \frac{1}{2}\boldsymbol{\sigma}\cdot\frac{d\boldsymbol{M}}{dT}\boldsymbol{\sigma} + \frac{d\boldsymbol{m}}{dT}\cdot\boldsymbol{\sigma} + \int_{T_0}^{T}\frac{C_{\boldsymbol{\sigma}}(\boldsymbol{0}, \nu)}{\nu}d\nu + s_0,$$

and the corresponding expression for the heat capacity at constant stress is

$$C_{\boldsymbol{\sigma}}(\boldsymbol{\sigma}, T) = C_{\boldsymbol{\sigma}}(\boldsymbol{0}, T) + T\left(\frac{1}{2}\boldsymbol{\sigma}\cdot\frac{d^2\boldsymbol{M}_T}{dT^2}\boldsymbol{\sigma} + \frac{d^2\boldsymbol{m}}{dT^2}\cdot\boldsymbol{\sigma}\right). \tag{7}$$

To sum up, according to the thermoelasticity model developed here, the initial eigenstrain affects neither the compliance (by assumption), the thermal expansion tensor, nor the heat capacity at constant stress.

4. The Gibbs potential of a heterogeneous material

4.1. Localization relations

We consider a representative volume (RVE) of a heterogeneous material whose constituents show (i) a thermoelastic behavior governed by the Gibbs potential (6) and the associated constitutive equations, and (ii) perfect bonding and thermal contact at their interfaces. Within the RVE, (i) the distribution of initial eigenstrains is assumed to be given, (ii) the temperature is assumed to be uniform, and (iii) the strain is assumed to be small. In view of the potential and moduli of interest here, the RVE is considered under only controlled macroscopic (average) stress $\overline{\boldsymbol{\sigma}}$ and temperature T, the macroscopic stress being controlled by applying uniform boundary tractions and the temperature by applying a uniform boundary temperature.

The stress $\boldsymbol{\sigma}(\mathbf{x})$ within the RVE can be written as

$$\boldsymbol{\sigma}\left(\mathbf{x};\overline{\boldsymbol{\sigma}},T\right)=\boldsymbol{B}\left(\mathbf{x};T\right)\overline{\boldsymbol{\sigma}}+\boldsymbol{b}\left(\mathbf{x};T\right). \tag{8}$$

In this equation, the first term on the right-hand side is the stress that develops in the RVE under $\overline{\boldsymbol{\sigma}}$ only, and the fourth-order tensor $\boldsymbol{B}$, which is independent of the eigenstrains, is the stress concentration tensor at temperature T. The second term corresponds to the stress that develops in the RVE under the eigenstrains and traction free boundary conditions. This latter term can be decomposed into two terms, viz.

$$\boldsymbol{b}\left(\mathbf{x};T\right)=\boldsymbol{b}_{\theta}\left(\mathbf{x};T\right)+\boldsymbol{b}_{0}\left(\mathbf{x};T\right),$$

which correspond to the stress developed under $\boldsymbol{m}_{\theta}\left(\mathbf{x};T\right)$ and $\boldsymbol{m}_{0}\left(\mathbf{x}\right)$, respectively. Although the stress $\boldsymbol{b}_{0}\left(\mathbf{x};T\right)$ arises from the constant (and nonuniform) eigenstrain $\boldsymbol{m}_0$, it also depends on the temperature because of the dependence on temperature of the compliance tensor.

4.2. Macroscopic Gibbs potential

The macroscopic Gibbs potential density G is defined as

$$G\left(\overline{\boldsymbol{\sigma}},T\right)=\langle g\rangle=\frac{1}{|V|}\int_V g\left(\mathbf{x};\boldsymbol{\sigma},T\right)\,\mathrm{d}V.$$

Here, $g\left(\mathbf{x};\boldsymbol{\sigma},T\right)$ is of the form (6) where $\boldsymbol{M}$, $\boldsymbol{m}$, $C_{\boldsymbol{\sigma}}$, s_0, and g_0 are made to depend also on position, and $\boldsymbol{\sigma}$ is the stress that develops in the RVE under the combined effects of $\overline{\boldsymbol{\sigma}}$ and $\boldsymbol{m}$.

Substituting the expression for g given in (6) and the stress localization relation (8) into the foregoing equation yields

$$\begin{aligned}G\left(\overline{\boldsymbol{\sigma}},T\right)=&-\frac{1}{2}\overline{\boldsymbol{\sigma}}\cdot\langle\boldsymbol{B}^{\mathsf{T}}\boldsymbol{M}\boldsymbol{B}\rangle\overline{\boldsymbol{\sigma}}-\langle\boldsymbol{B}^{\mathsf{T}}\boldsymbol{m}\rangle\cdot\overline{\boldsymbol{\sigma}}-\frac{1}{2}\langle\boldsymbol{m}\cdot\boldsymbol{b}\rangle\\&-\int_{T_0}^{T}\left(\int_{T_0}^{\xi}\frac{\langle C_{\boldsymbol{\sigma}}(\boldsymbol{\sigma}=\mathbf{0},\nu)\rangle}{\nu}\,\mathrm{d}\nu\right)\mathrm{d}\xi-\langle s_0\rangle\left(T-T_0\right)+\langle g_0\rangle.\end{aligned} \tag{9}$$

The derivation of this equation uses classical properties of the stress localization tensor $\boldsymbol{B}$ as can be found in [5].

5. Effective moduli

The macroscopic Gibbs potential (9) is that of a thermoelastic material. The macroscopic strain $\overline{\boldsymbol{\varepsilon}}=(\partial G/\partial\overline{\boldsymbol{\sigma}})_T$ takes the form

$$\overline{\boldsymbol{\varepsilon}}=\boldsymbol{M}^{\mathrm{eff}}\overline{\boldsymbol{\sigma}}+\boldsymbol{m}^{\mathrm{eff}}, \tag{10}$$

where, in view of (9), the effective compliance tensor $\boldsymbol{M}^{\mathrm{eff}}$ and thermal strain $\boldsymbol{m}^{\mathrm{eff}}$ are given by

$$\begin{aligned}\boldsymbol{M}^{\mathrm{eff}}&=\langle\boldsymbol{B}^{\mathsf{T}}\boldsymbol{M}\boldsymbol{B}\rangle,\\\boldsymbol{m}^{\mathrm{eff}}&=\langle\boldsymbol{B}^{\mathsf{T}}\boldsymbol{m}\rangle=\langle\boldsymbol{B}^{\mathsf{T}}\boldsymbol{m}_{\theta}\rangle+\langle\boldsymbol{B}^{\mathsf{T}}\boldsymbol{m}_0\rangle.\end{aligned}$$

From (10), the thermal expansion tensor $\boldsymbol{\alpha}^{\text{eff}}(\overline{\boldsymbol{\sigma}},T) = (\partial\overline{\boldsymbol{\varepsilon}}/\partial T)_{\overline{\boldsymbol{\sigma}}}$ is found to be given by

$$\boldsymbol{\alpha}^{\text{eff}}(\overline{\boldsymbol{\sigma}},T) = \frac{\mathrm{d}\boldsymbol{M}^{\text{eff}}}{\mathrm{d}T}\overline{\boldsymbol{\sigma}} + \boldsymbol{\alpha}^{\text{eff}}(\mathbf{0},T),$$

where $\boldsymbol{\alpha}^{\text{eff}}(\mathbf{0},T)$ is the free thermal expansion tensor,

$$\boldsymbol{\alpha}^{\text{eff}}(\mathbf{0},T) = \frac{\mathrm{d}\langle \boldsymbol{B}^{\mathsf{T}}\boldsymbol{m}\rangle}{\mathrm{d}T} = \frac{\mathrm{d}\langle \boldsymbol{B}^{\mathsf{T}}\boldsymbol{m}_\theta\rangle}{\mathrm{d}T} + \left\langle \frac{\mathrm{d}\boldsymbol{B}^{\mathsf{T}}}{\mathrm{d}T}\boldsymbol{m}_0 \right\rangle.$$

The macroscopic entropy $\overline{s}(\overline{\boldsymbol{\sigma}},T) = -(\partial G/\partial T)_{\overline{\boldsymbol{\sigma}}}$ can be written as

$$\overline{s} = \frac{1}{2}\overline{\boldsymbol{\sigma}}\cdot\frac{\mathrm{d}\boldsymbol{M}^{\text{eff}}}{\mathrm{d}T}\overline{\boldsymbol{\sigma}} + \frac{\mathrm{d}\boldsymbol{m}^{\text{eff}}}{\mathrm{d}T}\cdot\overline{\boldsymbol{\sigma}} + \frac{1}{2}\frac{\mathrm{d}}{\mathrm{d}T}\langle \boldsymbol{b}\cdot\boldsymbol{m}\rangle + \int_{T_0}^{T}\frac{\langle C_{\boldsymbol{\sigma}}(\mathbf{0},\nu)\rangle}{\nu}\,\mathrm{d}\nu + \langle s_0\rangle,$$

so that the effective heat capacity at constant stress $C_{\boldsymbol{\sigma}}^{\text{eff}}(\overline{\boldsymbol{\sigma}},T) = T(\partial\overline{s}/\partial T)_{\overline{\boldsymbol{\sigma}}}$ is given by

$$\begin{aligned} C_{\boldsymbol{\sigma}}^{\text{eff}}(\overline{\boldsymbol{\sigma}},T) &= C_{\boldsymbol{\sigma}}^{\text{eff}}(\mathbf{0},T) + T\left[\frac{1}{2}\overline{\boldsymbol{\sigma}}\cdot\frac{\mathrm{d}^2\boldsymbol{M}^{\text{eff}}}{\mathrm{d}T^2}\overline{\boldsymbol{\sigma}} + \frac{\mathrm{d}^2\boldsymbol{m}^{\text{eff}}}{\mathrm{d}T^2}\cdot\overline{\boldsymbol{\sigma}}\right], \\ &= C_{\boldsymbol{\sigma}}^{\text{eff}}(\mathbf{0},T) + T\left[\frac{1}{2}\overline{\boldsymbol{\sigma}}\cdot\frac{\mathrm{d}^2\boldsymbol{M}^{\text{eff}}}{\mathrm{d}T^2}\overline{\boldsymbol{\sigma}} + \frac{\mathrm{d}^2\boldsymbol{m}_\theta^{\text{eff}}}{\mathrm{d}T^2}\cdot\overline{\boldsymbol{\sigma}}\right] + T\left\langle \frac{\mathrm{d}^2\boldsymbol{B}^{\mathsf{T}}}{\mathrm{d}T^2}\boldsymbol{m}_0\right\rangle\cdot\overline{\boldsymbol{\sigma}}, \end{aligned}$$

where

$$\begin{aligned} C_{\boldsymbol{\sigma}}^{\text{eff}}(\mathbf{0},T) &= \langle C_{\boldsymbol{\sigma}}(\mathbf{0},T)\rangle + \frac{1}{2}\frac{\mathrm{d}^2}{\mathrm{d}T^2}\langle \boldsymbol{b}\cdot\boldsymbol{m}\rangle \\ &= \langle C_{\boldsymbol{\sigma}}(\mathbf{0},T)\rangle + \frac{1}{2}\frac{\mathrm{d}^2}{\mathrm{d}T^2}\langle \boldsymbol{b}_\theta\cdot\boldsymbol{m}_\theta\rangle + \frac{1}{2}\frac{\mathrm{d}^2}{\mathrm{d}T^2}\langle \boldsymbol{b}_0\cdot\boldsymbol{m}_\theta\rangle + \frac{1}{2}\frac{\mathrm{d}^2}{\mathrm{d}T^2}\langle \boldsymbol{b}_\theta\cdot\boldsymbol{m}_0\rangle + \frac{1}{2}\left\langle \frac{\mathrm{d}^2\boldsymbol{b}_0}{\mathrm{d}T^2}\cdot\boldsymbol{m}_0\right\rangle \end{aligned}$$

The following conclusions can be drawn:

- The initial eigenstrains have no influence on the effective compliance tensor.
- The initial eigenstrains affect the effective total eigenstrain.
- The initial eigenstrains affect the effective thermal expansion tensor. Their influence is made possible by the temperature dependence of the constituents' compliance tensors. If the constituents' compliance tensors are temperature independent, then the initial eigenstrains have no effect on the effective thermal expansion even if the constituents' thermal expansion tensors are temperature dependent. If the constituents' compliance tensors depend on temperature, then the initial eigenstrains affect the effective thermal expansion tensor even if the constituents' thermal expansion tensors are temperature independent.
- The initial eigenstrains affect the effective heat capacity at constant stress. Again, their influence is made possible by the temperature dependence of the constituents' compliance tensors. If the constituents' compliance tensors depend on temperature, then the initial eigenstrains affect the effective heat capacity even if the constituents' heat capacities are temperature independent.

References

[1] Chen B S and Tong L Y 2005 *Comput. Method. Appl. M.* **194** 1891–911
[2] Bobaru F 2007 *J. Appl. Mech.-T. ASME* **74** 861–74
[3] Becker T L, Cannon R M and Ritchie R O 2000 *Mech. Mater.* **32** 85–97
[4] Hill M R and Lin W-Y 2002 *J. Eng. Mater. Technol.* **124** 185–91
[5] Boussaa D 2011 *Mech. Mater.* **43** 397–407

Journal of Physics: Conference Series **419** (2013) 012008 doi:10.1088/1742-6596/419/1/012008

Precipitation phases at different processes and heat treat ments as well as their effects on the mechanical propertie s of super-austenitic stainless steel

Hunying Sun[1,2], Zhangjian Zhou[1], Man Wang[1], Shaofu Li[1], Liwei Zhang [1] and Lei Zou[1]

[1]School of Materials Science and Engineering, University of Science and Technology Beijing,Xueyuan Road 30, Haidian District, Beijing 100083, PR China

[2]Technology Center, Anyang iron &Steel Group Company, Anyang Iron and Steel Avenue 502, Yin Du District, Anyang 455004, PR China

Corresponding author: Zhangjian Zhou, E-mail: zhouzhj@mater.ustb.edu.cn

Abstract. A new type 1Cr30Ni30Mo2TiZr super-austenitic stainless steel has been developed. The microstructures, precipitation phases and mechanical properties of the steel under different deformation processes and heat treatment (solution, stabilized treatment) were investigated using X-ray Diffraction (XRD), scanning electron microscopy (SEM), transmission electron microscopy (TEM) as well as mechanical tests. The results indicate that coarse carbides such as Cr-rich $M_{23}C_6$, sigma (σ), and little chi (χ) phases were formed in the steel, and large α′ -Cr phases were also detected at three joint grain boundaries, and they were promoted by large strain. The precipitate phases were dissolved or transformed to intermetallic phase even at higher elevated temperature, and influenced the mechanical property obviously. These intermetallic compounds seriously reduced elongation of the rolled steel at room temperature and 700℃, but increased the forged one at 700℃. Impact absorbed energies of the stabilized specimens were lower than half of that solution status.

1. Introduction

Super-austenitic stainless steels are specifically designed by controlling blends of alloying additions and fine precipitates, they provide excellent corrosion resistance, high-temperature creep strength and oxidation resistance [1,2]. Because the large amount of alloying additions, it is common that extended use at elevated temperature results in precipitation of a number of fine second phases, including carbides, nitrides, and intermetallic phases [1-4]. The most commonly observed carbide: $M_{23}C_6$, MC and M_6C, intermetallic compound: sigma, chi, and Laves phases, and other phases such as α′ -Cr, R phase and Z phase, depending on the particular alloy and heat treatment carried out [1-4].

For a long time, many studies focused on determining second phases and the specific precipitate forms in the stainless steels with the local compositions, heating times and temperatures [5-7]. In recent years, several works has studied the interrelationship between the formation, morphology of the second phases and matrix microstructures in detail. For example, before $M_{23}C_6$ precipitation grain boundary (GB) occurs serrated and infliction a dramatic effect on its formation in austenitic stainless steels [8]. Precipitates form on serrated GBs were usually planar or faceted in shape and lower in quantity [8, 9], whereas those that formed on planar GBs tend to be triangular and were numerous. Furthermore, an increase in the misorientation can favor the formation of triangular carbides [9].
The intermetallic compounds σ, χ, R and Laves phases are dominate in the super stainless steels. It is difficult to distinguish them from each other, and usually they can transform to another intermetallic phase [10]. The nucleation and subsequent growth of the intermetallic phases were observed in different alloys widely, they are affected by the aging times and temperatures [11]. In addition to aging, the sigma phase can also form by radiation-induced segregation in Fe-Cr alloys [12].In spite of the very numerous past studies, and have continued apace, but now difference viewpoint still exist about this transform. For example, is on earth $M_{23}C_6$ carbide transforms to σ phase or on the contrary; σ phase transforms to χ or R phases or not?

As is known that second phases can cause the mechanical properties and corrosion resistance of highly alloyed steels decrease remarkably [1, 5, 11, 13]. Studies have shown that the impact toughness of the alloys is extremely sensitive to small amounts of precipitations, even after short-term annealing at high temperature in super-austenitic steels [14]. However, Shek [15] has found that when the distribution and morphology of the sigma phase are properly controlled through appropriate pretreatments, the creep strength, ductility, yield strength and tensile strength of a 25Cr–8Ni duplex stainless steel can be enhanced. Actually, the effect of second phases in super alloys on the properties has not a verdict. Although many researchers have researched the second phases in stainless steel and their effect on the properties, it seems that no consensus has yet been reached these days. In this study, we report the secondary phases at the different processes and heat treatments as well as their effects on the mechanical properties of the high-Cr, Ni super-austenitic stainless steel1Cr30Ni30Mo2TiZr.

2. Experimental Procedure

The material was received from super-austenitic stainless steel which was melted in the vacuum inductive furnace, then casted for cylinder and forged at high temperature (1000-1200℃) to a cuboid, at last hot rolled to 4mm. Measured nominal compositions of the alloy are given in Table 1. Some samples were deformation state. Some were heated to the temperature range from 1100 to 1180 ℃ and hold for about 30min, then quenched into water immediately. The stabilized treatments were carried out at temperature varied between 900 to 1050℃ and hold for 180 to 210min, cooled the samples with furnace.

All samples for characterization were prepared by standard metallographic techniques. In order to determine the structure of the matrix and the precipitates, XRD analyses of the heat treated steel samples at different conditions were carried out. The microstructures and resulting phases were analyzed with SEM, the chemical composition obtained from energy-dispersive spectroscopy (EDS). Some more definitive characterization of microstructure was carried out with TEM. Tensile tests were performed using dumbbell-shape specimens at room temperature and 700 ℃, respectively. The impact samples were tested at room temperature and -40℃using an half size Sharpy impact with a V- groove.

Table 1. Chemical composition of the tested steel (wt., %)

Ni	Cr	Si	Ti	Zr	V	W	Mo	C	Cu	Fe
30.4	31.5	0.65	0.20	0.2	0.15	0.20	2.1	0.10	0.12	34.8.

3. Experimental Results

3.1 Second phases

The XRD spectra showed that the matrix of experimental steels was a homogeneous austenite (figure 1). In the thermal forging state steel contained carbide and a small amount of sigma phase. After a solution heat treatment at 1150℃ for 30min, the carbide peak became narrow and smaller, the amount of carbide reduced greatly. However, in the stabilized treatment, the $M_{23}C_6$ peak increase and its width became larger visibly, as well as the diffraction peaks of Mo, Cr-rich sigma phase was observed. But diffraction peaks of other second phases were too teeny to be detected.

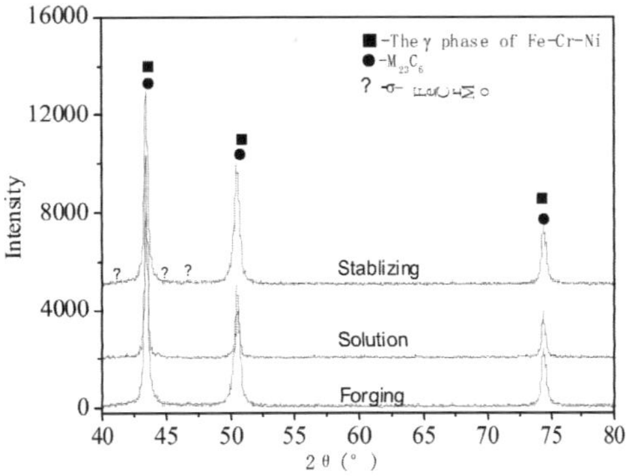

Figure. 1. XRD spectra of the forged steel at different treatments

Figure 2 showed SEM images of the different heat treatment specimens. In both the forged and hot rolled states, large amounts of second phases along GB were observed. In figure 2 (a), Some GBs presented serration which can be considered as an early term of precipitation (figure 2 (a) in the white arrows) [8], and only a few carbides with the shape of chain-like were observed on the GB. At a three-joint GBs lamellar α'-Cr had become coarsen in the result of higher energies. In rolled specimen, the amounts of the second phases increased greatly with the deformed strain increasing (see figure 2 (b)), and grew up into the matrix; some precipitates were seen within grains. Precipitates had a different morphology: the surface was planar, and the inner was a large number of coarse lamellar α'-Cr. Some white MC were discovered among the precipitates (arrow of P7 in figure 2 (b)), EDS showed its composition was C, Ti and Zr (as shown in table 2 P7). As figure 2 (c) shown, some second phases had not yet dissolved completely even in the solution-treated specimens. After were stabilized at 950℃ for 3.5 hours, the volume fraction of them increased rapidly, even formed a net along GBs off and on (see figure 2 (d)). Quite a number of precipitates dispersed to the intra-grains. Interestingly, some small precipitates came into being at large prior $M_{23}C_6$, EDS analysis showed they were sigma and chi phases (black arrows in figure 2 (d)).

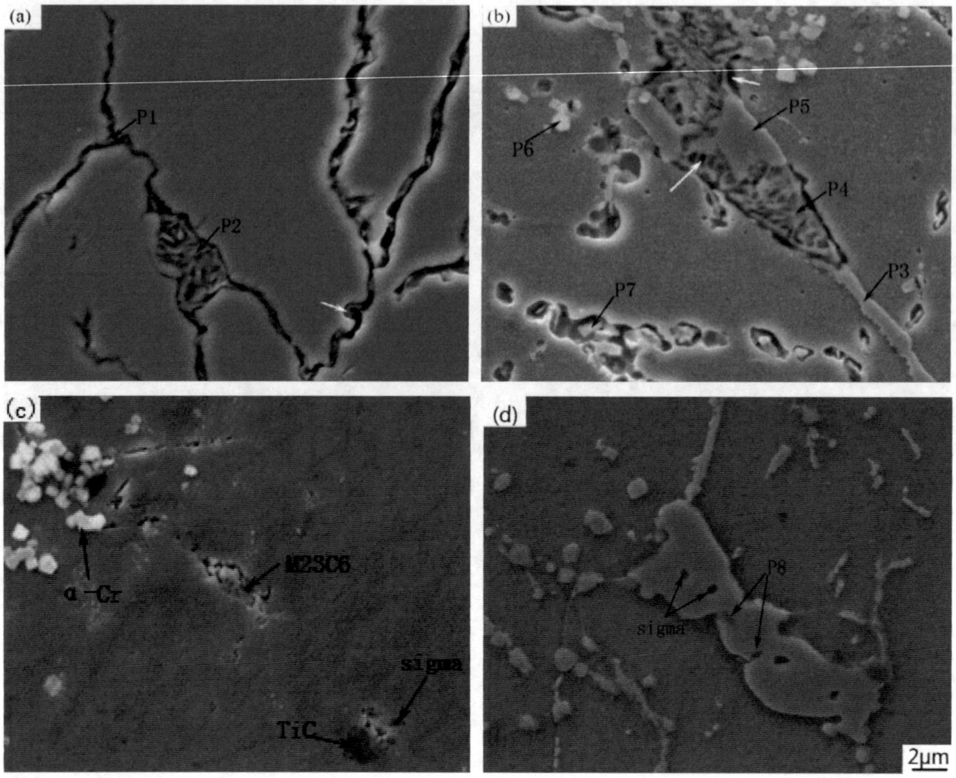

Figure. 2. SEM images of the tested steel at different states. (a) forging; (b) deformed to 50% at 1100℃; (c) solution of the forging steel; (d) stabilized treatment at 900 ℃ for 3.5h (pi(i=1-8) with arrow denote the precipitations which have been analyzed by EDS)

Table 2. EDS results of Fe, Cr, Mo and Ni of the labeled phases in figure 2 a-d

Precipitation	Wt., %					Phase
	C	Fe	Cr	Ni	Mo	
P1	5.05	16.38	73.71	4.87		Cr-rich
P2	3.11	14.74	74.66	7.4	4.47	Mo-rich-$M_{23}C_6$
P3	5.31	50.85	23.51	17.27		Fe-rich-$M_{23}C_6$
P4	3.24	15.63	68.80	7.14	5.19	$M_{23}C_6$
P5		39.4	46.15	10.44	4.98	Sigma
P6	0.49	15.55	75.18	3.01	6.33	α ' -Cr
P7	6.7	4.71	3.68	1.48	Ti33.29, Zr20.4	TiC, ZrC
P8		14.58	6.67	5.5	25.76,Ti 46.71	Chi

3.2 Mechanical properties

3.2.1 Tensile properties

Tensile test results at room temperature and 700 ℃ of the different processes and their own heat-treated samples were showed in figure 3. After heat treatment, the tensile strengths of specimens were lowered than their deformed states. On the contrary, the elongations of these specimens were larger than the deformed specimens. Comparing in figure 3 (a), the strengths of the rolling samples at room temperature were significantly higher than that of the forged specimens. After been heat treating, the results were inversed: the elongation of forging specimen were higher than that of rolled steel at room temperature (figure 3 (c)), and the results were on the contrary at high temperature except for the stabilizing state (see figure 3 (d)).

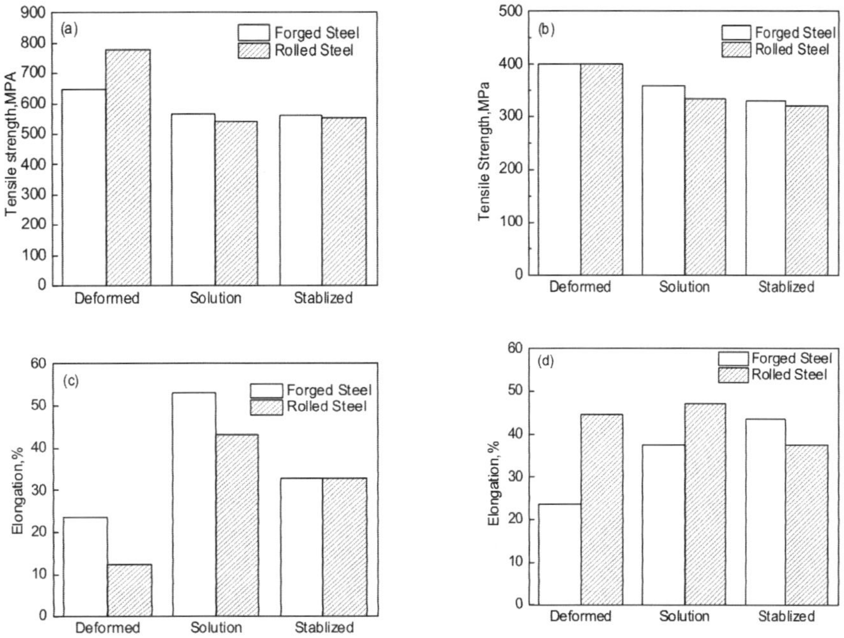

Figure. 3. Results of the tensile tests of the tested steel for forged and rolled steels
(a) tensile strength at room temperature, (b) tensile strength at 700℃,
(c) elongation at room temperature, (d) elongation at 700℃

3.2.2 Impact properties

The V-groove Charpy impact results were shown in figure 4. As a contrast, the data obtained from heat treated steels were also shown. Figure 4 notes that the impact energy of the tested steel was not sensitive to the temperatures, both the solution and stabilization treatment states. The proper solution-treated samples absorbed the more impact energy and did not completely fracture, even at the lower temperature (-40 ℃). But the impact ruptured of stabilizing treatment specimens at both the room temperature and -40 ° C; the impact energies were less than the half of the solution ones. Samples had become brittle during stabilizing treatment, so heat treatment specimen showed a significant drop in fracture toughness (figure 2 (b)) as studied by other studies [11, 13].

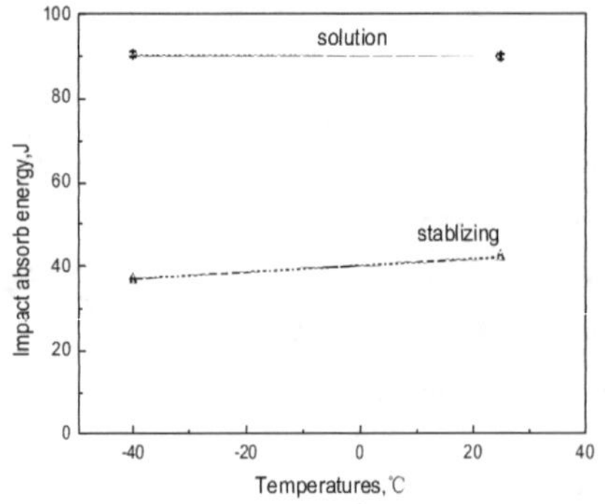

Figure. 4. Impact absorb energy of the tested steel

3.2.3 The effect of precipitates on the fracture mechanism

SEM observations of the fracture surfaces of solution specimens revealed a classical ductile fracture profile with many dimples. These dimples were very fine, shallow and uniformity (figure 5 (a) and figure 5 (b)). The low magnification SEM micrographs of the stabilizing treatment specimen exhibited more and larger cavities (marked with arrows) (figure 5 (c)). The magnified picture of the cavities showed as figure 5 (d). The nucleation of large cavities occurred at the position that existed coarse precipitates and there were some small cracks in them. The enlarged views of these cavities region showed some cleavage-like features. The fracture surfaces both of the impact and tensile specimens presented the brittleness characteristic with some coarse precipitates and visible cavities (see figure 5 (e)). Energy dispersive spectroscopy (EDS) analysis identified the particles indicating the difference in chemical composition between A and B, they were indicated as Cr-rich $M_{23}C_6$ and σ phase (see in figure 5 (f) and (g), respectively).

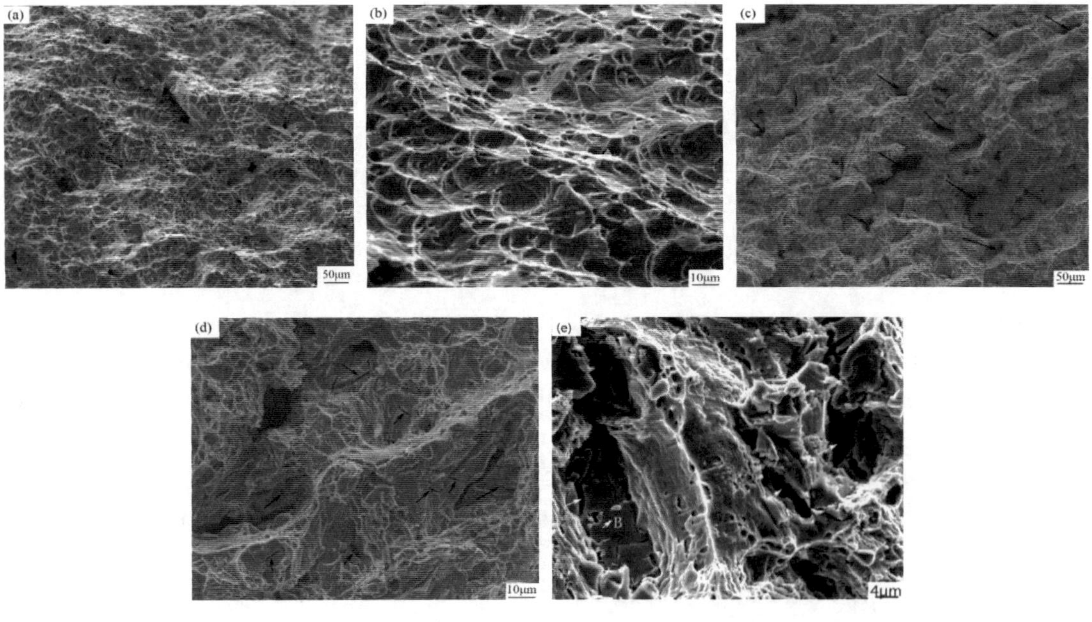

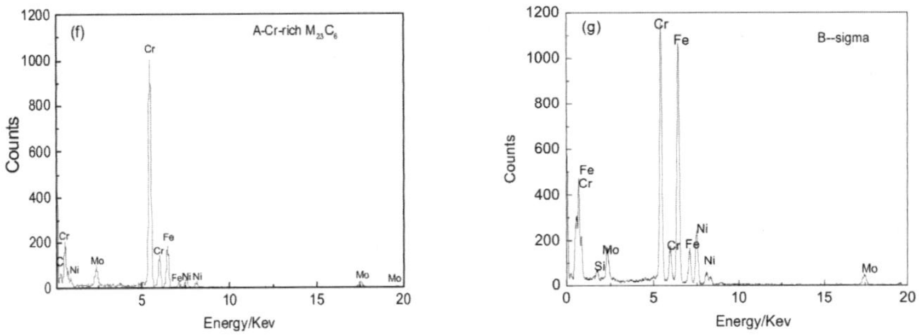

Figure. 5 SEM images of fracture surfaces at room temperature. (a) impact fracture surface of solution treatment specimen; (b) high magnification of (a); (c) impact fracture surface of stabilized treatment specimen at low magnification; (d) high magnification of (c); (e) tensile fracture surface of stabilized treatment specimen; (f) EDS analysis of A area in (d); (g) EDS analysis of B area in (e);

4. Discussion

4.1. Difference between sigma phase and $M_{23}C_6$ carbide, chi phase

For the super austenitic stainless steel, Cr content in Cr-rich carbide is in the range of 60-75% which is higher than that of other materials reported [17]. Cr-rich carbide and α′ -Cr are different in C contents, there are hardly any C for the latter. Moreover, α′ -Cr is lamellar, incompact, and there is some interstitial spaces away matrix (white arrows in figure 2 (b)). In the sigma phase of tested steel, the Mo content is higher than that in carbides, Cr and Fe contents are roughly equal in σ phase. Mo content in Chi phase is higher than that in the sigma; while the Cr content is lower in the chi phase. Usually, chi phase is spindly (figure 2 (d)). As shown in Table 2 (see figure2b the arrow P3), a Fe-rich $M_{23}C_6$ is observed in the super alloy, its color is slighter than Cr-rich $M_{23}C_6$. It is mentioned that the Fe-rich $M_{23}C_6$ is reported rarely, but Cr-rich of $M_{23}C_6$ has been reported frequently in stainless steel.

From figure 2 (a) and (b), $M_{23}C_6$ and lamellar α′ -Cr are prior precipitated on GBs, deformation or heat treatment make precipitates coarsen or transform to another phase. During α′ -Cr growing up to inner matrix, surface layer change and lamellar characteristics disappear, then the other coarse precipitate comes into being. Because of the large interspace between lamellar α′ -Cr and matrix, the coarse precipitate segments transformed from α′ -Cr tend to break off and α′ -Cr is exposed (as show in figure 2 (b)), EDS detect it is sigma phase. Moreover, vivid processing of carbide transform to intermetallic phases as is shown figure 2 (d), It is can be speculated that σ and χ phased nucleate at $M_{23}C_6$, this phenomenon is in accordance with the research by Tavares [18]. But this study can't determinate σ and χ which form firstly.

4.2 The effects of the second phases on the properties

Because the rolling sample contains more second phases and their strengthening effect (shown in figure 2 (a)-(d), the strength of rolled steel is higher than that of a forged state at room temperature. However, the second phases have coarsened, and the most are brittle and hard ($M_{23}C_6$), thus the elongation will decline sharply. During heat treatment processing, the second phases solute back to matrix, presence high temperature recoveries and a lot of dislocations move and counteract to each other, so significant reduction in tensile strength of the solution specimen was observed. Despite precipitation phase is harmful in the stabilization process, the second phases in forging specimen is relatively few and small, the process of reheat recovery and diffusion make second phases be dispersion, uniform in the intra-granular, which effectively improve the strength and elongation of the material at 700 ℃ . As shown in figure 4 (d), the effect of precipitates on the elongation at high temperature is harmless, even beneficial. This effect is consistent with the other researches [16]. It is believed that the fine and homogenous dispersion of ($\sigma + \gamma$) is conductive to tensile elongation, because large cavities may easily form at the boundaries of large austenite grains [15].

The hard and brittle sigma phase can result in harmful influence on the properties of the alloy [19]. As shown in figure 5 (e), the sigma phase cracks at about 550 MPa during room-temperature tensile tests. Nevertheless, Pohl et al. [20] has found that the sigma phase cracked at about 600 MPa during tensile tests at 700℃. The strength of sigma is lower than that at 700℃, so it is more inclined to occur crack at room temperature, and this can help to understand the elongations at both temperatures. Under load stress, it is impossible for coarse precipitates to absorb energy by their own strain and deformation, so they are prone to be nucleation sites for micro cracks (marked with an arrow pointing up in figure 5 (d)) or serve as for larger dimples or cavities with ductility matrix strain increasing. In the present case, these coarse precipitates react on the properties of the material greatly, especially the toughness and stretch. As shown in figure 3 (c) and (d), the elongations of stabilizing specimen are lower than those of solution one at room temperature and 700 ℃， moreover, the impact absorb energies are even lower than those half of solution one (see figure 4).

Possible factors related to this unexpected drop in fracture toughness and ductility might include location of the precipitation (along grain boundaries or not), type of precipitation (embrittling carbide or intermetallic), and shape (acicular or spheroidal) [10]. In this article, some carbide clusters in stabilizing specimen were discovered at the high angle GBs in both forged and rolled steels. They were triangle and aculeate, arranging trimly in parallel. At same time, some polygonal sigma phases were also observed in rolled steel (figure 6 (b)). These triangle and polygonal are deleterious to material properties, But planar carbide and spheroidal precipitations are beneficial, so it is necessary to control heat treatments parameters to obtain them [10, 21].

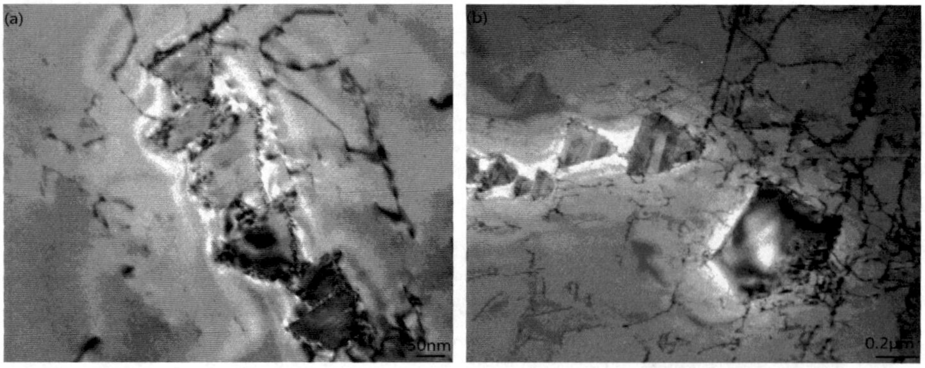

Figure. 6. Triangle carbides cluster in stabilizing specimens. (a) forged steel; (b)rolled steel

5. Conclusions

The main results obtain from this study are as follows:

(1) Many phases carbides ($M_{23}C_6$, MC), intermetallic phases (sigma, chi) were precipitated in the super austenitic steel. The precipitation of these phases was promoted significantly by hot-rolled process at elevated temperature.

(2) Carbide was a preferential precipitate than sigma and chi phase. These carbides transform into sigma phase, chi phase during short-term stabilized treatment or large deformation steel.

(3) After heat treatment, forging specimen and rolled steel showed different change law in elongation. The latter showed a more drastic decrease of impact energy due to containing the more precipitates.

(4) The sigma phase cracked at about 550 MPa during room-temperature tensile tests.

(5) In stabilized specimens, a series of triangle carbides clusters were observed in both forging and rolled steel. They were considered be more harmful to ductility of the materials.

Acknowledgements

The authors are grateful for the project supported by the National Basic Research Program of China (Grant No. 2007CB209800).

Reference

[1] Heino S 2000 *Mater. Trans.* A. **31** 1893
[2] Heino S, Knutson-Wedel M and Karlsson B 1999 *Mater. Sci. Forum* **318–320** 143
[3] Svoboda M, Kroupa A, Soousek J, Vrest'al J and Miodownik P 2004 Z Metallk. **95** 1025
[4] Lee T H, Kim S J and Jung Y C 2000 Metall. *Mater. Trans.* A **31** 1713
[5] Kim Y J, Chumbley L S and Gleeson B 2004 *Met. Trans.* A **35** 3377
[6] Kim Y J, Ugurlu O, Jiang C, Gleeson B and Chumbley L S 2007 *Meter. Trans.* **38** 203
[7] Kim Y J, Chumbley L S and Gleeson B 2008 *J. Mater. Eng. Perform.* **17** 32
[8] Hong H U and Nam SW 2002 *Mater. Sci. and Eng.* A **332** 255
[9] Hong H U, Rho B S, Nam S W, 2001 *Mater. Sci. Eng.* **A 318** 285
[10] Lo K H, Shek C H and Lai J K L 2009 *Mater. Sci. and Eng.* R **65** 39
[11] Muller C and Scott Chumbley L 2010 *J. Mater. Eng. Perform.* **19** 714
[12] Chakin V, Kazakov V, Goncharenko Yu, Ostrovsky Z, 1996 *J. Nucl. Mater.* **233–237** 573
[13] Chen TH, Weng KL and Yang J R 2002 *Mater. Sci. Eng.* A **338** 259
[14] Martins M, Rodrigues L and Forti N 2008 *Mater. Charact.* **59** 162
[15] Shek C H, Li D J, Wong K W and Lai J K L 1999 *Mater. Sci. Eng.* A **266** 30
[16] Han Y S and Hong S H 1997 Scripta Mater. **36** 557
[17] Xiang H L, He F S and Liu D 2009 *Acta Metall. Sini.* **45** 1456
[18] Tavares S S M, Moura V, Da Costa V C, Ferreira, M L R and Pardal J M, 2009 *Mater. Charact.* **60** 573
[19] Gojić M, Nagode, A, Kosec B, Kožuh S, Šavli, Š, Holjevac-Grgurić T and Kosec L 2011 Engineering Failure Analysis doi: 10.1016/j.engfailana 2011.07.016
[20] Pohl M, Storz O and Glogowski T 2007 *Mater. Charact.* **58** 65.
[21] Kim KJ, Ginsztker J and Nam SW 2005 *Mater. Lett.* **59** 1439

The fabrication and characterization of polymeric microcellular foams with designed gradient density

Qiang Shen, Yuanlu Xiong, Huan Yuan, Guoqiang Luo*, Xue Liang, Lianmeng Zhang

State Key Laboratory of Advanced Technology for Materials Synthesis and Processing,Wuhan University of Technology, Wuhan 430070, P.R. China

E-mail: luoguoqiang1980@sina.com

Abstract. The fabrication of polymeric foams with gradient density is a key to investigate the response of materials in quasi-isentropic loading. In this work, a non-traditional approach was proposed to fabricate the polymeric foam monolayer with designed density in a fixed volume by using the pre-mixture of expandable and expanded polymeric microballoons. After heating, the expandable microballoons expanded and bonded together to be integrated block with the final density from 0.1 g/cm^3 to 0.5 g/cm^3. The gradient density foams were fabricated by bonding with different monolayer. The microstructure and mechanical properties of each layer were investigated by SEM and compression strength tests. The results showed that the density of polymeric foams was matched with the design value very well by controlling the ratio of expandable and expanded polymeric microballoons. The uniform closed-cell structure with good bonding interface was shown from the SEM picture. And the mean cell size is less than 35μm. The polymeric foams exhibited ideal compressive behavior with the compressive strength as high as 9MPa. The Young's modulus and compressive strength nearly linear increased with the increasing of density. The extremely wide range of accessible densities should make this technology for broad application prospects.

1. Introduction

The quasi-isentropic compression technique has been an important way for measuring material's properties at high pressure, due to the accompanying temperature rise is much less severe than during shock-loading experiments [1,2]. It always used a density gradient material with increasing shock impedance away from the target material with light gas-gun [3,4]. This density gradient material is called impactor. The ideal impactor requires the densities ranging from 0.1 g/cm^3 to 15 g/cm^3 over a space of several millimeters [5]. The density gradient impactor under high and medium density region has been prepared successfully by tape casting method with different metal powder mixtures [2,6]. However, the impactor with low density region (from 0.1 g/cm^3 to 1.2 g/cm^3) is still hard to prepare by the same techniques.

The low density gradient material is usually composed with polymeric foams by stacking with different foam monolayer slices. Chen [7] has fabricated a multi-layer gradient foams material with the density range from 0.5 g/cm^3 to 1.0 g/cm^3 by stacking six epoxy resin foam monolayer together. Much lower density layer can not be made using the same preparation technique of mixing the epoxy resin matrix with plastic microballoons. Chemical or physical foaming method can easily prepare polymeric foams with much lower density [8]. However, it is hard to control the products with designed density due to the uncontrollable foaming speed. And the monolayer slices with required shape (less than 1mm) are difficult to process successfully by such a low density and large cell size foam block.

In order to fabricate the low density foam monolayer with designed density, a non-traditional approach was proposed to fabricate the polymeric foams in a fixed volume by using the pre-mixture of expandable and expanded polymeric microballoons together. After heating, it allows only expansion of the expandable powder to fill the interstices between the expanded microballoons [9,10]. The real density, microstructure and mechanical properties of the polymeric foams were investigated.

2. Experimental

The expandable microballoons (F-30D) and the expanded microballoons (FN-80SDE) were both purchased from Matsumoto Yushi-Seiyaku Co., Ltd., Japan. SEM images of the expandable powder and expanded powder are shown in Figure 1. The particle size of expandable microballoons is 10~20μm, and the expanded microballoons is 20~40μm. The packing density of F-30D and FN-80SDE was calculated by measuring the mass of the powder and the real volume occupied by the powder in a 100 ml graduated cylinder after tapping 50 times on the counter.

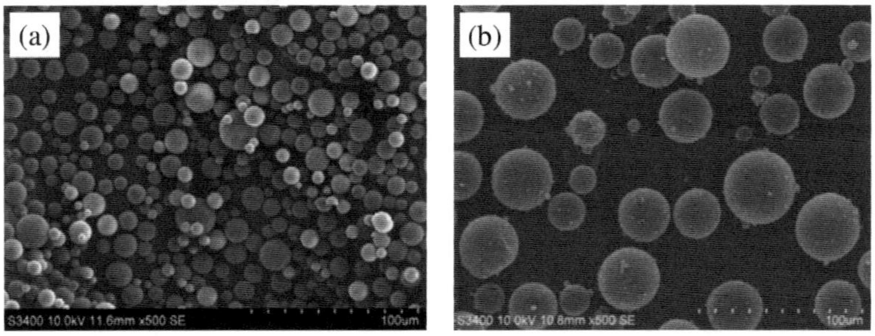

Figure 1 the SEM images of (a) expandable and (b) expanded microballoons

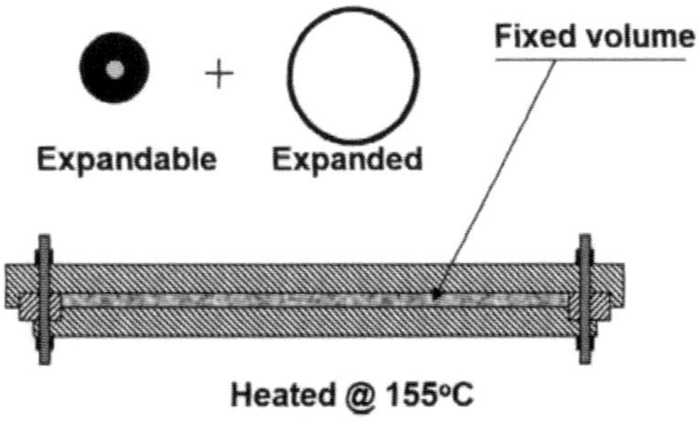

Figure 2 The schematic diagram of the fixed volume expansion technique

The expandable and expanded polymeric microballoons were mixed homogeneously together at a ratio such that the tap density of the mixed powder was the same as the final density of the foams. The designed density of each foam monolayers is varied from 0.1 g/cm^3 to 0.5 g/cm^3. After completely filling the mixed powder into the fixed volume mold, it was heated to 155 °C for 30 minutes in the oven. Therefore, the expandable microballoons expanded and bonded together as integrated foams. The schematic diagram is shown in Figure 2. The fixed volume is about ϕ50mm × 1mm for the preparation of monolayer slices. After totally cooling down, the sample was demoded. The real density was obtained by measuring the volume and weight of different monolayer. Samples were freeze fractured in liquid nitrogen and the fracture surface was sputter-coated with gold before SEM test. For mechanical strength test, the sample was fabricated in another mold with a fixed volume of ϕ20mm × 16mm.

3. Results and discussion

Results show that the packing density is 0.5050 g/cm^3 for expandable microballoons and 0.0086 g/cm^3 for expanded microballoons respectively. In a fixed volume, the design density is determined by the weight ratio between expandable and expanded microballoons. Figure 3 shows the tested densities and theoretical densities of the foams. The density decreased as the weight fraction of expanded powder increased. And the experimental data is very close to the theoretical value. It illustrates that the density of the polymeric foams is able to be controlled precisely by this fixed volume foaming method.

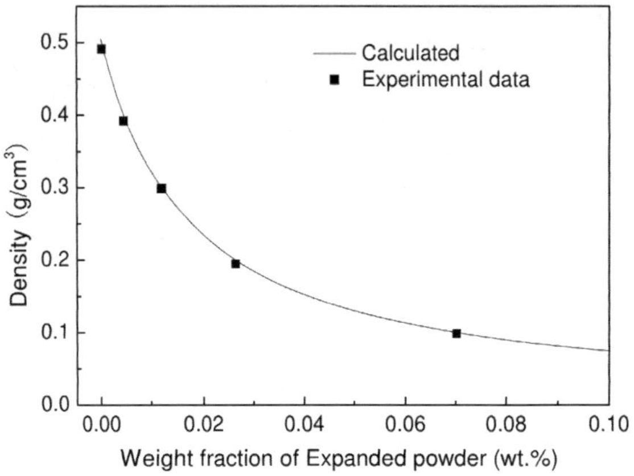

Figure 3 Tested and theoretical density value of monolayer at composition of different weight fraction of expanded powder

Figure 4 shows the SEM images of fracture surfaces of monolayer foam with different magnification and different density. It can be seen that the polymeric foams has a very smooth surface and neat shape with the thickness of 1mm. Much thinner monolayer can be obtained by adjusting the thickness of the fixed volume mold. So the density gradient foams could be easily prepared by tacking several monolayer slices together with epoxy resin as bonder. From the high magnification images shown in Figure 4(b) and (d), it is clear that the nice uniform closed-cell structure with the pore size less than 35 m was gotten. The polymeric foam with higher density has much smaller pore size. And the pore structure tended to deformed shown in Figure 4(c). It attributes to the expandable powder compete to grow up under heating. When the density of the foam was low, the pore structure can grow freely and leads a larger pore size in such a fixed volume which was shown in Figure 5(e).

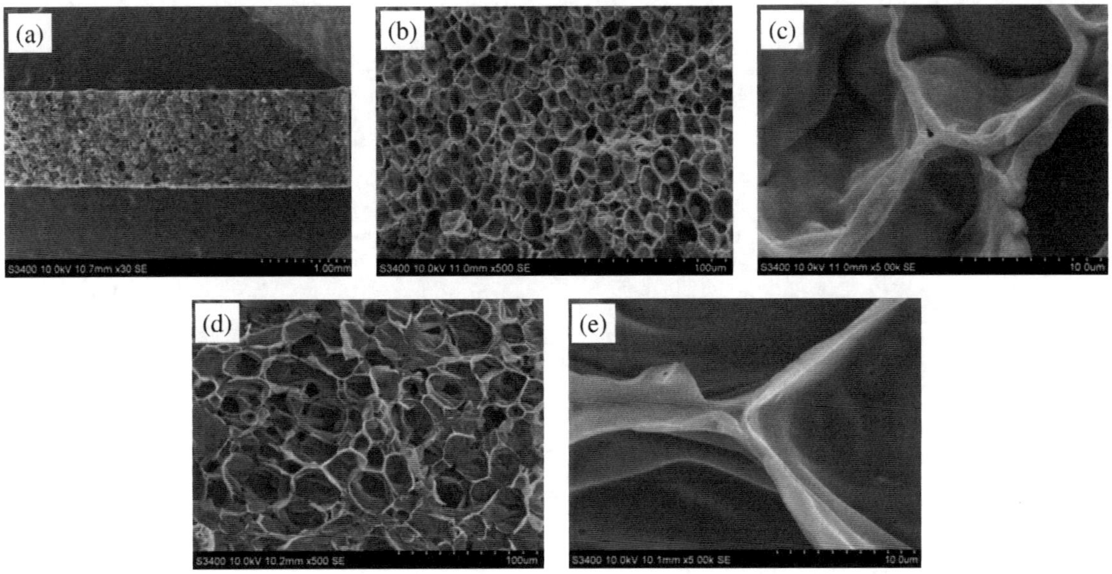

Figure 4 SEM images of fracture surfaces of monolayer foam with different density: (a),(b),(c) 0.4 g/cm^3; (d) (e) 0.1 g/cm^3

In addition, the foams show a very perfect bonding interface which implies the excellent mechanical properties according to the figure 4(c) and (d). Actually, those microballoons contain vinyl chloride which provides an activated shell surface. When the heating temperature is higher than the initial softening temperature (about 80~85°C for F-30D), the expandable microballoons will grow bigger and bigger due to the gasification phenomenon of the inside volatiles [11,12]. The shell surface is bonded together finally to be a foam block.

Figure 5(a) shows the compressive stress-strain curves of foams as a function of density varied from 0.1 to 0.5 g/cm^3, respectively. The typically linear regime and plateau regime of the stress-strain curves are very clear. The plateau stress is nearly constant during the plastic plateau regime which illustrates that the foam products exhibited nearly ideal compressive behavior. Figure 5(b) shows the change of Young's modulus and compressive strength as a function of density. The compressive strength can achieve as high as 9MPa when the density is about 0.5 g/cm^3. Both the Young's modulus and the compressive strength decreased linearly with the decreasing of density.

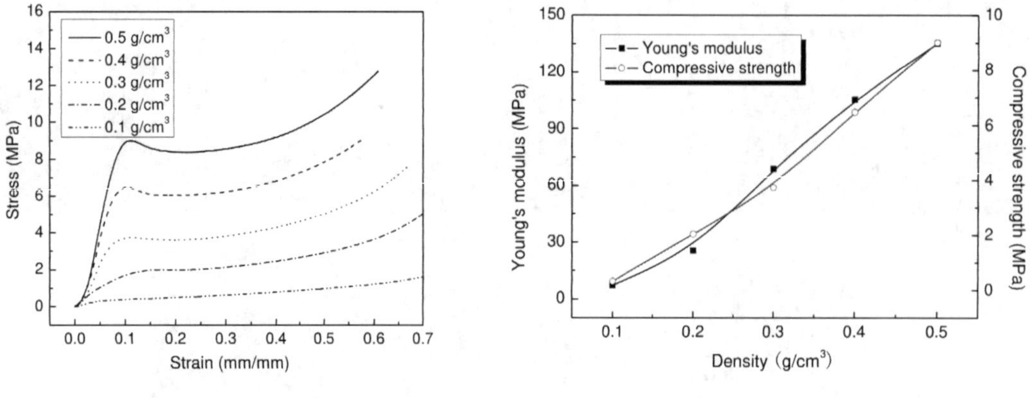

Figure 5 (a) the compressive stress-strain curves of foams; (b) the change of Young's modulus and compressive strength as a function of density

4. Conclusion

The foam monolayer slices with designed density could be fabricated by using the pre-mixture of expandable and expanded polymeric microballoons in a fixed volume mode under heating. The density can be controlled precisely varied from 0.1 to 0.5 g/cm^3. The cell size is less than 35μm. And it has a very perfect bonding interface which promotes the compressive strength up to 9 MPa. This work offers a simple method to get the low density gradient material by stacking monolayer slices with different fixed density.

Acknowledgements

This work is supported by the National Natural Science Foundation of China (51202175) and the 111 Project（B13035）.

References

[1] J.H. Nguyen, N.C. Holmes 2004 *Nature* **427** 339-342

[2] L.P. Martin, D. Orlikowski and J.H. Nguyen 2006 *Mat. Sci. Eng. A* **427** 83-91

[3] R.T. Krone, L.P. Martin, J.R. Patterson, D. Orlikowski and J.H. Nguyen 2008 *Mat. Sci. Eng. A* **479** 300-305

[4] H. Jarmakani, J.M. McNaney, B. Kad, D. Orlikowski, J.H. Nguyen and M.A. Meyers 2007 *Mat. Sci. Eng. A* **463** 249-262

[5] F.H. Streitz, J.H. Nguyen, D. Orlikowski, R. Minich, J.A. Moriarty and N.C. Holmes 2005, *Final Report 02-ERD-033, LLNL*, California

[6] L.P Martin, J.R. Patterson, D. Orlikowski and J.H. Nguyen 2007 *J. Appl. Phys.* **102** 23507-23510

[7] Z. Chen, Y. Qin, Z. Huang, Q. Shen and L. Zhang 2009 *Adv. Mater. Res.* **66** 284-287

[8] L.M. Matuana, O. Faruk and C.A. Diaz 2009 *Bioresource Technol.* **100(23)** 5947-5954

[9] L. Vaikhanksi, S.R. Nutt 2003 *Compos. A* **34** 755-763

[0] M. Jonsson, O. Nordin, E. Malmstrom and C. Hammer 2006 *Polymer* **47** 3315-3324

[11] H. Andersson, P. Griss and G. Stemme 2002 *Sensor. Actuat. B* **84** 290-295

[2] M. Jonsson, D. Nystrom, O. Nordin, E. Malmstrom 2009 *Eur. Polym. J.* **45** 2374-2382

Fabrication of Al-W Functionally Graded Impact Material via Vacuum Hot-Pressing Sintering Method

Pingan Chen, Guoqiang Luo*, Yinghu Liu, Qiang Shen, Lianmeng Zhang
State Key Lab of Advanced Technnology for Materials Synthesis and Processing,Wuhan University of Technology, 122 Luoshi Road, Wuhan 430070, P.R.China
E-mail:luoguoqiang1980@sina.com

Abstract Dense graded material as a type of functionally graded material (FGM) changes its wave impedance gradually along the thickness direction. In this investigation, Al-W functionally graded material was fabricated via vacuum hot-pressing sintering method (VHPS). The results showed that densified Al-W composite was fabricated at 550°C-300MPa-120min which the relative density was higher than 98.5% without intermetallic compounds. The density graded material of Al-W FGM was attained at the optimized parameters which the component of Al was from 10% to 100%. The microstructure of Al-W FGM composite indicated that W particles in single-layer composite were distributed homogeneously. The density of each layer in Al-W FGM composite was consistent with the design value.

1. Introduction

Functionally Graded Material is a kind of inhomogeneous composite, and its properties change gradually with structure and components which varies continually or quasi-continually [1-2]. Composites with graded wave impedance can be regarded as a special kind of functionally graded materials (FGMs), which create quasi-isentropic loading to target materials, and have come to show great potential for the application in dynamic high-pressure technology [3-6]. So the composites with graded wave impedance have attracted many attentions of researchers in the world.

Researchers have fabricated the simple binary functionally graded materials by decreasing components of the composites to control concisely the constituent and graded wave impedance variations [6-10]. F. H. Streitz and H. Nguyen [11] et al. have prepared Al-W functionally graded materials by powder metallurgy. The components of Al-W composites changed graded layer by layer and the polymer was added as binder in the composites. However, it was still not binary system because polymer was the third phase in the composites, and there was not research on the mechanism of densification. L. P. Martin [12] et al. have made Mg-W graded file-plates by type-casting to achieve the density quasi-continuous variations which varied from 1.74 $g\cdot cm^{-3}$ to 8.90 $g\cdot cm^{-3}$. By this method, the wave front could be controlled and the correlation between density and acoustic wave impedance was monotonous nearly linear which was consistent with the simulation. Therefore, the composites of Mg-W have achieved quasi-continuous variation for the wave impedance.

In this paper, we aimed to fabricate high densification Al-W composites with different components and Al-W functionally graded material which meets the design requirement. The hot-pressing sintering has been used to prepare Al-W composites with different components. The effects of W contents and processing parameters on the densification of Al-W composites have been studied to acquire the optimum processing parameters. Al-W functionally graded materials with density gradient have been fabricated successfully, and the structure and constituent of Al-W functionally graded materials have been characterized.

2. Experimental

In the present work, the starting W powders (particle size of 5 μm, purity>99.9%) were provided by Beijing Nonferrous Metal Research Institute, Al powders (particle size of 6 μm, purity > 99.9%) were provided by Xiamen Golden Egret Special Alloy Co. Ltd. The process was as follows：Al and W powders were weighed, and mixed in air in a shaker-mixer for 12h to produce Al-W powder blend. The long hours milling was to make distribution of the different powders more uniform. After the milling completed, the mixture was cold pressed into 10-mm-diameter and 3-mm-thick green compacts at a pressure of 400 MPa without any lubricant. The compacts were heated up in vacuum hot-pressing sintering furnace.

The density of sintered compacts was measured by the Archimedes principle. The theoretical density of the samples was calculated according to the rule of mixtures. Scanning electron microscopy (SEM, JSM-S3400N) was used to evaluate the microstructure and EDS of Al-W samples. X-ray diffraction (XRD, UltimaIII) for phase analysis was collected.

3. Results and discussion

3.1 Effect of sintering temperature on the phase of Al-W composites

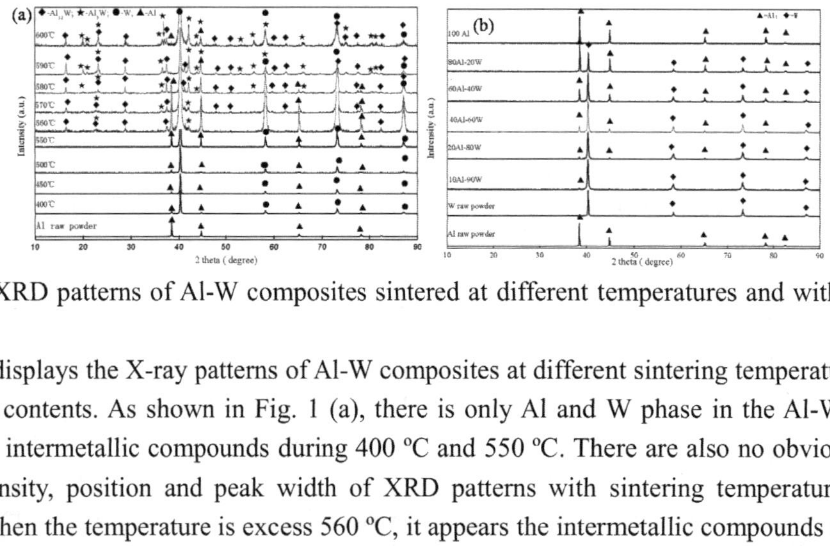

Fig. 1 The XRD patterns of Al-W composites sintered at different temperatures and with different W contents

Fig. 1 displays the X-ray patterns of Al-W composites at different sintering temperatures and with different W contents. As shown in Fig. 1 (a), there is only Al and W phase in the Al-W composites without any intermetallic compounds during 400 °C and 550 °C. There are also no obviously changes for the intensity, position and peak width of XRD patterns with sintering temperature increasing. However, when the temperature is excess 560 °C, it appears the intermetallic compounds of $Al_{12}W$ and Al_5W diffraction peaks besides Al and W peaks in XRD patterns. And the diffraction peak intensity of $Al_{12}W$ and Al_5W increase gradually in contrast to the intensity decreasing of Al and W diffraction peaks with increasing temperature. There is almost no Al phase diffraction peak at 600 °C. So Al is melted from 560 °C from which it forms intermetallic compounds. Because of the brittle of intermetallic compounds, the sintering temperature should be lower than 550 °C to avoid the formation of intermetallic compounds.

Fig. 1 (b) shows the XRD patterns of Al-W composites with different W contents at 550 °C. As it is shown, there is only Al and W phase diffraction peaks in the Al-W composites without any intermetallic compounds. With increasing W contents, the intensity of Al diffraction peak weaken, in contrast to that, the intensity of W diffraction peak makes stronger. And there is no position shifting or width variation for Al and W phase. It indicates that there is no obviously solid solution or formation of intermetallic compounds between Al and W at 550 °C for different W contents in the Al-W composites.

Therefore, in order to achieve high densified Al-W composites and ensure no formation of intermetallic compounds, the sintering temperature should be below 560 °C. And when the sintering temperature is 550 °C, there is only Al and W phase in the Al-W composites. So the sintering temperature is chosen to 550 °C to acquire densified Al-W composites.

3.2 Fabrication of Al-W functionally graded materials

For the Al-W functionally graded material, the front interface density of ρ_0 is the density of Al which is 2.70 g·cm^{-3}, the back interface density is 8.652 g·cm^{-3}. In this work, the total thickness for Al-W functionally graded material is designed as 3.5 mm, which the single layer thickness is 0.5mm. The processing parameter is 550°C-300MPa-120min.

3.2.1 The microstructure of Al-W functionally graded material

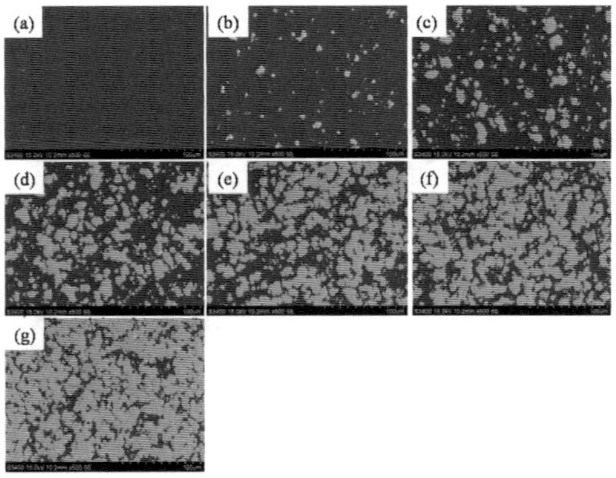

Fig. 2 the microstructures of the Al-W FGMs for each graded layer: (a) 100Al (b) 88.5Al-12.5W (c) 66Al-34.0W (d) 48.6Al-51.4W (e) 36.9Al-63.1W (f) 28.5Al-72.5W (g) 20.0Al-80.0W

Fig. 2 is the microstructure of each layer of Al-W functionally graded material. It can be shown that every layer of Al-W functionally graded material is densified, and Al and W are distributed homogeneously. With layers increasing, the W content increases while the Al content decreases. The first layer is pure Al. The W particles are almost distributed homogeneously in the Al matrix from the second to fourth layers. In the fifth and sixth layers Al and W are both distributed homogeneously, and in the seventh layer W particles are connected tightly and Al is filled into the pores between W particles.

3.2.2 EDS of Al-W density functionally graded material

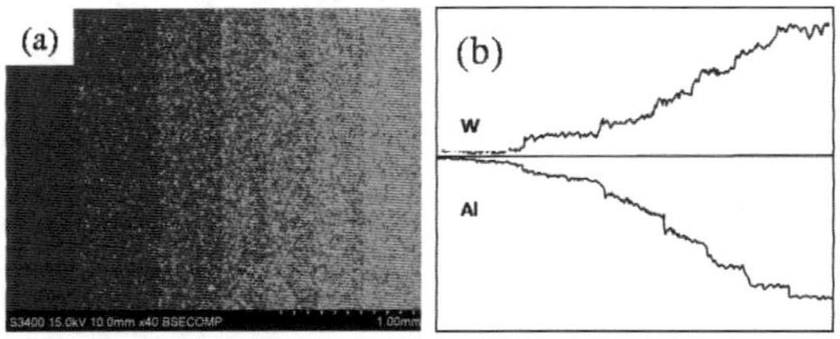

Fig. 3 the SEM image: (a) and EDS (b) of the Al-W FGMs

Fig. 3 shows the microstructure of Al-W functionally graded material with density gradient and EDS along the gradient direction. From Fig. 3 (a), it can be seen that the Al-W functionally graded material is nearly full-density; there is no obviously pores or any other defects in the composite. What's more, the transition layer interface is clear and the parallelism is good, which indicate that there is no fracture or dislocation for every layer during the sintering process. From Fig. 3 (b), it can be clearly seen that it appears good gradient variation along the thickness.

4. Conclusion

In this work, Al-W functionally graded material with density gradient is designed and the optimum processing parameters are acquired. And the Al-W functionally graded material which the constituent varies continually is fabricated under the optimum parameters.

1. The relative density of Al-W composites increases with temperature. When sintering temperature is below 550 °C, only Al and W phase can be detected, however, when sintering temperature exceeds to 550 °C, there is Al-W intermetallic compounds.

2. The transition layer interface of Al-W functionally graded material is clear and the parallelism is good. The components of Al and W exhibit well gradient distribution. Al-W functionally graded material is near full-density and Al and W particles are distributed homogeneously.

Acknowledgement

This work has been supported by National Natural Science Foundation of China (51202175) and self-determined and innovative research funds of WUT (2012-ZY-005) and the 111 Project (B13035).

Reference

[1] M.Niino, T.Hirai, R.Watanabe. Japanese Soci Comp 1987 13 257-64

[2] Kawasaki and R.Watanabe. Japanese Soci Powd Metall 1990 37 253-60

[3] L. M. Zhang, C. B. Wang, Q. Shen and J. G. Li. Ceramic Trans, 2001 114 667-9

[4] Q. Shen, L. M. Zhang, H. P. Xiong, J. S. Hua and H. Tan. Chinese Sci Bull 2000 45 1421-5

[5] C. B. Wang, L. M. Zhang and Q. Shen. Mater Sci Forum 2003 423-425 803-6

[6] G. Luo, J. Bai, H. Tan, et al. Metall Mater Trans A 2010 41 2389-95

[7] H. G. Tang, X. F. Ma, W. Zhao, et al. J Alloy Compd 2002 347 228-30

[8] M. M. Hukovic, N. Radic, Z. Grubac, et al. Electrochim Acta 2002 47 2387-97

[9] M. Stubicar, A. Tonejc, N. Radic. Vacuum 2001 61 309-16

[10] R. Tu, Q. Shen, J. S. Hua, et al. Proceedings of the 4th International Symposium on Functionally Graded Materials 1996 307-11

[11] F. H. Streitz, J. H. Nguyen, D. Orlikowski, et al. Final report 02-ERD-033 February 14, 2005

[12] L. P. Martin, D. Orlikowski and J. H. Nguyen. Mater. Sci Eng A 2006 427 83-91

[13] Jing Zhou. Master's degree thesis. Wuhan: Wuhan University of Technology. 2007

Molecular Dynamic Simulation of Failure of Ettringite

W Sun, D Wang and L Wang
3500 Transportation Plaza, Virginia Tech Transportation Institute, Blacksburg, VA, 24060, US

E-mail: sunwj@vt.edu, wangd@vt.edu, wangl@vt.edu

Abstract. Ettringite is an important component in the hydration products of cement paste. To better understand the failure modes under tensile loading of cement-based materials, mechanical properties of each individual hydration product needs to be evaluated at atomic scale. This paper presents a molecular dynamic (MD) method to characterize and understand the mechanical properties of ettringite and its failure modes. The molecular structure of ettringite is established using ReaxFF force field package in LAMMPS. To characterize the atomic failure modes of cement paste, MD simulations were conducted by applying tensile strain load and shear strain load, respectively. In each MD failure simulation, the stress-strain relationship was plotted to quantify the mechanical properties at atomic scale. Then elastic constants of the ettringite crystal structure were calculated from these stress-strain relationships. MD simulations were validated by comparing the mechanical properties calculated from LAMMPS and those acquired from experiments. Future research should be performed on bridging-relationships of mechanical properties between atomic scale and macroscale to provide insights into further understanding the influence of mechanical properties at atomic scale on the performance of cement-based materials at macroscale.

1. Introduction

Tensile strength is one of the major mechanical properties influencing the initiation and propagation of cracks in cement-based materials. Cracks exhibit more in cement paste with low tensile strength, consequently reducing durability of cement-based materials. Therefore, it is meaningful to understand how the mechanical properties influence the performance of hardened cement paste and other cement-based materials at atomic scale.

There are mainly four types of cement hydrates: calcium silicate hydrate (C-S-H), calcium hydroxide (CH), "alumina, ferric oxide, tri-sulfate" (AFm) and "Alumina, Ferric oxide, mono-sulfate" (AFt) phases (including ettringite, monosulphate, and monocarbonate, etc.). Figure 1 plots the volume percentage of each composition in hydrated Portland cement paste after the hydration of 14 months [1]. Among these cement hydration products, ettringite is a calcium aluminium sulfate mineral crystalizing in the trigonal system with the chemical formula of $Ca_6Al_2(SO_4)_3(OH)_{12}{\cdot}26H_2O$. It plays an important role in the crystallization development of cement hydration products and in the deterioration of hardened cement concrete due to sulphate attack [2-4].

This paper conducts series of molecular dynamic (MD) simulations using the ettringite molecular structure in Large-scale Atomic/Molecular Massively Parallel Simulator (LAMMPS). The mechanical properties of the ettringite molecular structure under strain loadings are investigated in MD simulations with the comparison with experimental results. A detailed understanding of the ettringite regarding to mechanical properties and failure modes is helpful for the prediction of crack failures in the long-term performance of hydrated cement concrete at macroscale.

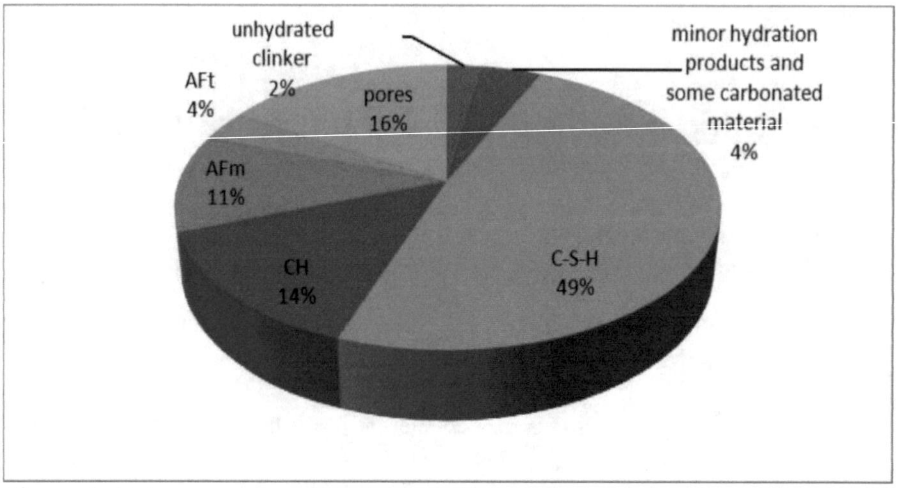

Figure 1. Volume percentage of phases in hydrated Portland cement paste, water-cement ratio=0.5, hydrated for 14 months (data are from [1]).

2. Computational theory

Ettringite molecular structure is established based on ReaxFF force field in LAMMPS; the optimized parameters are documented in a ffield file with atom positions in a data file from Liu et al.'s work [5]. The ReaxFF force field calculates the atomic interactions in ettringite with good accuracy. The total energy in ReaxFF is partitioned in different energy terms as defined by the following equation.

$$\begin{aligned} E_{\text{total}} &= E_{\text{bonds}} + E_{\text{angles}} + E_{\text{torsions}} + E_{\text{H-bonds}} + E_{\text{vdW}} + E_{\text{Coulomb}} \\ &+ E_{\text{lp}} + E_{\text{over}} + E_{\text{under}} + E_{\text{pen}} + E_{\text{coa}} + E_{C_2} + E_{\text{triple}} + E_{\text{conj}} \end{aligned} \tag{1}$$

where E_{bonds} is the sum of all bonded pairwise interactions directly derived from bond orders; E_{angles}, E_{torsions}, $E_{\text{H-bonds}}$, E_{vdW}, and E_{Coulomb} are angle, torsional, hydrogen bond, van der Waal, and Coulomb interactions respectively; E_{lp} is the sum of the long pair energy, a penalty for breaking up long pairs in O, N; E_{over} is the sum of overcoordination energy, penalty for overcoordinating atoms; E_{under} is the undercoordination energy, contribution for the resonance of π-electron between undercoordinating atoms; E_{pen} is the penalty for "allene"-type molecules ($H_2C{=}C{=}CH_2$); E_{coa} is the energy term due to the contribution of three-body angle conjugation in -NO_2 groups; E_{C2} is an energy correction that destabilizes the stability of C=C; E_{triple} is a triple bond stabilization energy in CO bonded pairs; E_{conj} describes the contribution of conjugation stability to the molecular energy.

A ReaxFF force field was developed for ettringite to simulate the first principles quantum mechanism interaction in hydrous calcium aluminium sulphate systems (H/Ca/Al/S/O) with good accuracy [5]. The geometries and energies of species and complexes in ettringite were calculated according to density functional theory (DFT); ReaxFF parameters were adjusted with the minimum differences between ReaxFF energies and DFT energies. Figure 2 plots a 2×2×1 molecular triclinic supercell structure, a basic calculation structure in the simulation using LAMMPS. The conformity and structure stability of proposed molecular structure using the aforementioned ReaxFF force field parameters were validated in an isothermal-isobaric (NPT) ensemble at room-temperature, and the predicted structures from MD simulations are in close agreements with experiments [5].

Figure 2. 2×2×1 supercell molecular structure of ettringite
(Green is the Oxygen atom; red is the Hydrogen atom; pink is the Calcium atom; Sulphate and Aluminium atoms cannot be seen from this point of view in the figure.)

3. Result and Discussion

MD simulations of tensile and shear failures of ettringite were performed on periodic boundary condition in three directions under strain loads to characterize the mechanical properties of ettringite. The stress-strain responses under different ambient temperatures were also investigated.

3.1 Elastic Tension failures at atomistic scale

An initial molecular structure of 10×4×4 ettringite supercell structure was adopted in a NPT ensemble at 1 atm of ambient pressure. Temperature was set at 280 K throughout the simulation. Figure 3 plots the uniaxial stress-strain relationship in x-direction of the ettringite molecules. A tensile load was applied at the constant engineering strain rate of 0.001 fs^{-1}. The ensemble was equilibriumed by performing an energy minimization of the system with a stoping tolerance for energy of 1×10^{-6} Kcal/mole and for force of 1×10^{-6} Kcal/mole-Angstrom. Timestep was set as 0.002 fs, and a total of 1000 fs was applied. The deformations of the ettrignite molecules in three directions and the stresses of the whole ensemble were documented for each timestep, and the last 400 fs was used for data collection. There was a tensile failure at the tensile strain of 8% along x-direction. According to Liu et al. [5], there is a dramatically decrease of the total number of Ca-O bonds after the tension failure.

Similiarly, figure 4 depicts the stress-strain relationship at both compression and tension in x-direction within a very small strain range of ±1%, where positive strain denotes tension and negative strain denotes compression. Figure 5 plots the shear stress-strain relationship under a velocity load of 0.002 Angstrom/ps in x-direction. The molecular strcutre failed at the shear strain of 1.4%.

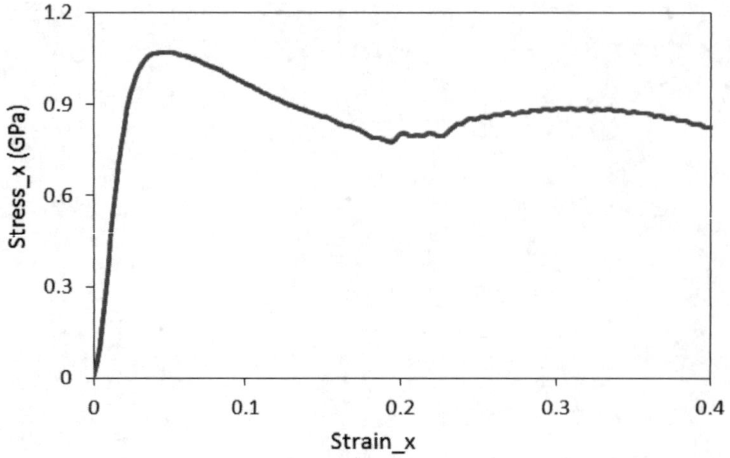

Figure 3. Tensile stress-strain relationship along x-direction due to the tensile strain load

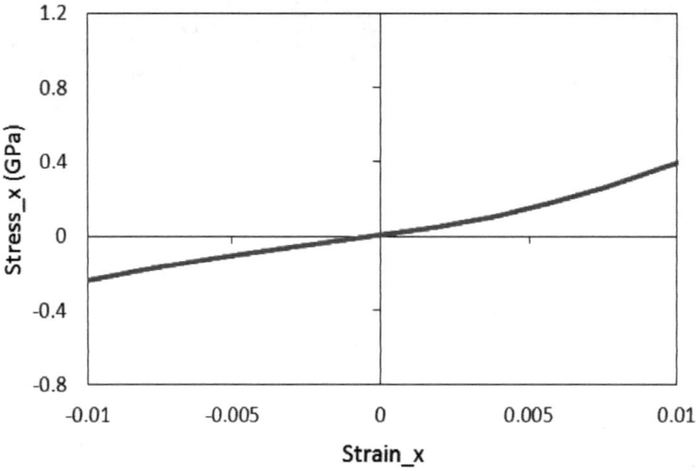

Figure 4. Stress-strain relationship due to strain load along the x-direction in the limit of small deformation

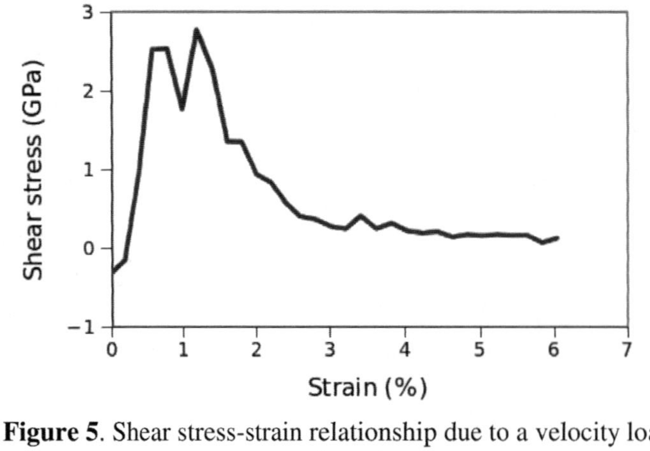

Figure 5. Shear stress-strain relationship due to a velocity load in x-direction

Table 1 tabulates the elastic constants of the ettringite crystal. Elastic constants were determined from the stress-strain relationships in LAMMPS simulations within very small strain ranges, where the stress-strain relationship is linear. These calculated elastic constants from LAMMPS simualtions are in good agreement with experimental results.

Table 1. Elastic constants of ettringite crystal.

Elastic constants (GPa)	LAMMPS	Experiment [6]
C_{11}	30	35.1 ± 0.1
C_{12}	20.1	21.9 ± 0.1
C_{13}	20	20.0 ± 0.5
C_{14}	1.5	0.6 ± 0.2
C_{33}	52	55.0 ± 1.0
C_{44}	12.2	11.0 ± 0.2
C_{66}	4.9	6.6 ± 0.2
Bulk modulus (K)	25.8	27.3 ± 0.9
Shear modulus (G)	12.2	9.5 ± 0.8
Poisson's ratio (v)	0.335	0.34 ± 0.02

3.2 Stress-strain relationship at different ambient temperatures

Figure 6 plots the tensile stress strain relationships in x-direction under varied temperatures of 1 K, 100 K, and 200 K respectively. As the ambient temperature increases from 1 K to 200 K, the critical stress, when tensile failures occur, decrease from 3.67 GPa down to 2.73 GPa. And the critical strain also shows a similar decrease trend. A possible reason is that atoms with higher kinetic energies at a higher temperature are more likely to break interatomic bonds at smaller deformations. According to the experimental results in [7], the elastic modulus of hydrated Portland cement with a water-cement ratio of 0.33 shows a increase trend, while the ambient temperature increases provided that the temperature is lower than 400 Fahrenheit (477.594 K).

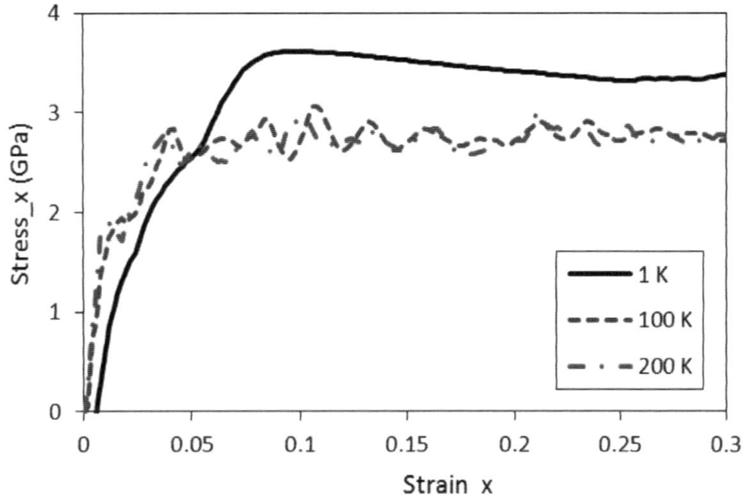

Figure 6. Tensile stress-strain relationships in x-direction under different ambient temperatures

4. Summary

MD simulations of the ettringite molecular structure were conducted in a NPT ensemble using LAMMPS. Tensile and shear failures were investigated to determine elastic constant from the linear stress-strain relationships at very small strain ranges. The calculated elastic constants showed good agreement with experimental results, which validates the accuracy of the MD simulations. The stress-strain relationship under different ambient temperatures showed that the critical stress and critical strain decreasesas the ambient temperature increases. A possible reason is that atoms with greater kinetic energies at a higher temperature are easier to escape from interatomic bonds.

Future research should be conducted on the establishment of corresponding bridging-relationships of mechanical properties between atomic scale and macroscale. These bridging-relationships can provide insights into understanding how the mechanical properties and crack failures at atomic scale influence crack failures of cement paste and other cement-based materials at macroscale.

Acknowledgement
The authors want to express their sincere thanks to Andres Jaramillo-Botero and Williams A. Goddard, III for the help on the information share and the establishment of ettringite molecular structure.

References

[1] Bye G C 1999 *Portland cement* (second edition) (Thomas Telford, London, GB) p131
[2] Santhanam M, Cohen M D and Olek J 2001 *Cem. Concr. Res.* **31** 845
[3] Mehta P K and Monteiro P J M 2006 Concrete: Microstructure, Properties and Materials, third edition (New York: McGraw-Hill)
[4] Linklater C M, Albinsson Y, Alexander W R , Casas I, Mckinley I G and Sellin P 1996 *J. Cont. Hydrol.* **21** 59
[5] Liu L, Jaramillo-Botero A, Goddard, III W and Sun H 2012 *J. Phys. Chem. A* **116** 3918
[6] Speziale S, Jiang F, Mao Z, Monteiro P J M, Wenk H-R, Duffy T S and Schilling F 2008 *Cem. Concr. Res.* **38** 885
[7] Harmathy T Z and Berndt J E 1966 *Journal of American Concrete Institute* **63** 93

Preparation and properties of ATO films and their effects on the TiO_2/ATO system

J. A. Galaviz-Pérez[1a], Fei Chen[a], J. R. Vargas García[b], Qiang Shen[a], Lianmeng Zhang[a]
[a] State Key Laboratory of Advanced Technology for Materials Synthesis and Processing, Wuhan University of Technology, Wuhan 430070, PR China
[b] Dept of Materials and Metallurgical Eng., National Polytechnic Institute, México 07300 DF México

E-mail: jorgegalavizperez@gmail.com

Abstract. Multilayered TiO_2/ATO thin films were prepared by the spin-coating method to evaluate the electronic interactions and the effect of the experimental conditions on optical transparency, electric resistivity and carrier density. ATO chemical composition, number of layers and annealing temperature were modified to study their effect on the global properties of the TiO_2/ATO system. TiO_2/ATO films exhibited an optical transparency as high as 81 % in the visible wavelength and a strong absorption within the UV spectra. Electric resistivity as low as $3.4x10^{-1}$ Ω·cm was measured for TiO_2/ATO films. Optical properties were influenced mainly by the presence of upper TiO_2 layer and ATO film thickness. The effect of ATO film thickness, composition and annealing temperature on TiO_2/ATO transparency and resistivity are discussed.

1. Introduction

Transparent conductive oxides (TCO´s) are important component for solar cell applications and have drawn strong interest for improving their performances [1,2]. The invention of Grätzel cells opened the interest to develop cheaper TCO´s due to the promissory application of this kind of solar cell [3–5]. After monocrystalline silicon solar cells technologies, the most important material are TCO´s and among them the indium doped tin oxide (ITO) has been selected as the best option for industrial applications, however, indium is an expensive and rare material, increasing ITO costs and therefore solar cell ones. Antimony doped tin oxide (ATO) is a promissory substitute for ITO because Antimony relatively low cost and elevated abundance in earth crust [6]. Nowadays many ATO synthesis routes are under study in order to met the requirements for solar cell applications [7–11]. TiO_2 is a fundamental material for Grätzel cells in combination with TCO´s of elevated efficiency [12,13]. Among the preparation techniques on use to prepare TCO´s, the sol-gel technique is well known for its readiness and low cost, which could encourage the preparation of such films to overcome the comparative low resistivity of ATO films in comparison with ITO. This work evaluates the effect of annealing temperature, microstructure and morphology of ATO/quartz substrate on the electrical resistivity and UV-vis transparency of TiO_2/ATO/quartz films.

[1] To whom any correspondence should be addressed.

2. Experimental

Details on ATO sols preparation has been reported in a previous paper [14] and details on TiO_2 sol preparation can be found elsewhere [15]. Three ATO sols were prepared using different acetylacetone (acac) ratios and ethylene glycol (EG) as hydrolysis modifiers named ATO1, ATO4 and ATOG for acac ratio of 1, 4 and 1 with addition of EG, respectively. Films were prepared by spin-coating onto quartz substrates at ambient temperature and at a spin speed of 5000 rpm by 10s. ATO films with 1 and 3 layers were prepared named as follows: ATO 11, ATO 13, ATO 41, ATO 43, ATOG1 and ATOG3. TiO_2/ATO films consisted in one layer of TiO_2 over ATO film. These films are named as TATO 11, TATO 13, TATO 41, TATO 43, TATO 1EG1 and TATO 1EG3. Films were annealed at 450, 550 and 650 °C to evaluate the microstructure. Crystal structure of films was analyze by XRD (rigaku ULTIMA III, CuKα). Surface morphology and film cross section were analyzed by FE-SEM (HITACHI S-4800, 10 kV). Optical and near UV transparency was measured by UV-Vis from 200 to 800 nm in wavelenght (Shimadzu UV-2550). The electrical resistivity and Hall measurments was measured by using the van der Pawn method (Accent Optical HL5500PC) at ambient temperature.

3. Results and discussion

XRD results (not shown) revealed only the presence of cassiterite and anatase crystalline phases. The morphology of films was influenced by the solutions composition, as show by figure 1, which compares the morphology of TATO 13, TATO 43 and TATO 1G3 films annealed at 650 °C. The use of acetylacetone and ethylene glycol as hydrolysis modifiers changed noticeably the morphology from granules of about 20 nm in diameter (figure 1a) using an acetylacetone molar ratio of 1, to small granules of less than 10 nm (figure 1b) with an acetylacetone ratio of 4. The granules of film TATO 13 have coalesced forming voids between the interconnected granules. TATO 43 film consisted in smaller granules interconnected due to coalescence and forming smaller voids. TATO 1G3 film exhibited highly coalesced particles leading to a smooth surface with quasi-linear voids. Films morphology was generally smooth. The annealing temperature or the number of layers did not affect noticeably the morphology of films.

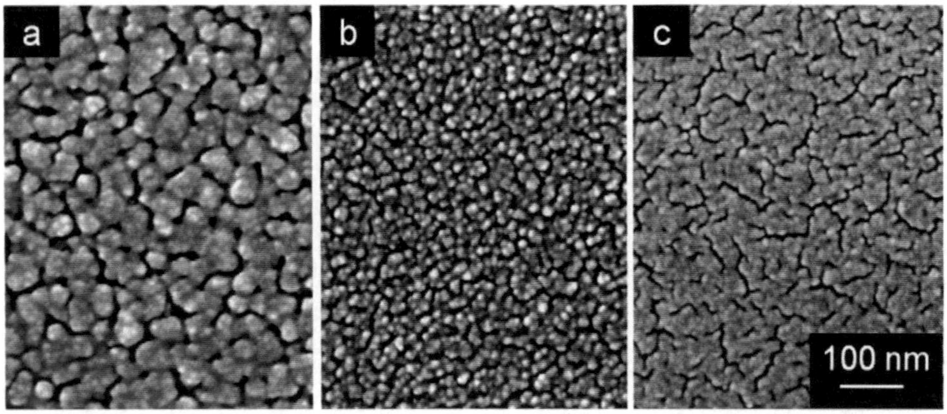

Figure 1. FE-SEM images of TATO13 (a), TATO43 (b) and TATOG3 (c) films annealed at 650 °C. The scale is the same for the three images

Figure 2 shows the transparency of films annealed at 650 °C within the UV-vis wavelength (from 200 to 800 nm). Transparency of ATO and TATO films with 1 and 3 layers are presented in figure 2a and detailed spectra for TATO films are showed in figure 2b. ATO 11, ATO 41 and ATO 43 films exhibited an elevated transparency above 90% from 350 and 800 nm in wavelength, under 350 nm transmittance decreased rapidly to 30% at 200 nm. ATO film 13 exhibited a maximum transmittance of about 80%. TATO films exhibited a different behaviour than ATO films with maximum transparency of 80% for TATO 41 film and 70% for TATO 1EG3. A strong absorption between 300 and 380 nm is observed for all TATO films. Evidently, the TiO_2 layer absorbs the UV wavelength completely below 380 nm. The UV radiation strongly interacts with crystalline TiO_2, especially the anatase phase [16], despising the low thickness of about 100 nm but remarking the highly crystallinity of TiO_2 above 400 °C and the apparent absence of rutile phase. In this case, the nature of the ATO solution not noticeably affected the transparency of films.

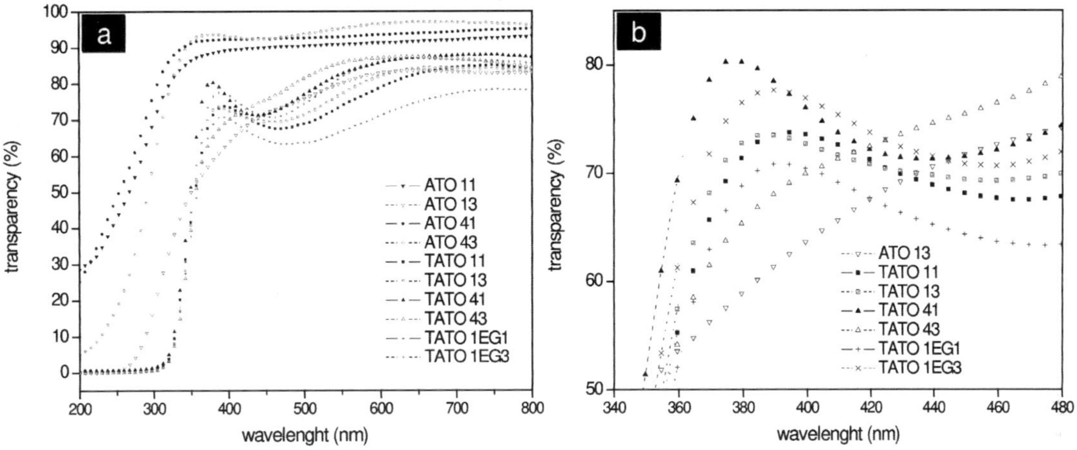

Figure 2.- UV-vis transparency spectra for ATO and TATO films annealed at 650 °C.

Table 1 sumarizes the Hall measurement results of ATO and TATO films at different annealing temperatures. Resitivity of ATO and TATO films decreased with annealing temperature, but resistivity values for TATO films are up to 100 times above of correspondent ATO films which is spected for TiO_2 films, which electric resistance commonly range 10^{19} Ω·cm [17,18]. However carrier mobility is not highly affected by the TiO_2 layer and for TATO 13, TATO 41 and TATO 43 films carrier mobility is slightly enhanced. This last could be cause by the enhancement in bandgap by TiO_2 layer, which limits the recombination of free electrons in the interface [19].

ATO film morphology consisted on nanoparticles in a varable size by the use of hydrolisis modifiers [14] and a thickness about 300 nm. The small particle size and low thickness reduced absorption in the visible spectrum and limited absorption in the UV (200-400 nm) region. The addition of a TiO_2 layer onto the ATO film not changed noticeably the morphology and thickness was increased in just about 100 nm, which not increased considerably the opacity of film. Thus, the increased absorption in the UV spectrum is completely related to TiO_2 layer at the point that transparency down to 0% at about 250 nm in wavelenght. A fundamental absorption process has been reported between 200-300 nm [20] that explains the differences in transparency between ATO and TATO films. TATO 41 film exhibited the highest transparency at 380 nm (80%) while TATO 43 showed a decreasing transparency in UV, but a slow absorption rate. The number of layers clearly affected the films transparency but sol composition (thus morphology) is the most noticeable feature that affected the final transparency of TATO films, after the fundamental absorption process. TATO 1 and TATO 1EG films showed a noticeable coalescence that reduced the transparency in the visual as well as the UV spectra. TATO 4 films were formed by smaller nanoparticles of less than 10 nm, leading to a better transparency over 350 nm in wavelenght. The differences between TATO films transparencies were as much as 10% above 350 nm.

Table 1. Hall measurement results for ATO and TATO films.

		resistivity 10^{-1} (Ω·cm)			carrier mobility 10^{-1} ($cm^2 \cdot V^{-1} \cdot cm^{-1}$)		
	thickness	annealing temperature (°C)			annealing temperature (°C)		
sample	(nm)	450	550	650	450	550	650
ATO 11	100	8.8	2.5	2.8	4.25	2.2	9
TATO 11	200	36	9	13	1.16	1.5	1.3
ATO 13	300	0.28	0.049	0.97	2.61	3.5	3.6
TATO 13	400	0.38	0.59	0.5	6.5	6.1	5
ATO 41	100	2.4	1.37	0.59	5.3	0.23	4.5
TATO 41	200	35	40	71	35	13	27
ATO 43	300	0.47	0.17	0.025	2.6	2.1	1.7
TATO 43	400	12	0.34	2.2	3.26	7.85	3.13

4. Conclusions

TiO_2/ATO films were successfully prepared using the spin coating method. Films morphology was heavily influenced by the addition of acetyl acetone and ethylene glycol as hydrolysis modifiers. Transparency of films was strongly affected by the presence of TiO_2 and by the hydrolysis modifier contents leading to a strong absorption in the UV-vis region. Films with an acetylacetone contents ratio of 4 exhibited the best optical and electric properties. A light enhancement in carrier mobility, due to the presence of the TiO_2 layer, is possibly caused by the enhancement in band gap and the subsequent limitation in recombination of free electrons for TiO_2/ATO films. Annealing temperature influenced electric properties, exhibiting the lowest resistivity at 550 °C.

5. Bibliography

[1] Chopra KL, Paulson PD, Dutta V. Progress in Photovoltaics Research and Applications 2004;92:69.
[2] de Graaf A, van Deelen J, Poodt P, van Mol T, Spee K, Grob F, Kuypers A. Energy Procedia 2010;2:41.
[3] Grätzel M. Journal of Photochemistry and Photobiology C: Photochemistry Reviews 2003;4:145.
[4] Lenzmann FO, Kroon JM. Advances in OptoElectronics 2007;2007:1.
[5] Ngamsinlapasathian S, Sreethawong T, Yoshikawa S. Thin Solid Films 2008;516:7802.
[6] Laznicka P. Economic Geology 2006;101:1445.
[7] Muto Y, Oka N, Tsukamoto N, Iwabuchi Y, Kotsubo H, Shigesato Y. Thin Solid Films 2011;520:1178.
[8] Boltz J. Sputtered Tin Oxide and Titanium Oxide Thin Films as Alternative Transparent Conductive Oxides. Aachen University, 2011.
[9] Coste S, Lecomte A, Thomas P, Merle-Mejean T, Champarnaud-Mesjard JC. Journal of Sol-Gel Science and Technology 2006;41:79.
[10] Chatelon JP, Terrier C, Roger JA. Journal of Sol-Gel Science and Technology 1997;10:55.
[11] Liu S, Ding W, Chai W. Physica B 2011;406:2303.
[12] Vigil E, González B, Zumeta I, Docteur S, Peiró AM, Gutiérrez–Tauste D, Domingo C, Domènech X, Ayllón J a. Journal of Crystal Growth 2004;262:366.
[13] Chen X, Mao SS. Chemical reviews 2007;107:2891.
[14] Galaviz Pérez JA, Chen F, Qiang S, Vargas García JR, Zhang L. Inorganic Thin Films and Coatings 2012;In press.
[15] Li Y, Hagen J, Schaffrath W, Otschik P, Haarer D. Solar Energy Materials and Solar Cells 1999;56:167.
[16] Viseu TMR, Almeida B, Stchakovsky M, Drevillon B, Ferreira MIC, Sousa JB. Thin Solid Films 2001;401:216.
[17] Mardare D, Rusu G. Materials Science and Engineering B 2000;75:68.
[18] Yildiz a., Lisesivdin SB, Kasap M, Mardare D. Journal of Non-Crystalline Solids 2008;354:4944.
[19] Fotsa-Ngaffo F, Caricato AP, Romano F. Applied Surface Science 2009;255:9684.
[20] Travina T, Mukhin Y. Russian Physics Journal 1966:40.

Acknowledgments

This research was funded by the project 111 (B13035). Author J. A. Galaviz-Pérez wish to thank to National Council of Science and Technology of Mexico (CONACYT) for the support for the postdoctoral stage.

 doi:10.1088/1742-6596/419/1/012013

High-temperature oxidation behavior of ZrO_2-$ZrSiO_4$/NiCr composites and its applicability for functionally graded materials

K Wu, A Rosin, H Park, M Willert-Porada

Chair of materials processing, university of Bayreuth, 95447 Bayreuth, Germany

Email: kangjian.wu@uni-bayreuth.de

Abstract. In the present study, high-temperature oxidation behavior of pressurelessly sintered ZrO_2-$ZrSiO_4$/NiCr8020 samples with different volume fractions of ceramic is investigated with the aim to develop oxidation resistant ZrO_2-$ZrSiO_4$/NiCr functionally graded materials out of them. Sintered samples with 0, 25, 50 and 75 vol% ceramic were oxidized at 1000 °C for 36 h in synthetic air. The oxidation kinetics depends strongly upon the phase connectivity of NiCr in the ceramic matrix, which also governs the crack pattern formed due to the volume expansion of NiCr particles. While nearly full oxidation of the NiCr particles occurs in the 75 vol% ceramic composite with a relatively low phase connectivity of NiCr, the 25 vol% ceramic composite with a highly percolative NiCr phase exhibits a metal-like oxidation behavior and its overall oxidation was limited to the surface near area.

1. Introduction

Functionally graded materials (FGMs) have the potential for multifunctionality combining different materials and providing a spatial variation in composition, structure and properties that cannot be achieved by conventional homogeneous materials [1]. Metal-ceramic FGMs are promising candidates for future high-temperature applications, with the ceramic component offering thermal barrier effects and protecting the metal from melting, corrosion and oxidation, while the FGMs are toughened and strengthened by the metallic component [2]. ZrO_2/NiCr FGMs are typical FGMs of thermal barrier type and they have already received fairly intensive study [3-8] in the recent 20 years. However, to the best knowledge of authors, few studies [3,4] have dealt with the high-temperature oxidation behavior of this type of FGM. Moreover, we have found no reports yet concerning the characteristic oxidation behavior of individual constituent layers of such FGMs.

Borchert [3] found that the oxidation resistance of ZrO_2/NiCr FGMs is able to be improved significantly by adding a relatively low volume fraction of $ZrSiO_4$ in the ZrO_2 matrix as inclusion. In order to develop oxidation resistant ZrO_2-$ZrSiO_4$/NiCr FGMs, high-temperature oxidation behavior of ZrO_2-$ZrSiO_4$/NiCr composites with different volume fractions of ceramic is investigated based on their oxidation kinetics and microstructure in the present study.

2. Experimental

2.1. Starting materials

Ceramic powders used in the present study were 8 mol% yttria-stabilized zirconia powder (TZ-8YS, D_{50}=0.54 μm, Tosoh, Japan) and $ZrSiO_4$ powder (D_{50}=0.85 μm, Helmut Kreutz GmbH, Germany). The ZrO_2 powder was delivered in the form of spray-dried granules. The applied metal powder was irregularly shaped NiCr8020 powder (D_{50}=13.80 μm, H.C. Starck, Germany) with 20 wt% Cr as alloying element. Fig. 1 shows the typical microstructure of the starting materials.

Figure 1. SE images of starting materials: (a) TZ-8YS; (b) $ZrSiO_4$; (c) NiCr8020.

2.2. Sample preparation
For the high-temperature oxidation experiments, ZrO_2-$ZrSiO_4$/NiCr samples with different volume fractions of ceramic were fabricated via die and isostatic pressing followed by pressureless sintering. In the fabrication process, 5 vol% $ZrSiO_4$ was selected as the inclusion phase to be added into the respective ZrO_2 matrix. Metal-ceramic powder mixtures with 75, 50 and 25 vol% ZrO_2-$ZrSiO_4$ (5 vol%) ceramic, respectively, were firstly prepared by weighing the powders according to their volume ratios and mixing them by planetary ball milling for 3 h using ethanol as solvent. Pure NiCr powder (0 vol% ceramic) was mixed with an aqueous PVA solution (5 wt%) in order to improve the green strength of the sample. After drying, the mixtures were sieved through a screen with a mesh size of 200 μm to crush the soft agglomerates. Subsequently, they were filled into a steel die with an inner diameter of 8 mm and compacted just with hand as pre-compaction. After demoulding, green bodies were pressed cold-isostatically under 300 MPa as post-compaction, followed by pressureless sintering at 1350 °C for 3 h under argon in a high-temperature alumina tube furnace (Thermal Technology, USA).

2.3. High-temperature oxidation
Sintered samples had a dimension of about 7 mm (Φ) × 2.5 mm. A hole with a diameter of 1 mm was drilled in the center of the samples, so that they can be hung on a platinum wire as shown in Fig. 2, in order to ensure a homogeneous oxidation of the samples. Subsequently, they were heated with 5 °C/min up to 1000 °C under nitrogen followed by oxidation at this temperature for 36 h under

flowing synthetic air (N_2/O_2 80/20), with a gas flowing rate of 40 ml/min. Weight gain of the samples during the oxidation was registered by using a TG/DSC analyzer (STA 499 C/4/G, Netzsch, Germany).

Figure 2. Sample arrangement for the oxidation experiments.

2.4. Characterization

Relative densities and porosities of the as-sintered samples were determined by the Archimedes' method. Specific surface area of the composite samples was measured using a physisorption analyzer (ASAP 2010, Micromeritics, Germany). In order to improve the measuring accuracy, at least 10 identical samples were used for each individual measurement. However, surface area of the pure metal samples was still too low to be measured correctly by the physisorption method. Thus, specific surface area of the pure metal sample was deduced by fitting the initial linear portion of the measured parabolic oxidation curve with that taken from the literature [9], in which Ni-19.4 %Cr samples were oxidized at 1000 °C in static air. Additionally, microstructure and element distribution of the oxidized samples were studied by using a scanning electron microscope (JSM-840A, Jeol, Japan) equipped with a dispersive X-ray spectrometer (Oxford Instruments, UK).

3. Results

3.1 Sintered samples prior to oxidation

In the present study, ZrO_2-$ZrSiO_4$/NiCr samples with 75 to 0 vol% ceramic are termed as 75C to 0C. Table 1 shows the basic properties of the as-sintered samples. We note that the samples exhibit similar relative densities ranging from 86.40 to 90.41 %. Specific surface areas of the metal-ceramic composite samples are comparable among each other, whereas that of the pure metal sample is significantly lower than those of the composite ones.

Table 1. Basic properties of the as-sintered non-FGM samples.

Phase	Relative density (%)	Porosity (%)	Open porosity (%)	Specific surface area (cm^2/g)
75C	86.40	13.60	6.59	1517
50C	90.41	9.59	5.12	843
25C	88.94	11.06	9.31	711
0C	87.73	12.27	9.29	9

Microstructure of the sintered samples with different volume fraction of ceramic is shown in Fig. 3. We note that NiCr particles are distributed homogeneously in all the composite phases. As the volume fraction of metal increases from 25 to 75 vol%, matrix phase of the composite transits gradually from ceramic matrix with dispersed NiCr particles to metal matrix with an interpenetrating ceramic network. For the 0C sample, several small dark gray spots can be observed. These spots are composed of Ni–Cr_xC_y [10], which was formed due to an incomplete debindering.

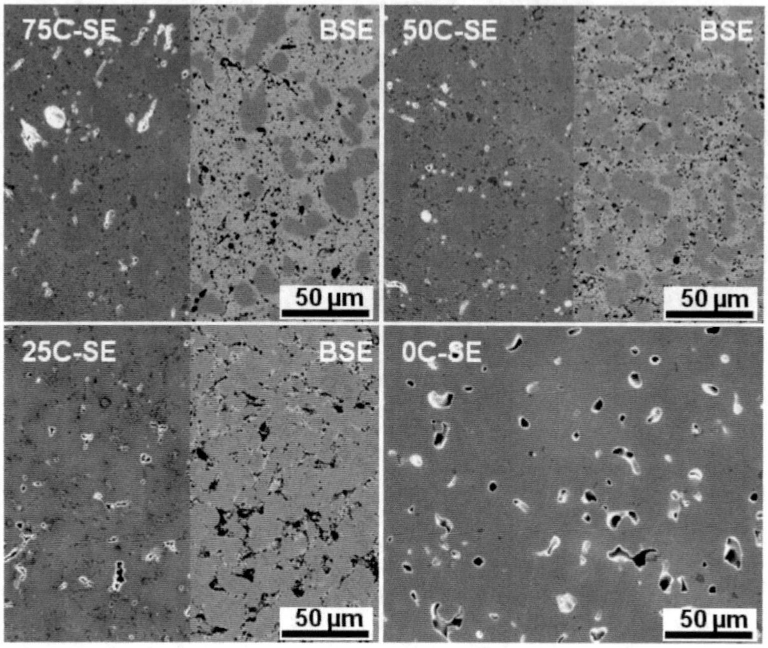

Figure 3. SE and BSE images of the ZrO_2-$ZrSiO_4$/NiCr samples.

3.2 Samples after oxidation

Since all the samples were still porous after sintering, their surface area can vary substantially during the oxidation process, thus application of the parabolic law assuming a constant surface area during oxidation for data analysis is not suitable for this case. Instead, we use the degree of oxidation to describe the oxidation kinetics, which was calculated by dividing the weight gain of the sample by the total weight gain, when the sample is totally oxidized. Fig. 4 shows the degree of oxidation as a function of oxidation time for the different ZrO_2-$ZrSiO_4$/NiCr samples. We note that for all the phases the degree of oxidation increases with increasing oxidation time, although the increase for 0C is difficult to be identified from the figure. During the first 10 h of oxidation, the three composite phases exhibit similar degrees of oxidation. From t=10 h up to the end of the process, the degree of oxidation decreases with increasing volume fraction of metal at a certain oxidation time. Additionally, it is noteworthy that 75C was oxidized almost completely after an oxidation time of 36 h.

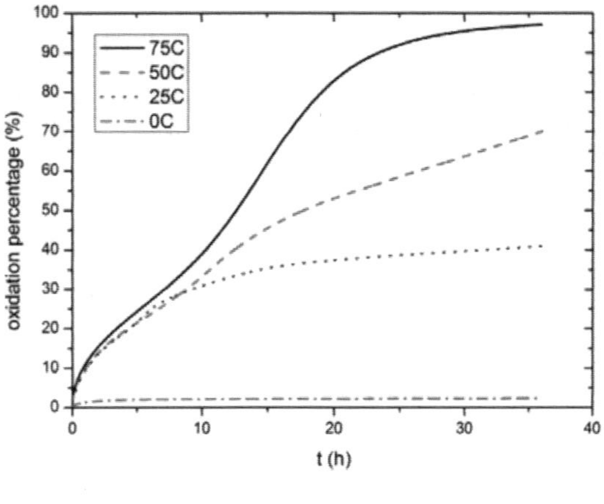

Figure 4. Influence of time on degree of oxidation for the ZrO_2-$ZrSiO_4$/NiCr samples.

Fig. 5 shows the kinetic constants of oxidation for the ZrO_2-$ZrSiO_4$/NiCr samples. In the first five hours, kinetic constants of the composite phases decrease rapidly from about 8.5 to 2.5 % h^{-1}, while that of 0C is much smaller and it decreases from 0.93 to 0.05 % h^{-1}. Afterwards, kinetic constant of 75C increases intensively to 5.17 % h^{-1} at t=15 h, decreases then rapidly again to a minimum value of 0.17 % h^{-1} at t=35 h, agrees well with its degree of oxidation reaching nearly 100 % at t=36 h as shown in Fig. 4. Similarly, kinetic constant of 50C increases slightly to 2.79 % h^{-1} at t=10 h, decrease then gradually to a value of 0.91 % h^{-1} at t=35 h. For the metal-rich phases, kinetic constant of 25C decreases monotonically up to 0.21 % h^{-1} at t=35 h, while that of 0C remains nearly constant at a low value of ~0.01 % h^{-1} up to the end of the process.

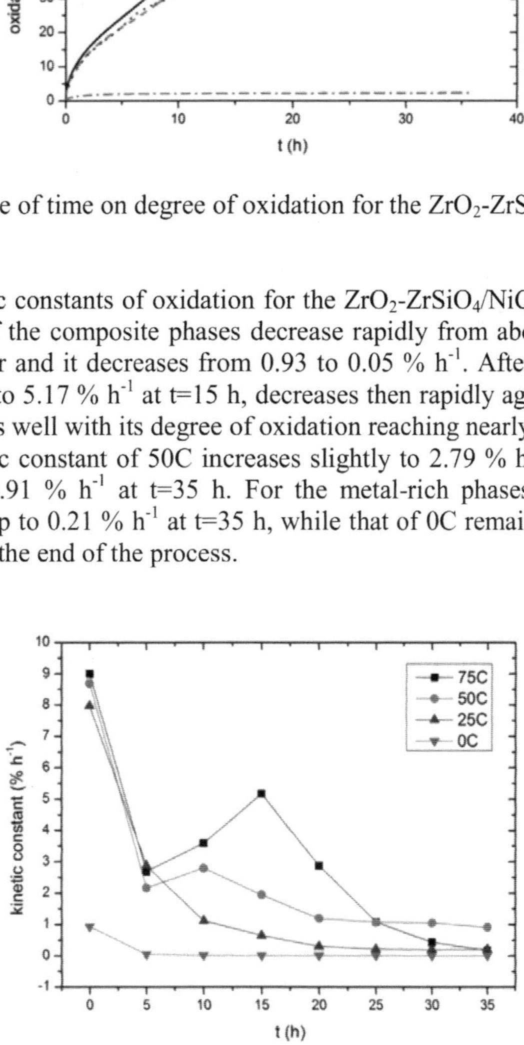

Figure 5. Kinetic constants of oxidation for the ZrO_2-$ZrSiO_4$/NiCr samples.

Fig. 6 shows microstructure and element mapping obtained from the center of the oxidized 75C sample. We observe from Fig. 6(a) and (b) that all the NiCr particles were oxidized and voids were generated in most of the particles, indicating that the whole sample has been oxidized almost completely. In addition, a network of oxide interpenetrating in the ceramic matrix can be observed. When comparing the element mapping in Fig. 6(c) and (d), we note that Cr in the NiCr particles has been depleted completely and a dense Cr_2O_3 scale is formed around each NiCr particle. Nickel phase of the metal particles was also strongly oxidized since oxygen was able to be detected over the whole particles. When comparing Fig. 6(d), (e) and (f), we may deduce that the interpenetrating network of oxide is composed of NiO, even though no network of oxygen could be identified from Fig. 6(c) due to the presence of oxygen detected from the ZrO_2-$ZrSiO_4$ (5 vol%) matrix.

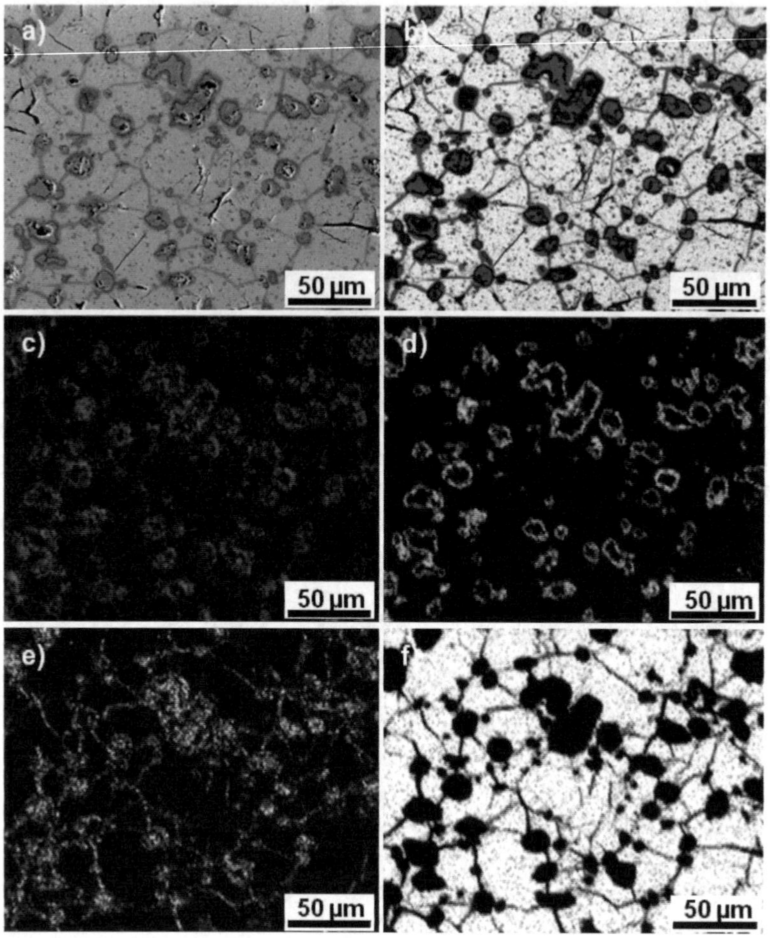

Figure 6. Microstructure and element mapping obtained from the center of the oxidized 75C sample: (a) SE image; (b) BSE image; (c) oxygen (red); (d) chromium (green) (e) nickel (blue) (f) zirconium (white). All elements were recorded from their K_α line.

Fig. 7 shows microstructure of the oxidized 50C sample, which was obtained from near the surface and the center of the sample, respectively. The subsurface zone where the NiCr particles were strongly oxidized reaches a thickness of ~620 µm. From Fig. 7(b), we observe that a relatively finer network of NiO than that in 75C was formed in this zone. In contrast, the center of the sample was just slightly oxidized, with a thin oxide scale around each metal particle and the inside of the particles remaining unoxidized (Fig. 7(c) and (d)). Additionally, no network of NiO was found in the ceramic matrix.

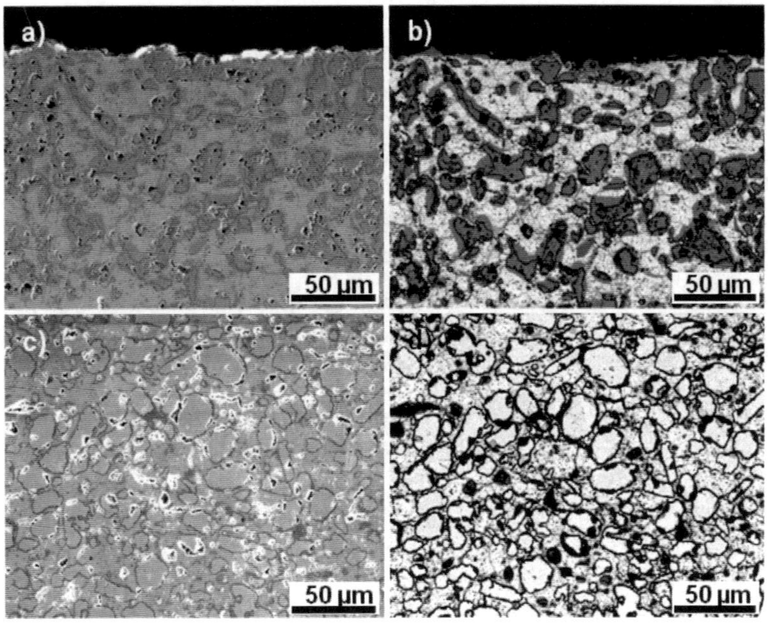

Figure 7. Microstructure of the oxidized 50C sample with (a) SE and (b) BSE images obtained near the surface and (c) SE and (d) BSE images obtained from the center of the sample.

Microstructure of the oxidized 25C sample is shown in Fig. 8. The strongly oxidized subsurface zone reaches a thickness of about 110 µm. A finer network of NiO than that in 50C can also be observed in the subsurface zone, with the amount of the NiO-network significantly less than those in 75C and 50C. Dissimilar to the 75C and 50C samples as described above, a continuous dense oxide scale along the surface of the sample was formed, which can reduce the surface area and protect the sample from further oxidation effectively. From the EDX result (spectrum 1 in Fig. 8(a)) with 43.69 at% oxygen and 56.31 at% nickel, we may deduce that this oxide scale is also composed of NiO.

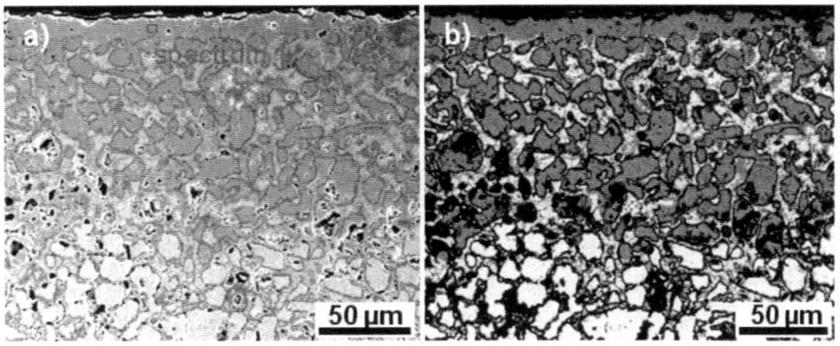

Figure 8. Microstructure of the oxidized 25C sample: (a) SE image; (b) BSE image.

Fig. 9 shows microstructure of the oxidized 0C sample. We note that after the oxidation process all the open pores near its surface were blocked by oxide scales, which are dark gray and marked by using arrows in Fig. 9(a). Moreover, a continuous oxide scale along the surface of the sample was formed. EDX results detected on the light gray and dark gray phases are listed in Table 2. In spectrum 1, the element Cr and O have a mol ratio of about 2:3 with the Ni content of only 0.92 at%, indicating that the dark gray phase is mainly composed of Cr_2O_3. In spectrum 2, no oxygen was detected and the elements Cr and Ni have a mol ratio of about 1:9, lower than the nominal ratio of the NiCr8020 alloy, indicating that the light gray phase is the unoxidized NiCr phase with reduced content of Cr. Fig. 9(b) shows the microstructure obtained from the center of the 0C sample. Expect for several dark gray spots composed of Ni-Cr_xC_y as mentioned above, no oxide scales can be observed.

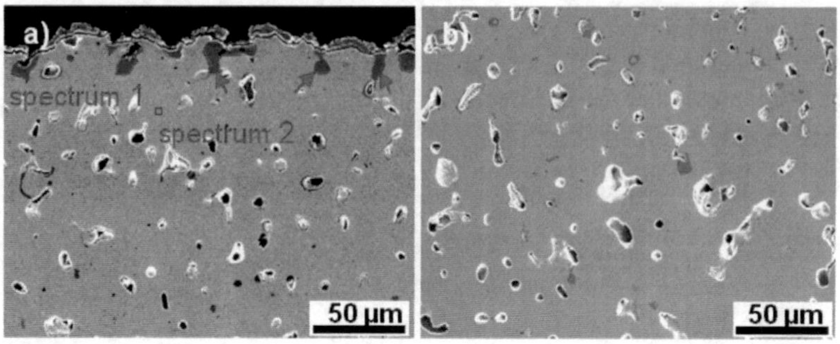

Figure 9. Microstructure of the oxidized 0C sample with (a) SE image obtained near the surface and (b) SE image obtained from the center of the sample. Blocked open pores near the surface are marked by using arrows.

Table 2. EDX results detected on the cross section of the 0C sample.

	Element (at%)		
	O	Cr	Ni
Spectrum 1	54.19	44.89	0.92
Spectrum 2		11.02	88.98

4. Discussion

Although dispersed NiCr particles are observed from the microstructure of 75C and 50C (Fig. 3), a percolative network of NiCr was formed in both samples. Percolation theory [11] tell us that a random distribution of inclusions form branching clusters of touching inclusions and at a critical volume fraction (percolation threshold) of ~0.16, one of the clusters forms a percolative network of inclusions that spans the entire volume of the specimen. For elongated inclusions, the percolation threshold can be dramatically reduced [12]. Lange et al. [13] found that more than 95% of the inclusions will be part of the same network when the volume fraction of inclusion reaches 0.20. After sintering, volume fractions of NiCr particles based on the total volume in 75C, 50C and 25C were 21.60, 45.21, 66.71 vol%, respectively, indicating that a percolative network of NiCr was formed in all the composite samples, but the phase connectivity of NiCr still increases with increasing volume fraction of metal, as shown in Fig. 3.

In the first 5 h of the oxidation process, the composite phases exhibit similar degrees of oxidation and kinetic constants. Their kinetic constants decrease from about 8.5 % h^{-1} to 2.5 % h^{-1} in this period, which can be attributed to the oxidation of Cr near the surface of the samples by inward diffusion of oxygen through both open pores and oxygen ion-conducting ZrO_2 matrix. Simultaneously, the oxidation of Cr resulted in a volume expansion of the metal particles, since the volume of Cr_2O_3 is about twice as large as that of Cr according to the Pilling-Bedworth ratio [14]. However, due to the relatively low content of Cr of 20 wt% in each metal particle, volume expansion during this period is considered to be relatively slight without crack formation in the ceramic matrix, since existing pores in the ceramic matrix can help to accommodate slight volume expansion of metal particles [15]. Additionally, surface areas of the composite phases were probably reduced to a certain extent by the volume expansion of NiCr particles. For the pure metal phase, kinetic constant of 0C also decreases during this period from 0.93 to 0.05 % h^{-1}, indicating that oxide scales were formed and they grew gradually in the open pores near the surface of the sample, resulting in a reduction of the surface area. The kinetic constant of 0C is much smaller than those of the composite phases, since the latter exhibit significantly higher specific surface areas as compared to the former after sintering.

After depletion of Cr near the surface of the composite samples, nickel phase in the NiCr particles near the surface began to be oxidized. Since Ni is the dominant element in each metal particle and the volume ratio between NiO and Ni reaches a relatively high value of 1.65 [14], the oxidation of Ni can thus give rise to a more intensive volume expansion of NiCr particles than that of Cr. Moreover, nickel cations diffuse more quickly through oxide scales than oxygen anions [14], i.e. NiO tends to grow outwards with voids generated inside the NiCr particles, which further enhances the effect of volume expansion. When internal stress intensity caused by the volume expansion reached the fracture toughness K_{IC} of the ceramic matrix, cracks were initiated from several sides of each metal particle, propagated in the ceramic matrix and were arrested by adjacent metal particles. Cracking of the ceramic matrix led to an increase in the surface area and an improved oxidation kinetics. An increase in the kinetic constants of the 75C and 50C samples was thus observed after an oxidation time of 5 h (Fig. 5), whereas the extent of increase for 50C is much lower than that for 75C. From the microstructure of oxidized 75C and 50C (Fig. 6 and 7), we note that the cracks in 50C is also much finer than that in 75C. These observations can be attributed to the higher phase connectivity of NiCr and resulting higher fracture toughness of 50C than those of 75C. Kinetic constant of 25C decreases monotonically during this period of time. Moreover, it exhibits the lowest extent of cracks in 25C among the composite phases. The reason is, on the one hand, the phase connectivity of NiCr in 25C is the highest among the composite phases. On the other hand, the simultaneous formation of a continuous NiO scale on the surface helps to reduce the surface area and to improve the oxidation resistance, as shown in Fig. 8.

After cracks were initiated, remaining unoxidized nickel cations diffused outwards through oxide scales into the cracks and filled the cracks with NiO, resulting in the formation of an interpenetrating network of NiO in the ceramic matrix, which reduced the surface area and the oxidation kinetics again. For the 0C sample, its kinetic constant keeps at a nearly constant value of only about 0.01 % h^{-1} from t=5 h up to the end of the process, indicating that all the open pores near the surface had been blocked by the formation of oxide scales. The porous 0C sample began to be oxidized like a dense metal from this point of time.

From the high-temperature oxidation behavior of the ZrO_2-$ZrSiO_4$/NiCr samples with different volume fraction of ceramic, we may conclude that in order to improve the oxidation resistance of ZrO_2-$ZrSiO_4$/NiCr FGMs processed by powder metallurgy, high relative density and low surface area are very beneficial on the one hand. On the other hand, fracture toughness of the ceramic-rich side of FGM should be improved, by for example adding high-temperature resistant fibers or applying elongated NiCr particles to increase the phase connectivity of metal.

5. Conclusions

High-temperature oxidation behavior of pressurelessly sintered ZrO_2-$ZrSiO_4$/NiCr samples with different volume fractions of ceramic is investigated in the present study. The oxidation kinetics depends strongly upon the phase connectivity of NiCr in the ceramic matrix, which also governs the crack pattern formed due to the volume expansion of NiCr particles. In order to obtain an oxidation resistant ZrO_2-$ZrSiO_4$/NiCr FGM, fracture toughness of the ceramic-rich side of FGM should be improved.

Acknowledgements
The authors would like to thank the graduate school 1229 of German Science Foundation for the financial support. Ingrid Otto, Peter Kostolansky and Philipp Ponfick are also gratefully acknowledged for their technical supports.

References

[1] Kawasaki A and Watanabe R 1987 Finite element analysis of thermal stress of the metal/ceramic multi-layer composites with controlled compositional gradients *J. Jpn. Inst. Met.* **51** 525
[2] Uemura S 2003 The activities of FGM on new application *Mater. Sci. Forum* **423** 1
[3] Borchert R 1998 Process development for the fabrication of metal-ceramic functionally graded materials by microwave sintering (Göttingen: Cuvillier)
[4] Khor K and Gu Y 2000 Thermal properties of plasma-sprayed functionally graded thermal barrier coatings *Thin Solid Films* **372** 104
[5] Zhu J, Lai Z, Yin Z, Jeon J, Lee S 2001 Fabrication of ZrO_2–NiCr functionally graded material by powder metallurgy *Mater. Chem. Phys.* **68** 130
[6] Kawasaki A and Watanabe R 2002 Thermal fracture behavior of metal/ceramic functionally graded materials *Eng. Fract. Mech.* **69** 1713
[7] Lia J, Zeng X, Tang J, Xiao P 2003 Fabrication and thermal properties of a YSZ–NiCr joint with an interlayer of YSZ–NiCr functionally graded material *J. Eur. Ceram. Soc.* **23** 1847
[8] Jin X, Wu L, Sun Y, Guo L 2009 Microstructure and mechanical properties of ZrO2/NiCr functionally graded materials *Mater. Sci. Eng.* A **509** 63
[9] Choe H and Dunand DC 2004 Mechanical properties of oxidation-resistant Ni–Cr foams *Mater. Sci. Eng.* A **384** 184
[10] Neukam M 2011 Process development for metal–ceramic composites on the basis of cellular nickel-base alloys (in German) (Aachen: Shaker Verlag)
[11] Zallen R 1983 *The physics of amorphous solids* (New York: Wiley) pp 135–204
[12] Holm EA and Cima M 1989 Two-Dimensional Whisker Percolation in Ceramic Matrix-Ceramic Whisker Composites *J. Am. Cerum. Soc.* **72** 303
[13] Lange F, Atteraas L, Zok F 1991 Deformation consolidation of metal powders containing steel inclusions *Acta Metall. Mater.* **39** 209
[14] Bürger R 2001 *Handbook for high-temperature material engineering* (Wiesbaden: Vieweg) pp 272-275
[15] Morales-Rodriguez A et al 2006 Influence of oxidation on the high-temperature mechanical properties of zirconia/nickel cermets *Scripta Mater.* **54** 2087

Microstructure Deformation and Fracture Mechanism of Highly Filled Polymer Composites under Large Tensile Deformation

Zhang Jiang Tao[1], Song Dan Ping, Zhang Mei, Zhai Peng Cheng

State Key Laboratory of Advanced Technology for Materials Synthesis and Processing, Wuhan University of Technology, Wuhan 430070, P.R.China

[1] zhjiangtao@whut.edu.cn

Abstract. The microstructure deformation and fracture mechanisms of particulate-filled polymer composites were studied based on microstructure observations in this paper. By using in-situ tensile test system under scanning electron microscopy, three different composites composed of polymer binder filled by three different types of particles, namely Al particles, AP particles and HMX particles, with the same total filler content were tested. The roles of initial microstructure damage and particle type on the microstructure deformation and damage are highlighted. The results show that microstructure damage starts with the growth of the initial microvoids within the binders or along the binder/particle interfaces. With the increase of strain, the microstructure damages including debonding at the particle/binder interface and tearing of the binder lead to microvoid coalescence, and finally cause an abrupt fracture of the samples. Coarse particles lead to an increase of debonding at the particle/binder interface both in the initial state and during the loading process, and angular particles promote interface debonding during the loading process.

1. Introduction

The particulate-filled polymeric composites (FPCs) constitute a broad class of structural and functional materials, which are highly packed particulate composite materials containing hard particles embedded in a polymeric binder. Such as the polymer concretes, the dispersion-hardened thermoplastics, syntactic foam and the solid rocket propellants, they are widely used in various engineering fields[1,2].

When the material is strained, irreversible microstructure damage in the form of microvoids in the binder or along the binder/particle interface occurs. As the applied strain increases, the growth of damage takes place as successive nucleation and coalescence of the microvoids or as polymer binder tearing[3]. These damage processes are microstructure dependent, and which is responsible for the microstructure dependence of the macromechanical behaviour of FPCs[4]. In order to develop an accurate analysis method to characterize the macromechanical properties of FPCs, the evolution of different microstructural damage mechanisms of FPCs should be correctly understood and their effects on macromechanical properties should be measured.

During the last decades, several techniques have been developed to directly observe the microstructural damage process of FPCs, such as Moiré interferometry[5, 6], digital speckle photography[7, 8], white-light digital image cross-correlation (DICC) analysis[9], and in-situ tensile tests under scanning electron microscope(SEM)[10]. Among these techniques, the last one is the most frequently used technique. By using the in-situ SEM observation, Zheng et al[10] investigated the crack initiation, propagation, and termination on the plerosphere/PP composites, and concluded that the toughness and strength of the polymer can be improved due to large numbers of microcracks caused by the particles. Kim and a coworker[11] studied the cavitation in rubber toughened polymers and the debonding in particulate-filled polymers, and found that the cavitation process plays an important role in promoting the shear yielding of the matrix ligaments between microvoids, which is the main mechanism of absorbing the energy. The debonding process has also been studied. Many factors, such as the filler content[13], filler size[13] and interaction between the filler particles and the matrix[12] have been proven to affect the debonding process. The debonding process can affect the mechanical properties of particulate-filled polymers. Homogeneous debonding leads to tough fracture behavior, and local debonding leads to brittle behavior[14].

In this work, a series of quasi-static tests by using the in-situ SEM observation were carried out to study the microstructure deformation and fracture behavior of a propellant simulation material, which is a highly-particle-filled composite in which the total content of the fillers are 74%, with a few percent polymer binder. The short, rectangle-shaped specimens were machined from the original bulk materials. The in-situ SEM observation was used to monitor the process of microstructure deformation to observe the initiation and propagation of microcracks, and reveal the failure mechanisms of the specimens.

2. Materials and Experiments

The particulate-filled polymeric composites (FPCs) used in the experiments are propellant simulation materials consisting three kinds of filler particles, HMX, AP and Al particles. The total content of the filler particles is 74%. The contents and average particle sizes of the fillers are given in table 1.

Table 1. Contents and average sizes of HMX, AP and Al particles.

sample	Al% (14 μm)	AP%(146 μm)	HMX% (18 μm)
1	59	15	0
2	0	15	59
3	18	5	51

Figure 1. The in-situ loading device DEBEN Microtest-2000 placed in the chamber of Scanning Electron Microscopy(SEM)

Journal of Physics: Conference Series **419** (2013) 012014 doi:10.1088/1742-6596/419/1/012014

In order to investigate the process of microstructure deformation and damage of the propellant simulation material under tensile loading, the in-situ tensile tests under scanning electron microscopy(SEM) is adopt to observe the microstructure evolvement process. As shown in figure 1, the in-situ loading device of DEBEN Microtest-2000 is used to apply tensile loads on samples, and SEM is used to observe the microstructure deformation and damage modes. The correlation between components and microstructural deformation mechanism was investigated with a series of in-situ SEM observation.

The samples for micromechanical deformation observations were prepared as follows. The samples with a rectangle section of 10mm×1mm and the length of 20mm were cut out from the bulk materials. Before the samples were fixed on the tester, their surfaces were golden-coated for SEM observations. The failure strain of the tested simulation materials is higher than 250%. Due to the restriction of the space of the test system, the maximum applied displacement of the loading device is only about 10mm, so the effective length for observation is limited to 3mm(the distance between two clamps) to ensure that the samples can be broken in the tensile tests.

Tensile specimens of the simulation materials were stretched in the in-situ loading device at a crosshead speed of 0.4 mm/min. At several given elongations, the crosshead was stopped and sustained for several minutes to scan the deformed regions of the samples by SEM, and the typical SEM pictures were captured for further analysis of the microstructure deformation.

3. Results and Discussions

3.1. Initial microstructures

The initial microstructures of the samples are shown in figure 2. The large particles with smooth surface are AP particles, and their average size is 146μm. The bright spherical particles in figure 2 (a) and (c) are Al particles, and their average size is 14μm. The dark and irregular particles in figure 2 (b) and (c) are HMX particles and their average size is 18μm. It can be observed from the SEM photography that there are two kinds of initial damage in propellant simulation materials mainly shown as follows

- Broken particle: mainly occurred in large AP particles.
- Microcrack and void: mainly appeared at the interfaces among HMX, large AP particles and polymer binder.

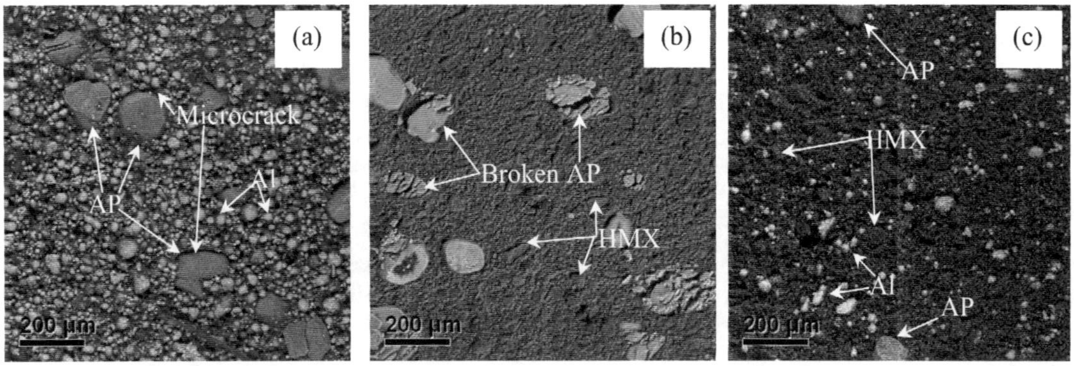

Figure 2. SEM photography of initial microstructure. (a) Sample 1(59% Al, 15% AP), (b) Sample 2 (15% AP, 59% HMX) and (c) Sample 3(18% Al, 5% AP, 51% HMX)

3.2. Tensile fracture process

In the simulation materials, the stiffness and strength of filler materials are much higher than those of the binder material. As results, the deformation mainly takes place in the binder materials, and the initial defects of microvoids and interface debonding grow obviously even at the initial stage of the tensile test. With the increase of deformation, it is observed that the interfacial debonding firstly occurs along the large AP particles, and then the large irregular HMX particles during the tests for samples containing HMX particles.

Owing to the development of the microvoids in matrix and along the interfaces, the microvoids homogeneously develop over the entire surface of the sample with the increase of deformation in the initial stage of deformation. When the displacement is around 3mm, the microstructures develop into homogeneous void structures over the entire sample. When the displacement attains about 5mm, as shown in figure 3, a local cracklike structure can be observed that crosses to both sides of the samples perpendicular to the loading direction. In the cracklike regions, large deformation of the binder ligaments can be seen, and the interconnected voids have been formed. The deformed binder shows a network structure. Some fractured matrix ligaments that have developed large deformation can also be seen. It can be seen that the debonding along large particle interfaces are comparatively severe. Many voids can be seen on both sides of the particles in the direction parallel to the applied strain that are formed by large deformation of the matrix between particles after the debonding process, especially around the large AP and irregular HMX particles.

Figure 3. SEM photography of microstructure at the deformation of 5mm. (a) Sample 1(59% Al, 15% AP), (b) Sample 2(15% AP, 59% HMX) and (c) Sample 3(18% Al, 5% AP, 51% HMX)

At the last stage of tension, the process of interface debonding is completed and very few interface debonding occurs[15]. The deformation mainly takes place by the growth of voids and stretch of binder ligaments in the cracklike structure. Finally, the fracture of binder ligaments leads to the coalescence of voids and results in the fracture of sample from the cracklike structure regions.

It can be concluded from the above experiment results that, the damage mechanisms of the tested propellant simulation material mainly include the growth and coalescence of initial defects, interfacial debonding along large particles and the fracture of binder ligaments. At low deformation levels, the dominant mechanism of damage accumulation is the growth of initial defects and interfacial debonding. The interface debonding of large irregular particles precedes that of small smooth particles. While at high values of tensile deformation, the dominant mechanism is the fracture of binder ligaments between filler particles.

3.3. Effects of filler particles

As shows in figure 3(a) of sample 1, the damage mechanisms are mainly interfacial debonding along the large AP particles, large voids are formed on both sides of the particles in the direction parallel to the applied stress. Al particles are strongly bonded with the polymer binder, and very few interface dedonding occurs along the Al particles. The binder ligaments with large deformation are presented between Al particles. The stretch and fracture of binder ligaments between Al particles can be observed in the cracklike regions.

For sample 2, large AP particles debonded first, then interfacial debonding around large irregular HMX particles occurred, and debonding appeared first at sharp corners. The binder formed a network structure as shown in figure 3(b).

For sample 3, large AP particles tended to interfacial debonding first, followed with a large amount of interface debonding along HMX particles. In some regions with severe damage, HMX particles were almost apart from the binder completely. In this sample, due to the contact of Al particles and HMX particles, a small quantity of Al particles debonded from the contact places as shown in figure 3(c). It indicates the HMX particles may affect the polymer binder coating of Al particle and the dedonded particles are not fully coated by the polymer binder.

By comparison, the interfaces along large AP particles easily debonded. For samples with high volume fraction of AP (sample 1 and 2), it is easy to cause severe local damage which act as the fracture source leading to sample failure. For samples with comparatively low volume fraction of AP, the development of deformation inside the sample is comparatively uniform under large range of deformation, the polymer binder around the small particles can deform more fully, which leads to homogeneous interfacial debonding of small particles as shown in figure 3(b), especially for irregular HMX particles.

Figure 4. Fracture surfaces of the samples. (a) Sample 1(59% Al, 15% AP), (b) Sample 2(15% AP, 59% HMX) and (c) Sample 3(18% Al, 5% AP, 51% HMX)

Figure 4 illustrates the fracture photographs. For all samples, the surfaces of AP particles were quite smooth without residual binder, which indicates the wettability and adherence of the polymer binder and AP particles are insufficient, and interfacial debonding can occur completely. For AP particles, the debonding is also highly related to particle distribution and surface coating. Although the surface of AP particle is smooth, some large AP particles clustered or the surfaces were not coated by binder, which lead to the debonding of AP particles completely. The debonding of small Al and HMX particles is invisible during the overall tensile process, which showed that these particles were well coated by polymer binder, and the bonding strength was comparatively high. The shape of HMX particles is extremely irregular, large HMX particles debonding started from sharp corners. The homogenous debonding of large quantity of HMX particles leads to the uniform deformation of sample and improve the fracture toughness.

3.4. Force-displacement curves

Figure 5 shows the typical load-displacement curves recorded during the tensile process. The fluctuation of the curve in figure 5 (a) caused by the pause of tension for SEM observation. It can be seen that for each time of crosshead suspension, it could lead to stress relaxation. But after certain deformation when it started to reload, the loads can recover to the original magnitude measured by continuous loading as shown in figure 5(b), which is the load-displacement curve measured by continuous tensile loading

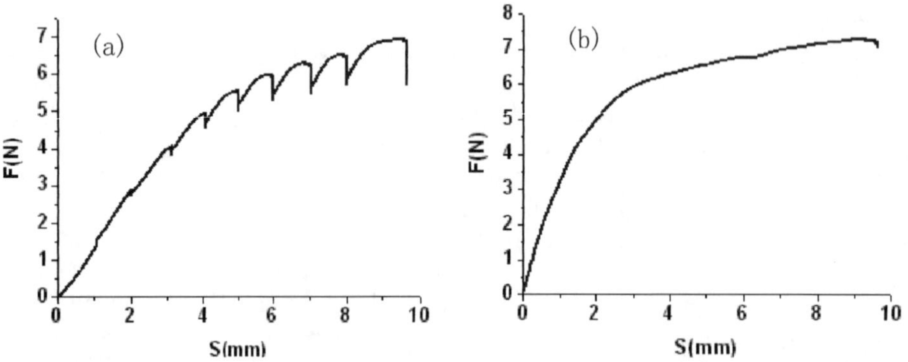

Figure 5. Typical load-displacement curves of sample 3. (a) Recorded with crosshead suspension for SEM observation, (b) Recorded by continuous loading

The load-displacement curves in figure 5 can be divided into two stages: the nonlinear elastic stage and hardening stage. In elastic stage, the damage was mainly the elastic expansion of initial voids, and it can recover after unloading. While at the beginning of hardening stage, the interface debonding of large particles and inelastic expansion of voids (or the inelastic deformation and damage of binder) occurs. With the increasing of deformation, the damages continuously accumulated and the microcracks began to expand, which cause the degradation of mechanical properties of the stimulation materials. But the loading capacity of sample still slowly increases due to the strengthening effect of particles and strengthening deformation of binder. When the deformation reached the maximum magnitude, the sample cracked immediately due to the macrocracks caused by the fracture of local matrix tearing.

It should be noted that during the tensile process, obvious deformation and sliding occur near the clamps due to large deformation of the samples, thus the recorded displacement-time curve is not precise, only the record of loads is accurate. Thus, these tests cannot be used directly to calculate the stress-strain curves of samples. The macroscale experiments are still needed to investigate the effect of particle fraction and types on the mechanical properties of the simulation materials.

4. Conclusions

The in-situ tensile tests under SEM is adopt to observe the microstructure evolvement of the propellant simulation materials under tensile loading, and the effect of particle characters on the microstructure damage mechanisms are analyzed. Three different composites composed of polymer binder filled with the same filler content of Al particles, AP particles and HMX particles were tested. The results show that at low deformation levels, the dominant damage mechanism is the growth of initial defects, and then the interfacial debonding along the large AP particles. While at the large deformation, the dominant mechanism is the fracture of binder ligaments between filler particles. Al particles are strongly bonded with the polymer binder, and very few interface dedonding occurs along the Al particles. The wettability and adherence of the polymer binder and AP particles are insufficient. The interfacial debonding can occur completely along AP particles and large voids are formed on both sides of the particles in the direction parallel to the applied strain. The shape of HMX particles is extremely irregular, large HMX particles debonding started from sharp corners. In the sample with high content of HMX, the debonding of large quantity of HMX particles leads to the uniform deformation and improvement of the fracture toughness. While in this sample, the contact of Al particles and HMX particles causes a small quantity of Al particles debonded from the contact places.

Acknowledgments
This work was supported by the National Natural Science Foundation of China under Grant No. 11202154, the Fundamental Research Funds for the Central Universities of China under Grant Nos. 2012-Ia-021 and 2011-IV-090, and by Open Foundation of State Key Laboratory of Explosion Science and Technology (Beijing Institute of Technology, China) under Grant No. KFJJ11-9Y.

References

[1] Zezin Yu P 1994 *Mech. Compos. Mater.* **30** 131
[2] Saini D R, Nadkarni V M, Nigam K D P, Grover P D 1987 *J. Compos. Mater.* **21** 782
[3] Gdoutos E E and Papakaliatakis G 2001 *Fatigue Fract. Eng. M.* **24** 637
[4] Pierard O, LLorca J, Segurado J, Doghri I 2007 *Int. J. Plasticity* **23** 1041
[5] Rae P J, Palmer S J P, Goldrein H T, Field J E, Lewis A L 2002 *Proc. R. Soc. Lond. A* **458** 2227
[6] Chen P W, Xie H M , Huang F L , Huang T, Ding Y S, 2006 *Polym. Test.* **25** 333
[7] Larsson L, Mikael S O and Fredrik T 2004 *Opt. Laser. Eng.* **41** 767
[8] Icardi U 2003 *Mech. Mater.* **35** 35
[9] Rae P J, Palmer S J P, Goldrein H T, Lewisb A L, Field J E 2004 *Opt. Laser. Eng.* **41** 635
[10] Zheng Y H, Shen Z G, Wang M Z, Ma S L, Xing Y S, 2007 *J. Appl. Polym. Sci.* **106** 3736
[11] Kim, G M and Michler G H 1998 *Polymer* **39** 5699
[12] Fang Z P, Hu Q L and Angew M 1999 *Chem.* **265** 1
[13] Dubnikova I L, Muravin D K, Oshmyan V G 1997 *Polym. Eng. Sci.* **37** 1301
[14] Yang J H, Zhang Y and Zhang Y X 2002 *J. Appl. Polym. Sci.* **85** 1207
[15] Zhang M, Zhang J T, Zhai P C, Liu L S, Shi H J 2011 *Int. J. Mater. Prod. Tec.* **42** 98

Fracture analysis of a viscoelastic functionally graded materials strip with a central crack

Z H Wang[1]**, L Zhang**[2]**, L C Guo**[3]
[1,2]Department of Astronautic Science and Mechanics, Harbin Institute of Technology, Harbin 150001, P.R. China
[3]Department of Astronautic Science and Mechanics & Department of Material Science and Engineering, Harbin Institute of Technology, Harbin 150001, China

E-mail: guolc@hit.edu.cn

Abstract. The fracture mechanical behavior of a crack in viscoelastic functionally graded materials (FGMs) is investigated. Firstly, the extensional relaxation modulus is assumed as separable form in space and time. Then, the crack problem of the viscoelastic FGMs can be turned into an elastic crack problem of FGMs by using correspondence principle. Applying Fourier transfom and boundary conditions, the elastic crack problem of FGMs is reduced to singular integral equations. The stress intensity factors (SIFs) can be obtained by solving the singular integral equations. Base on correspondence principle and Laplace transform, the SIFs of viscoelastic FGMs can be determined by inverting the transformed SIFs of correspondence elastic crack problem.

1. Introduction

One of the primary advantages of viscoelastic functionally graded materials (FGMs) is that FGMs can resist high temperature. It should be point out that materials will exhibit significant creep and stress relaxation behavior at high temperature which must be considered. Thus, the crack problem of viscoelastic FGMs has attracted a significant amount of attention. The correspondence principle is revisited and established for a class of FGMs [1-2]. The problem of a propagating semi-infinite mode III crack in an infinite inhomogeneous viscoelastic body is analyzed [3]. A crack in a viscoelastic strip of a functionally graded material FGM is studied under tensile loading conditions [4]. The viscoelastic fracture of multiple cracks in a functionally graded strip is studied [5]. The plane thermomechanical behavior of a crack in a viscoelastic FGMs coating with arbitrary material properties bonded to a homogeneous substrate is studied [6]. A meshless method based on the local Petrov-Galerkin approach is proposed for continuously nonhomogeneous linear viscoelastic solids [7]. To the authors' knowledge, it should be pointed out that some further important work needs to be done for the crack problem in viscoelastic FGMs. This paper presented a method to investigate the fracture behavior of a crack in functionally graded materials.

2. Basic equations

[1] Email:wasngzhihai@126.com
[2] Email:19666894@hit.edu.cn
[3] Email: guolc@hit.edu.cn

A nonhomogeneous viscoelastic FGMs strip containing a crack of length $2c$ ($2c = b - a$), as shown in Fig.1. Without loss of generality, it is assumed that the crack line lies on the x axis of the coordinate system (x-y), from a to b. The x coordinates of upper boundary and lower boundary are 0 ($x = 0$) and L ($x = L$), respectively. The basic equations of quasi-static viscoelasticity of FGMs are
The strain-displacement relationship

$$\varepsilon_{ij} = (u_{i,j} + u_{j,i})/2 \tag{1}$$

The equilibrium equations

$$\sigma_{ij,j} = 0 \tag{2}$$

and the boundary conditions are

$$\sigma_{ij} n_j = f_i \tag{3}$$

$$u_i = \Delta_i \tag{4}$$

in which σ_{ij} ($i, j = x, y$) are stresses, ε_{ij} are strains, u_i are displacements, $f_1(x,t)$ and $f_2(x,t)$ are known functions depicting the loading conditions on the crack face and can be determined from actual loading conditions. The viscoelastic constitutive law can be written as

$$s_{ij} = \int_0^t G_1(x, y, t-\tau)\frac{de_{ij}}{d\tau}d\tau \tag{5}$$

$$\sigma_{kk} = \int_0^t G_2(x, y, t-\tau)\frac{d\varepsilon_{ij}}{d\tau}d\tau \tag{6}$$

with

$$e_{ij} = \varepsilon_{ij} - \frac{1}{3}\varepsilon_{kk}\delta_{ij} \tag{7}$$

$$s_{ij} = \sigma_{ij} - \frac{1}{3}\sigma_{kk}\delta_{ij} \tag{8}$$

where s_{ij} and e_{ij} are deviatoric components of the stress and strain tensors, respectively. δ_{ij} is the Kronecker delta, $G_1(x, y, t)$ and $G_2(x, y, t)$ are the relaxation functions, and t denotes the time. Laplace transform of the basic equations (1)-(6) reduce to

$$\bar{\sigma}_{ij,j} = 0 \tag{9}$$

$$\bar{\varepsilon}_{ij} = (u_{i,j} + u_{j,i})/2 \tag{10}$$

$$\bar{s}_{ij} = p\bar{e}_{ij}\bar{G}_1(x, y, p) \tag{11}$$

$$\bar{\sigma}_{kk} = p\bar{\varepsilon}_{ij}\bar{G}_2(x, y, p) \tag{12}$$

$$\bar{\sigma}_{ij}\bar{n}_j = \bar{f}_i \tag{13}$$

$$\bar{u}_i = \bar{\Delta}_i \tag{14}$$

in which p is the Laplace transform variable. The over bar means the Laplace transform. Thus, the crack problem of the viscoelastic FGMs can be investigated by analyzing equations (9)-(14). It should be mentioned that the basic equations (9)-(14) of the Laplace transform homogeneous viscoelasticity have a form identical to those of nonhomogeneous elasticity. Therefore, the crack problem of the viscoelastic FGMs is turned into an elastic crack problem of FGMs. The elastic crack problem of FGMs with arbitrary properties can be turned into a group of singular integral equations [8].

$$\frac{1}{\pi}\int_a^b \left[\frac{4\mu_{n0}}{1+k}\frac{1}{x-u} + k_{n1}(u, x) + \pi h_{n2}(u, x)\right] g(u)du = -e^{\delta_n x}\sigma_0(x) \tag{15}$$

More detailed information on singular integral equations can be found in [8]. By solving the singular integral equations, the SIFs of elastic problem can be evaluated [8-11]. The SIFs for elastic problem can be expressed as

$$K_I^e(a,p) = \lim_{x\to a} \sqrt{2(a-x)}\sigma_{yy}(x,p) \quad (16)$$

$$K_I^e(b,p) = \lim_{x\to b} \sqrt{2(x-b)}\sigma_{yy}(x,p) \quad (17)$$

$$K_{II}^e(a,p) = \lim_{x\to a} \sqrt{2(a-x)}\sigma_{xy}(x,p) \quad (18)$$

$$K_{II}^e(b,p) = \lim_{x\to b} \sqrt{2(x-b)}\sigma_{xy}(x,p) \quad (19)$$

The SIFs for viscoelastic FGMs can be calculated by applying the Laplace numerical inversion technique [12]. The SIFs for the viscoelastic FGMs can be written as

$$K_1(a,t) = \sum_{m=0}^{l} C_m P_m^{(0,\beta)}(2\exp(-\gamma t)-1) \quad (20)$$

Similarly, the SIFs $K_1(b,t)$, $K_{II}(a,t)$ and $K_{II}(b,t)$ can be determined by the same method. More detailed information on Laplace numerical inversion technique can be found in [12].

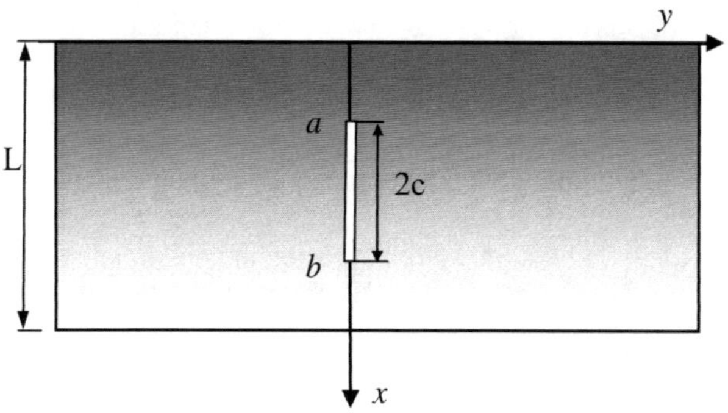

Fig.1 Geometry of the viscoelastic FGMs

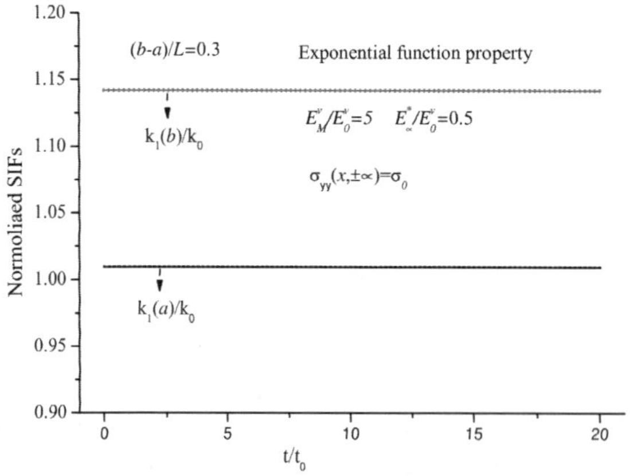

Fig.2 Variations of SIFs with time

As shown in Fig.2, The curves of SIFs with time t/t_0 are given when $\sigma_{yy}(x,\pm\infty)=\sigma_0$, the extensional relaxation functions $E(x,t)=E_0\left[E_\infty^*/E_0^*+\left(1-E_\infty^*/E_0^*\right)\exp(t/t_0)\right]\exp\left[x\ln(E_M^v/E_0^v)\right]$, $E_\infty^*/E_0^*=0.5$, $E_M^v/E_0^v=5$, $t_0=1$, $v(x)=0.3$ and $k_0=\sigma_0\sqrt{(b_0-a_0)/2}$. It can be observed that the SIFs decrease first with time t/t_0, and become a constant when $t/t_0\to\infty$.

3. Conclusions

In this paper, a new model is proposed to investigate the fracture mechanical behavior of a crack in viscoelastic functionally graded materials. The stress intensity factors for elastic FGM crack problem is first obtained by using the Fourier transform method and boundary conditions and the correspondence principle between the elastic and the Laplace transformed viscoelastic equations is used to obtain SIFs for viscoelastic FGMs. Finally, the SIFs for viscoelastic FGMs are realized by inverting the transformed SIFs. Basic Formulae for SIFs are given.

Acknowledges

The work is supported by NSFC (11072067), NCET-08-0151, Doctoral Fund Ministry of Education of China, Heilongjiang Science Fund for Distinguished Young Scholar, SRF for ROCS (SEM), Heilongjiang Science Fund for ROCS.

References

[1] Paulino G H and Jin Z H 2001 Correspondence principle in viscoelastic functionally graded materials. *J. Appl. Mech.-Trans. ASME* 68, 129-132

[2] Mukherjee S and Paulino G H 2003 The elastic-viscoelastic correspondence principle for functionally graded materials. *J. Appl. Mech.-Trans. ASME* 70, 359-363

[3] Herrmann J M and Schovanec L 1994 Dynamic steady-state mode-III fracture in a nonhomogeneous Visocoelastic body. *Acta Mech 106*, 41-54.

[4] Jin Z H and Paulino G H 2002 A viscoelastic functionally graded strip containing a crack subjected to in-plane loading. *Eng. Fract. Mech.* 69, 1769-1790.

[5] Pan F Y, Li W J, Wang B L and Zhang X H 2009 Viscoelastic fracture of multiple cracks in functionally graded materials. Comput. *Meth. Appl. Mech. Eng* 198, 2643-2649

[6] Cheng Z Q, Meguid S A and Zhong Z 2010 Thermo-mechanical behavior of a viscoelastic FGMs coating containing an interface crack. *Int. J. Fract.* 164, 15-29

[7] Sladek J, Sladek V, Zhang C and Schanz M 2006 Meshless local Petrov-Galerkin method for continuously nonhomogeneous linear viscoelastic solids. *Comput Mech* 37, 279-289.

[8] Guo L C and Noda N 2007 Modeling method for a crack problem of functionally graded materials with arbitrary properties - piecewise-exponential model. *Int. J. Solids Struct.* 44, 6768-6790.

[9] Guo L C and Noda N 2008 Fracture mechanics analysis of functionally graded layered structures with a crack crossing the interface. *Mech. Mater.* 40, 81-99.

[10] Guo L C Wu L Z, and Ma L 2004 The interface crack problem under a concentrated load for a functionally graded coating-substrate composite system. *Compos. Struct.* 63, 397-406.

[11] Guo LC Wang Z H Zhang L 2012 A fracture mechanics problem of a functionally graded layered structure with an arbitrarily oriented crack crossing the interface. *Mech. Mater.* 46, 69-82.

[12] Miller M K and Guy W T 1966 Numerical inversion of the Laplace transform by use of Jacobi polynomials. *SIAM Journal on Numerical*

The influencing mechanism of modification layer on the performance of SiC_{3D}/Al multi-function gradient composite

Li Bowen[1,2], Ma Zhuang[1,2*], Wang Yangwei[1,2] ,Wang Fuchi[1,2]
[1] School of Materials Science and Engineering, Beijing Institute of Technology, 100081, Beijing, China, [2] National Key Laboratory of Science and Technology on Materials under Shock and Impact, Beijing, 100081, China
*Corresponding author e-mail: hstrong929@bit.edu.cn.

Abstract. SiC skeleton surface was oxidized in this paper. Vacuum-pressure infiltration method is used to prepare SiC_{3D}/Al composite. The effects of the thickness of the interface modification layer were investigated. The results showed that the thickness of SiO_2 layer increases with the prolonged time of the skeleton oxidation. The brittle phase basically disappeared at the interface of the composite with 9 hours pre-oxidized, which lead to the high interface bonding strength. As a result, fracture morphology of the oxidized composite is mainly composed with plastic toughening of pure aluminum. Therefore, the static compressive strength of the composite raises up to 1165.2Mpa.

Keywords: SiC_{3D}/Al composite; thickness of the interface modification layer; fracture mechanism;

1. Introduction

Due to the unique network interpenetrating structure of the SiC_{3D}/Al composite, with high specific strength of the SiC as well as plastic deformable ability of aluminum, the SiC_{3D}/Al composite is widely used in many fields, such as electronics, mechanics, aviation and etc[1].

The best way to produce completely infiltrated SiC_{3D}/Al composite is vacuum pressure infiltration, which fullfilled with most of standards to create shape-perfect, structure compact composite [2]. Therefore, this method was used to prepare the composite in this paper.

During the special artistry in producing SiC_{3D}/Al with vacuum pressure infiltration, the SiC structure and Al matrix experienced a long directly contacting period under high temperature and high pressure. This lead to the formation of fragile phase along interfaces[3-5]. Therefore, the interface cannot efficiently conduct and distribute stress with applied load on the composite. Consequently, the composite property will be discounted. The interface bonding is a key fact to elevate performance of this composite material. One of the solutions is to reconstruct the outer shell of the SiC structure by oxidization, creating a compact oxygen layer of SiO_2, to detach the SiC from A1 during infiltration[5,6]. However, if the thickness of SiO_2 layer is insufficient to block the direct contact of Al from SiC, Al_4C_3 will formed. So, the thickness of the SiO_2 layer is very important to the vacuum pressure infiltration. In this paper, a series of SiC_{3D}/Al composites with different SiC pre-oxidation processes were investigated and the performance-microstructure relationship was discussed.

2. The experiment procedure and method

2.1. Experimental materials

The three-dimensional connected porous SiC with the density of $2.5g/cm^3$, relative porosity of 21.88% was used as the ceramic phase of the experimental composite, while industrial pure aluminum was used as the metal phase during the preparation.

The ZYQ250/400-2.1000 model vacuum pressure infiltration furnace was used to produce SiC_{3D}/Al composite material at temperature between 750 ~ 800 ℃, and infiltration pressure of 5-6.5 Mpa. The infiltration sketch was shown in Fig.2.

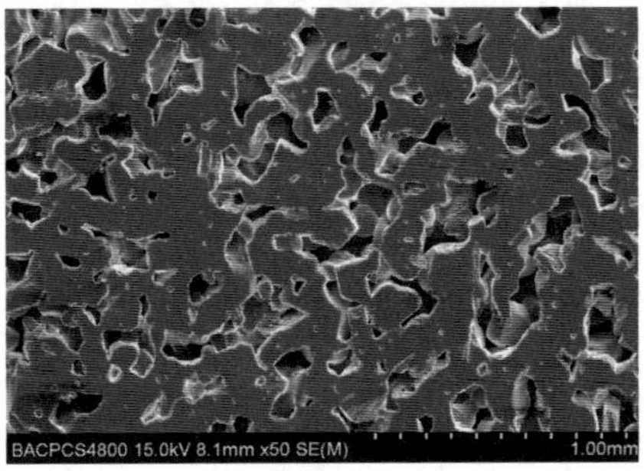

Figure 1 Skeleton microstructure.

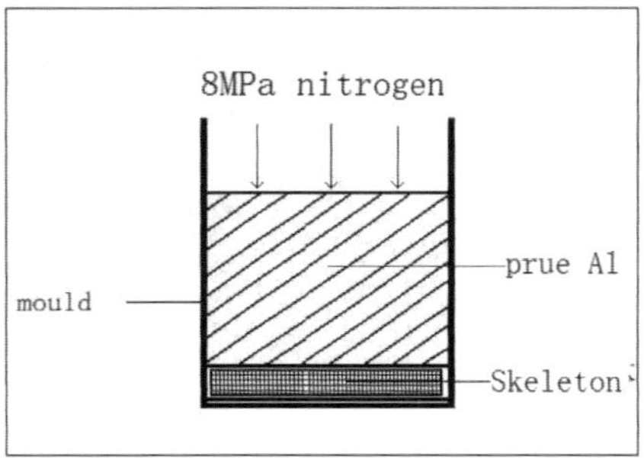

Figure 2 Infiltration sketch.

2.2. SiC skeleton surface oxidation

To estimate the effects of the thickness of SiO_2 layer on the properties of the composites, this experiment was carried out under 1200℃ in an oxygen sufficient environment where the SiC surface oxidation would take place. The oxidation time under 1200℃ were 3h, 6h, 9h respectively.

2.3. Composite performance and organization structure characterization.

The microstructural morphology of the polished composites was observed using ice emission scanning electron microscope (S4800). The Rigaku X-ray diffraction analyzer was used to analyze the phase composition of the reaction product (Cu target, K, Ni filter, scanning range 15 ~ 90; tube voltage 40 kv, tube current 50 mA; Slit size DS = 1, RD = 0.15, SS = 1, scanning speed 5 / min).

The compression properties of the composite were determined by WDW - E100D tester, with the strain rate controlled in the range of 10^{-4} s^{-1} ~ 10^{-3} s^{-1}.

The composite was etched for 30s using solution of HNO_3(33 vol.%) [5]with corrosion voltage of 20V, corrosion current of 1.5A, so as to observe the surface morphology and phase composition.

3. Results and analysis

3.1. Microstructure morphologies and mechanical properties

Figure 3 shows the microstructure of composites with skeletons treated by different pre-oxidation procedures. The samples shown in figure 3a, 3b, 3c, 3d were pre-oxidized for 0 hr, 3 hrs, 6 hrs and 9hrs, respectively. It is evidently in figure 3 that the integrities of all the samples were very good. The density of these composite are above 99%.

In order to study the influence of pre-oxidation, the static compressive strength of composites was tested in this paper. The results are shown in Table 1.

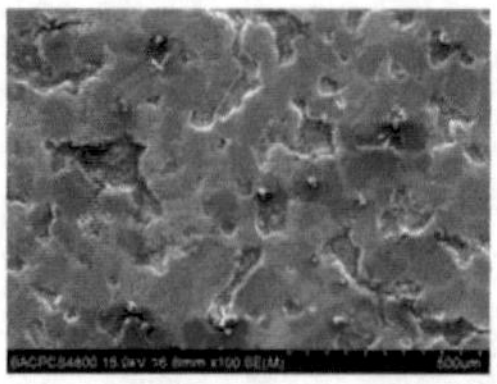

a skeleton un-oxidized

b skeleton oxidized for 3hrs

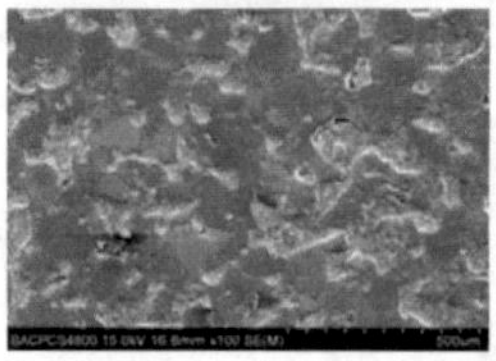

c skeleton oxidized for 6hrs

d skeleton oxidized for 9hrs

Figure 3 Microstructure of composites.

Table 1 The compressive strength of composites with un-oxidized and oxidized skeleton.

Sample No.	un-oxidized	3hrs oxidation	6hrs oxidation	9hrs oxidation
	300.3	414.0	1096.1	1074.0
compressive strength	365.7	737.6	401.6	1119.8
	300.5	593.1	433.2	1301.9
average	322.5	505.8	643.6	1165.2

The compressive strengthes of the composite with oxidized skeleton were significantly higher than those with skeleton unprocessed. Besides, the compressive strength improved with the increasing oxidation time and reached its highest level when the oxidation time prolonged to 9 hours.

Many investigations have been done for the SiO_2 layer on the skeleton surface, in which the existence of SiO_2 layer is considered to suppress the formation of Al_4C_3 and increase the wettability between SiC skeleton and Al matrix[7]. But the formation of Al_4C_3 could not be completely prevented if the thickness of SiO_2 layer is insufficient. Therefore, the SiO_2 layer thickness plays an important role in influencing of the interface reaction during the infiltration process.

3.2. The role of the oxide layer thickness

The thickness of SiO_2 layer increases with the prolonged time of the skeleton oxidation. As analyzed before, the thickness of the layer is a sensitive factor that influences the property of the composite. Hence, the thickness of the layer should be quantitatively analyzed first. In this paper the experience formula [8] is used to calculate the thickness of the SiO_2 layer.

$$h = \frac{\Delta W}{100 \cdot s} \cdot M_{SiC} / \left[\left(M_{SiO_2} - M_{SiC} \right) \cdot \rho_{SiC} \right]$$

where h is the thickness of the oxidized layer, S represents the specific surface of SiC skeleton (cm^2/g), ρ_{SiC} represents the SiC density ($3.2 g/cm^3$); $\triangle W$ the weight gain (%), being M_{SiC} and M_{SiO2} the molecular masses of SiC and SiO_2, respectively.

Table 2 The thickness of SiO_2 layer

Sample No.	3hrs oxidation	6hrs oxidation	9hrs oxidation
thickness/nm	121.1	179.6	223.4

a skeleton oxidized for 3hrs b skeleton oxidized for 6hrs c skeleton oxidized for 9hrs

Figure 4 Micro morphology of SiC/Al composite with oxidized skeleton

Figure 4 shows the microstructural morphology of the skeletons surfaces which were obtained from the electrolytic solution etched composites sample. Al phase was removed in etching process. The hexagonal crystals attached on the surface of skeleton in Figure 4.(a) is analyzed by EDS, the result is shown in figure 5.

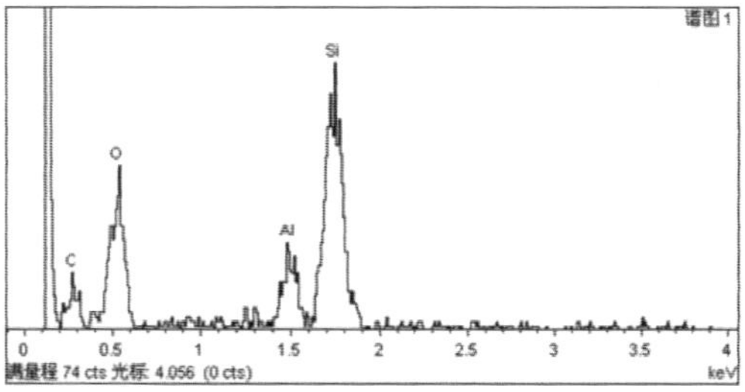

Figure 5 The EDS analysis of SiC skeleton.

It can be concluded from the EDS results that the hexagonal substance in figure 4a is Al_4C_3. The XRD results for all the skeleton surface are shown in figure 6, which further demonstrated the existence of Al_4C_3. Thus, the interface reaction product is Al_4C_3. It can be concluded from the intensities of the peaks that amounts of Al_4C_3 in the skeletons decreased with the pre-oxidation time increasing, which is also accordance with the morphologies shown in figure 4. As shown in figure 4a, numerous brittle hexagonal crystals generated on the interface of the composite with 3 hours of pre-oxidized skeleton. There was a significant reduction of brittle crystals in the composite with 6 hours pre-oxidized skeleton, which is exhibited in Figure 4b. Al_4C_3 crystal was barely generated in the composite with 9 hours pre-oxidized skeletons.

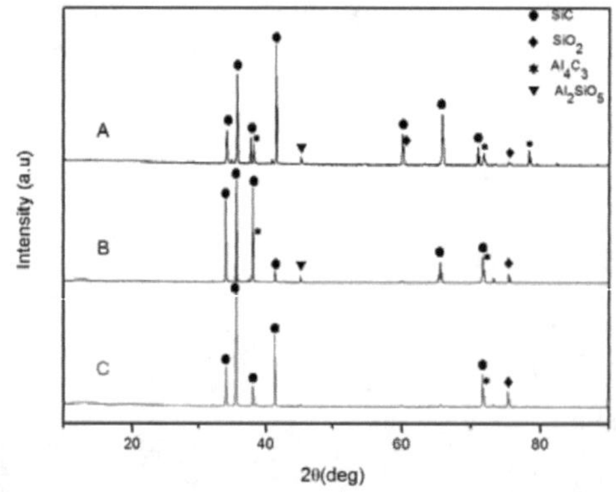

Figure 6 The XRD analysis of SiC skeleton with different oxidation time

The existence of hexagonally stacked Al_4C_3 would cause the irregular arrangement of the atoms at the interface and also induced the stress concentration in the deformation area, which finally lead to the growth of the cracks and influenced the mechanical property of the composite. Based on the results above, it can be concluded that the existence of SiO_2 layer could suppress the formation of Al_4C_3 and elevate the mechanical property of composite.

3.3. Fracture morphology of the composite

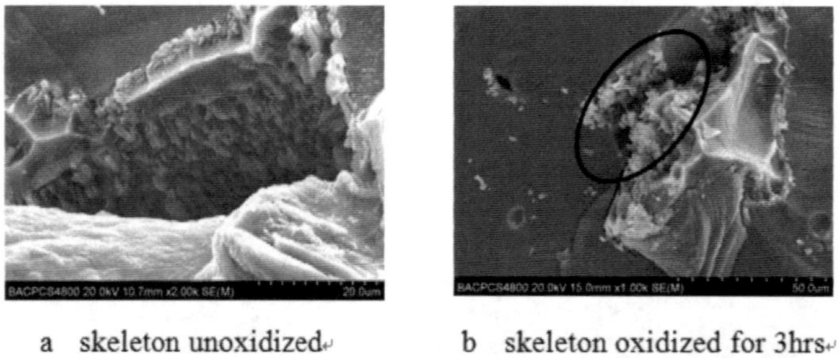

a skeleton unoxidized b skeleton oxidized for 3hrs

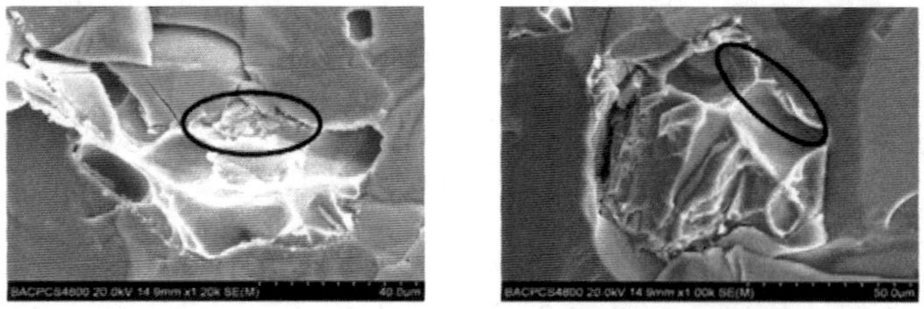

c skeleton oxidized for 6hrs d skeleton oxidized for 9hrs

Figure 7 Fracture morphology of the composite.

Figure 7 shows the fracture morphology of composites. The relatively dark and regular phase is SiC, while the light and rugged phase is Al or interface reaction product. The figure shows that the primarily fracture mode of the composite without oxidation treatment is interfacial brittle cracking. The brittle phase stacking at the interface leads to the interfacial premature cracks, so the plastic deformation capacity of Al phase cannot be well performed. Based on the analysis above, the mechanical properties of the composites was degraded. From the labeled zone in figure 7(b), (c), (d), the interface of the composite bonded closely and the fractured mechanism of the composites pre-oxidized was Al phase ductile tearing. The content of the brittle phase decreased due to the increasing of oxidation time. After 9 hours of pre-oxidation, the brittle phase disappeared essentially, which was in accordance with the analysis results mentioned before. There was obviously dimple in the Al phase surface. Pure Al played an important role in improving the mechanical properties of composite.

4. Conclusions

The thickness of SiO_2 layer increased with the prolonged time of the skeleton oxidation. The brittle phase basically disappeared at the interface of the composite with 9 hours pre-oxidized skeleton, which lead to the high interface bonding strength and plastic toughening of pure aluminum, as a result, the static compressive strength of the composites raised up to 1165.2Mpa.

The fracture morphology of the unprocessed composite is mainly constituted with brittle cracking in the interface, while the fracture morphology of the oxidized composite is mainly composed with plastic toughening of pure aluminium. Hence, the mechanical property of composite improves.

References

[1] Zhang Zhijin, Wang Yangwei, Yu Xiaodong, Wang Fuchi, Li Kai, Luan Zhiqiang, The Preparation of 3D-Reticulation SiC Reinforced Al Matrix Composite, Rare Metal Materials and Engineering, Suppl.2, 38(2009), 499-501

[2] WANG Zhi- tai, YU Huan, XU Zhi- feng, CAI Chang- chun, YAN Qing- song, Fabricating process of high volume fraction SiCp/ Mg composites by vacuum pressure infiltration, Journal of Nanchang Hangkong University(Natural Science), 21(2007) 1-5

[3] Z L Shi, J M Yang. The interfacial characterization of oxidized SiC(p):2014 Al Composites [J]. Materials Science and Engineering A, 2001, (303): 46-53.

[4] G.G. Sozhamannan, S. Balasivanandha Prabu. Influence of interface compounds on interface bonding characteristics of aluminium and silicon carbide[J]. Materials characterization 60 (2009) 986 – 990.

[5] Youngman Kima, Jae-Chul Lee. Processing and interfacial bonding strength of 2014Al matrix composites reinforced with oxidized SiC particles[J]. Materials Science and Engineering A 420 (2006) 8–12.

[6] L. Salvo, G.L. Espérance, M. Suéry, J.G. Legoux. Interfacial reactions and age hardening in Al - Mg - Si metal matrix composites with SiC particles[J]. Mater. Sci. Eng. A, A177 (1994), pp. 173 - 183.

[7] J.C. Lee, J.P. Ahn, Z. Shi, Y. Kim, H.I. Lee. Modification of the interface in SiC/Al composites[J]. Metall. Mater. Trans., 31A (2000), pp. 2361–2368

[8] Urena A, Martinez E E, Rodrigo P, et al. Oxidation treatments for SiC particles used as reinforcement in aluminium matrix composites[J]. Composites Science and Technology, 2004, (64): 1843-1852.

Preparation and characterization of PMMA graded microporous foams via one-step supercritical carbon dioxide foaming

Huan Yuan, Junguo Li, Yuanlu Xiong, Guoqiang Luo*, Qiang Shen, Lianmeng Zhang
State Key Laboratory of Advanced Technology for Materials Synthesis and Processing,Wuhan University of Technology, Wuhan 430070, P.R. China
E-mail:luoguoqiang1980@sina.com

Abstract. Supercritical carbon dioxide ($ScCO_2$) foaming which is inexpensive and environmental friendly has been widely used to prepare polymer-based microporous materials. In this paper, PMMA graded microporous materials were foamed by PMMA matrix after an unstable saturation process which was done under supercritical condition of 28MPa and 50℃. The scanning electron microscopy (SEM) was utilized to observe the morphology of the graded foam. A gas adsorption model was proposed to predict the graded gas concentration in the different region of the polymer matrix. The SEM results showed that the solid and foam region of the graded foam can be connected without laminated layers. With the increasing thickness position of the graded microporous foam, the cell size increased from 3.4 to 27.5 μm, while the cell density decreased from 1.04×10^9 to 1.96×10^7 cells/cm^3. It also found that the gradient microporous structure of the foam came from graded gas concentration which was obtained in the initial saturation process.

1. Introduction

Supercritical fluids have received much more attention in science and technology in recent years for its distinctive properties. They can be used as foaming agent due to their many advantageous, such as high diffusion coefficient, strong solvent power and the plasticization of glassy polymers[1,2]. Among so many supercritical fluids, supercritical carbon dioxide ($ScCO_2$) has widely used because their properties of environmental friendly and low critical condition(31.1℃ and 7.38MPa)[3,4]. Microporous polymer which is foamed from polymer using $ScCO_2$ as blowing agent has smaller cell size and higher cell density compared to traditional foams. These qualities made them a high compressive strength or tensile strength[5]. In addition, the foam products prepared by supercritical CO_2 foaming technique can be well controlled and calibrated.

Functionally graded materials (FGMs) which have non-homogeneous composition and graded microstructure can apply to the fields of aerospace, biomedical and automotive industries[6,7]. There are so many methods to fabricate FMGs, such as chemical vapor deposition(CVD)[8], which is common used to prepare thin film FGMs. Powder metallurgy is another method to produce both thin film and bulk FGMs. But this method often needs many steps including the mixing, compaction, sintering and subsequent welding[9]. The research on polymer-based FGMs is quite limited compared to ceramic- and metal-based ones. Changchun Zhou's team had presented a two-step foaming method to prepare functionally graded microporous foams using $ScCO_2$ as blowing agent[10]. They first put the polymer sample in a pressure vessel to absorb $ScCO_2$, then the sample was taken out from the vessel and foamed using a modified hot press device. The foaming process was mainly controlled by the saturation time, foaming temperature, and foaming time.

In this paper, we present a one-step foaming method to fabricate graded microporous foams using $ScCO_2$ as blowing agent. Microstructure gradient was created by controlling the gas adsorption process. The carbon dioxide will form concentration gradient inside the polymer during the adsorption process if the saturation time is appropriate. This gas concentration gradient will induce cell density gradient after foaming. This paper presents the fabrication and structural characterization of the polymer-based functional graded microporous foams.

2. Experimental

2.1. Materials

Polymethyl methacrylate (PMMA) was used as polymer matrix, acquired from Nantong Rayon Chemical Co., Ltd., China. The density of the material was 1.19 g/cm^3 and the glass transition temperature (Tg) was 106℃.PMMA solid matrix was cut into 2mm×15mm×15mm rectangular samples. The foaming agent CO_2 (>99.95%) was kindly supplied by Wuhan Yousheng Gas Limited Liability Company.

2.2. Foaming process

The PMMA sample was firstly absorb CO_2 under a supercritical condition at 28MPa and 50℃ for an hour. Then the pressure was rapidly released in 2s to introduce a thermodynamically unstable state. The CO_2 in polymer nucleates and foams. After foaming for 60s, the sample was shaped by cooling in ice and water.

2.3. Sample characterization

Scanning electron microscope (SEM) was used to characterize the morphology of the foam sample. The foam samples were freeze-fractured in liquid nitrogen and the fracture surface ware sputter coated with gold. The cell size and cell density were calculated according to the SEM images. Cell density and cell size analysis was conducted using ImageJ (National Institute of Health) and more than 100 cells were tested. The cell density was calculated using the following equation (1):

$$N=\left(\frac{nM^2}{A}\right)^{3/2} \tag{1}$$

Where $\boldsymbol{N}$ is the cell density (cells/cm^3), $\boldsymbol{A}$ is the area of the SEM image and $\boldsymbol{M}$ is the magnification factor, $\boldsymbol{n}$ is the number of cells counted from the SEM image[5].

3. Results and discussion

3.1 The morphology of uniform microporous foams

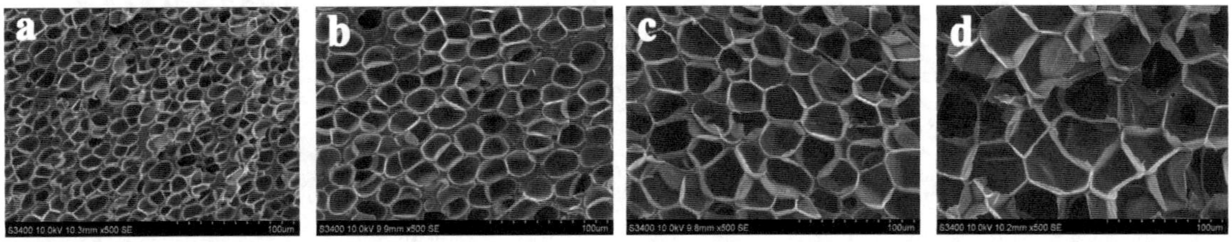

Figure 1. Morphology of microporpos uniform foams after saturating at 20MPa (a)65℃(b)80 ℃(c)110 ℃(d)140 ℃ for 6h

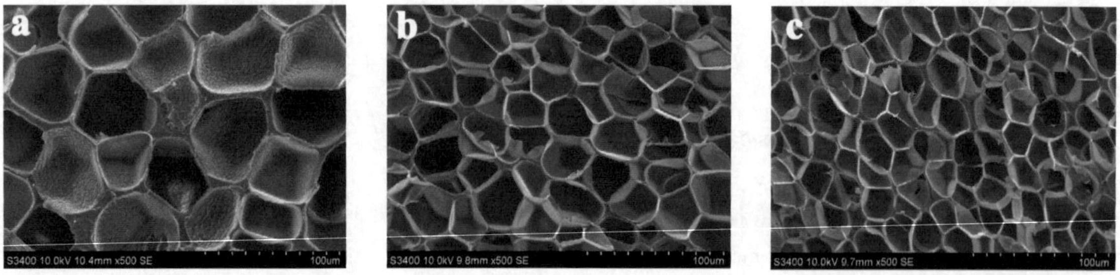

Figure 2. Morphology of microporpos uniform foams after saturating at 110℃ (a)12MPa; (b)20MPa; (c)24MPa for 6h.

In order to investigate the saturation temperature on the morphology of the microporous foams, we do many experiments at different saturation temperature and the results were showed in figure 1. The cell size was increasing from 12.4 to 43.5 μm with the temperature. The cells began to deform at high temperature. Then we prepare uniform foams at different pressure. The results were showed in figure 2. It was clear that the cell size was decreasing from 48.5 to 22.4μm with the saturation pressure. In conclusion, the foams with smaller cell size and good shape can be prepared at low temperature and high pressure. So we choose to prepare the graded microporous foams at 28MPa and 50℃.

3.2. The morphology of graded microporous foams

Figure 3 shows the morphology of graded microporous foams after saturating at 28MPa and 50℃ for one hour. It can be seen that gradient structure can be successfully gotten at a high pressure and low temperature condition. The diffusion depth of CO_2 in PMMA matrix is about 1386μm. In the region near the cortex of the foam sample, more CO_2 gas nucleates and foams that induce a high cell density and a thin pore wall. However, the cell size gradually increased toward the region inside the foam sample. Less CO_2 nucleates and foams in that region, so it has a low cell density and a thick pore wall. Unlike other functional graded materials which are made by lamination method[11,12], there is no need to use a bonding agent to prepare a graded microstructure in this study. The solid and foam region can be bonded gradually without laminated layers.

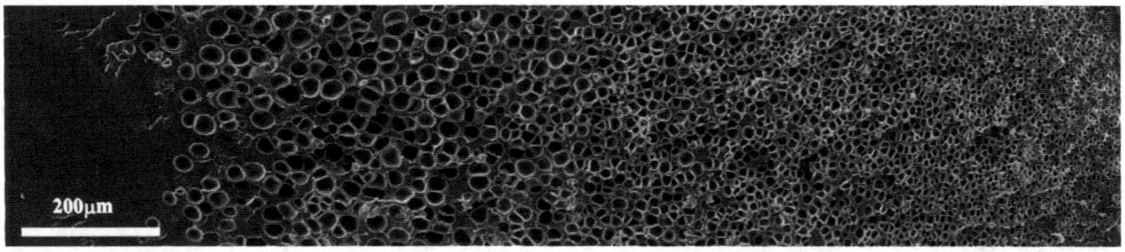

Figure 3. Morphology of graded microporpos foams after saturating at 28MPa and 50℃ for 1 hour

3.3. Cell density and cell size analysis

The cell size and cell density were calculated from the SEM image of graded microporous foams. The image was cut into different regions according to the thickness position. The foamed region represents the low thickness position while the solid region represents the high thickness position. The average cell density and cell size were measured in each region.

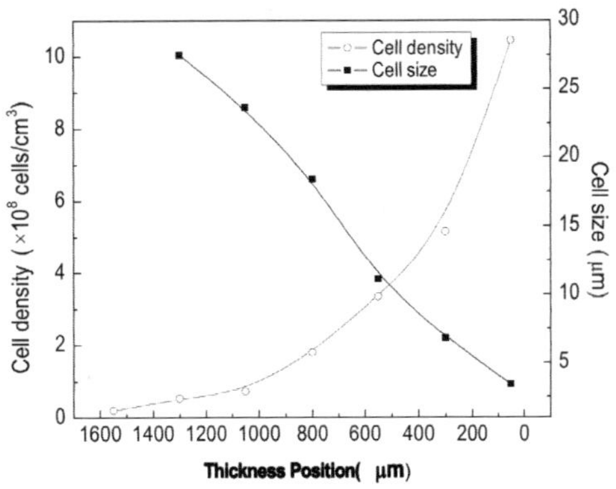

Figure 4. The relationship between the location of the bubbles and the cell size or cell density after saturated at 28MPa and 50℃ for one hour

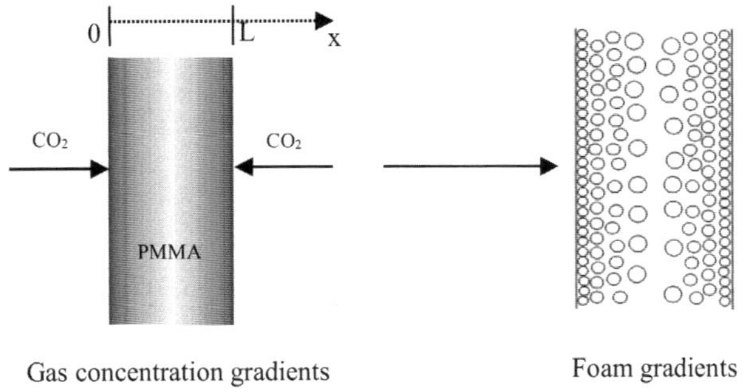

Figure 5. The formation of gas concentration gradients and foam gradients

Figure 4 shows the relationship between the location of the cells and the cell size or cell density. The cell size was changed along with the position of the graded foam. When the cell was near the surface of the foam, the size was small and the smallest size was about 3.4μm. While the size of the biggest cell which located in the deep position of sample was about 27.5 μm, almost 8 times bigger than the smallest one.

The cell density was also gradually changed along with the thickness position which demonstrated an opposite variation of the cell size. In low temperature, the deepest thickness position showed the lowest cell density, as low as 1.96×10^{7}cells/cm^3. The cell density reached 5.15×10^{7}cells/cm^3 at the position of 300 μm and got the highest value in 50 μm, about 1.04×10^{9}cells/cm3, almost 2 orders higher than that in the deeper position.

3.4. The CO_2 adsorption and the formation of concentration gradients

Figure 5 shows the gradient structure formation process. First, the $ScCO_2$ saturated in PMMA solid matrix in which will form a concentration gradient. The final diffusion depth was L/2. Then, after a quick pressure relief, the gas in the saturated sample became unstable and tended to foam. So the concentration gradient will induce a foam gradient. The gas concentration in the surfaces of the sample was quiet high so that there were a large amount of cells in the surfaces and the cell size was small. However, the gas concentration in the centre of the saturated sample was lower, so the cells became bigger and the number of cells was limited.

In addition, the CO_2 can reduce the polymer's glass transition temperature (T_g)[14], the higher the gas concentration, the lower the glass transition temperature. If the temperature is higher than T_g, the polymer tend to foam. In this study, the gas concentration in different position of the sample was different. In the surfaces of the solid matrix, it was easy for the gas immersion. The gas concentration was high there, so it induced a lower T_g and it was easy for that region to foam. However, in the centre of the solid matrix, it took a long time to make the sample fully saturated. The gas concentration low, so the T_g was higher there. It is hard for the sample to foam in that region.

4. Conclusion

（1）The polymer-based functionally graded materials were prepared through a one-step supercritical foaming method which using CO_2 as blowing agent.

（2）The gradient structure can be successfully gotten at a high pressure and low temperature condition. The cell size and cell density were changed gradually along with the thickness position.

（3）The graded microporous structure of the foam came from graded gas concentration which was obtained in the initial saturation process.

Acknowledgements

This work is supported by the National Natural Science Foundation of China (51202175) and the 111 Project（B13035）.

References

[1] A.I. Cooper 2000 *J. Mat. Chem.* **10** 207-234
[2] I. Tsivintzelis, A.G. Angelopoulou and C. Panayiotou 2007 *Polymer* **48** 5928-5939
[3] C. Jiang, J. Pi and Q. Pan 2009 *Poly. Mat. Sci. and Eng.* **25** 162-165
[4] P.G. Jessop and W. Leitner1999, *Chemical Synthesis using Supercritical Fluids*(Wiley,VCH, Weinheim)
[5] L. Chen, L. S. Schadler and R. Ozisik 2011 *Polymer* **52** 2899-2909
[6] Bandyo padhyay A, Krishna B, Xue W and Bose S. 2008 *Sci. Mater. Med.* doi: 10.1007/s10856-008-3478-2.
[7] Wang B.L and Mai Y.W 2007 *J. Therm. Stresses.* **30(6)**:523－58.
[8] Groves J and Wadley H 1997 *Compos. B* **28(1-2)** 57-70
[9] Kawasaki A. and Watanabe R. 1997 *Ceram. Int.* **23(1)**73－83
[10] C.C. Zhou, P.Wang and W.Li 2011 *Compos. B* **42** 318-325
[11] Gupta N. 2007 *Mater. Lett.* **61(4–5)** 979–82
[12] Chen Z, Yan Q, Huang Z, Shen Q and Zhang L 2009 *Adv. Mater. Res.* **66** 284–7
[13] Wang H and Li W 2008 *J. Manuf. Sci. Eng. Trans. ASME* **130(2)** 41–49
[14] J.A.R.Ruiz，Jean M.T., M. Pedros and M.Dumon *J. of Supercritical Fluids.* **57** *87-94*

Fabrication of W/Cu FGM By Aqueous Tape Casting

Shulong Liu[1], Qiang Shen[2], Guoqiang Luo[3], Meijuan Li[4], Lianmeng Zhang[5]

[1,2,3,4,5]*Country State Key Lab of Advanced Technology for Materials Synthesis and Processing, Wuhan University of Technology, Wuhan 430070, P.R.China*

liushulong19820314@gmail.com, luoguoqiang1980@sina.com, sqqf@263.net, meijunli@whut.edu.cn, lmzhang@whut.edu.cn

Abstract. Tungsten Copper-based metals (W/Cu) were extensively used as electrical contact materials in switching systems for the electric power industry. In this paper, a novel investigation to prepare Tungsten Copper-based metal composite materials according to functionally graded material (FGM) concept and the method of tape casting was reported. Cu-coated W powders with different Cu weight fraction were synthesized via electricless plain in methanol-water solvent. The green tapes with different composition and thickness were laminated and then sintered to prepare W-Cu functionally graded materials. XRD, EDS, SEM and metallographic analyses were used to characterize the material microstructure and combination between different layers. The results showed that the Cu-coated W powders had grate compressibility leading to wettability of powders. The parallelism and flatness of intermediate layer were good and the combination was tight.

1. Introduction

Electrical contact material in a variety of electrical applications was very important. Because of high strength and high conductivity coupled with low thermal expansion coefficient, W-Cu alloys had been widely used in the field of power engineering and plasma technology, such as contact materials for medium and high voltage interrupters as well as welding electrode [1, 2]. However, the fabrication of W-Cu composites proved to be difficult due to their mutual insolubility of W and Cu, in addition the high melting temperature of W (3410℃) compared to that of Cu (1083℃) [3, 4]. Traditional powder metallurgy (PM) technology often resulted in drawbacks such as low densification, microstructure inhomogeneity and limited composition variation. In recent years, many investigations had been published to develop processing of W-Cu composites. It had been found that electrical properties of contact materials were strongly influenced by high purity, high relative sintered density and a fine homogeneous microstructure [5-13].Actually, temperature and current distributions form the contact surface to bulk were non-uniform during a switch action, which implied that the electrical contact material should be inhomogeneous to match with such non-uniform property distributions [14]. Obviously, a contact material with ideal operating parameters was very difficult to fabricate by conventional manufacturing technique [15]. Functionally graded material (FGM) concept was suitable to prepare a good electrical contact material [16].

In this paper, A potential and possible method for W-Cu electric contact materials that were prepared by electricless plain technology, tape casting technology and vacuum heating-press sintering technology was reported, and some characteristics were investigated simply.

2. Experimental

In the tape casting process, powders, binders and plasticizers were dispersed in a solvent to form slurry. The slurry was layered onto a moving belt and passed beneath the blade edge that shaped the slurry into a green tape of steady thickness. When the solvents evaporated, the particles coalesce into a relatively dense, flexible dried tape. The dried tape was then cut to size for next process. After layers had been stacked, the multilayer was sintered.

However, the tape casting of metal matrix composites was in its early stages. Cu-coated W powders were synthesized via electricless plain. SEM (Fig.1a) examination revealed an irregular and polyhedral shape of the metal crystal with a clean and smooth surface of grains. The particles had a size distribution in the range of 6-20 μm approximately. The copper plating solution was composed of copper sulfate solution (0.08M $CuSO_4 \cdot 5H_2O$, 0.24M $C_4H_4O_6KNa \cdot 4H_2O$) and formaldehyde solution (20M HCHO, 30 wt .% NaOH). Plating was carried out at 40℃ in a container with magnetic stirring. During the reaction process, the pH value of the system was kept as a constant (11) by adding NaOH solution. After plating, the powder was washed with distilled water to remove chemical solutions. And then dried and weighed to ensure that the nominal value of Cu was deposited on W.

The binder conferred its mechanical strength to the tape by forming organic bridges into the tape during solvent evaporation. Polyvinyl alcohol (PVA1788) and a low boiling point were used as binder in the slurry preparation. The dispersant agent used was acrylic ammonium salt polymer enables to wet Cu-coated W powders and leaded to a good dispersion of the powders in the slurry by developing repulsive forces between particles. Glycerin was used as the plasticizer enabled to make the tape more flexible by decreasing the glass transition temperature of the PVA skeleton.

After formulation, the slurry was mixed 12 hours at 20 rpm using planetary milling. Once been homogenized, the mixture was degassed 5 minutes and then directly layered onto a Mylar silicon film using two motorized doctor blade at the desired thickness. After solvent evaporated at room temperature, the as-cast green tape could be cut to any desired shape. The Cu-coated W composite tape layer was then to be stacked and the sample was sintered by vacuum heating-press sintering at 930°C, 60 MPa. The powders, the tapes and the sintered samples were characterized by SEM, XRD, EDS.

3. Results and discussion

Tungsten powders were coated with copper by electroless plating method using formaldehyde solution as a reducing agent. The Cu layer covering a W particle was controlled by the content of $CuSO_4.5H_2O$. The Cu-precipitation is proceeding according to the following reaction:

$$Cu^{2+}+2HCHO+4OH^{-} \rightarrow Cu^{0}+H_2+2H_2O+2HCO_2^{-}$$

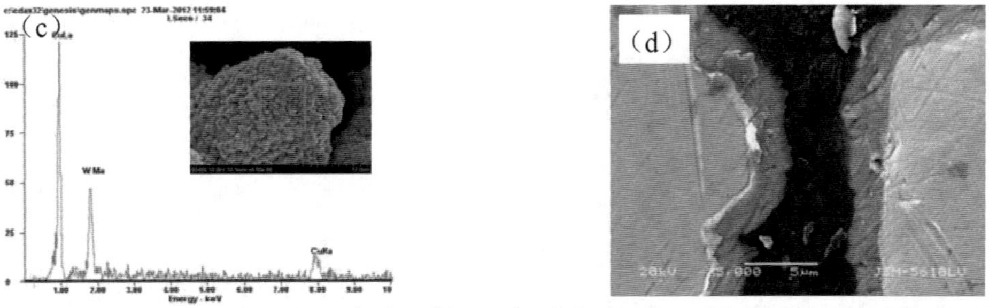

Figure 1. SEM images and EDS analysis of investigated powders: (a) SEM image of pure W, (b) SEM image of Cu-coated W powder, (c) EDS analysis of surface, (d)SEM image of cross-section.

The surface and cross-section (SEM) micrographs of the investigated powders of Cu-coated W powder are shown in Fig.1b and d, respectively. It is evident that the surface of the W particle is coated with a layer of tiny copper. The deposited Cu-coating layers consisted of particles with dome clusters or egg-shaped tops [17]. The Cu-coated W particles of Cu-loading are analyzed using EDS. Spectra of both W and Cu were in evidence, as illustrated in Fig.1b and c.

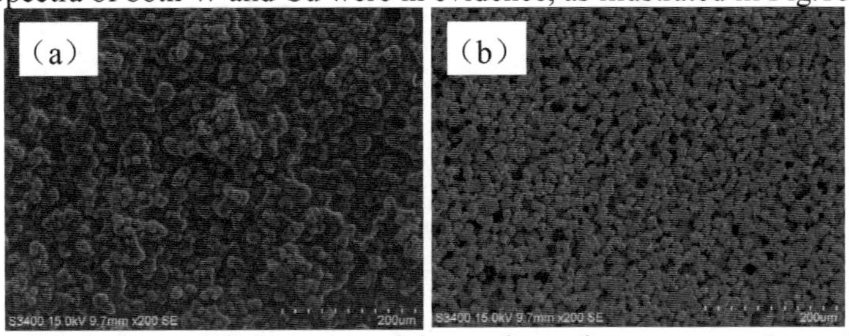

Figure 2. The surface of green tape SEM micrograph: (a) upper surface, (b) bottom surface.

Fig.2 shows SEM images of the surface of green tapes. It can be seen that the particle distribution in slurry is very uniform, but there are many small uniform holes, which are mainly caused by evaporation of the solvent.

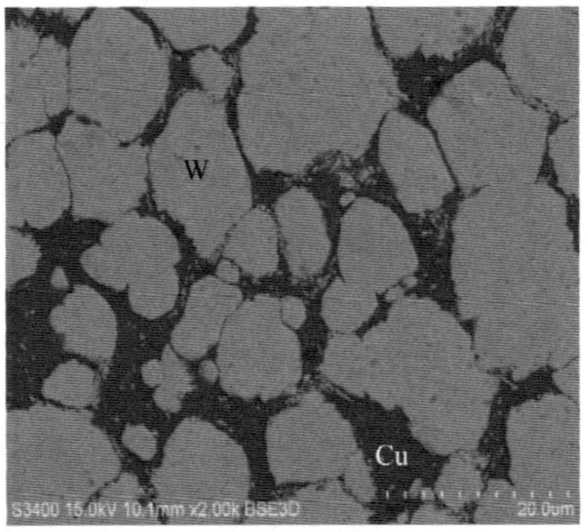

Figure 3. SEM images of W-40wt.% Cu sintered samples.

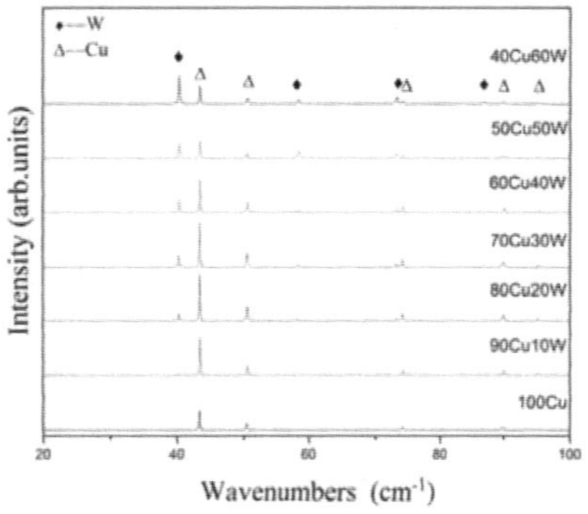

Figure 4. XRD patterns of sintered samples.

Because of a high dihedral angle in the W-Cu system, the W-W network was retained, even after Cu melt formation and the resulting tungsten skeletal rigidity restricts capillary force induced densification [17-18]. On the other hand, the micrograph of coated composites shows in Fig.3, reveal that the Cu coating separated the tungsten particles, thereby greatly reducing the degree of contact between the tungsten particles, even though the Cu content is low. In case of coated composites, due to the existence of Cu-Cu contacts, their sintering before the onset of Cu liquid formation behave like homogeneous sintering system in which densification is obtained by volume and grain boundary diffusion mechanisms. Thus, a highly contiguous microstructure is expected. Tungsten coated with copper composites might have a reduced level of solid state sintering between W-W contacts, but more densification would occur by rearrangement. The microstructures of coated W-Cu composites reveal homogeneous structure. The XRD result of W-Cu composites, only peaks related to copper and tungsten are seen and no compound between these two elements has been formed in Fig. 4.

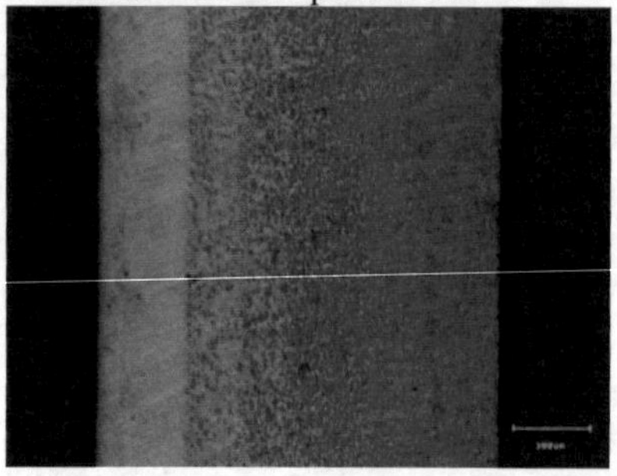

Figure 5. Metallic phase microstructure analysis of W-Cu density graded materials with 40-100wt.%Cu.

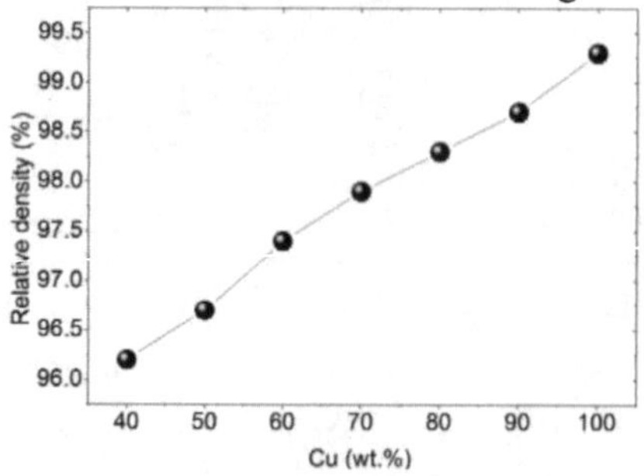

Figure 6. Variation of the relative density of W-Cu composites with varying Cu-coated W weight fraction.

Fig. 5 shows a representative microstructure of a graded W/Cu composite. A one dimensional gradient ranging from Cu (left) to W (right) is clearly visible within a distance of about 1.4mm. The gradient is continuous and accompanied by an increase in the size of W dark areas. The parallelism and flatness of intermediate layer are good and the combination is tight. The interface diffusion of different layers is obvious. The relative densities of the final W-Cu samples could achieve to 96.2%-99.3% as Cu weight fraction varying from 40 wt. % to 100 wt. % as shown in Fig. 6.

4. Conclusions

(1) W-Cu composition powder with core-shell structure was successfully prepared by electricless plain. Only two spectra of both W and Cu were fond by using EDS. Cu was highly dispersed throughout the whole W.

(2) The green tapes of W-Cu composition had a smooth surface and homogenous structure.

(3) The individual compositions were fabricated into monolithic tapes for characterization by laminating multiple layers together, hot-pressing to nearly-full density. The excellent planarity and parallelism between the layers were separated from each other by clearly defined interfaces, indicating that the initial design could be kept stable.

Acknowledgments

This work had been supported by National Natural Science Foundation of China (51202175), the Electronics Information Functional Materials Key Laboratory of Ministry of Education, the 111 Project (B13035) and self-determined and innovative research funds of WUT (2011-IV-038).

References

[1] L. J. Kecskes, M. D. Trexler, B .R. Klotz, K. C. Choand and R. J. Dowding 2001 *Metall. Mater. Trans.* **32** 2885.
[2] B.Thomas, K.V.Walter, M.Roland and K.E.Sacegr 1991 *IEEE Trans.* **14** 8.
[3] Cheng J, Lei C, Xiong E, Jiang Y, Xia Y. 2006 *J. Alloys Comp* **421** 146-50.
[4] Cheng J, Song P, Gong Y, Cai Y, Xia Y. 2007 *Mater Sci Eng A* **461** 1-5.
[5] J.Kaczmar 1989 *Powder Metall* **32** 171.
[6] B.J.Ding, Z.Yang and X.Wang 1996 *IEEE Trans. on CPMT* **19** 76.
[7] Wenge CHEN, Bingjun DING and Hui ZHANG 2002 *The Chinese J. Nonferrous Metals* **18** 1224.
[8] H. Zhang, Z. Yang and B. Ding 1999 *IEEE Trans. on CPT Part A* **22** 455.
[9] Y. Wang and B. J. Ding 1999 *IEEE Trans. on CPT Part A* **22** 467.
[10] F. Wang, F. Zhuge, H. Zhang and B. Ding 2003 *Mater. Res. Bull.* **38** 629.

[11] Y. Wang, C. Zhang, H. Zhang, B. Ding and K. Lu 2003 *J. Phys. D: Appl. Phys.* **364** 2649.
[12] B. Yang 1994 *Advances in powder metallurgy and particulate materials* **5** 17-27.
[13] S. H. Hong, B. K. Kim. 2003 *Mater Lett* **57** 2761-2767.
[14] Y. Wang, F. Zhang, L. Z. Qi, L. Wang and L. C. Cheng. 2005 *Key Engineering Materials* **280-283** 1917-1920.
[15] L. Wang, Y. Zhao, X. T. Zhu and Y. Wang 2007 *Key Engineering Materials* **336-338** 2616-2618
[16] A. Mortensen and S. Suresh 1995 *Int Mater. Rev.* **40** 239
[17] Ozkal B, Upadhyaya A, Ovecoglu ML 2004 *Euro Powder Metall* **40** 1-6.
[18] Yoon ES, Lee JS, Oh ST, Kim BK. 2002 *J Refract Met Hard Mater* **20** 201-206.

Influence of Al_2O_3/YSZ micro-laminated coatings on high temperature oxidation and spallation resistance of MCrAlY alloys

Junqi Yao, Yedong He and Deren Wang
Beijing Key Laboratory for Corrosion, Erosion and Surface Technology, University of Science and Technology Beijing, 100083 Beijing, PR China

E-mail: htgroup@mater.ustb.edu.cn

Abstract. Al_2O_3/YSZ micro-laminated coatings with different layers were prepared on MCrAlY alloys by magnetron sputtering and characterized by high-resolution field emission scanning electron microscopy (FE-SEM) and X-ray diffraction (XRD). Results indicated that the laminated structures of Al_2O_3 and YSZ layers were observed with compact microstructure and the thickness at sub-micron level each layer. High-temperature cyclic oxidation test at 1000℃ in air was performed to investigate the oxidation and spallation resistance of the coatings on MCrAlY substrates. Result shows that the coatings exhibit more excellent oxidation and spallation resistance with the increase of the layers, which can be attributed to the increase of stress tolerance and fracture toughness in the laminated coatings by the thinner layers and crack deflection toughening.

Keywords: Al_2O_3; YSZ; micro-laminated coatings; oxidation and spallation resistance; fracture toughness

1. Introduction

Due to the formation of a slowly growing protective alumina scale, MCrAlY-coatings (M=Ni, Co) are commonly used in the hottest sections of gas turbine components [1]. In industrial gas turbines, coatings of MCrAlY type are used as overlay coatings and as bond coats for thermal barrier coatings (TBCs) [2]. Constituents in a TBC system include superalloy substrate, superalloy bond coat (which is generally MCrAlY or Pt-modified nickel aluminide), ceramic coating (which is typically yttria-stabilized zirconia, YSZ), and thermally grown oxide (TGO) formed at the interface between the bond coat and ceramic coatings during high temperature operation [3]. It is generally accepted that the growth rate, morphology, microstructure and adherence of the aluminum-based TGO scale during high-temperature service is of crucial importance for TBC life [4]. After long term exposure, the alumina scales have been shown to be prone to cracking and spallation due to the thermal expansion mismatch between the oxide and the metallic substrate [5]. So the processes of cracking and spallation are the key factors influencing the service lifetime [6]. In order to circumvent the drawback, it is a practical way to add a transition layer between MCrAlY and the ceramic top coating to form a more effective obstruction of oxygen diffusing and improve the interface situation [7].

The design of laminated composite coatings has been accepted as an effective way to improve the fracture toughness generally. Clegg and his assistants prepared the SiC/C laminated composite ceramic; the fracture toughness reached 15 MPa • $m^{-1/2}$ and the fracture energy reached 4625 J/m^2, which are far greater than that of conventional materials [8]. It's reasonable to deduce that the laminated composite coatings can exhibit beneficial effects in the high-temperature application. He Y D and his co-workers developed the Al_2O_3/YSZ micro-laminated coatings in various preparation methods, which exhibit more excellent oxidation and spallation resistance than the coating with single phase [9,10]. Tomaszewski et al. demonstrate that the crack deflection in alumina layer was found to be the main mechanism responsible for an increase in mechanical properties of the layerd zirconia/alumina composites [11]. So it's necessary to investigate the influence of laminated coatings on the high-temperature oxidation and spallation resistance in more detail.

As an effective technology to prepare coatings, the magnetron sputtering (MS) succeeds in application of many fields [12-14]. MS has sorts of comparative advantages such as higher deposition rate, more excellent density and bonding property. In this paper, thinner Al_2O_3/YSZ laminated composite coatings were prepared on MCrAlY substrates by magnetron sputtering. When verifying the number of layers, the influence of Al_2O_3/YSZ coatings on the high-temperature oxidation and spallation resistance was investigated and the mechanisms of such excellent performance were discussed.

2. Experimental procedure

Laminated composite coatings were sputtered onto the surface of MCrAlY by magnetron sputtering, with the layers of Al_2O_3 prepared by reactive magnetron sputtering. In the procedure of preparing Al_2O_3, DC electrical source was adopted with some influencing factors as 120℃ of the substrate temperature, 0.5Pa of Ar and O_2 as working gas with the proportion of 7:1, 540V of voltage, 160mA of current flow, and 70mm of distance between target and samples. During the preparation of YSZ, RF electrical source was used with 0.5Pa of Ar as working gas, the same substrate temperature and distance between target and samples. The scale of substrate is 10mm×10mm×3mm. The ingredient of MCrAlY is Ni-39.5, Co-32, Cr-20, Al-8, Y-0.5. All the surface of samples were ground to #2000 abrasive, followed by ultrasonic cleaning with acetone and ethanol, and then cleaned by ionized water.

The high-temperature behavior of coatings was investigated by cyclic oxidation at 1000℃ in air. After sputtered, the test samples were exposed in high-temperature environment induced by tube type resistance furnace 10 hours every time in quartz crucibles. After every break the samples and quartz crucibles were cooled to the room temperature by natural cooling, then the total weight and the weight of quartz crucible without sample were both measured by electronic balance with an accuracy of 10^{-5}g. With a series of mentioned data, the curve of weight gain and spallation per unit area can be drew, where we can get some useful information about high-temperature behavior.

Si substrate was used to observe the cross-section morphology of the composite coatings due to its electronical conductivity and brittle fracture characteristic. The cross-section morphology and surface image of coatings were obtained by high-resolution field emission SEM. Phases of the laminated coatings and oxide scales formed after oxidation were both analyzed by X-ray diffraction (CuKα, λ=0.15406 nm, step wise of 0.02°, continuous scanning).

3. Results

3.1 Characterization of Al_2O_3/YSZ laminated composite coatings with different layers

The FE-SEM images of Al_2O_3/YSZ laminated composite coatings with different layers prepared by MS were showed in the figure 1. It's clearly revealed in figure 1(a) that the coatings are very dense and comprised with uniform nano-particles. Figure 1(b) shows the cross-section of the Al_2O_3 single coating, and figure 1(c)~1(f) shows the cross-sections of the laminated composite coatings with three layers, five layers, seven layers and nine layers. For convenience, they are simplified as 1-layer, 3-layer, 5-layer, 7-layer and 9-layer. The total thickness of each component is mainly constant while the aggregate thickness is about 6 μm. The interfaces between layers are distinct without any cracks or flaws due to the compact bond from the magnetron sputtering.

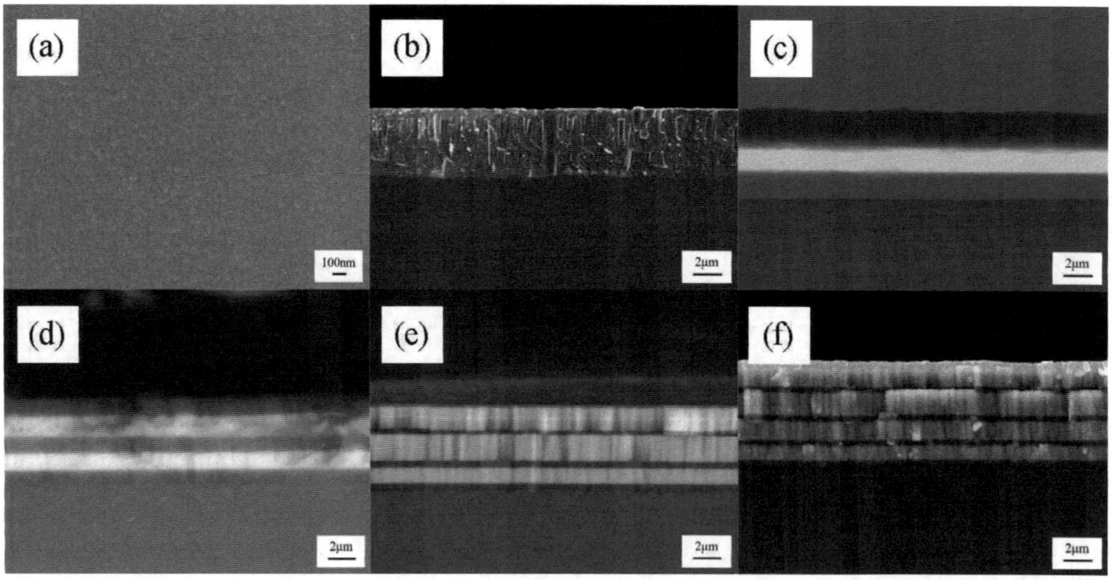

Figure 1. FE-SEM images of Al_2O_3/YSZ laminated composite coatings: (a) surface of the coatings; (b) 1-layer coating; (c) 3-layer coatings; (d) 5-layer coatings; (e) 7-layer coatings; (f) 9-layer coatings.

3.2 High-temperature cyclic oxidation test of the micro-laminated coatings on MCrAlY substrates

Figure 2 shows the experimental results of cycle oxidation test of different samples at 1000℃ in air for 400h. It can be illustrated from figure 2(a) that all the samples with different layers have smaller weight gain than the blank sample (1.34 mg/cm^2). And the oxidation resistance of samples with different layers is improved greatly as the layers increase. The 9-layer laminated coatings exhibit the best oxidation resistance with 0.26 mg/cm^2 weight gain per unit area. Figure 2(b) reveals that the spallation resistance of samples with different layers exhibits an ascensive trend when increasing the number of the layers. The spallation of the sample with nine layers is as few as 0.09 mg/cm^2, which is much smaller than that of the blank sample and the sample with 1-layer coating. With a same total thickness, more layers of Al_2O_3 and YSZ means more interfaces and each layer becomes thinner. From these oxidation kinetic curves of samples, there is a trend summarized that the more layers of Al_2O_3 and YSZ at a certain total thickness, the better high temperature oxidation and spallation resistance the specimen has.

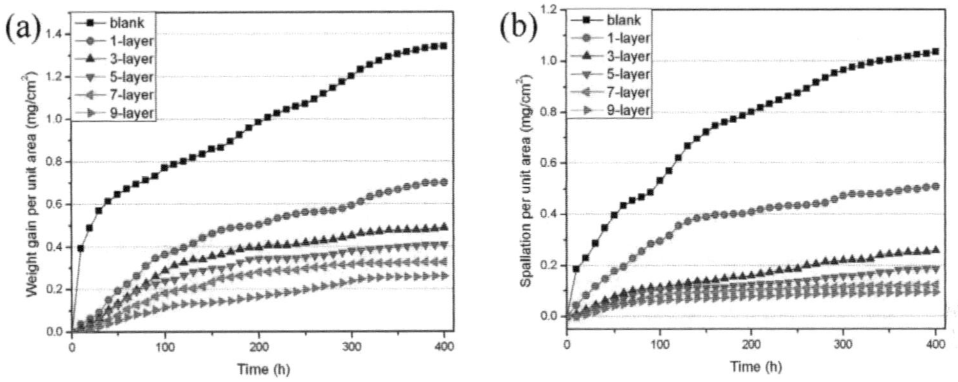

Figure 2. Cyclic oxidation kinetic curves of samples with laminated coatings at 1000℃ for 400h (a) weight gain per unit area versus time; (b) spallation per unit area versus time.

3.3 Characterization of samples after high-temperature cyclic oxidation test

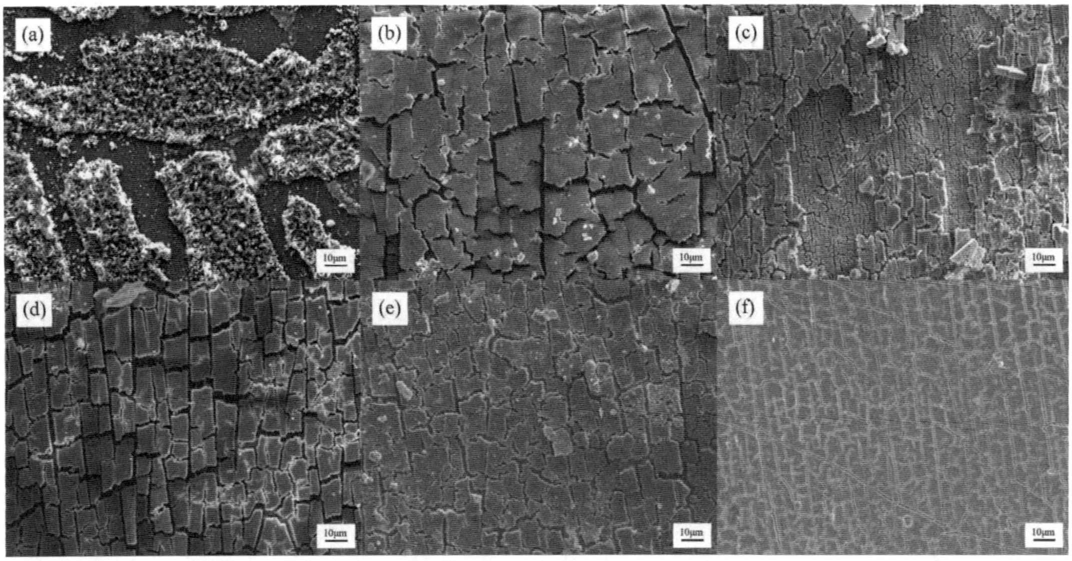

Figure 3. FE-SEM surface images of different samples after cyclic oxidation: (a) blank; (b) 1-layer coating; (c) 3-layer coatings; (d) 5-layer coatings; (e) 7-layer coatings; (f) 9-layer coatings.

Figure 3 illustrates the surface morphologies of the samples after high-temperature cyclic oxidation test. The blank sample exhibits a rough surface with mainly needle-like morphologies of θ-Al_2O_3 as shown in figure 3(a). It can be found obviously that the scale of all the microcracks existed in different coatings is rather slight. The top layer in figure 3(b) has been fragmented, which means large area of spalling. Compared to others, figure 3(f) shows a more smooth and intact surface morphology with cracks at the smallest size. After high-temperature cyclic oxidation test for 400h, the composite coatings with nine layers maintained the laminated structure comparatively completely, where the smallest residual stress can be inferred. Put in another way, the 9-layer coatings show excellent spallation resistance in the high-temperature environment. Figure 4 shows the cross-section images of sample with 9-layer coatings after cyclic oxidation and that of the blank sample as a comparison. The 9-layer laminated coatings were reserved integrally with some cracks throughout one layer and thin TGO after cyclic oxidation. Besides, the blank sample was oxidation severely with thicker TGO contained many cracks and voids. In figure 5, the XRD results show that, after cyclic oxidation, t-ZrO_2 and α-Al_2O_3 are identified as stable phases at the room temperature which could ensure the excellent oxidation and spallation resistance of the laminated coatings.

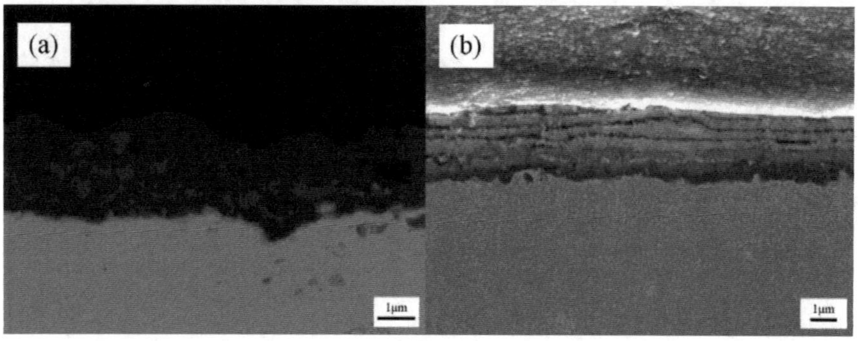

Figure 4. Cross-section images of samples after cyclic oxidation: (a) blank; (b) 9-layer coatings.

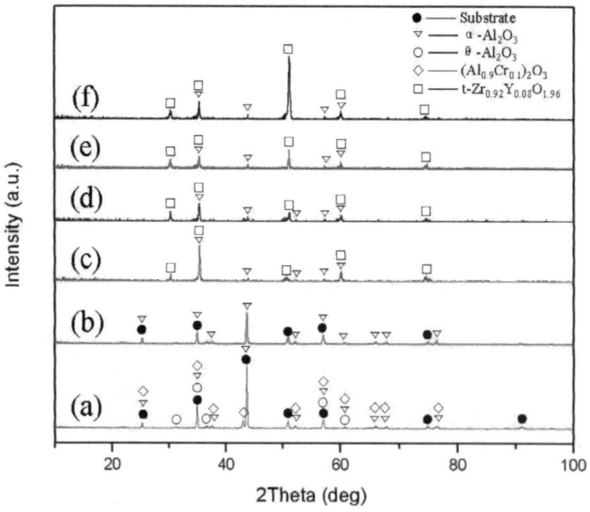

Figure 5. XRD spectra of different samples after cyclic oxidation at 1000℃ for 400h: (a) blank; (b) 1-layer coating; (c) 3-layer coatings; (d) 5-layer coatings; (e) 7-layer coatings; (f) 9-layer coatings.

4. Discussion

It has been acknowledged that the excellent oxidation and spallation resistance of the laminated composite coatings are attributed to fundamental conditions: (1) thermodynamically stable constituent phases [15-17]; (2) entirely sealed structure of the component with low oxygen diffusion coefficient (here is Al_2O_3 [18]) on the substrate; (3) special mechanical properties which benefit from the as-designed laminated structures. The excellent oxidation and spallation resistance of the Al_2O_3/YSZ laminated composite coatings acquired by increasing the number of layers can be analyzed as follows.

(1) The increase of the stress tolerance in the laminated structures

The basic formula about fracture toughness of materials is shown as follows [19]:

$$K_{IC} = \sigma_c \sqrt{a} Y \tag{3}$$

Where K_{IC} is the critical stress intensity factor (fracture toughness), σ_c is the critical stress for crack propagation, a is the half length of crack and Y is a geometrical factor related to the shape and loading modes. K_{IC} and Y are constant in certain conditions. When the total thickness is constant, the thickness of each layer will decrease with the increase of the number of layers. The propagation processes of the cracks in vertical direction could be limited in a shorter length, no more than the thickness of one layer. Thus, it can be inferred from the above formulas that the coatings could stand the higher critical stress, which means the mechanical properties of the laminated coatings could be improved significantly. The shorter micro-cracks guarantee the maximum sealed function as shown in the figure 3(f), which avoids further oxidation of the substrate. So the oxidation resistance can be improved relatively by increasing the number of layers.

(2) The increase of the fracture toughness in the laminated structures

The toughening effect of the laminated structures can be attributed to the energy release mechanisms in various forms, such as crack deflection or crack bifurcation, while there is no macroscopic damage occurred in the coatings. As the cracks in vertical direction can be limited in a shorter length, the cracks deflection will occur more easily, benefited from the more interfaces, which means the cracks propagation path would be extended as shown in figure 6 [20]. And then more strain energy would be released, which could improve the fracture toughness of laminated composite coatings. Thus, with the

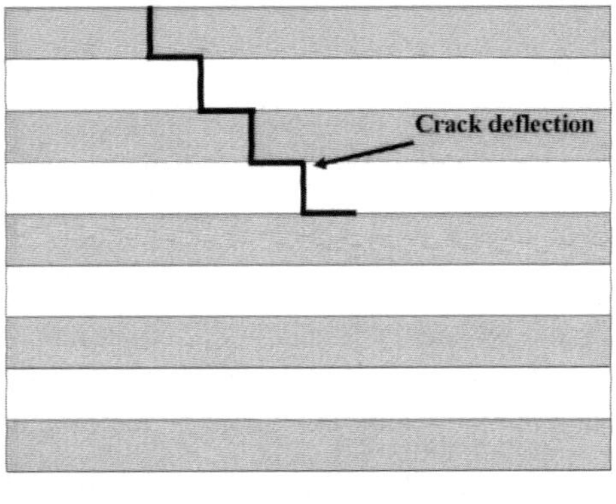

Figure 6. Toughening mechanism of crack deflection in laminated composite coatings

increase of layers, the more excellent mechanical properties can bring in the better spallation and oxidation resistance of micro-laminated coatings.

5. Conclusions

In this paper, the oxidation and spallation resistance of Al_2O_3/YSZ micro-laminated coatings with different layers were investigated. The conclusions can be drawn as follows:

(1) The oxidation and spallation resistance of Al_2O_3/YSZ micro-laminated coatings prepared by magnetron sputtering on MCrAlY substrates could be improved significantly by increasing the number of layers while keeping the total thickness constant.

(2) The beneficial effects could be attributed to the increase of stress tolerance and fracture toughness in the laminated coatings by the thinner layers and crack deflection toughening.

Acknowledgements

Authors gratefully acknowledge the financial support from the Chinese National Nature Science Foundation (grant 51071030).

References

[1] Goward G W 1998 Progress in coatings for gas turbine airfoils *Surf. Coat. Technol.* **108-109** 73-79

[2] Rickerby D S and Matthews A 1991 *Advanced Surface Coatings: a Handbook of Surface Engineering* (London: Springer)

[3] Li Y, Li C J, Zhang Q, Yang G J and Li C X 2010 Influence of TGO composition on the thermal shock lifetime of thermal barrier coatings with cold-sprayed MCrAlY bond coat *J. Therm. Spray Techn.* **19** 168-177

[4] Busso E P, Lin J, Sukurai S and Nakayama M 2001 A mechanistic study of oxidation-induced degradation in a plasma-sprayed thermal barrier coating system.: Part 1: model formulation *Acta Mater.* **49** 1515-28

[5] Evans H E 1995 Stress effects in high temperature oxidation of metals *Int. Mater. Rev.* **40** 1-40

[6] Evans H E 1989 Cracking and spalling of protective oxide layers *Mater. Sci. Eng. A.* **120** 139-146

[7] Gao J G, He Y D and Wang D R 2010 Preparation of YSZ/Al_2O_3 micro-laminated coatings and their influence on the oxidation and spallation resistance of MCrAlY alloys *J. Eur. Ceram. Soc.* **31** 79-84

[8] Clegg W J, Kendall K and Alford N M 1990 A simple way to make tough ceramics *Nature* **347** 445-447

[9] Yao M M, He Y D, Wang D R and Gao W 2005 Oxidation resistance of Fe-13Cr alloy with micro-laminated (ZrO_2-Y_2O_3)/(Al_2O_3-Y_2O_3) films *J. Rare Earth.* **23** 559

[10] Gao J G, He Y D and Wang D R 2010 Fabrication and high temperature oxidation resistance of ZrO_2/Al_2O_3 micro-laminated coatings on stainless steel *Mater. Chem. Phys.* **123** 731-736

[11] Tomaszewski H, Weglarz H and Boniecki M 2000 Effect of barrier layer thickness and composition on fracture toughness of layered zirconia/alumina composites *J. Mater. Sci.* **35** 4165-76

[12] Kelly P J and Arnell R D 2000 Magnetron sputtering: a review of recent developments and applications *Vacuum* **56** 159-172

[13] Kelly P J, Hisek J, Zhou Y, Pilikington R D and Arnell R D 2004 Advanced coatings through pulsed magnetron sputtering *Surf. Eng.* **20** 157-162

[14] Cooke K E, Hampshire J, Southall W and Teer D G 2004 Industrial application of pulsed DC bias power supplies in closed field unbalanced magnetron sputter ion plating *Surf. Eng.* **20** 189-95

[15] Ma X X, He Y D, Wang D R and Zhang J 2012 Superior high-temperature oxidation resistance of a novel (Al_2O_3-Y_2O_3)/Pt laminated coating *Appl. Surf. Sci.* **258** 4733-40

[16] Medraj M, Hammond R, Parvez M A, Drew R A L and Thompson W T 2006 High temperature neutron diffraction study of the Al_2O_3-Y_2O_3 system *J. Eur. Ceram. Soc.* **26** 3515-24

[17] Lakiza S M and Lopato L M 1997 Stable and Metastable Phase Relations in the System Alumina-Zirconia-Yttria *J. Am. Ceram. Soc.* **80** 893-902

[18] Heuer A H 2008 Oxygen and aluminum diffusion in α-Al_2O_3: How much do we really understand? *J. Eur. Ceram. Soc.* **28** 1495-1507

[19] Griffith A A 1921 The phenomena of rupture and flow in solids *Phil. Trans. R. Soc. A* **221** 163-198

[20] Sbaizero O and Lucchini E 1996 Influence of residual stresses on the mechanical properties of a layered ceramic composite *J. Eur. Ceram. Soc.* **16** 813-18

Numerical Simulation in relation to Adiabatic Shearing Behaviors in Titanium Alloy

Gao Ruihua[1,2],Fan Qunbo[1,2*], Wang Fuchi[1,2]
[1] School of Materials Science and Engineering, Beijing Institute of Technology, 100081, Beijing, China, [2] National Key Laboratory of Science and Technology on Materials under Shock and Impact, Beijing, 100081, China

*Corresponding author e-mail: fanqunbo@bit.edu.cn

Abstract. The adiabatic shearing phenomena are commonly found in titanium alloys, but they are rarely simulated in the microstructure scale. In the current study, the macro-dynamic compression is implemented and the microstructures are successfully embedded into the macro model by introducing a multiscale simulation technique, thus help to reveal the adiabatic shearing deformation mechanism of titanium alloy. The simulation results show that for the equiaxed titanium alloy, the adiabatic shearing process is determined by the phase proportion outside the shear band instead of the phase proportion inside; the study further shows that within a certain proportion of α phase, with the increase of α phase proportion, the adiabatic shear sensitivity decreases.

1. Introduction

Titanium is an adiabatic shear-sensitive material, and it is easy to produce significant localized deformation under the condition of high strain rate and strain state, resulting in the formation of the adiabatic shear bands[1-3]. Adiabatic shear bands generated by thermal instability, when the effect of thermal softening predominates over the effect of strain hardening and strain rate hardening, give rise to plastic deformation instability into uniform plastic deformation localized deformation, forming of adiabatic shear band. The initiation of strain shear localization corresponding the highest point in the stress - strain curve, which indicates the loss of the carrying capacity of the titanium alloy materials, and the occurrence of plastic failure. The adiabatic shear deformation is an important aspect of the mechanical response of titanium alloy under high strain rate [4].

The adiabatic shear property of titanium alloy is closely related to its composition and microstructure. For the titanium alloys of the same composition, different microstructure morphology is associated with different adiabatic shear performances. On the titanium alloy of the relationship between the microstructure and the adiabatic shear performance, Zhang Wangfeng [5] adopted the linear regression method to study the impact of its anti-adiabatic shear properties of the titanium alloy sheet organization characteristic parameters, and obtained a function that could reflect the relationship between lamellar titanium anti-adiabatic shear properties, β grain size D, the α cluster size d, and α lamellar thickness b; based on Bayesian neural networks and fuzzy logic, S.K.Kar successfully implemented the prediction of the relationship between the microstructure characteristic parameters and the mechanical properties of theTi-6Al-4V titanium alloy. However, these methods are not able to reveal the intrinsic mechanism of titanium alloy deformation process in the evolution of the microstructure, and the micro-level deformation information can not be obtained. Therefore, it is particularly important to establish the relationship between microstructure and macro-dynamic compression, so as to study the effect of the characteristics on the titanium alloy adiabatic shear performance. In this paper, the LS-DYNA developed by Ansys and the microstructure oriented finite element software (MOF) developed by our group [6,7] are adopted to establish a multiscale simulation method for the titanium alloy adiabatic shear deformation, so as to study the key factor that dominates the adiabatic shearing sensitivity of equiaxed titanium alloy, and identify the influence of phase proportion on adiabatic shear sensitivity.

2. Multiscale simulation of titanium alloy adiabatic shear deformation

In the current study, the multiscale simulation method of titanium alloy adiabatic shear deformation is proposed by establishing the macro finite element model first, extracting the boundary conditions from a specified micro region subsequently, exerting the boundary conditions on the finite element model of the microstructure, finally calculating with LS-DYNA.

2.1. Macro-dynamic compression

The macro-dynamic compression finite element model is shown in Figure 1. The specimen is a 2D axisymmetric finite element model, subdivided by 2000 elements, and element size is set to 25μm. As shown in Figure 1, the radius and the hight are 5mm, respectively . The upper edge is applied by a rigid wall, the lower edge is constrained in the vertical direction, and the friction coefficient between the upper edge and the rigid wall is 0.08. In order to ensure the strain rate of approximately 3400/s, the rigid wall is loaded by the speed of 17m/s, and the time of macro-dynamic compression is 80μs. Since large local shearing deformation will happen along the 45° diagonal direction for a macro-dynamic compression sample, serious element distortion will happen if with the traditional Lagrange algorithm. Therefore, the Abitrary-Lagrangian-Eulerian algorithm (ALE algorithm) is adopted in this current study [8], which optimizes those distorted elements accordingly, thus keeping the calculation stability.

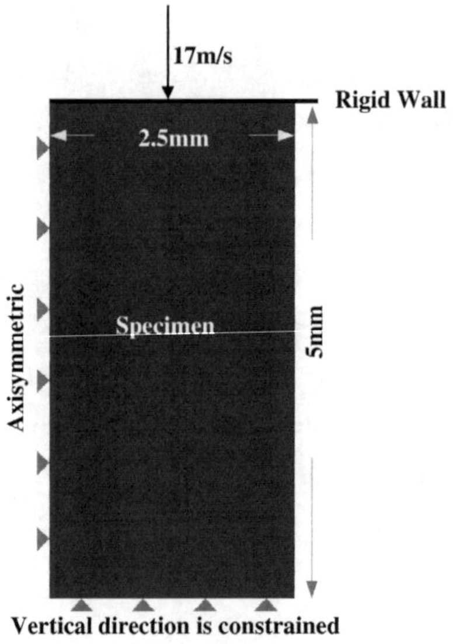

Figure 1. Titanium alloy macro-dynamic compression finite element model

2.2. The numerical simulation of adiabatic shear deformation for titanium alloy

The numerical simulation of adiabatic shear deformation for titanium alloy is shown in Figure 2. Firstly, a 300μm ×300μm micro region to be studied is selected from the macrostructure finite element model, while the model is fully discretized in order to accurately track the nodal displacements history data within this region; then, finite element mesh generation principle combined with digital image processing approach has been used to generate the equiaxed finite element model based on the actual microstructural images. In this paper, the model is subdivided by 40000 elements, and element size is set to 1.5μm. While the micro model is loaded by the boundary conditions that extracted from the macro mode, and a secondary analysis calculation is carried out, in order to achieve the deformation information inside the micro region on a smaller scale.

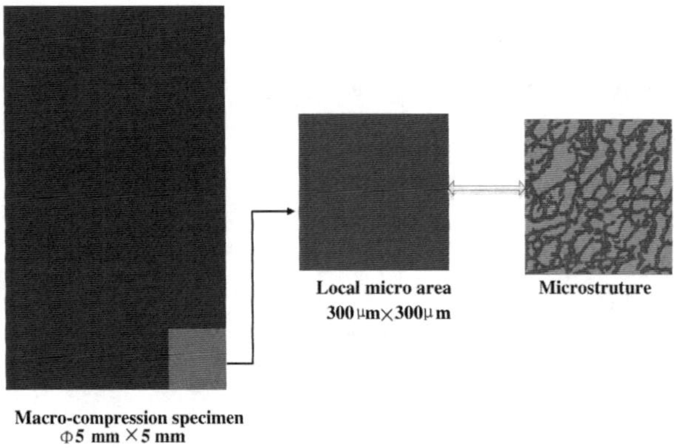

Figure 2. The numerical simulation of adiabatic shear deformation for titanium alloy

3. The influence of equiaxed microscopic characteristics on titanium alloy adiabatic shear sensitivity

Adiabatic shear sensitivity of titanium alloy is closely related to many factors, such as the phase proportion, phase size, and phase morphology. In this current study, however, the phase proportion is focused on. The phase proportion contains the phase proportion inside the adiabatic shear band, and the phase proportion outside adiabatic shear band. These factors influence each other, thus it is difficult to determine which one is the key factor. Meanwhile, the strength of β phase is stronger than α phase, and the α phase has a better uniform plastic deformation ability than the β phase has. Therefore, these issues will be focused on.

3.1. The influence of the phase proportion within the shear band on adiabatic shear sensitivity of equiaxed titanium alloy

Because plastic deformation is mainly concentrated in a narrow adiabatic shear band in the process of adiabatic shearing deformation, it seems that the phase proportion within the shear band dominates the adiabatic shear sensitivity. Therefore, the effect of the α phase proportion within the adiabatic shear band on the adiabatic shear sensitivity is discussed firstly. Three equiaxed titanium alloys with different microstructure characteristics are designed, as is shown in Figure 3 (a)~(c), where the blue is α phase, the red is β phase. Based on the micro model, the contours of strain are achieved by further calculation, as is seen in Figure 3 (d)~(f). The statistics show that α phase proportions within the shear bands are 60.6%, 22.5%, 14.3%, respectively.

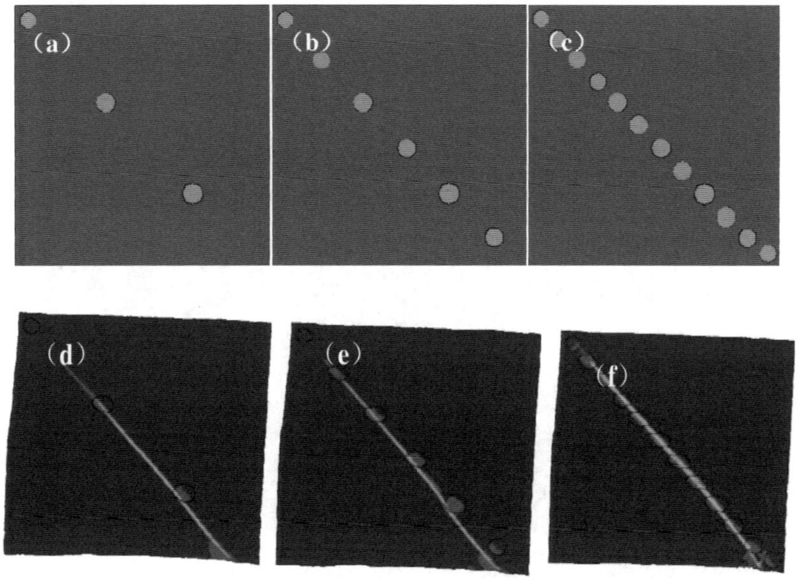

Figure 3. The equiaxed titanium alloys with different α phase proportions within shear bands

(a) (d), 14.3%; (b) (e), 22.5%; (c) (f), 60.6%

The calculation method of Section 2.2 is adopted to calculate crack initiation time, and the result is shown in Table 1. As is seen in Table 1, little change is found when the α phase proportion within the shear band is varied from 60.6% to 14.3%, and the maximum difference is only 0.8μs, which indicates that the phase proportion within the shear band has little influence on adiabatic shear sensitivity for equiaxed titanium alloys. Therefore, it is not the key factor that dominates the adiabatic shear sensitivity.

Table 1. The crack initiation time of equiaxed titanium alloys with different α phase proportions within the shear bands

α phase proportion (%)	Crack initiation time (μs)
14.3	42.5
22.5	43.3
60.6	42.7

3.2. The influence of the a phase proportion outside the shear band on adiabatic shear sensitivity of equiaxed titanium alloy

In the present work, a series of equiaxed titanium alloys with different microstructure characteristics are constructed. The α phase proportion distributed in the direction of 45^0 is kept constant, while the α phase proportion outside the shear band is continuously changed. As is shown in Figure 4 (a)~(f), the α phase proportions outside the shear bands of equiaxed titanium alloys are 13.8%, 20%, 27%, 34.7%, 42%, 46%, respectively.

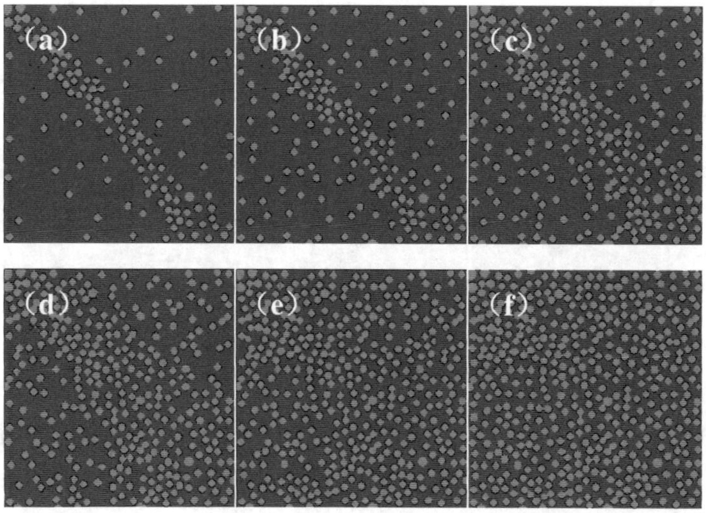

Figure 4. The equiaxed titanium alloys with different α phase proportions outside the shear bands

(a) 13.8%; (b) 20%; (c) 27%; (d) 34.7%; (e) 42%; (f) 46%

As is shown in Figure 5, when the calculation time is 44μs, the strain contours corresponding with different α phase proportion are presented accordingly.

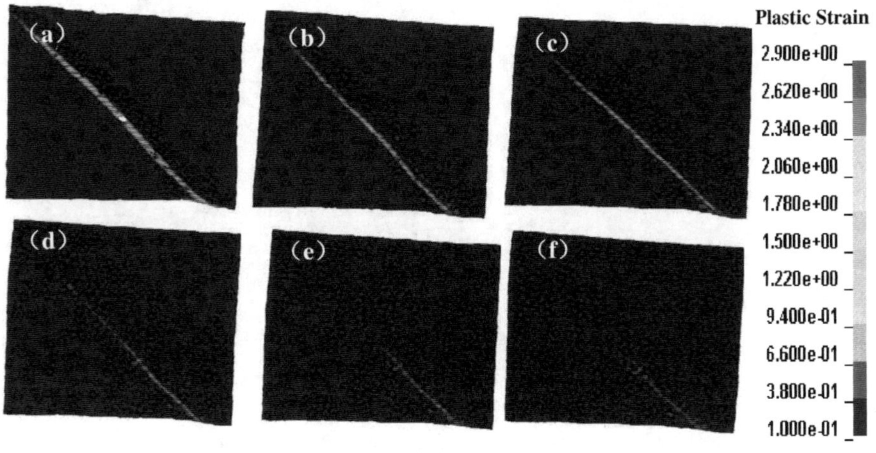

Figure 5. The effective plastic strain contours with the calculation time is 44μs

(a) 13.8%; (b) 20%; (c) 27%; (d) 34.7%; (e) 42%; (f) 46%

As is seen in Figure 5, the strain localization region is readily observed in the direction of 45 °, leading to the formation of the adiabatic shear band. When the α phase proportion outside the shear band is varied from 13.8%~46%, with the increase of the α phase proportion, the strain localization region decreases. For the α phase proportion of 13.8%, it has the largest strain localization region, the highest strain is more than 2.9, and the initiation of apparent cracks are observed; however, when the α phase proportion increases to 46%, the highest strain is lower than 0.7, and the strain localization region almost disappears. In order to further study the phenomenon, influence of the α phase proportion outside the shear band on adiabatic shear sensitivity is calculated and shown in Figure 6. The results show that as the α phase proportion changes from 13.8% to 46%, the crack initiation time is varied from 40.1μs to 46.2μs, and the difference of crack initiation time is up to 6.1μs, which indicates the α phase proportion outside the shear band is the key factor that dominates the adiabatic shear sensitivity of equiaxed titanium alloys and the adiabatic shear sensitivity decreases with increasing α phase proportion within a certain proportion of α phase. The research further shows that since the α phase has a better uniform plastic deformation ability than the β phase has, with the increase of the α phase proportion, the uniform plastic deformation ability increases, resulting in a decreased tendency of strain localization, thus the adiabatic shear sensitivity decreases[9-11].

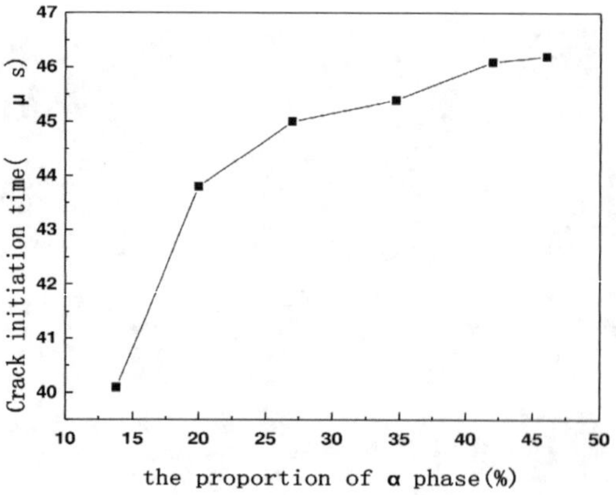

Figure 6. The crack initiation time of equiaxed titanium alloys with different α phase proportion

4. Conclusions

In the current research, the macro-dynamic compression is implemented and the microstructures are successfully embedded into the macro model by introducing a multiscale simulation technique, so as to study the influence of the phase proportion on the adiabatic shear sensitivity of equiaxed titanium alloy. The main conclusions are as follows:

(1) The phase proportion within the shear band has little effect on the adiabatic shear sensitivity of equiaxed titanium alloy.

(2) For the equiaxed titanium alloy, the phase proportion outside the shear band is the key factor that dominates the adiabatic shear sensitivity.

(3) Within a certain proportion of α phase, with the increase of the α phase proportion outside the shear band, the adiabatic shear sensitivity decreases.

5.Reference

[1] X.Liu and C Tan 2009 *J. Mate. Sci. Eng.* **501** 30-6

[2] J peris and P Verleysen 2010 *J. Inter. Impa. Eng.* **37** 703-14

[3] L W Meyer and K T Sommer 2008 *J. Mech. Time-Depend. Mater.* **12** 237-47

[4] R W Armstrong and S.M.Walley 2008 *J. Int. Mate. Revl.* **53** 105-128

[5] Zhang Wangfeng and Cao Chunxiao 2009 *J. Rare. Met. Mater. Eng.* **38** 972-5

[6] Shen Wei 2009 *J. Comput. Mater. Sci.* **38** 972-5

[7] Shen Wei 2010 *J. Surf. Coat. Technol.* **204** 3376-81

[8] Ye hengkui 2005 *J. Ship. Mate. Mechan.* **9** 11-2

[9] E Lee 2004 *J. Mate. Sci. Eng.* **24** 11-2

[10] J Y Kim and I O Kim 2007 *J. Mate. Sci. Forum.* **539** 2269-74

[11] D Lee and S Lee 2004 *J. Mate. Sci. Eng.* **366** 25-37

Microstructural characterization of the Mg/Cu/Al diffusion bonded joint

Jian Zhang[1], Qiang Shen[1], Guoqiang Luo[1*], Yiyu Wang[1], Meijuan Li[1], Lianmeng Zhang[1]

[1]State Key Lab of Advanced Technology for Materials Synthesis and Processing, Wuhan University of Technology, Wuhan 430070, China

[*]Corresponding author: Guoqiang Luo

E-mail: luogq@whut.edu.cn

Abstract. A vacuum hot-pressed diffusion bonding method was used to prepare an Mg/Cu/Al laminated composite. Both the Mg/Cu and Al/Cu interfaces were investigated by means of scanning electron microscopy, electron probe microanalysis, X-ray diffraction spectrometer system and Vickers microhardness test. The results showed that two kinds of intermetallic compounds, Al_4Cu_9 adjacent to the Cu side and Al_2Cu adjacent to the Al side, were formed in the interface of Al-Cu. Meanwhile, Mg_2Cu was formed at the interface of Mg/Cu. The maximum value of shear strength is 13.1 MPa and the fracture of the joints had taken place at the Mg-Cu interface. The microhardness of the interface increased due to the formation of the intermetallic compounds, which is the main cause leading to poor bond properties.

1. Introduction

Magnesium and aluminum alloys are low density metals, which have been used in different forms by automotive and aerospace industries to produce lightweight structures [1-2]. The use of composite structure made of Mg/Al dissimilar alloys is a popular method [3]. However, dissimilar welding of magnesium and aluminum is very difficult. The major metallurgical problem of joining magnesium with aluminum is due to the formation of much more Mg-Al intermetallic compounds (IMCs) [4-5]. Eliminating or improving Mg-Al IMCs has become an issue to achieve a reliable welding of Mg/Al dissimilar alloy.

Various joining methods have been used for joining of Mg with Al, such as cold metal transfer welding [6], friction-stir bonding [7], and diffusion bonding [8]. Among these, diffusion bonding is always adopted. During diffusion bonding, the bare metal surfaces are placed in intimate contact with each other to be joined under pressure and at elevated temperatures with minimum macroscopic deformation. Currently, the application of interlayer is commonly applied in diffusion bonding since the interlayer is utilized to minimize or eliminate problems caused by special chemical or metallurgical issues of the alloys [9]. In this paper, Cu foil as a interlayer for diffusion bonding aluminum and magnesium was investigated for the first time. The interface structure and mechanical properties of the joints are studied. The study provides a theoretical reference and selection method for Mg/Al dissimilar metals connection.

2 Experimental procedure

Samples of pure Al (≥ 99.9wt%) and pure Mg (≥ 99.9wt%) were used in the present study and were cut into size of ∅25 mm×5 mm. The pure copper (≥ 99.9wt%) of 100 μm was chosen as a interlayer. Before welding, the specimen surfaces were prepared by conventional grinding techniques with final grinding on 1200 # emery paper. The specimens were ultrasonically cleaned in an acetone bath to remove adhered contaminants and then dried in air. Diffusion bonding was carried out under a constant bonding pressure of 1 MPa and at the bonding temperature of 430-470 °C for a bonding time of 10 to 90 min. Vacuum pressure was less than 6×10^{-3} Pa. In the bonding process, heating rates of the experiment were kept at 15 °C/min, the assemblies were cooled in the processing chamber under vacuum. The bonding specimen was sectioned by a cutting machine. The microstructure and BSE image of the joints were characterized by S-3400 scanning electron microscope(SEM), the concentration profiles across the interfaces were evaluated by JXA-8230 equipped with Oxford Inca X-Act EDS, the phase constitution on the fracture surface were identified using a Rigaku ultima III X-ray diffraction system (XRD). The shear strength of the joints was tested using QJ-210A universal machine. Vickers microhardness profiles of the joint were determined along the sample with a 10g load and a loading time of 15 s using a MicroMet-2104 microhardness tester.

3 Results and discussion

3.1 Microstructural characterization

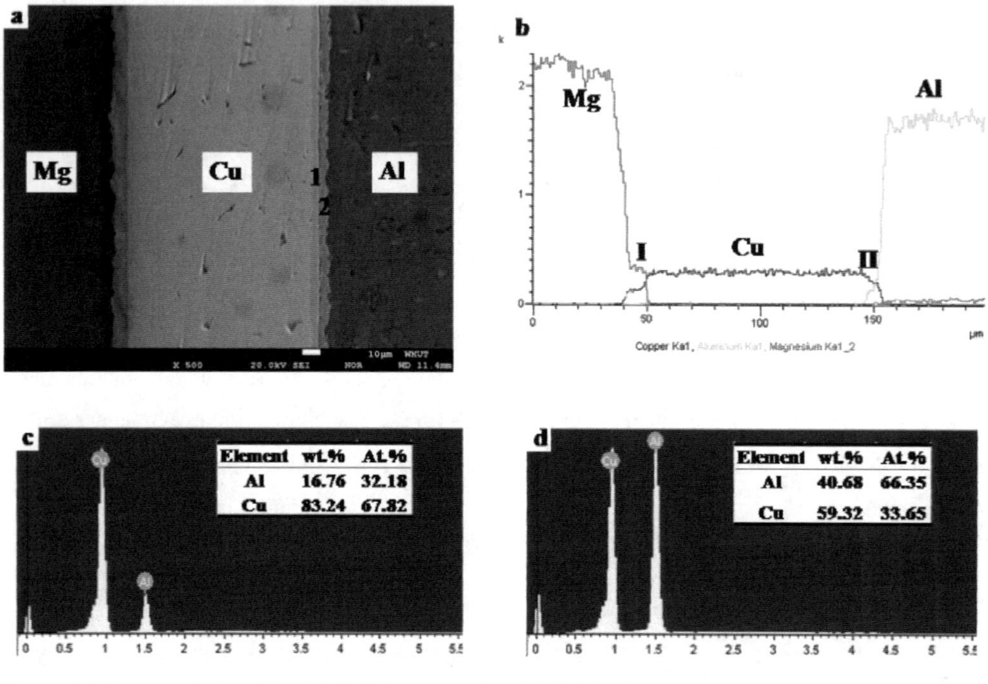

Figure 1 Micro-analysis of the Al/Cu/Mg composite: (a) SEM micrograph; (b) line analysis of elements; (c) EDS spectrum of 1 in Figure 1(a); (d) EDS spectrum of 2 in Figure 1(a)

Figure 1(a) is the SEM micrograph of the cross-section of the Al/Cu/Mg joint, under the diffusion bonding process with the bonding temperature of 450 °C, holding time of 60 min and a pressure of 1 MPa. There are two diffusion bonding regions that can be recognized, Mg-Cu interface and Al-Cu interface, respectively. Figure 1(b) shows the line analysis of elements Mg, Cu, Al in the Mg/Cu/Al joint. It is found that two obvious diffusion layers in the joint (region I and II in Figure 1(b)), indicating the joint is formed through mutual diffusion of Mg-Cu and Al-Cu under heating and pressure. Figure 1(c) and (d) are the EDS spectra of points 1 and 2 in Figure 1(a), respectively. The results show that in point 1, Al content of 32.18 at.% and Cu content of 67.82 at.% are measured, it should be Al_4Cu_9 phase according to the phase diagram [10]. Point 2 consists of 66.35 at.% Al and 33.65 at.% Cu, so it is determined to be Al_2Cu phase.

The phases of the Mg/Cu/Al fracture interface is analyzed by XRD and showed in Figure 3. According to the XRD peaks related with Al or its intermetallic compounds are not detectable, inferring results, inferring that the fracture has taken place somewhere at the interface of the Mg-Cu rather than at the Al-Cu interface, indicating that the Mg-Cu interface is the weak point at the weak region within Mg/Cu/Al layered composite.

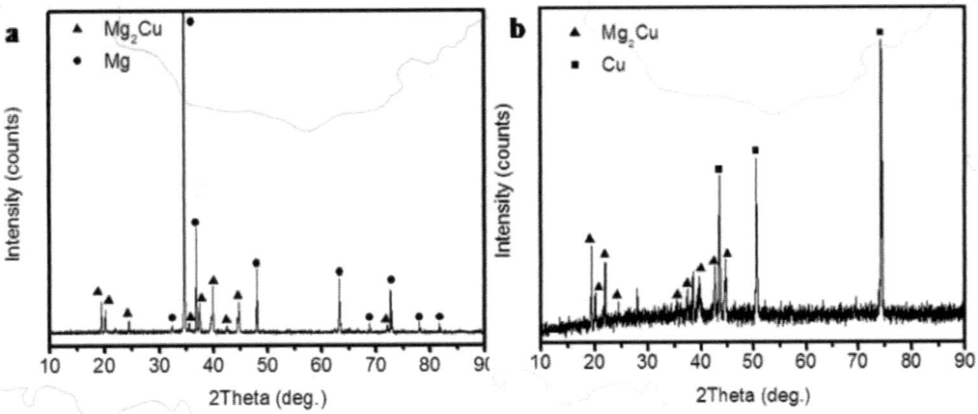

Figure 2 XRD pattern of the Mg/Cu/Al fracture surface: (a) Mg side and (b) Al side

3.2 Mechanical characterization

No matter what kind of intermetallic compounds are formed in the interface, the hardness of the interface will increase [4]. Figure 3 shows the results of the microhardness test of the Mg/Cu/Al joint at 440 °C, 450 °C, and 460 °C for 60 min. In the hardness profiles, the Mg-Cu and Al-Cu interfaces exhibit the hardness peaks due to the formation of intermetallic compounds during bonding. Moreover, the hardness at the Mg-Cu interface is higher than that at the Al-Cu interface. The peak hardness of the Mg-Cu interface approaches 301 Hv, whereas this is about 235 Hv at the Al-Cu interface.

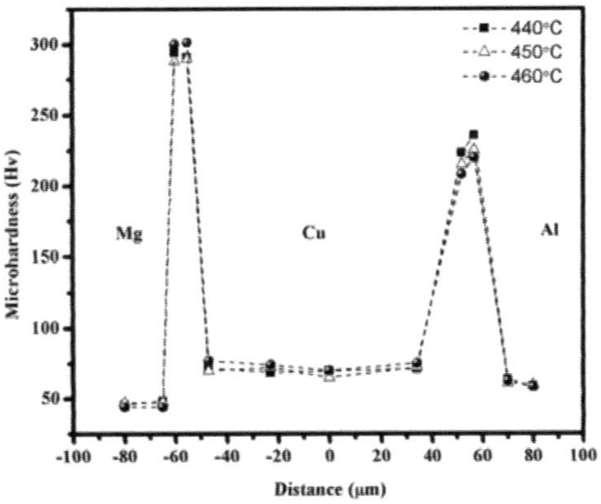

Figure 3 Microhardness profile of the Mg/Cu/Al joint

The interfacial bond strength of the Mg/Cu/Al joints is evaluated by a shear strength test. The average shear strength results are displayed in table. 1. For all the specimens with bonding temperature of 420 °C and the specimens at bonding temperature of 440 °C with holding time lower than 30 min, no bonding was achieved between the mating surfaces. This is because the insufficient temperature to cause diffusion of atoms or insufficient holding time to achieve a sound bond [11-12]. The results also show that the shear strength increases with increasing bonding time which can be attributed to the diffusion of atoms and the formation of intermetallic compounds at the joint interface during the bonding process. The intermetallic formation produces metallurgical bond at the interface. The maximum shear strength is 13.1 MPa obtained at 460 °C for 60 min.

Table 1 Experimental conditions and joint shear strength results

Conditions	Temperature (°C)	420		440			460			
	Time (min)	30	90	10	30	60	90	10	30	60
Shear Strength (MPa)		-	-	-	5.5	8.9	12.2	4.3	8.4	13.1

4 Conclusions

(1) Mg/Cu/Al laminated composite were prepared by means of diffusion bonding at the temperature of 430~470 °C for a bonding time of 10-90 min.

(2) Two kinds of intermetallic compounds, Al_4Cu_9 and Al_2Cu were formed in the interface of Al-Cu. Meanwhile, the Mg_2Cu phase was formed in the interface of Mg/Cu. Fracture occurs in the Mg-Cu interface rather than in the Al-Cu interface.

(3) The shear strength of the joints increased with the increasing in temperature, and the highest shear strength obtained 13.1 MPa at 460 °C for 60 min under 1MPa.

(4) The microhardness of the interface increased due to the formation of the intermetallic compounds. The peak hardness was at the Mg-Cu interface which approached 301 Hv.

Acknowledgements

The authors gratefully acknowledge the sponsorship from the National Natural Science Foundation of China (No. 51202175) and the 111 Project (B13035).

References

[1] Li X R, Liang W, Zhao X G, Zhang Y, Fu X P and Liu F C 2009 *J. Alloy. Compd.* **471** 408
[2] Choi D, Ahn B, Lee C, Yeon Y, Song K and Jung S 2011 *Intermetallics.* **19** 125
[3] Yang H Y, Guo X W, Wu G H, Wang S H and Ding W J 2011 *Surf. Coat. Technol.* **205** 2907
[4] Zhang J, Luo G Q, Wang Y Y, Shen Q and Zhang L M 2012 *Mater. Lett.* **83** 189
[5] Liu P, Li Y J, Geng H R and Wang J 2005 *Mater. Lett.* **59** 2001
[6] Shang J, Wang K H, Qi Z, Zhang D K, Huang J and Li G L 2012 *Mater. Des.* **34** 559
[7] Kwon Y J, Shigematsu I and Saito N 2008 *Mater. Lett.* **62** 3827
[8] Zhao L M and Zhang Z D 2008 *Scripta. Mater.* **58** 283
[9] Alhazaa A N and Khan T I 2010 *J. Alloy. Compd.* **494** 351
[10] Mitsuo N and Yukihisa Y 1977 *Wire. J.* **4** 71
[11] Voropai N M and Shinyaev A Y 1967 *Met. Sci. Heat. Treat.* **9** 926
[12] Chen S D, Ke F J, Zhou M and Bai Y L 2007 *Acta. Mater.* **55** 3169

Study for double-layered AZO/ATO transparent conducting thin film

Miaomiao Cao, Yudong Li, Jing Yang and Yigang Chen*

Department of Electronic Information Materials, School of Materials Science and Engineering, Shanghai University, 149 Yanchang Road, Shanghai 200072, China

E-mail: yigangchen@shu.edu.cn

Abstract. The purpose of this study is to provide transparent conductive composite films, used for such as dye-sensitized solar cells (DSSCs). In this work, transparent conductive oxide films with double-layer structure, ATO(antimony doped tin oxide)films covered on AZO(aluminum doped zinc oxide)films, were prepared on glass substrates by RF magnetron sputtering method. Subsequently the films were post-annealed at different temperature. The structure, surface morphology, optical and electrical properties of the films were investigated as a function of annealing temperature. Our results indicate that the composite films can maintain good electrical and optical properties at a temperature higher than 450°C compared to that of single-layer film.

Keywords: Transparent conductive oxide, composite, optical properties, electrical properties, thermal stability

1. Introduction

Transparent conductive oxide film, hereinafter referred to as TCO, is a functional thin film with low resistivity and high transmittance in the visible light range. TCOs are mainly used for optoelectronic devices such as flat panel displays [1] and photovoltaic fields [2-4] including inorganic and organic devices and dye-sensitized solar cells (DSSCs). The improvement of electrical and optical properties of TCO materials will greatly increase the efficiency and performance of these devices. The composite film technology can improve the shortcomings of the monolayer or achieve monolayer unattainable performance, which has become a hot research area of the film surface technology. Because optical, electrical and chemical properties of each layer material of the composite films are not identical, these composite films will have some unique characteristics. For example, the double-layer composite films,

FTO (fluorine doped tin oxide) films deposited on ITO (indium-tin-oxide) films coated on glass substrates, have excellent performance and make up their own defects as well as improve the efficiency of dye-sensitized solar cells [5].

The main preparation methods of the composite films include physical vapor deposition (PVD) [6,7] and chemical vapor deposition (CVD) [8]. The former contains evaporation, sputtering, ion plating, arc plating, plasma plating, ion cluster beam (ICB) and molecular beam epitaxy (MBE), etc., and the latter covers vapor deposition, liquid phase deposition, electrolytic deposition, glow discharge deposition and the metal organic chemical vapor deposition (MOCVD), etc. There are many unique preparation methods such as ion implantation and laser assist deposition, etc. Among these techniques, RF magnetron sputtering is an attractive technique for the growth of composite films because it has advantages of high deposition rate, low substrate temperature, good adhesion of the film and large area coating capability [9].

In this work, the double-layer composite films, ATO (antimony doped tin oxide) films deposited on AZO (aluminum doped zinc oxide) films coated on glass substrates, were developed without direct substrate heating by RF magnetron sputtering. The reason of growing double-layer structure is that the heat resistance of ATO films is better than that of AZO films in the high temperature process. In this article, we reported a detailed study about the temperature dependence of the structure and properties of these composite films.

2. Experimental details

2.1 Film preparation

AZO and ATO thin films were successively deposited on super white glass substrate ($15\times15\times3mm^3$) without direct heating by RF magnetron sputtering from the stoichiometric AZO and ATO ceramic targets (99.999% purity) of 2.4 in. diameter. The AZO ceramic target contained 2 wt% of Al2O3 and the composition of ATO ceramic target was 6 wt % of Sb2O3. As the base pressure was smaller than $8 \times 10{-}4$ Pa, the Ar and O2 were introduced into the chamber. The total working pressure of films growth was fixed at $7 \times 10{-}1$ Pa and the oxygen partial pressure ratio ($P_{O2}/(P_{O2}+P_{Ar})$) was about 10-2 during AZO deposition and about 10-1 during ATO deposition. The applied RF power of composite films was 100 W and 80W, respectively. Then, the AZO/ATO bilayer films were annealed in a rapid heating furnace at 150°C, 250°C, 350°C, 450°C, 550°C and 650°C, respectively, in the vacuum for 30 minutes in order to study the temperature dependence of the structure and properties of these composite films.

2.2 Film characterization

The film thickness was measured by a XP-2 step profiler. The crystal structure of the films was characterized by X-ray diffraction (XRD) using Cu Ka radiation. The surface morphology and cross-section of the films were studied using scanning electron microscope (SEM), the acceleration voltage was 20 kV. The transmittance in the 200–800 nm regions was performed using a UV-VIS spectrophotometer. Concentrations and mobility of charge carriers and resistivity of the samples were determined using Hall effect measurement system (Accent HL5500PC).

3. Results and discussion

Figure 1 shows the cross sectional SEM micrograph of the AZO/ATO composite transparent conductive film. As can be seen from the figure each monolayer including the first layer of the AZO film and the second layer of the ATO film is clearly seen. Because of the difference of the sputtering process parameters, the film thickness of the two layers in the AZO/ATO films is different. The thickness of AZO film is confirmed as approximately 380 nm and ATO has a thickness of about 75 nm as measured with XP-2 step profiler, which is in good agreement with the SEM result. It also can be seen from the figure that the ATO films are deposited tightly on the AZO films, which illustrates the conductive composite films could be prepared by this method.

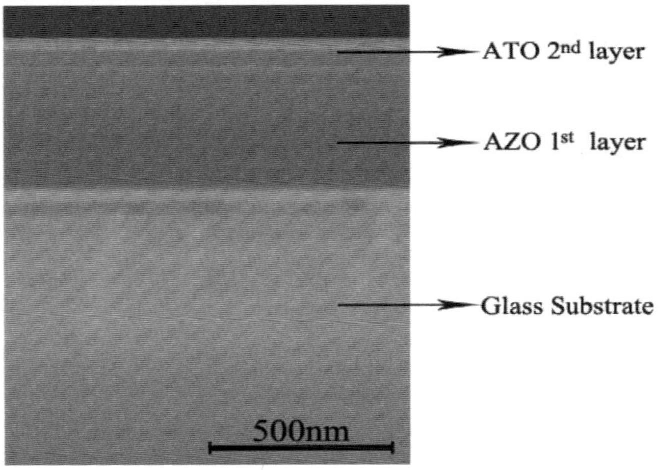

Figure 1. A cross sectional SEM image of the double-layered film composed of AZO (1st layer) and ATO (2nd layer).

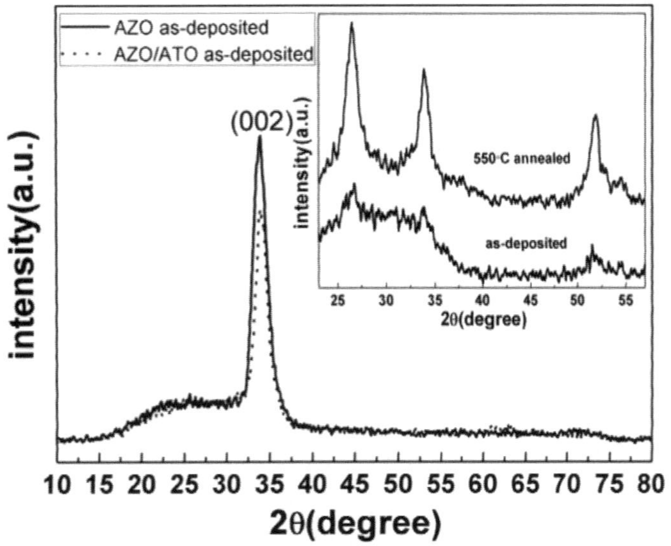

Figure 2. XRD patterns of bare AZO and 70 nm ATO-coated AZO films deposited at room temperature on the glass substrate. The inset shows XRD scan of as-deposited and 550°C annealed ATO (70 nm) film.

Figure 2 shows the XRD spectra of the bare AZO and the AZO/ATO films. The figure clearly presents the (002) diffraction peak of ZnO in the bare AZO films, which suggests good crystallized AZO films with wurzite hexagonal structure can be obtained. The diffraction peaks of the AZO/ATO films are dominated by the wurzite hexagonal characteristic peaks of AZO and do not show ATO characteristic peaks, which can be explained by the inset in Figure 2. Although the diffraction peaks of the SnO_2 in the ATO films deposited at room temperature are not observed, the inset shows the appearance of the tetragonal rutile peaks (110), (101), and (211) of pure SnO_2 when the ATO thin films are annealed at 550°C. This explains why AZO/ATO composite films deposited at room temperature mainly show the wurzite hexagonal characteristic peaks of pure ZnO: the AZO film deposited at room temperature is crystalline and ATO film deposited at room temperature is amorphous, even more, the thickness of AZO film is thicker than that of the ATO film.

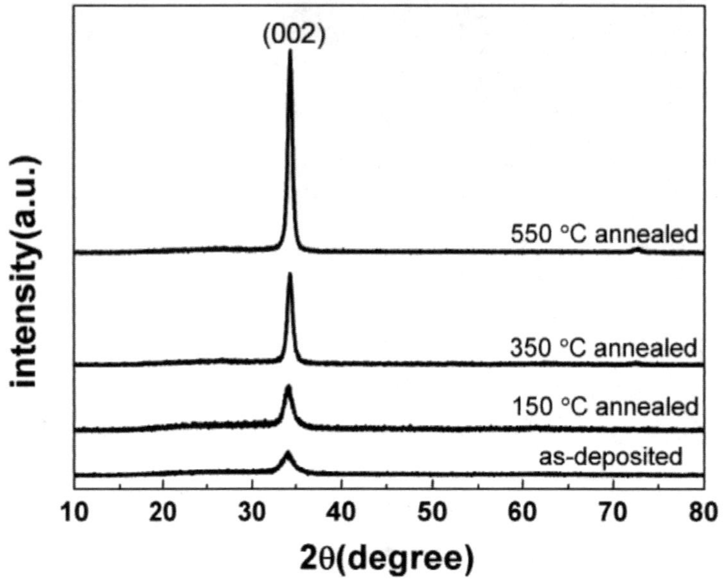

Figure 3. XRD patterns of AZO/ATO composite films annealed in vacuum at different temperatures.

Figure 3 shows the XRD patterns of AZO/ATO composite films annealed in vacuum at different temperatures. The films were annealed at 150, 350 and 550°C for 30 minutes, respectively. The figure shows that the AZO/ATO as-deposited and annealed films are polycrystalline films with wurzite hexagonal structure. The strongest diffraction peak is the (002) direction. Because ATO films' thickness is too thin, the diffraction figures do not show the ATO films' characteristic peaks. In a temperature range of 150~550°C, with the annealing temperature increasing, the intensity of (002) peak increases gradually and the FWHMs of the peak are reduced accordingly, which reveals that the crystallization of the as-deposited AZO/ATO films could be improved after annealing. The crystal sizes of the films are estimated using Scherrer formula [10] to be about 21.2, 23.1, 26.2 and 30.8 nm for the samples as-deposited and annealed at 150, 350 and 550°C, respectively. The above results show that the crystal sizes of the films become larger and the crystallization becomes better with the increasing of the annealing temperature.

Figure 4 shows the optical transmittance spectra of the AZO/ATO films deposited at room temperature and annealed at different temperatures as a function of wavelength. The oscillations of the interference peaks in the visible region are observed in all the deposited and annealed AZO/ATO films due to the interference effect with respect to the increase in annealing temperature. Moreover, the optical transmittance increases slightly with increasing the annealing temperature. Particularly near the absorption edge, the transparency of the annealed films increases obviously, compared to that of films deposited at room temperature. Furthermore, the blue shift of the absorption edge with increasing the annealing temperature is mainly attributed to the Moss-Burstein effect, since the absorption edge of a degenerate semiconductor is shifted to shorter wavelengths with increasing carrier concentration [11]. The experimental results show that the optical properties of the bilayered films could be improved after this high temperature treatment. According to the preliminary analysis, this could arise from the corresponding optical band gap changes caused by structural changes after annealing. The results indicate that the composite films have good optical transparency.

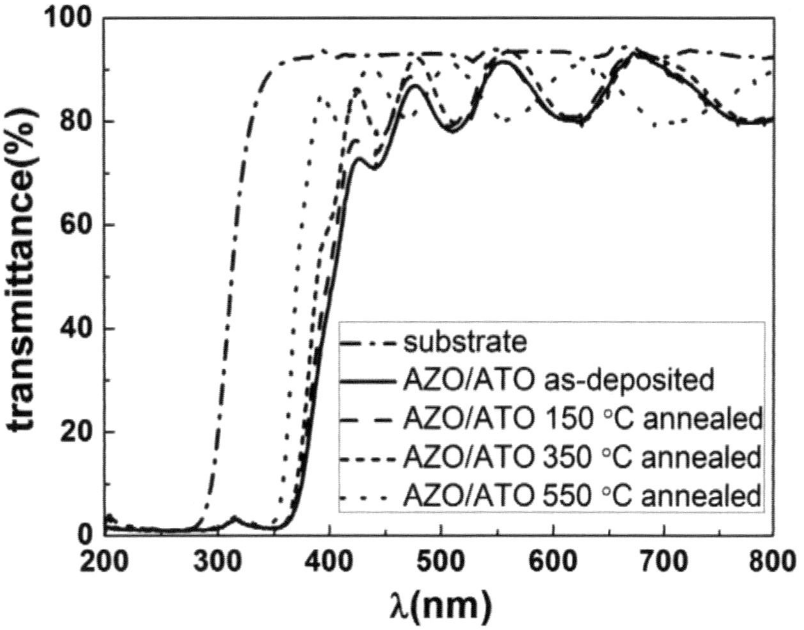

Figure 4. Optical transmittance spectra of glass substrate and AZO/ATO composite films before and after annealing at different temperatures in vacuum.

Figure 5 shows the resistivity, carrier concentration and Hall mobility of the AZO/ATO composite films as a function of the annealing temperature. According to quantum free electronic theory, the conductivity of the films depends on both the charge carrier concentration and transferring. The carrier comes mainly from oxygen vacancies and doping metal ions, while mobility are mainly originated from (derives mainly from) both the grain boundary scattering and ionized impurity scattering. As can be seen, within the temperature range from 350°C to 650°C, the carrier concentration increases with annealing temperature increasing, which is attributed to two different reasons: Sb^{3+} extinguishing and oxygen desorption. With increasing the annealing temperature, the mobility first increases and then decreases. The maximum mobility value is obtained for 450°C annealed sample, which is because once the charge carrier concentration is high enough to overcome the structural barriers, the ionized impurity scattering arises along with a decrease of the mobility. With the annealing temperature increasing, the resistivity of the AZO/ATO composite films declines sharply and the lowest resistivity of 6.1×10^{-3} Ω·cm is obtained at 550°C, while as the annealing temperature increases to higher than 550°C, the resistivity of the composite films begin to rise again. Some literature reported that this phenomenon was considered to be due to the dispersion and the deviation of the grain caused by overhigh annealing temperature [12]. Some other literature speculated that hyperplasia and mobile mechanism of grain boundaries existed in the high annealing temperature, which promoted the splitting of the large grain and decreased the average grain size, and finally led to the deterioration of the conductive performance [13]. According to the figure 5, when annealing between 450 ℃and 600 ℃, the composite films have good conductivity, so the AZO / ATO composite conductive films can be used to prepare the dye-sensitized solar cells.

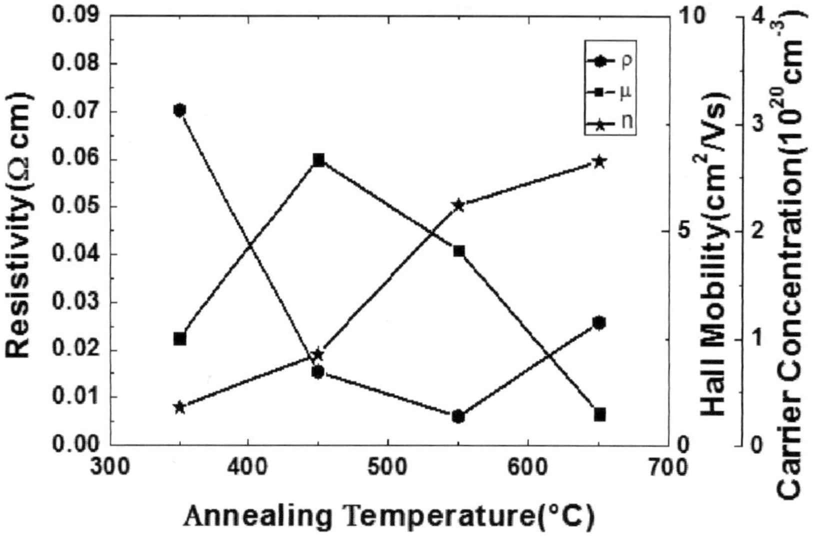

Figure 5. Resistivity, Carrier concentration and Hall mobility of AZO/ATO composite films annealed at different temperatures in vacuum.

4. Conclusions

AZO/ATO transparent conductive composite films have been successfully deposited on glass substrates by RF magnetron sputtering at room temperature. Post-annealing has a significant impact on crystallization as well as optical and electrical properties of the composite films. The transmittance in the visible region (400–800 nm) of the composite transparent conductive film is about 80~93%. Moreover, the optical transmittance increases slightly and the absorption edge is shifted to shorter wavelengths with increasing in the annealing temperature. An ATO protective layer could provide a fine thermal stability in terms of electrical conductivity at high temperatures and the lowest resistivity of 6.1×10^{-3} Ω·cm was obtained at 550°C. In summary, the AZO / ATO composite conductive films can be used to prepare the dye-sensitized solar cells.

Acknowledgments

The authors thank for the financial support of Eastern-scholar program and Shanghai Pujiang program, China.

References

[1] Kwak D J, Park M W and Sung Y M 2008 Discharge power dependence of structural and electrical properties of Al-doped ZnO conducting film by magnetron sputtering (for PDP) *Vacuum* **83** 113–8

[2] Tak Y H, Kim K B, Park H G, Lee K H and Lee J L 2002 Criteria for ITO (indium–tin-oxide) thin film as the bottom electrode of an organic light emitting diode *Thin Solid Films* **411** 12–6

[3] Granqvist C G 2007 Transparent conductors as solar energy materials: a panoramic review *Sol. Energy Mater. Sol. Cells* **91** 1529–98

[4] Batzill M and Diebold U 2005 The surface and materials science of tin oxide *Prog. Surf. Sci.* **79** 47–154

[5] Kawashima T, Matsui H and Tanabe N 2003 New transparent conductive films: FTO coated ITO *Thin Solid Films* **445** 241–4

[6] Jia C L, Wang K M, Wang X L, Zhang X J and Lu F 2005 Formation of c-axis oriented ZnO optical waveguides by radio-frequency magnetron sputtering *Opt. Express* **13** 5093–9

[7] Yang S M *et al.* 2010 Microstructural investigation of ZnO films grown on (111) Si substrates by plasma-assisted molecular beam epitaxy *J. Cryst. Growth* **312** 1557–62

[8] Yoshii N, Nakamura A, Hosaka S and Temmyo J 2009 Investigation of morphology and crystallinity of ZnO crystal formed by side-flow-type MOCVD *J. Electrochem. Soc.* **156** K117–K120

[9] Yang W F,Liu Z G,Peng D L,Zhang F,Huang H L,Xie Y N and Wu Z Y 2008 Room-temperature deposition of transparent conducting Al-doped ZnO films by RF magnetron sputtering method *Appl. Surf. Sci.* **255** 5669-5673

[10] Selmi M, Chaabouni F, Abaab M and Rezig B 2008 Studies on the properties of sputter-deposited Al-doped ZnO films *Superlattices Microstruct.* **44** 268-75

[11] Fang G J, Li D J and Yao B L 2002 Fabrication and vacuum annealing of transparent conductive AZO thin films prepared by DC magnetron sputtering *Vacuum* **68** 363-72

[12] Dong J H 2002 *Preparation and Properties Study of Transparent Conductive In_2O_3:Sn and ZnO:Al Thin Films* (Chongqing: Chongqing University) pp 20–53

[13] Feng T H, Lu J X, Zhang Y X, Yang S E, Li R, Jin R M, Wang H Y and Gao X Y 2005 Study of preparing polycrystalline silicon thin films by RTA *Electronic Components & Materials* **24** 26-28

Microstructure of Diffusion-Bonded Mg-Ag-Al Multilayer Composite Materials

Yiyu Wang, Guoqiang Luo*, Jian Zhang, Qiang Shen, Lianmeng Zhang

State Key Laboratory of Advanced Technology for Materials Synthesis and Processing, Wuhan University of Technology, Wuhan 430070, PR China

* Corresponding Author e-mail address: luogq@whut.edu.cn

Abstract: Mg-Al bonded composite materials expand Mg and Al alloys' applications by combining their unique performances together. However, the formation of Mg-Al intermetallic compounds in interface zone of Mg/Al directly-bonded joint seriously obstructs its further development. To solve this problem, Mg-Ag-Al multilayer composite materials have been successfully prepared by diffusion bonding technology. The effect of key process parameter (bonding temperature) on microstructure of this material has been mainly investigated. The results show that Mg and Al were well bonded by using silver interlayer when the bonding temperature exceeded 370°C. But $Mg_{17}Al_{12}$ and Mg_2Al_3 compounds were formed in the interface zone at temperatures higher than 420°C. By means of controlling the bonding temperature (380 °C–420 °C), silver interlayer effectively restrained the generation of Mg-Al intermetallic compounds, and Mg-Ag intermetallic compounds (Mg_3Ag, MgAg) were formed in the interface zone instead.

1. Introduction

Welding composite materials play an important role in manufacturing and construction industries for their unique performances [1, 2]. Mg/Al bonded composite materials with their excellent properties together, such as low density, high strength and high corrosion resistance have been widely used in many fields, especially in aerospace, automobile and electronic products by largely reducing the weight of the structural components and saving the costs [3, 4].

Diffusion bonding method is an efficient way to produce composite materials, which has been widely used in bonding dissimilar metals [5-7]. However the formation of brittle Mg-Al intermatellic compounds at the interface of directly-bonded Mg/Al joint has become an obstacle in its development [8, 9]. Interlayers such as Zn alloy interlayer and Ti interlayer inserted between Mg and Al have been proved to be an efficient way to solve the problem above by blocking the inter-diffusion of Mg and Al atoms [10, 11]. Silver interlayer with high ductility and good metallurgical property has been successfully used in welding similar and dissimilar metals, even ceramics [12, 13]. Bonding temperature as the most important parameter in diffusion bonding process has a great effect on the microstructure and mechanical properties of the joints [14, 15].

In the present work, Mg and Al metals have been successfully jointed at different bonding temperatures by using silver interlayers. The effect of bonding temperature on the variation in the microstructure of diffusion-bonded Mg-Ag-Al multilayer composite materials has been investigated particularly, which provides an important guidance for understanding the fundamental law of welding science deeply.

2. Experimental procedures

Pure magnesium Mg1 and commercial pure Al 1060 plates with dimension of Φ25mm×5mm were used in this work. Table 1 shows the chemical composition of the materials above. Pure silver (>

99.99%, wt %) was used as the magnetron sputtering target material.

Table 1. Chemical composition of Mg1 and 1060Al

Materials	Chemical composition (Wt %)								
	Mg	Al	Cu	Fe	Si	Zn	Ti	Mn	Ni
Mg1	Bal	0.0056	0.0007	0.0025	0.012	0.003	0.005	-	0.0006
Al 1060	0.02	Bal	0.03	0.05	0.10	0.02	0.01	0.005	-

In all cases, Mg1 and Al 1060 were ground flat by grit SiC paper to 2000# and a diamond polishing agent to 0.05 μm, and then cleaned in ethanol and acetone before being silver sputtered. Then silver interlayers were deposited onto the clean surfaces of Mg1 and Al 1060 using magnetron sputtering technology. The thickness of the silver interlayer on each side is 2.5μm. Then these samples were fitted together and introduced into a vacuum furnace (10^{-4}–10^{-3}Pa). To investigate the effect of bonding temperature on microstructure of the Mg-Ag-Al multilayer composite materials, samples at different bonding temperatures (370 °C–450 °C for 30min) were prepared respectively. The load used in this investigation was 5 MPa for all conditions. The microstructure test samples were cut from the bonded materials by wire-cutting.

Microstructures at the joint interfaces were characterized by using scanning electron microscope (SEM, S-3400N HITACHI). The main element concentration profiles and composition of the reaction layers near the interfaces were analyzed by JXA-8230 electron probe micro analyzer (EPMA) equipped with Oxford Inca X-Act energy dispersive spectrometer (EDS). The appearance and phase compositions of fracture surfaces after strength tests were measured by SEM and a Rigaku Ultima X-ray diffraction system (XRD). All tests were carried out on the polished surfaces, without etching.

3. Results and discussion

The experimental results of the joints under different bonding temperatures were presented in Table 2. Several inferences were obtained as follows: (1) when the bonding temperature was lower than 380 °C, Mg and Al cannot be jointed because of the low interdiffusion speed of the atoms at low temperature (Fig. 1(a)). (2) When the bonding temperature was higher than 440 °C, the bonding of Mg and Al failed due to the melting of Mg base caused by the eutectic reaction at high temperature (Fig. 1(c)). (3) When the bonding temperature was controlled from 380 °C to 420 °C, strongly-combined Mg/Ag/Al joints were produced without the formation of Mg-Al compounds at the interfaces (Fig. 1(b)). At 430 °C and 440 °C, Mg-Al compounds occurred at interfaces, which was harmful to the strength of the joints.

Table 2. Experimental conditions and the interfacial compound results

No.	Bonding process parameters			Bonding condition	Compounds at the interface	
	Temperature/°C	Time/min	Pressure/MPa		Mg-Ag compounds	Mg-Al compounds
1	370	30	5	Failed	—	—
2	380	30	5	Completed	Yes	No
3	390	30	5	Completed	Yes	No
4	400	30	5	Completed	Yes	No
5	410	30	5	Completed	Yes	No
6	420	30	5	Completed	Yes	No
7	430	30	5	Completed	Yes	Yes
8	440	30	5	Completed	NO	Yes
9	450	30	5	Failed	—	—

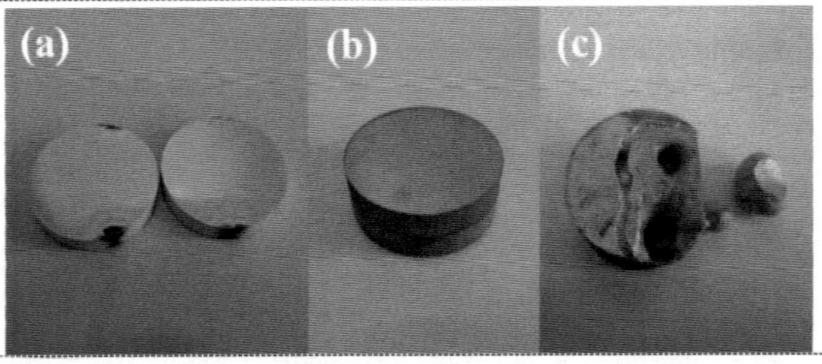

Figure 1. Photographs of the joints bonded at (a) 370°C, (b) 400 °C, and (c) 450 °C.

A diffusion transition zone was formed between Mg and Al base metals during diffusion bonding process. The interfacial microstructures of the Mg-Ag-Al joints bonded at 400 °C and 420 °C for 30 min were investigated by SEM, shown in Fig. 1. It can be seen that Mg and Al have been successfully jointed without porosity, crack under current technological conditions. Fig. 2(a) shows that two reaction layers different from the parent metals were formed at the joint interface, which demonstrates that the activity of Ag is high enough to form the reaction product layers. The reaction phases marked by layers A and B were identified according to the quantitative analyses of the compositions by EPMA. Mg-Ag intermetallic compounds Mg_3Ag (Mg 74.93, Al 25.07, at. %) and MgAg (Mg 50.68, Al 49.32, at. %) were the phases of layers A and B respectively. At Mg-Ag-Al interface, the interdiffusion between Mg and Ag resulted in the formation of the layered structures (layers A, B). No Mg-Al or Ag-Ag compounds were found at the interface. When the bonding temperature rose to 420 °C, the microstructure of the joint interface changed a lot compared with that at 400 °C. Fig. 2(b) shows that a three-layered structure was observed instead of two layers. The phase compositions of layers C and D marked in Fig. 1(b) were still the Mg-Ag compound: Mg_3Ag, MgAg respectively. However, the thickness of layer C increased largely with the rising temperature compared with that of layer A at 400 °C, which is the consequence of the grain growth. On the contrary, layer MgAg (D) reduced for the continuous diffusion of Mg and Ag. The phase of the newly-grown layer E close to Al base was classified as the Mg-Ag-Al ternary compound: $AgMg_6Al_4$ (Mg 53.59, Al 36.46, Ag 9.65, at. %). The formation of $AgMg_6Al_4$ means that Mg has successfully transferred into Al side overcoming the blocking effect of silver interlayer at the current process. Mg-Al compounds can be avoided when the bonding temperature increases beyond 420 °C. So the optimized bonding temperature plays an important role in controlling the interfacial phase composition of the joints.

The line and mapping distributions of the main element Mg, Ag and Al cross the interfacial zone of the joints bonded at 410 °C for 30 min were measured by EPMA, presented in Fig. 3. Fig. 3(b) and (e) show that the amount of Mg decreased from Mg base to Al base. The distribution of Ag concentrated into two parts at the interface which was in line with the two-layered structure in Fig. 3(a). The interface contained almost no Al because of the low interdiffusion of Ag and Al at 410 °C. The generation of the thin layer close Al base resulted in the diffusion of Mg, Ag and Al. According to the data of the distribution of Al, it can be observed that silver interlayers play an important role in avoiding the formation of Mg-Al intermetallic compounds with high brittleness and hardness by preventing Al atoms diffusing from Al base to Mg base.

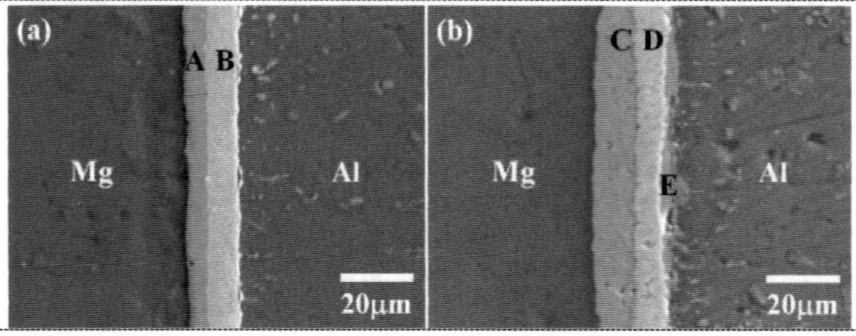

Figure 2. Microstructure of the joints bonded at (a) 400°C, (b) 420 °C investigated by SEM.

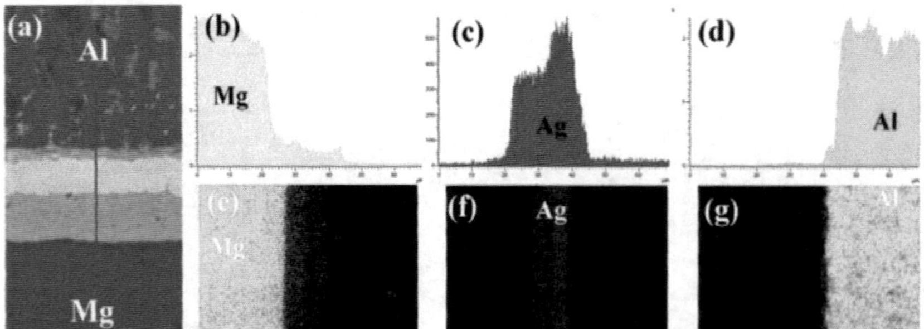

Figure 3. EPMA analyses of the joint bonded at 410 °C: (a) BSED image; Line distribution of Mg (b), Ag (c) and Al (d); Mapping distribution of Mg (e), Ag (f) and Al (g).

In order to understand better the changes at the interface of the joints, the interfacial microhardness of the Mg-Ag-Al joints bonded at 410 °C for 30 min was characterized by Vickers hardness, shown in Fig. 4. The average hardness value of Mg and Al base were about 45 HV and 75 HV respectively. The hardness value in the diffusion zone increased largely with a maximum value of 240 HV due to the formation of Mg-Ag compounds, which was considered to be a disadvantage to the joints. The higher hardness means the lower ductility and ability to resist deformation of the joints. The reduction of the joint interface hardness becomes a way worthy of consideration to improve the bonding condition of the joints in subsequent studies.

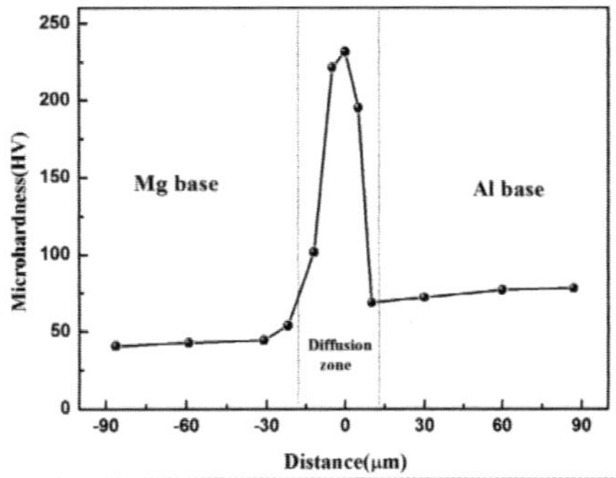

Figure 4. Vickers hardness of the Mg-Ag-Al joints bonded at 410 °C for 30 min.

4. Conclusions

Mg-Ag-Al multilayer composite materials have been successfully prepared without the formation of brittle Mg-Al intermetallic compounds at the interfaces at 380 °C-420 °C for 30 min by diffusion bonding technology.Mg_3Ag close to Mg base and MgAg close to Al base are the main phases at the interface of Mg-Ag-Al joints. The addition of silver interlay has blocked the interdiffusion of Mg and Al atoms resulting in the absence of the Mg-Al compounds by controlling the bonding temperature.The microhardness of the joint interfaces increased largely compared with the value of Mg and Al parent metals due to the formation of Mg-Ag intermetallic compounds.

Acknowledgments

The authors gratefully acknowledge the financial support from the National Natural Science Foundation of China (No. 11072228), the Fundamental Research Funds for the Central Universities in China (No. 2011-IV-038) and the 111 Project (B13035).

References

[1] Sohn W H, Bong H H and Hong S H 2003 Mater. Sci. Eng. A. **355** 231
[2] Elthalabawy W and Khan T 2011 J. Mater. Sci. Technol. **27** 22
[3] Liu L M, Wang H Y and Zhang Z D 2007 Scr. Mater. **56** 473
[4] Liu P, Li Y J, Geng H R and Wang J 2005 Mater. Lett. **59** 2001
[5] Kundu S and Chatterjee S 2010 Mater. Sci. Eng. A. **527** 2714
[6] Lee J G, Kim G H, Lee M K and Rhee C K 2010 Intermetallics. **18** 529
[7] Alhazaa A N and Khan T I 2010 J. Alloy Compd. **494** 351
[8] Dietrich D, Nicke D, Krause M, Lampke T, Coleman M P and Randle V 2011 J. Mater. Sci. Technol. **46** 357
[9] Yang H Y, Guo X W, Wu G H, Wang S H and Ding W J 2011 Surf. Coat. Technol. **205** 2907
[10] Zhang J, Xiao Y, Luo G Q, Shen Q and Zhang L M 2012 Mater. Lett. **66** 113
[11] Nami H, Halvaee A, Adgi H and Hadian A 2010 J. Mater. Process Technol. **210** 1282
[12] Barrena M I, Gómez de Salazar J M, Merino N and Matesanz L 2008 Mater. Charact. **59** 1407
[13] Atasoy E and N Kahraman 2008 Mater. Charact. **59** 1481
[14] Joseph Fernandus M, Senthilkumar T and Balasubramanian V 2011 Mater. Des. **32** 1651
[15] Chen S D, Ke F J, Zhou M and Bai Y L 2007 Acta. Mater. **55** 3169

Modeling the Dynamic Damage Process of the SiC_{3d}/Al Interpenetrating Phase Composites

Li Guoju[1,2], Fan Qunbo[1,2*], Wu Zheng[1,2], Zhang Xu[1,2], Wang Yangwei[1,2], Wang Fuchi[1,2]

[1] School of Materials Science and Engineering, Beijing Institute of Technology, 100081, Beijing, China,
[2] National Key Laboratory of Science and Technology on Materials under Shock and Impact, Beijing, 100081, China

*Corresponding author e-mail: fanqunbo@bit.edu.cn

Abstract. In the current study, a 3D mesoscopic structure FE-Model of interpenetrating SiC_{3d}/Al composite is built based on the digital image-based modeling technique together with optimized methods of three-dimensional mesh generation. Subsequently, the finite element method is proposed to simulate the dynamic damage process of the interpenetrating phase composites SiC_{3d}/Al under dynamic axial crushing. The cracking process in micro 3D space is clearly presented in the current study. It is shown that the cracks initialization and propagation mainly appear in the region of interface between ceramic and metallic phase. Moreover, the ceramic phase attributes to the model's damage predominantly. The method proposed in this paper would be of help in the microstructure design of Interpenetrating Phase Composites.

1. Introduction

Interpenetrating Phases Composites (IPC) are characterized by the ceramic skeleton and metallic matrix both continuous in three-dimensional space, and this special structure makes full use of the ceramic's high specific strength stiffness and the metal's high toughness. In the military materials filed, the IPC are receiving great attention on the research of the relationship between the mesoscopic structure characteristics and mechanical properties not only because it can effectively block the projectile's penetration, efficiently absorb and reflex the impact energy, but also because it can maintain a certain macro-structural integrity.

To describe the deformation or damage behaviour of IPC, the representative volume element (RVE) model by computational mesomechanics [1, 2] was employed. Some researchers [3~5] used the idealized unit cell model as the RVE model to simulate the elastic and plastic behaviour of the IPC under uniaxial compression/tension. In such cases, however, the detail features such as shape of cross-section and three-dimensional orientations inside the cell were ignored. Leon[6] employed the RVE model based on the digital image-based (DIB) modelling technique together with voxel array based methods of three-dimensional mesh generation to simulate the overall mechanical behaviour of the IPC. Though such model could reflect the real structure compared with the idealized unit cell model, the model was composed of identically sized hexahedron elements generated by voxel array based methods, leading to the phenomenon of jaggies in the interface between two-phase, thus influencing the calculation stability unavoidably.

In the current study, a 3D mesoscopic structure FE-Model of SiC_{3d}/Al with smooth interface is built based on the digital image-based modeling technique together with optimized methods of three-dimensional mesh generation. In addition, Johnson-Holmquist and Johnson-Cook constitutive models [7] are employed to describe the SiC ceramic phase and Al metallic phase, respectively. Subsequently, the model is imported into the commercial code LS-DYNA [8] to investigate the damage initiation and propagation under dynamic axial crushing.

2. Computation and modelling

2.1. CT data acquisition

3-D image scanning is performed using the Sky-Scan1172 Micro-CT. 50 kV and 800 μA are imposed respectively for the Voltage and the Current. In addition, the transmission X-ray image is set for 180 degrees of rotation. It is the active scanning mode that the sample as well as the sample stage is fixed, and the X-ray tube rotates on the sample stage plan's axis. Transmission X-ray images are recorded at 0.4° rotational steps for 180° of rotation. Subsequently, the transmission X-ray images are then reconstructed into cross-sectional images (voxel size of 2.2 × 2.2 × 2.2 μm) images using Sky-Scan's cluster reconstruction software (NRecon).

2.2. Image segmentation and mesh generation

The process of the SiC_{3d}/Al 3D FE model generation based on Micro-CT scanning images (the above mentioned cross-sectional images) is presented as figure 1. Micro-CT scanning images (see figure 1 (a)) are imported into ScanIP module [9] and thresholded for values from 65 to 255 to create the respective masks (see figure 1 (b)). Then the ceramic phase 3D image model is built by 3D view operation as figure 1(c) presents. At the same time, by duplicating and inverting the above masks, the metallic phase 3D image model (see figure 1 (d)) is also created. As a result, the SiC_{3d}/Al 3D image model is generated as figure 1 (e) shows. Subsequently, the SiC_{3d}/Al 3D image model is imported into ScanFE module [10] and meshed with a mixture of hexahedral and tetrahedral elements. Generated SiC_{3d}/Al 3D FE model (see figure 1 (f)) is constructed with 480512 tetrahedral elements as well as 906577 brick elements.

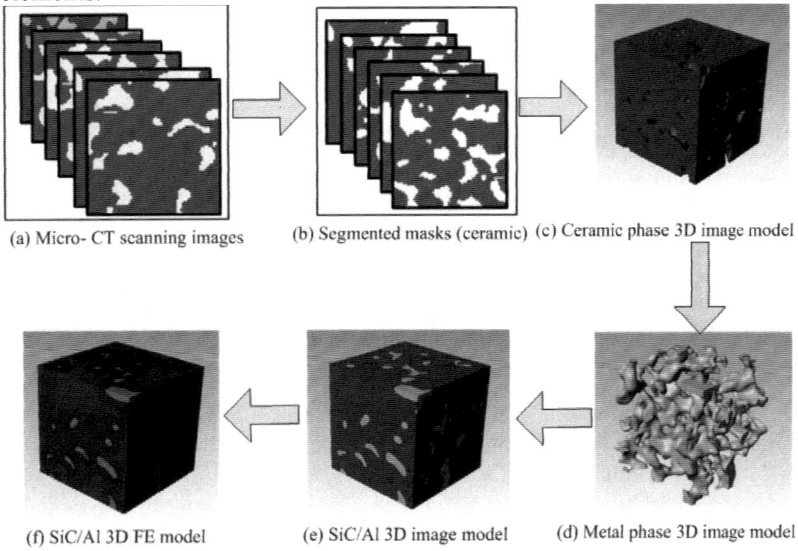

Figure 1. The process of SiC_{3d}/Al FE model generation based on Micro-CT scanning images.

2.3. Boundary conditions and loads
In this investigation, the external load is applied on the nodes uniformly on plane z=H in the z-direction (see figure 2) by prescribing the velocity with a constant strain-rate of 3500/s. Subsequently, the displacements of nodes on plane z=0 are fixed (see figure. 2). The nodes on plane x=0, are free to displace in the plane but were all constrained to have the same normal displacement u during the deformation in order to maintain a flat surface. The nodes on plane x= L, y=0, y=W are defined in the similar way, which meet equation (2)~(4)respectively.

$$u(0, y, z) = u(0,0,0) \tag{1}$$

$$u(L, y, z) = u(L,0,0) \tag{2}$$

$$v(x,0, z) = v(0,0,0) \tag{3}$$

$$v(x,W, z) = v(0,W,0) \tag{4}$$

L, W and H is the the length, width and height of the model respectively.

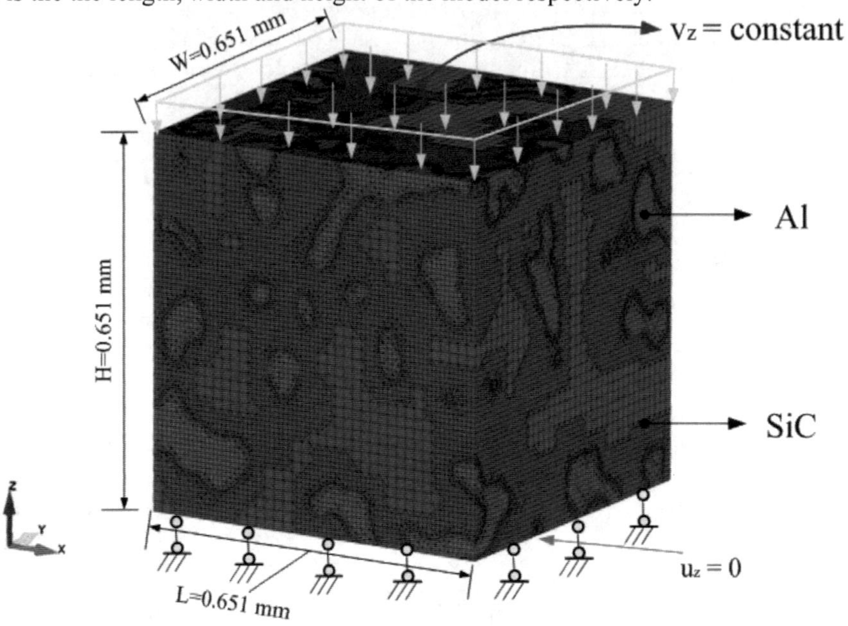

Figure 2. Boundary and loads of the SiC_{3d}/Al FE model.

3. Results and discussions

3.1. The damage analysis of the model
In this investigation, the average stress and strain in the model can be obtained from:

$$\varepsilon_{\text{ave}} = \Delta H / H \tag{5}$$

$$\sigma_{\text{ave}} = F/A \tag{6}$$

Where ΔH is the is overall deformation of model along Z-direction, H is the length of model in z direction, F is the sum of reaction forces at nodes on the surface Z=H and A is the area of that surface.

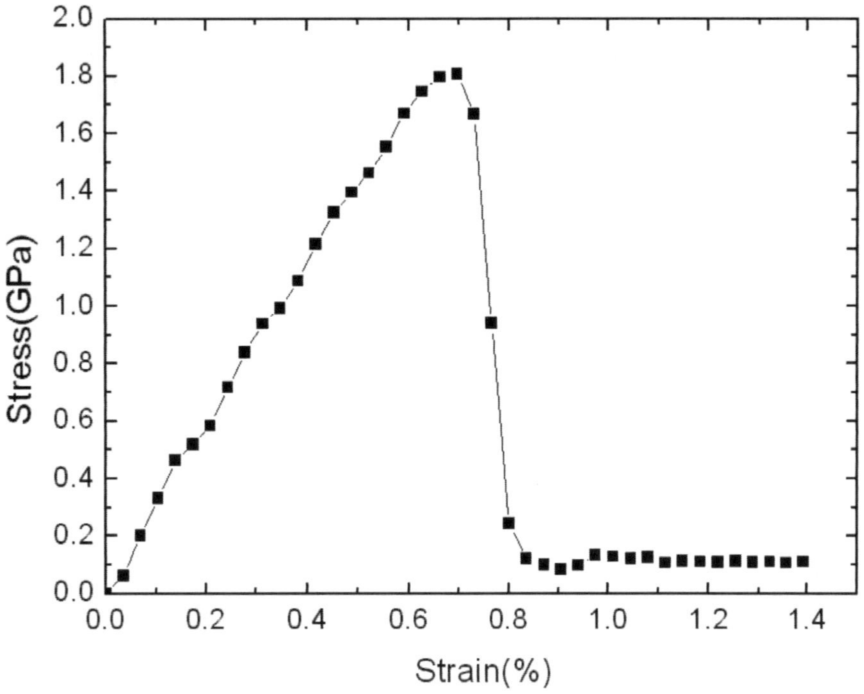

Figure 3. The average Stress/strain curves of model under dynamic axial compression.

Figure 3 and figure 4 show the average stress-strain curve of the model and damage process respectively. It is seen that the average stress increases linearly from 0 to 0.939GPa with the average strain increasing from 0 to 0.3%; at the same time, the crack initiation appears at the region near the interface between the ceramic and metallic phases (see figure 4 (a)). The average stress of the model still increases lineally until to 1.81GPa when the average strain reaches to 0.68; subsequently, the average stress collapses rapidly due to the crack propagation, bridging and accumulation.

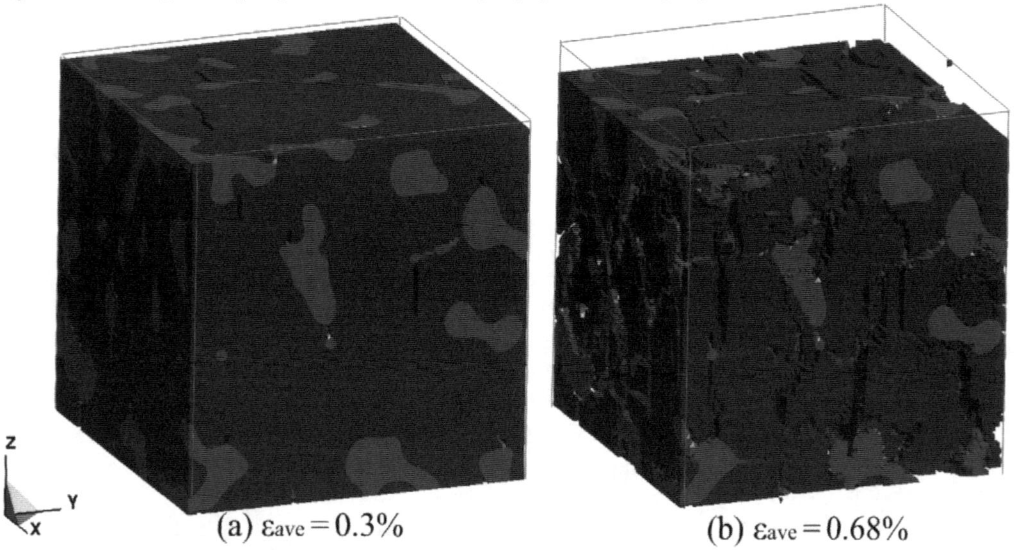

Figure 4. Thc overall damage process of the model.

3.2. Damage mechanisms

In order to investigate further the role of ceramic and metallic phase in the model under dynamic axial compression, the curves of residual mass ratio versus strain of the two phases are presented in figure 5. To describe the damage process, the residual mass ratio β in this investigation is defined as:

$$\beta = 1 - m_{\text{erode}} / m_0 \tag{7}$$

Where m_{erode} is the mass of the eroded elements of ceramic phase or metallic phase in the model, and m_0 is the original mass of the corresponding phase.

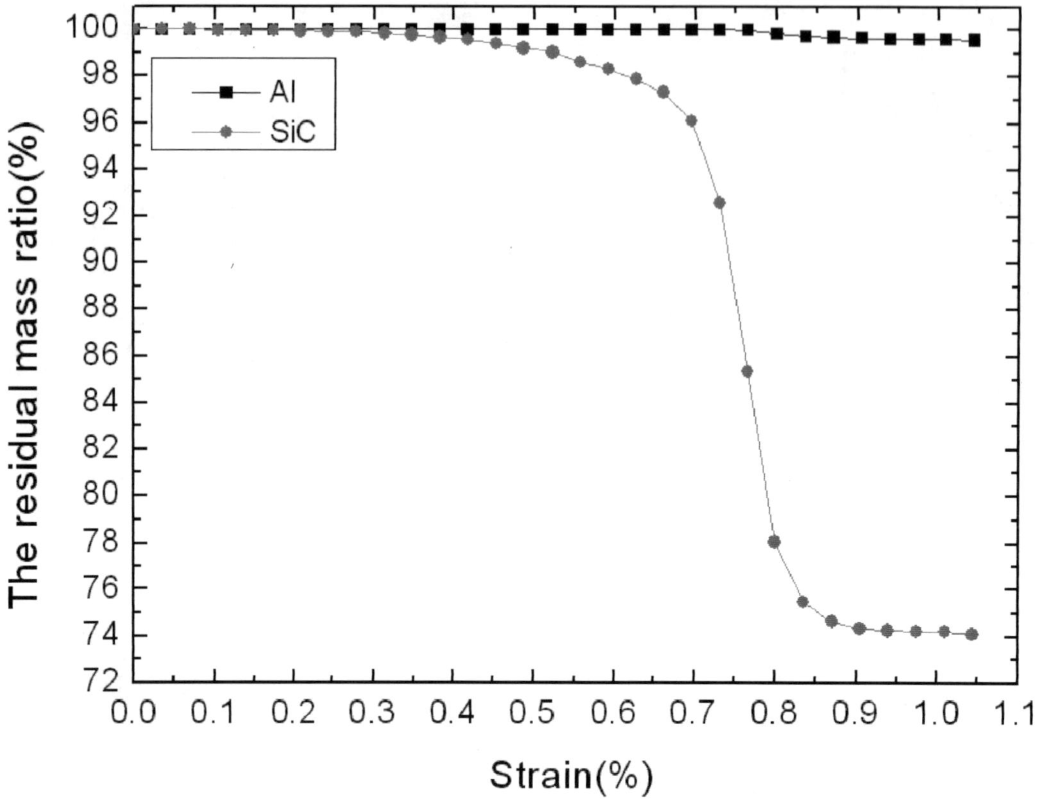

Figure 5. The residual mass ratio of ceramic phase as well as metallic phase in the model.

It can be seen from figure.6 that both the mass of ceramic and metallic phases do not decrease when the average strain increases from 0 to 0.3%; subsequently, the residual mass ratio of ceramic phase decreases gradually to 96% while the one of metallic phase is still 100% when the average strain reaches to 0.68, which corresponds to the stress peak presented in figure 3. However, with the strain increasing to 0.87%, the mass of ceramic phase decreases sharply to 73%, while the mass of metallic phase changes little. Comparing with figure 3, it is evident that the ceramic phase attributes to the model's damage predominantly. The underlying reason lies in the fact that the ceramic phase is with high strength but unavoidable intrinsic brittleness, while the metallic phase make little contribution to the model's damage because of its toughness.

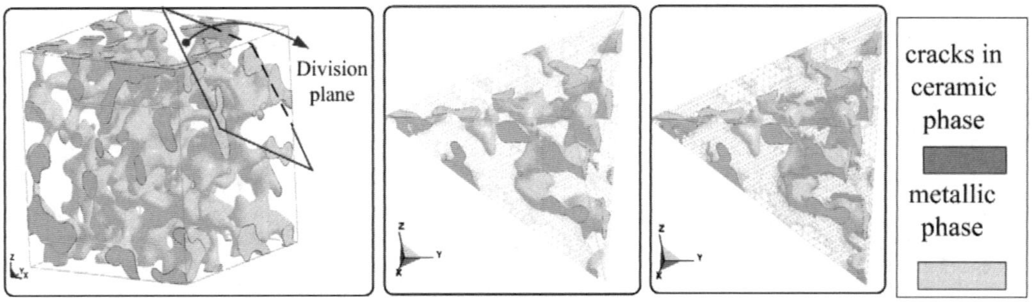

(a) Selecting a part of the model (b)$\varepsilon_{ave} = 0.30\%$ (c)$\varepsilon_{ave} = 0.68\%$

Figure 6. The damage process in the ceramic phase at different strain values

In this study, a part of the model is cut by a division plane to further investigate the damage process of ceramic phase as shown in figure 6, To illustrate the crack initiation and propagation locations of the ceramic phase, only the metallic phase and the cracks in ceramic phase are displayed, colored with green and pink, respectively. It can be seen from figure 6 (b) that the cracks are mainly initiated near the interface between the ceramic and metallic phase at the strain of 0.3%. When the strain increases to 0.68%, it is found that the cracks start to bridge and accumulate along the interface When the cracks are accumulate to a certain amount, they will propagate inside the ceramic phase, thus leading to the collapse of the model finally.

4. Conclusions

In this study, a 3D mesoscopic structure FE-Model of interpenetrating SiC_{3d}/Al composite **is** established by using Micro-CT scanning technique. Subsequently, the finite element modeling is applied to investigate its dynamic damage process under axial crushing. The results show that, it is the silicon ceramic phase that makes the dominant contribution to the model's damage due to its intrinsic brittleness, and the cracks initiate and propagate mainly appear in the region of interface between the ceramic and metallic phases.

5. References

[1]Dawson P R and Needleman A 1994 *Mater. Sci. Eng. A.* **175** 43
[2]Baskes M I, Hoagland R G and Needleman A 1992 *Mater . Sci. Eng. A.* **159** 1
[3]Wegner L D and Gibson L J 2000 *INT. J. MECH. SCI.* **42** 925
[4]Mills N J 2007 *INT. J. SOLIDS. STRUCT.* **44** 51
[5] Jhaver R and Tippur H 2009 *Mater Sci Eng A.* **499** 507
[6]Mishnaevsky Jr L L 2005 *Mater Sci Eng A.* **407** 11.
[7] Duane S 2003 4th Europan *LS-DYNA User Conference* Elsevier **44**51
[8]LS-DYNA, Livermore Software Technology Corporation, Las Positas Road Livermore, CA American, http://www.lstc.com
[9]ScanIP.Simpleware Ltd, Innovation Centre, Rennes Drive,EX4 4RN, UK, http://www.simpleware.com.
[10]+ScanFE. Simpleware Ltd, Innovation Centre, Rennes Drive, EX4 4RN, UK, http://www.simpleware.com.

Mechanical properties of tantalum-tungsten interlayer between tungsten tile and thimble to prevent helium leak from He-cooled divertor

Pingping Zhang, Weiping Shen [1], Yanan Zhou, Qingling Zhang
Institute of Nuclear Materials, School of Materials Science and Engineering, University of Science and Technology Beijing, Beijing 100083, China

E-mail: shenwp@ustb.edu.cn

Abstract. The tungsten parts made of pure tungsten tile and dispersion strengthened tungsten thimble with 3 mm interlayer of tantalum-tungsten alloy are fabricated by Spark Plasmas Sintering (SPS). The process of SPS is that the temperature is raised to 1700 °C at a rate of 100 °C/min and kept for 3 min, under a constant pressure of 50MPa along the Z-axis. The mechanical properties of the interlayer with different percent of tantalum are measured. The results show that with increasing percent of tantalum, the hardness first increases and then decreases; and as the indentation on the sample is closer to dispersion strengthened tungsten, the value of Vickers hardness is much higher. The Vickers hardness of interlayer is the highest when the content of tantalum is 50% and the indentation is next to dispersion strengthened tungsten. Bending strength drops with increasing content of tantalum, when the content of tantalum is 100% the value of bending strength is the lowest. The fracture toughness is highest as the content of tantalum is 25%, the value is 9.89MPa•m$^{1/2}$. The toughening tungsten-tantalum interlayer between tungsten tile and thimble would better prevent helium leak from He-cooled divertor for DEMO.

1. Introduction

Different concepts of He-cooled divertor have been studied, which shapes are finger (HEMJ or HEMS), T-tube and plate [1]. The concept of HEMJ (He-cooled modular divertor with multiple-jet cooling) was developed at KIT several years ago. Selection of suitable materials and technologies to manufacture the HEMJ is the most difficult, because divertor design requires surviving at cyclic heat load of 10-15 MW/m^2. During operation, the temperature of the tile is raised to 1000℃, while it remains 600-700℃ at thimble/steel joining area [2]. To satisfy this situation, many kinds of materials have been used, such as pure W, La_2O_3-W, brazing alloy, etc. The other difficulty is manufacture, many methods have been researched, such as PM (powder metallurgy), EDM (electro discharge machining) and conventional machining (turning, milling, grinding). Each of these methods has its own advantages and disadvantages.

To make use of He-cooled divertor, tungsten is regarded as the most suitable armor material, because it possesses a good thermal conductivity, high melting point, and high strength. And the PM is regarded as a suitable method to manufacture He-cooled divertor. But the micro-cracks between tile

[1] Corresponding author. Tel.: +86 62334951; fax: +86 62334951.
E-mail address: shenwp@ustb.edu.cn.

and thimble should be took care, from the previous experiments, it was found micro-cracks at bottom of the tile, and the micro-crack will go through the tile and thimble because low thermal fatigue resistance of pure tungsten tile, very thin brazing filler alloy and dispersion-strengthened tungsten thimble in high-heat-flux (HHF) [3,4,5]. To solve the problem of cracking, a new design of He-cooled divertor such as adding a few millimeter interlayer should be studied.

In this paper, pure tungsten and dispersion-strengthened tungsten (W-2% TaC) were also used for tile and thimble respectively, Spark Plasma Sintering (SPS) for the fabrication of He-cooled divertor. The characteristic of this method are SPS could make the high densities and the interlayer was used to block crack propagation. Different interlayer will have different mechanical properties. This paper will discuss the mechanical properties of samples with different interlayer to find the optimizations.

2. Experimental procedure

Two kinds of powders were used for the interlayer, tungsten and tantalum. Tungsten was supplied by Xiamen Golden Egret Special Alloy Co. Ltd, the size of tungsten powders is 2μm and the content of oxygen is 0.045%. Tantalum was commercial powder, the nominal size of which is 48μm. Tantalum carbide (TaC) was synthesized by a liquid precursor route, the average size of particles is about 50nm with 5% free carbon. The progress of fabrication contains two steps, the first step was mixing the powders of tungsten and TaC by ball milling, the second step was sintering the samples with the interlayer by SPS.

Ball-milling process was performed at room temperature in a horizontal planetary ball mill [4] using tungsten rods of φ8mm×10mm as grinding media. The raw materials were mixed with the powers of tungsten and nano-TaC in a weight ratio of 10:1 (total mass of 100 g), followed by ball-milling at a rotating speed of 150 r/min under Ar atmosphere for 6 h.

And then, the mixed powders were charged into an SPS graphite die of 20 mm in diameter. W-TaC powders were at bottom of graphite die, W-Ta powders were at the middle of graphite die, and pure tungsten powders were at top of graphite die. Graphite paper was filled between the graphite die and powder blend. Then put the die into the SPS furnace to sintering. After the SPS chamber was evacuated (about 10Pa), the temperature was raised to 1700 °C at a rate of 100 °C/min and kept for 3 min, under a constant pressure of 50MPa along the Z-axis.

The samples were cut in size of 13mm×3mm×4mm by EDM to test bending strength and fracture toughness with a notch in interlayer about 3mm. The microstructure of fractures was observed using a scanning electron microscope (SEM), the bending strength and fracture toughness was measured using three-point bending of 10mm span, and the Vickers hardness of samples was determined at a load of 100g and for 10s. The measured hardness value was calculated at an average of 5 different points on the same line of sample.

3. Results and discussion

3.1. Bending strength of the interlayer

The bending strengths of interlayer with different content of Ta are shown in Fig.1. It is apparent that the bending strengths decrease with increasing content of tantalum in the interlayer. When the content of tantalum is 0, the bending strength is the highest, reaches about 820MPa. When the content of tantalum is 100%, the bending strength is the lowest, decreases to about 140 MPa. The force impending dislocation motion decreases when the content of tantalum increases. There are two reasons to explain the value of bending strength is low.

First, with increasing content of tantalum, the size of grain becomes larger and the number of grain boundary becomes much fewer. Fig. 2 shows the microstructure of the interlayer with different content of tantalum. Due to grain boundaries become much fewer, the force impending dislocation motion will decrease.

And then, the other reason for the falling bending strengths is the lattice constant. Lattice constant increases linearly when the concentration of tantalum increases [6]. According to dislocation theory, slip plane spacing increases with an increase of lattice constant. It means that the width of the dislocation will be wider, which reduces the resistance to lattice dislocation motion. So the bending strength shows a downward trend with increasing the content of tantalum (Fig.1).

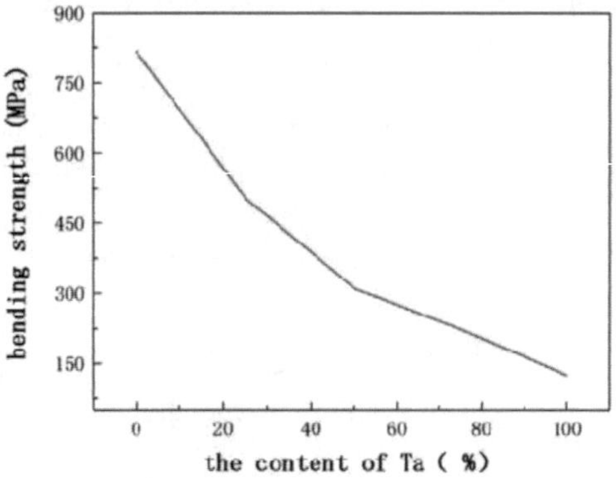

Fig.1. Bending strength of interlayer with different content of tantalum at room temperature

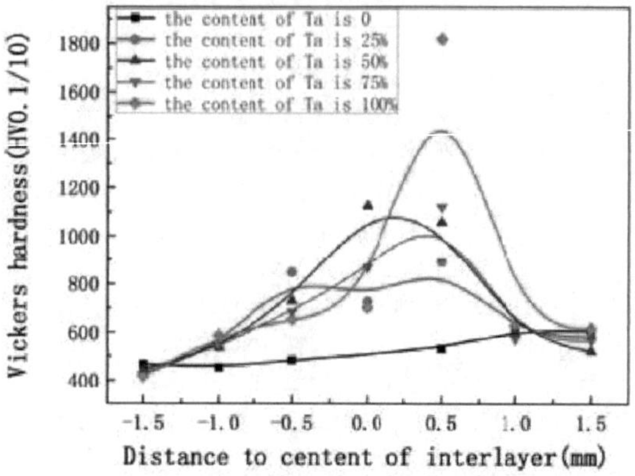

Fig.3. Vickers hardness of W-Ta interlayer with different content of Ta in SPS samples

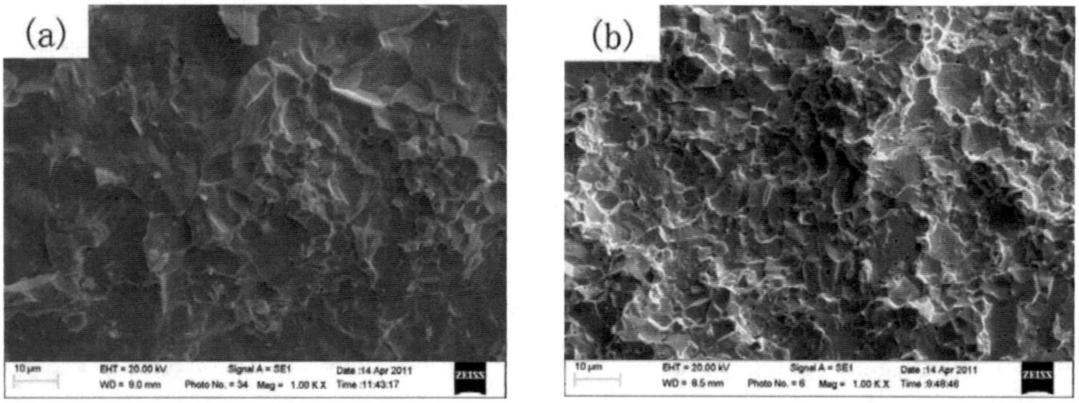

(a) Coarse granular fracture of 100% tantalum (b) Fine granular fracture of 100% tungsten

Fig.2. The microstructure of different interlayer cross-section by SEM

3.2. Vickers hardness of the interlayers

The Vickers hardness of different content of Ta in samples (Fig.3) clearly shows the Vickers hardness depending on the content of Ta and the distance to the dispersion strengthened tungsten. In Fig.3, the left presents pure tungsten and the right presents dispersion strengthened tungsten with 2% TaC. It shows that the Vickers hardness of dispersion strengthened tungsten is higher than the pure tungsten. The Vickers hardness of interlayer are around 500~1100HV0.1/10. As the content of tantalum increases, the Vickers hardness first increases and then decreases. So the value of hardness is highest when the content of Ta is 50%. And the Vickers hardness is much higher when the indentation is closer to the dispersion strengthened tungsten. For the samples, when the indentation is away from the center about 0.5 mm, the hardness is the highest. So when the content of tantalum is 50%, at the indentation away from center about 0.5 mm, the Vickers hardness is the highest, the value is about 1100 HV0.1/10. But as the content of tantalum is 100%, the Vickers hardness is highest at the indentation away from center at 1mm.

The Vickers hardness is always affected by the microstructure of samples, the hardness is mainly depending on the solid solution and grain boundary hardening. The contribution of grain boundaries can be described by Hall-Petch equation [7].

$$\sigma=\sigma0+kd-0.5$$

where $\sigma0$ is the intrinsic hardness of the material, d is the diameter of grain and k is a material constant. When the content of Ta was a little, Ta will strengthen the materials as solid solution composition, so the hardness increases when the content of Ta increases.

For solid solution hardening of W in Ta, the hardening effect can be calculated using the linear hardening model based on the elastic size misfit interaction with screw dislocations [7, 8].

$$\sigma=0.1274c$$

where c is the atomic concentration of W. For the equation, Vickers hardness will decreases as the content of Ta increases. Because of these two mechanisms, the boundary hardening plays a major role when the content of tantalum is low; the solid solution plays a major role when the content of tantalum is high. So with increasing content of tantalum, the Vickers hardness first increases and then decreases. The Vickers hardness will increase as the indentation is close to dispersion strengthened tungsten. It indicates that free carbon in the nano-TaC powders diffuses into the W-Ta layer to form tungsten carbide or tantalum carbide, as the Vickers hardness of tungsten carbide and tantalum carbide is higher than tungsten and dispersion strengthen tungsten, so the hardness of interlayer is higher than tungsten and dispersion strengthened tungsten. The different content of carbon diffused into W-Ta and then forms different phases when the distance is different. The content of carbon diffused in W-Ta and dispersion strengthened tungsten shows in Fig.4. The left of sample is the interlayer of W-Ta, the right is dispersion strengthened tungsten. It might be well to point out that the content of carbon in the right is more than it in the left, and then the content of tungsten carbide and tantalum carbide are more, so the right Vickers hardness is higher.

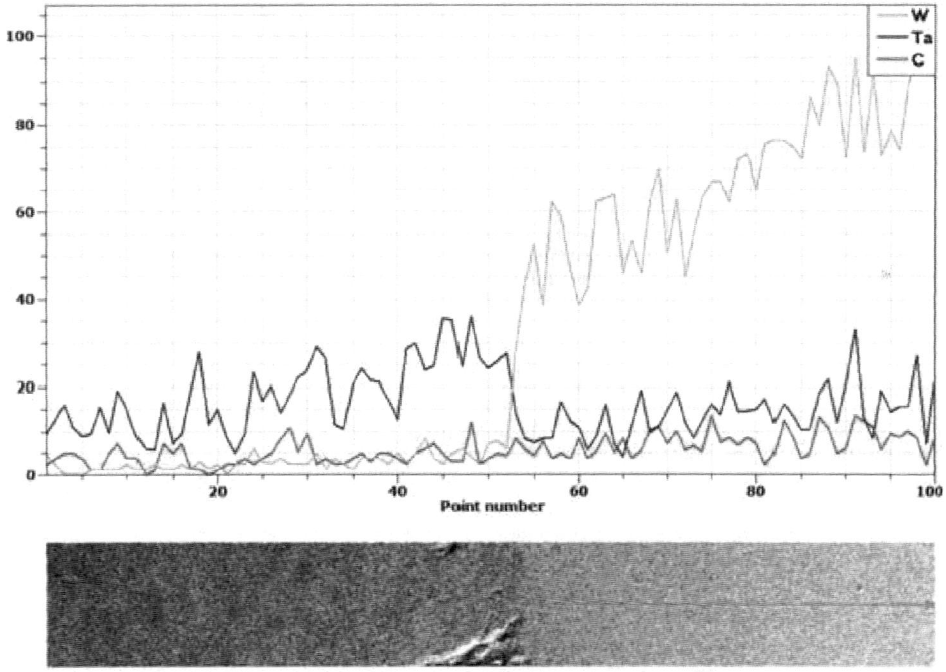

Fig.4. Carbon diffused in a sample from W-Ta interlayer (left) to dispersion strengthened tungsten (right).

3.3. Microstructure and fracture toughness of the interlayer

The thermal cracking is regarded as the important problem to be solved in He-cooled divertor of DEMO, so the new design of adding the interlayer of tungsten-tantalum is studied. The fracture surface of different content of Ta in samples (Fig.5) clearly shows the size of grain gets much larger and from brittle intergranular fracture to transgranular cleavage fracture appears in the fracture surface with increasing content of tantalum. It suggested that the bending force between grains is different with different content of tantalum. It may be that a solid solution has been formed between tungsten and tantalum. And then the cracking can't go along the grain boundaries, it have to need much more energy to go through the grain. But the number of grain boundary gets fewer with increasing content of tantalum, the energy to go through the grain will get low. So it can be said the interlayer of tungsten-tantalum alloy can block crack propagation when the content of tantalum is proper.

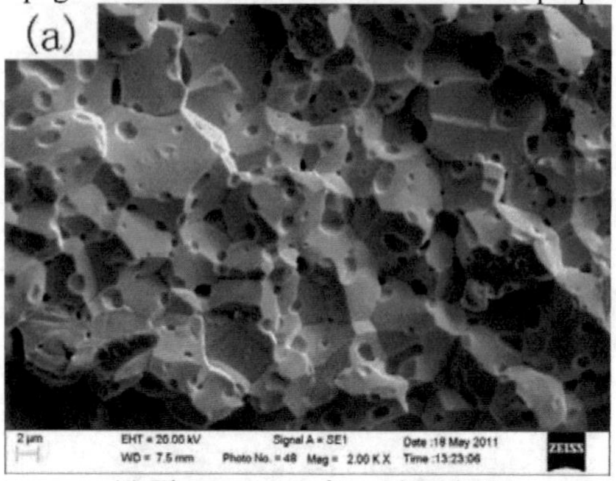

(a) The content of tantalum is 0;

(b) The content of tantalum is 25%

(c) The content of tantalum is 50%;

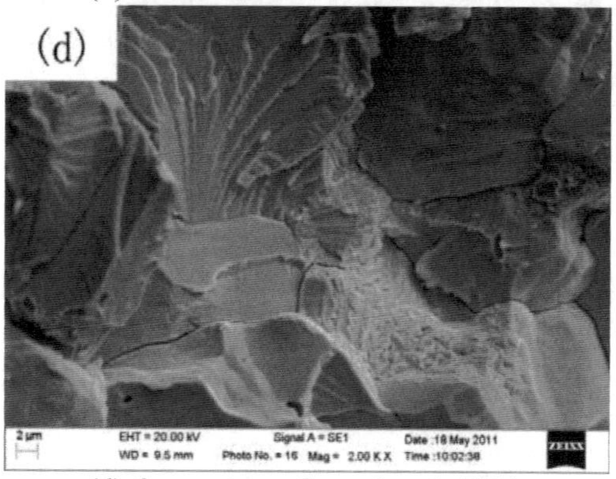

(d) the content of tantalum is 75%;

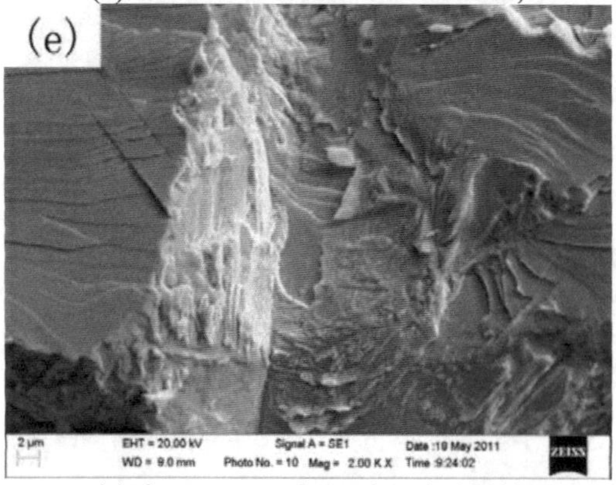

(e) the content of tantalum is 100%

Fig.5. Microstructure of fracture surface in interlayer with different content of tantalum by SEM:

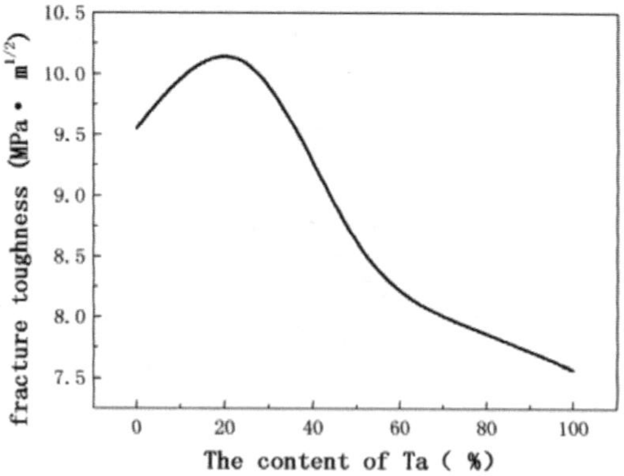

Fig.6. Fracture toughness of interlayer with content of the tantalum at room temperature

Fracture toughness is the most important mechanical properties of materials to describe the ability of blocking crack propagation. Fig.6 shows the fracture toughness of interlayer with different content of Ta. It was shown that the fracture toughness is rise and fall. The fracture toughness is highest when the content of tantalum is 25%, the value is 10.08MPa•m$^{1/2}$. There are three categories of possible mechanisms affecting the fracture toughness. The first one is the effects of flaws, the second one is the effects of interfaces, and the last one is related to the effect of interfaces involving the deformation mechanisms [9].

For these reasons, the most important actor is the force impeding grain boundary sliding. With increasing content of tantalum, the forces of impeding grain boundary are different. When the content of tantalum is 25%, the flows and interfaces are much relatively, it means that the grain boundary strength and the force are large, so the value of fracture toughness is the highest.

Some phenomenons can be found through compare Fig.5 and Fig.6.The brittle intergranular fracture of coarse grains and pores in Fig.5 (a) and then the interlayer has lower fracture toughness. Transgranular cleavage and intergranular fracture of fine grains make more crack deflections in Fig.5 (b). And the highest fracture toughness, and fracture toughness reduce gradually with increasing grain size of transgranular cleavage fracture in Fig.5(c)-(e). By the same reason, content of about 25% Ta in W, proper grain size and moderate interlayer thickness have better He leak-proof between tungsten armor and thimble of HEMJ by thermal shock.

4. Conclusions

Pure tungsten and dispersion strengthened tungsten with interlayer of W-Ta alloys with Ta composition from 0 to 100% were fabricated by Spark Plasmas Sintering (SPS) and their Vickers hardness, bending strength and fracture toughness were measured. Conclusions derived from the tests are given as follows:

(1) With increasing content of tantalum, Vickers hardness first increase and then decrease. When the content of tantalum is 50%, the Vickers hardness is the highest. In the interlayer, the Vickers hardness is much higher when the indentation is closer to the dispersion strengthened tungsten. So when the content of tantalum is 50% the Vickers hardness at the indentation between W-Ta and dispersion strength is the highest, the value is 1100 HV0.1/10.

(2) The plasticity of tantalum is higher than tungsten, and the hardness is lower than tungsten, so with increasing content of tantalum from 0 to 100%, the bending strength dropped precipitously from 820MPa to 120MPa.

(3) The transgranular cleavage fracture appears in surface when the content of tantalum increases, it means that cracking need more energy to go through the materials with increasing content of about 25% tantalum.

(4) The fracture toughness is first increasing and then decreasing, with increasing content of tantalum. When the content of tantalum is 25%, the fracture toughness is the highest, the value is 9.89 MPa•m$^{1/2}$. When the content of tantalum is 100%, the fracture toughness is the lowest, the value is 7.57MPa•m$^{1/2}$. Presumably the tungsten-tantalum alloy interlayer with proper tantalum content between tungsten tile and thimble of mock-up can preferably prevent helium leak from He-cooled divertor.

5. Acknowledgments

This project was supported byNational Magnetic Confinement Fusion Program (No.2010GB109000); Project supported by the State Key Development Program for Basic Research of China (Grant No. 2008CB717802).

6. References

[1] M.S. Tillack, A.R.Raffray, X.R.Wang, et al., Recent US activities on advanced He-cooled W-alloy divertor concepts for fusion power plants, Fusion Engineering and Design.2011,86(1):71-98.

[2] P. Norajitra, L.V. Boccaccini, E. Diegele, et al., Development of a helium-cooled divertor concept: design-related requirements on materials and fabrication technology, Journal of Nuclear Materials, 329–333 (2004): 1594.
[3] P.Norajitra, S,Antusch, R.Giniyatulin, et al. Progress of He-cooled divertor development for DEMO, Fusion engineering and design.86 (2011): 1656–1659
[4] Y. Chen, T. Halstead, J.S. Williams, Influence of milling temperature and atmosphere on the synthesis of iron nitrides by ball milling, Materials Science and Engineering. A206 (1996): 24-29.
[5] Prachai Norajitra, Radmir Giniyatulin, Vladimir Kuznetsov, et al., He-cooled divertor for DEMO: Status of development and HHF tests. Fusion Engineering and Design, 85 (2010): 2251–2256
[6] Zhang Tingjie, Zhang Deyao and Ding Xu. Study on Mechanical Properties of Several Tantalum Alloys with Higher Tungsten Content. Rare Metal Materials and Engineering, 1996, 25 (4):5~10.(in Chinese)
[7] C.L. Wang, M. Zhang, J.P. Chub et al., Structures and nanoindentation properties of nanocrystalline and amorphous Ta–W thin films, Scripta Materialia 58 (2008): 195–198.
[8] W.J. Tomlinson, R. Rushton. The fabricability and hardness of concentrated binary tantalum alloys containing titanium, hafnium, vanadium, molybdenum, tungsten, rhenium and palladium, Journal of the Less Common Metals, 1986,115(1):L1-L4.
[9] Z. Zak Fang , Xu Wang, Taegong Ryu, et al., Synthesis, sintering, and mechanical properties of nanocrystalline cemented tungsten carbide–A review, Int. Journal of Refractory Metals & Hard Materials, 27 (2009): 288–299.

The influence of ALN-Al gradient material gradient index on ballistic performance

Youcong Wang[1],Qiwen Liu[1],Yao Li[1],Qiang Shen[2].

[1]Department of Engineering Structure and Mechanics, Wuhan University of Technology, Wuhan 430070, P. R. China
[2]State Key Laboratory of Materials Synthesis and Processing, Wuhan University of Technology, Wuhan 430070, P. R. China

E-mail:360326203@qq.com

Abstract: Ballistic performance of the gradient material is superior to laminated material, and gradient materials have different gradient types. Using ls-dyna to simulate the ballistic performance of ALN–AL gradient target plates which contain three gradient index (b = 1, b = 0.5, b = 2). Through Hopkinson bar numerical simulation to the target plate materials, we obtained the reflection stress wave and transmission stress wave state of gradient material to get the best gradient index. The internal stress state of gradient material is simulated by amplification processing of the target plate model. When the gradient index b is equal to 1, the gradient target plate is best of all.

1. Introduction

In the double composite armor system, due to the differences of elastic modulus, hardness, density between ceramic panel and toughness panel, it result in layer acoustic impedance mismatching and stress concentration, Under the impact of projectile, the reflection tensile wave cause ceramic panel serious damage, which greatly limits the target plate ballistic performance. In the 1990 s, American scientists put forward a new gradient armor concept, and it gradually becomes the development trend of armor materials. The content of gradient armor varied with the thickness (step change) in the middle of the transition layer, not only reserved the ceramic material resistance to penetration of the superior performance, but also having good toughness metal materials, that can improve the composite target plate ability to resistance to penetration[1]. The study of gradient material impact behavior, there are two ways to describe gradient material constitutive. One way is that determining to use JH2 or JC constitutive equation by the volume fraction of ceramic. We choose using JH2 constitutive when volume fraction of ceramic is higher than fifty percent, otherwise using JC constitutive. Like Douglas W. Templeton, he uses interpolation to describe the middle intermediate layers material properties, and results shows an increase of approximately 15% in the ballistic performance of the simulated FGM when compared to double laminated armor [2]. Another way is using a new constitutive to describe gradient material. Like Y. Li and K. T. Ramesh, they put forward an equation to describe the stress-strain relationship of mixed material, and get several gradient structure stress states [3][4]. But they have no further analysis to get the optimization of gradient type. And theoretically for gradient materials are also deduced. Hugh A. Bruck use one-dimensional model to get a time delay phenomenon, which despond on the composition gradient and the difference in base material properties. This is presented by time-history profiles of reflected stress waves[5]. Therefore, this paper not only studies AlN-Al gradient armor gradient index for the influence on ballistic performance, but also to get the optimization of gradient index by the time-history profiles of reflected stress waves.

2. Numerical simulation of ballistic performance to gradient armor

2.1 Numerical simulation model

Figure1 computational model

The whole model take 1/4 modeling, and the radius of projectile is 0.5 cm, length is 5 cm. The total thickness of target plates is 3 cm, and we choose six layers to ensure the accuracy and calculated amount about gradient material. Each layer thickness is 0.5 cm, and radius is 10 cm, and the nodes are tied between layer and layer.

2.2. Material parameters

Table1 material parameters of projectile

Density	Elastic modulus	Poisson ratio	Yield stress	Tangent modulus	Beta
ρ	E	v	σ_S	σ_t	β
(g/cm^3)	(GPa)		(GPa)	(GPa)	
7.65	200	0.29	1.6	80	1.0

Considered three target plates gradient model and the variation about ceramic volume fraction of ALN with the thickness are shown as follows equation 1:

$$f(z)=\left(\frac{z}{h}\right)^b \tag{1}$$

b —interpolation index;
h — target plate thickness;
z —along the thickness direction coordinate values;
We take b is 2, 1, and 0.5, as shown in figure 2:

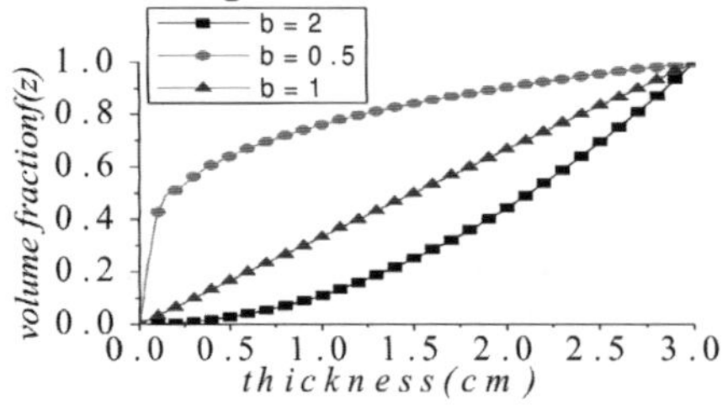

Figure 2 ALN volume fraction varied with the thickness of the target

Target plate material parameters (b=1) is shown in figure 3[3]:
The Constitutive model of layer 1 is the AlN JH2 material model, layer 6 is Al JC material model, and the middle layer material parameters are obtained by the equation 1.
When b = 1, layer1, 2, 3 is JH2 material model, and layer 4, 5, 6 is JC material model; When b= 0.5, layer 1, 2, 3, 4 is JH2 material model, and layer 5, 6 is JC material model; when B = 2, layer 1, 2 is JH2 material model, and layer 3,4,5,6 is JC material model.

The top three layer material model parameters of double laminated are pure ceramics, and the others are pure metal aluminum.

	Layer 1	Layer 2	Layer 3	Layer 4	Layer 5	Layer 6
Density, ρ (Kg/m^3)	3226	3134	3043	2951	2860	2768
Shear Modulus, G (GPa)	127	107	87	66	46	26
Tensile Strength, T (GPa)	0.50	0.75	1.00			
Intact Strength, σ_i (GPa)	4.31	3.50	2.80			
Maximum Strength, s_{max} (GPa)	5.50	4.50	3.50	2.50	1.50	0
Maximum Failed Strength, s_{max} (GPa)	0.20	0.16	0.12			
Bulk Modulus, K1 (GPa)	201	176	151	127	102	77
Pressure Constant, K2 (GPa)	260	234	207	181	154	128
Pressure Constant, K3 (GPa)	0	25	50	75	100	125
Damage Constant, D1	0.16	0.56	0.63			
Damage Exponent, N	1.00	1.26	1.47			
Specific Heat, c (J/kg-°C)	735	763	791	820	848	876
Yield Stress, C1 (GPa)				1.25	1.0	0.5
Pressure Coefficient, C4				0.83	0.5	0
MaximumStrength, s_{max} (GPa)				2.5	1.5	0
Fracture Constant, D1				0	0	0.14
Fracture Constant, D2				0.16	0.22	0.14
Fracture Constant, D3				-2.1	-2.0	-1.5

Figure3 target plate material parameters (b=1)

2.3 ballistic performances results

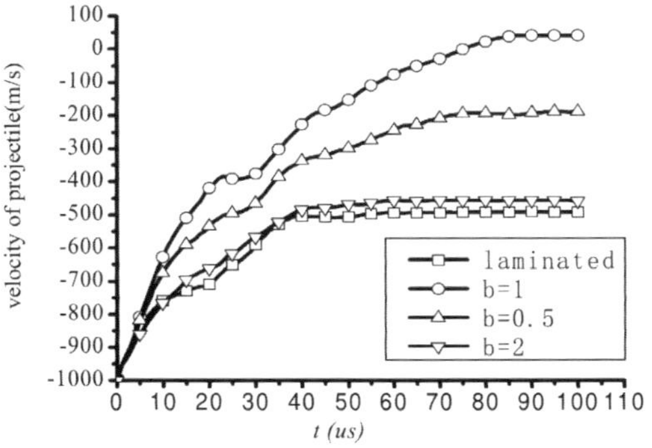

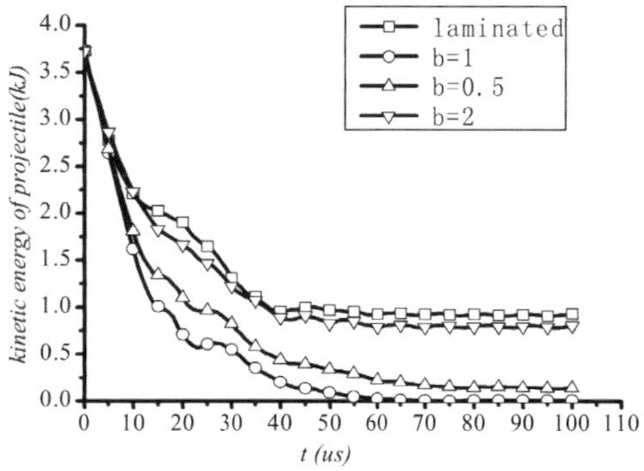

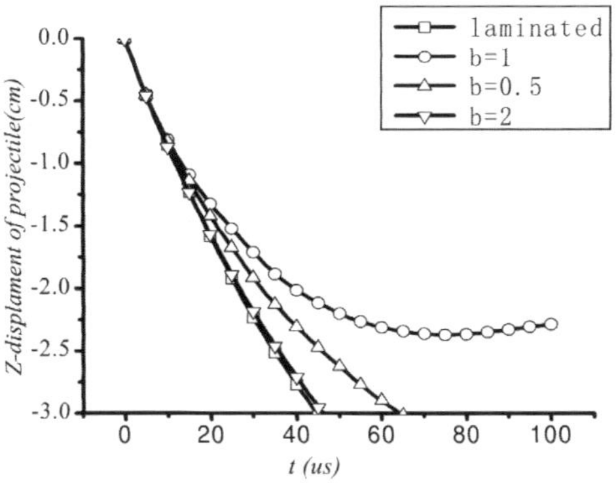

Figure 4 velocity, kinetic energy and Z-displacement of projectile in four kinds of target plate

From figure 4, we can get the residual velocity of the projectile, the kinetic energy of the projectile, and penetration depth of the target plates about four kinds of target. And it can be seen that the ballistic performances of target (the gradient index b = 1) is best, and the double laminate composite is worst.

3. Numerical simulation of Hopkinson bar

According to the theory about one dimensional stress wave, the peak value of target plate material reflection wave is smaller, then the ballistic performance of target plate is better [6].

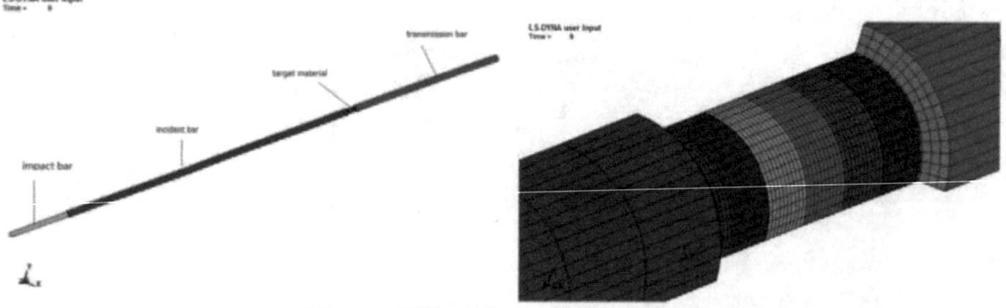

Figure 5 Hopkinson bar model

The whole model contains impact bar, incident bar, target plate material, transmission bar, and the impact bar impact incident bar as 32 m/s speed. The modeling data of each part are shown as table 2:

Table 2 model size

part	radius(*mm*)	length(*mm*)	material mode
impact bar	5	200	elastic steel model
incident bar	6.5	1000	elastic steel model
target plate material	5	30	JC,JH2 interpolation
transmission bar	6.5	500	elastic steel model

The peak values of reflection and transmission stress wave about four kinds of target plates are shown as table 3:

Table 3 reflection and transmission wave peak value about the target plates

target plate	Incident wave (*GPa*)	reflection wave (*GPa*)	transmission wave (*GPa*)
laminated	0.610	0.395	0.505
b=1	0.610	0.370	0.512
b=0.5	0.610	0.353	0.514
b=2	0.610	0.384	0.510

It can be seen from table 3: the reflection wave peak value of double laminated target plate is higher than that of gradient target. And the reflection wave peak value of gradient index b = 0.5 is minimum, But according to figure 4, ballistic performance of target (b=1) is better than that of target (b = 0.5), and this will be explained in section 4.

4. Stress wave in the target plate material internal propagation

Due to the low thickness of the target plate material (only 3 cm) and the velocity of stress wave (5000*m/s*), it is difficult to extract the internal stress wave propagation curve from Hopkinson bar simulation or experiments. By using numerical simulation technology, it is easy to obtain the internal stress wave propagation curve by the thickness of model expanded 100 times (3m). Initial loading conditions for the gradient material is applying a 2 us pressure wave on the pure ceramic surface, which is as shown in figure 6. And the reflection wave peak value about four kinds of target plates are shown in table 4.

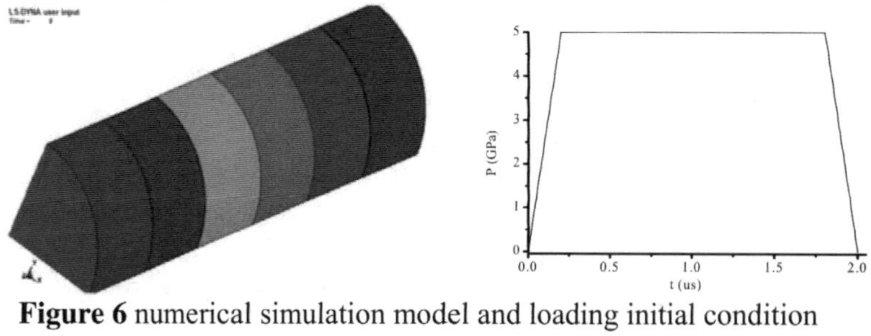

Figure 6 numerical simulation model and loading initial condition

Table 4 reflection wave peak value

	b=1	laminated	b=0.5	b=2
layer1	0.00252	0.00107	0.00243	0.00205
layer2	0.00246	0.00110	0.00244	0.00252
layer3	0.00280	0.00180	0.00288	0.00264
layer4	0.00260	0.00112	0.00288	0.00242
layer5	0.00225	0.00101	0.00245	0.00205
layer6	0.00191	0.00113	0.00170	0.00213

From table 4 we can see that, in the target plate internal stress state, the reflection wave peak value of double laminated is minimum. But the tensile strength of pure ceramic is less than that of gradient material ceramic phase. When the peak values of reflection stress wave are at the same level, the third ceramic layer of double laminated is failure first of all, and the failure spreads to the whole ceramic plate quickly. But gradient armor ceramic phase layer have just partial failure. Though the comparison the gradient material ceramic phase of b = 1 with b = 0.5, it can be seen that the reflection wave peak values are the same in layer 3 and layer 4 (ceramic phase) when the gradient index b = 0.5. And when b = 1, the reflection wave peak value of layer 3 ceramic phase is less than that of b = 0.5 slightly, but the tensile strength is a little stronger because of the gradient index interpolation. Therefore, the ballistic performance of gradient index b = 1 is better than that of b = 0.5.

5. Conclusions

Through the numerical simulation of AlN-Al gradient material by LS-dyna, the conclusions are as follows:

1. We use interpolation to describe the intermediate layers material properties and get ballistic performance of four kinds of target, and b=1 is the best gradient index can be seen from figure 4;
2. Through the Hopkinson bar numerical simulation, double laminated target plate reflection wave peak value is higher than gradient, gradient index b = 0.5 minimum value of the reflection wave;
3. By expanding the model of target 100 times, the internal stress wave propagation of target can be obtained. And it has proved that b=1 is better than b=0.5 by the reflection wave peak value about layer 3 and layer 4.

Acknowledgments

The work was supported by Fundamental Research Funds for the Central Universities

References

[1] Yang-wei Wang, Fu-chi Wang, Xiao-dong Wang, Zhuang Ma 2007 research advancement on graded ceramic-metal armor composites *acta armamentaria* **28 No.2**

[2] Douglas W. Templeton, Tara J. Gorsich and Timothy J. Holmquist 2006 computational study of a functionally graded ceramic-metallic Armor *23rd international symposium on ballistics*

[3] Y. Li, K.T. Ramesh 1998 Influence of particle volume fraction ,shape and aspect ratio on the behavior of particle-reinforced metal-matrix composites at high rates of strain *Acta Materialia* **46(16),** 5633-5646

[4] Y. Li, K.T. Ramesh and E.S.C. Chin 2001 dynamic characterization of layered and graded structures under impulsive loading *International Journal of Solids and Structures* **38** 6045-6061

[5] Hugh A. Bruck 2000 A one-dimensional model for designing functionally graded materials to manage stress waves *International Journal of Solids and Structures* **37** 6383-6395

[6] Lisheng Liu, Qingjie Zhang 2003 Propagation Characteristics of Impact waves across aplanar interface between a ceramic layer and a composite graded layer of ceramic/metal graded materials journal of wuhan university of technology 25 No.8

[7] LSTC 2007 LS-DYNA KEYWORDUSER'S MANUAL VOLUMES I AND II. VERSION 971. MAY. *http://www.lstc.com/pdf/lsdyna_971_manual_k.pdf*

Layered Composite of TiC-TiB_2 to Ti-6Al-4V in Graded Composition by Combustion Synthesis in High-gravity Field

Xuegang Huang, Zhongmin Zhao[1], Long Zhang

Department of Vehicle and Electrical Engineering, Mechanical Engineering College, Shijiazhuang 050003, China

E-mail:emei127@126.com

Abstract. By taking combustion synthesis to prepare solidified TiB_2 matrix ceramic in high-gravity field, the layered composite of TiC-TiB_2 ceramic to Ti-6Al-4V substrate in graded composition was achieved. XRD, FESEM and EDS results showed that the bulk full-density solidified TiC-TiB_2 composite was composed of fine TiB_2 platelets, TiC irregular grains, a few of α-Al_2O_3 inclusions and Cr alloy phases, and α'-Ti phases alternating with Ti-enriched carbides constituted the matrix of the joint in which fine TiB platelets were embedded, whereas some C, B atoms were also detected at the heat-affected zone of Ti-6Al-4V substrate. The layered composite of the solidified ceramic to Ti-6Al-4V substrate in graded composition with continuous microstructure was considered a result of fused joint and inter-diffusion between liquid ceramic and surface-molten Ti alloy, followed by TiB_2-Ti peritectic reaction and subsequent eutectic reaction in TiC-TiB-Ti ternary system.

1. Introduction

Ceramic-to-metal seals and joints are assuming increasing importance in today's technology, and it is safe to predict that their uses will increase in the future as the advantages of ceramics as construction material are realized. While ceramics possess low density, good high temperature strength and creep resistance, their fracture toughness and thermal shock resistance are poor, limiting their use in demanding applications. Metallic materials, on the other hand, possess high fracture toughness and excellent thermal shock resistance while their poor strength-related properties at high temperature generally make them unattractive. Joining ceramics to metals, to combine inherent advantages of these two kinds of materials, has been pursued to meet the material requirements in many applications.

[1] Corresponding author. Fax: +86 311 87994737.
E-mail address: zhaozm2007@yahoo.cn (Z. Zhao).

Hence, various ceramic-to-metal joining techniques have been developed and improved over the past 50 years, such as mechanical joining, adhesive joining, friction joining, high eney beam welding, microwave joining, ultrasonic welding, explosive welding, reaction joining, combustion reaction joining, field-assisted bonding, brazing, diffusion bonding, transient liquid phase bonding (TLPB) and partial transient liquid phase bonding (PTLPB) and so on. Among these solid state diffusion bonding and reactive metal brazing have been explored most extensively, whereas the PTLPB with the advantages of brazing and diffusion bonding is a promising technology [1-3].

However, thermal residual stresses arising from cooling after high temperature processing, and problems due to poor chemical compatibility are barriers to the wide spread used of ceramic-metal joints. The common problem has been the large difference in thermal expansion characteristics of ceramics and metals that are of interest to structural applications. The stresses that arise after joining of many ceramic-metal combinations exceed the fracture strength of ceramic, particularly in regions close to free surface near the interface, usually resulting in either cracking of the ceramic or poor strength of the joints. As a result, functionally graded joints between ceramics and metals offer a solution to the thermal stress problem and have created wide interest recently [4, 5].

Recently, large-bulk fined-grained $TiC-TiB_2$ composites are achieved by a novel process called as combustion synthesis in high-gravity field [6, 7], combustion synthesis in high-gravity field is a new method for forming a layered composite of ceramic-metal in graded composition. In this paper, based on taking combustion synthesis in high-gravity field to prepare solidified $TiC-TiB_2$ composites, and by introducing the plates of stainless steel and Ti-6Al-4V alloy at the bottom of mixed reactive powder, the experiment is renewed to obtain the joint of the ceramic and the metal, so that a novel layered composite is achieved by joining the ceramic to the metal for the applications in the field of special engineering.

2. Experimental

Raw materials 30-63 μm in particle size were prepared from high purity (>99.9%) CrO_3 and Al powders, while purity (>97%) B_4C powder with particle size < 3 μm and high purity (>99%) Ti powder with particle size < 34 μm. The molar ratio (Ti to B_4C) of 3:1 was chosen as the starting composition, so the composition of the solidified $TiC-TiB_2$ composite was determined by Eq. 1. In order to obtain full-liquid reaction products in combustion reaction, the adiabatic temperature of the whole combustion system was designed as 3200°C, and the (CrO_3+Al) subsystem was added as the activators for increasing the adiabatic temperature according to Eq. 2. In order to protect the ceramic of $TiC-TiB_2$ liquid from the oxidation, CaF_2 powders (density of 3.18 g · cm^{-3} and a melting temperature of 1423 °C) were used as flux and were mixed with the thermit of (CrO_3+Al), the mass fraction ratio of the primary system ($Ti+B_4C+CrO_3+Al$) to the flux-mixed system (CaF_2+CrO_3+Al) was determined as 1:1.

$$3Ti+B_4C \rightarrow TiC+2TiB_2 \tag{1}$$

$$CrO_3+2Al \rightarrow Al_2O_3+Cr+1094kJ \tag{2}$$

After mixing the above powders, a Ti-6Al-4V alloy plates with 10 mm in thickness and 150 mm in diameter were putted at the bottom of the crucibles, then, the crucibles were filled with the reactive blends of the above powders at constant pressure and inserted into two combustion chambers at the end of the rotating arms of the centrifugal machine. The whole combustion system were firstly ignited with the W wire (diameter of 0.5 mm) at the top of the reactive blends after the centrifugal machine had provided a high-gravity acceleration in the combustion chambers greater than 500g ($g=9.8 m \cdot s^{-2}$). As the combustion chambers were cooled to ambient temperature, the centrifugal machine was stopped and the crucibles were taken out of the combustion chambers. Finally, TiC-TiB_2 ceramic of 4~5 mm in thickness on Ti-6A1-4V alloy substrate were obtained after the samples were taken out of the crucibles and Al_2O_3-CaF_2 flux layer at the top of the sample was eliminated by grinding.

3. Results and discussion

XRD results showed that the ceramic was composed of fine TiB_2, TiC, a few of Cr metallic binder and Al_2O_3 inclusions, as shown in figure 1. Hence, TiC and TiB_2 in ceramic prepared in current experiment are both close to their respective stoichiometry, which promises the achievement of stoichiometric TiC-TiB_2 by combustion synthesis in high-gravity field. FESEM images and EDS results showed a large number of randomly-orientated, fine TiB_2 platelets (presented by the dark areas in figure 2) were uniformly embedded in the irregular TiC grains (presented by the grey areas in figure 2) and Cr metallic phases (presented by the white areas in figure 2), and a few of black inclusions of α-Al_2O_3 were also observed, as shown by the arrow in figure 2. Physical and Mechanical properties showed that density, relative density, microhardness and fracture toughness of TiC-TiB_2 ceramic were 4.3 $g \cdot cm^{-3}$, 98.5%, 25.1 GPa and 14.5 ± 3.5 $MP \cdot m^{-0.5}$, respectively. Meanwhile, FESEM images of crack propagation paths showed that crack deflection and crack bridging are considered the main interaction mechanisms between the crack and TiB_2 platelets, resulting in the achievement of pull-out toughening by TiB_2 platelets, as shown by the arrow in figure 3. Thus, greater crack-opening displacement within the bridging zone is achieved, so the stress concentration around crack tip is relieved due to the presence of the closure stress behind the crack tip, resulting in greater resistance to crack propagation in the ceramic. Hence, high fracture toughness of TiC-TiB_2 ceramic results from the coupled toughening mechanisms of crack deflection, crack bridging and pull-out by TiB_2 platelets along with plastic deformation by Cr metallic binder.

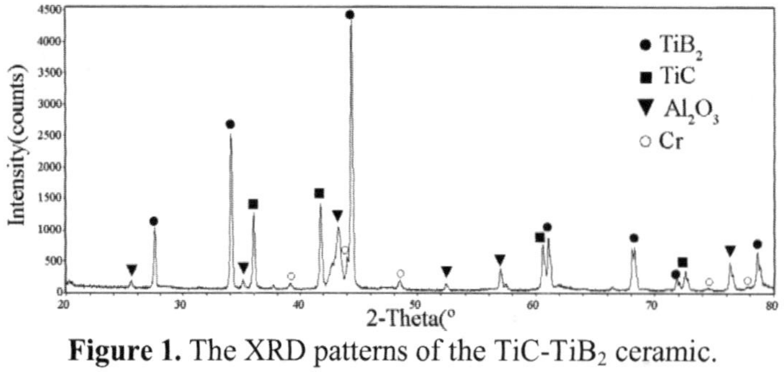

Figure 1. The XRD patterns of the TiC-TiB_2 ceramic.

According to the literature [6-8], the procedure of preparing bulk solidified TiC-TiB_2 composites by combustion synthesis in high-gravity field is described in brief as follows: by selecting (B_4C+Ti+C) as the primary system and the thermit of (CrO_3+Al) as the subsystem, and designing the adiabatic temperature above the melting point of all products, combustion synthesis was conducted in the presence of a high-gravity acceleration larger than 500g, finally, near-full-density TiC-TiB_2 composites with fine even ultrafine microstructures were achieved through rapid solidification of liquid TiC- TiB_2 ceramics. Moreover, it can be concluded that the new processing for preparing the solidified TiC-TiB_2 composites involves three stages: formation of the immiscible liquids consisting of Ti-Cr-C-B liquid and oxide liquid following thermit reaction and the decomposition of B_4C during the first stage, formation of the layered melt in high-gravity field due to density difference between the immiscible liquids of Ti-Cr-C-B liquid and oxide liquid during the second stage, and formation of TiC-TiB_2 solidified fine even ultrafine microstructures due to rapid coupled growth of TiC and TiB_2 phases from Ti-Cr-C-B liquid during third stage. Hence, similar to the preparation of bulk solidified TiC-TiB_2 composite, the formation of the ceramic on Ti alloy substrate is a result of the rapid deposition of TiC-TiB_2-Cr liquid, followed by rapid solidification of the liquid ceramic on the top surface of Ti alloy substrate,.

Figure 2. FESEM microstructure of TiC-TiB_2 ceramic.

Figure 3. FESEM image of crack propagation path at polished ceramic.

FESEM images showed that there was only the microstructure transformation rather than a clear joint interface between the ceramic and Ti alloy, as shown in figure 4. Meanwhile, compared with the ceramic microstructure far away from the joint, the microstructure of the ceramic nearby the joint was obviously refined, the ultrafine-grained microstructure characterized by the average thickness of fine platelets smaller than 1 μm was achieved, as shown in figure 4(b) and figure 4(c). In addition, a few of isolated Al_2O_3 inclusions were observed only at the area of the ceramic, as shown in figure 4(c).

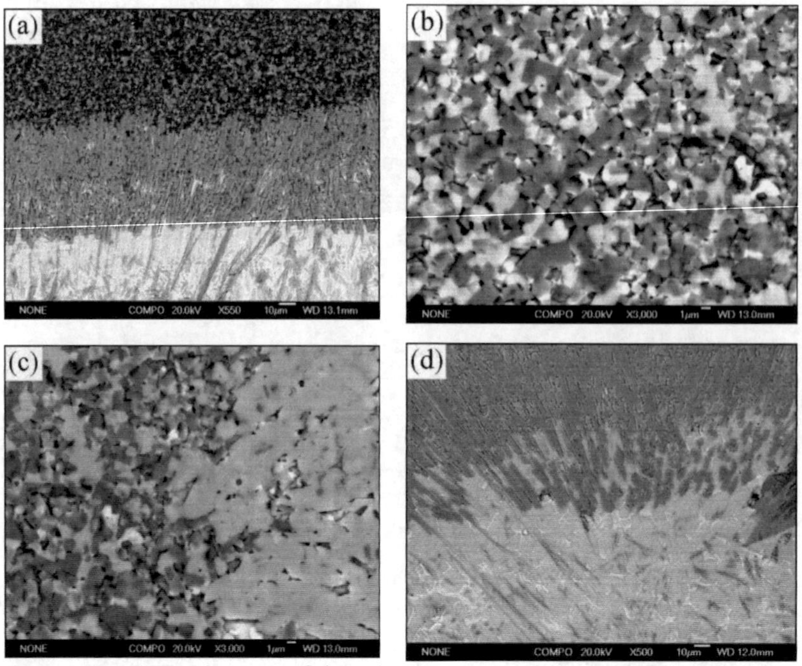

Figure 4. FESEM images of the microstructures from TiC-TiB_2 ceramic to Ti alloy substrate. (a) joint between the ceramic and Ti alloy substrate (b) the ceramic nearby the joint (c) joint area (d) Ti alloy substrate far away from the joint

The XRD patterns and FESEM images show that the joint area was composed of martensitic Ti, TiB, TiB_2 and Ti-enriched carbides, as shown in figure 5 and figure 6. A considerable amount of ultrafine hexagonal-pillar-shaped primary TiB grains (diameter of 0.5 to 1.0 μm and length of 0.5 to 1.5 μm) and fine spherical Ti-enriched carbides are distributed in the α'-Ti martensitic matrix, what's more, some coarse TiB_2 platelets are also obtained, as shown in figure 6.

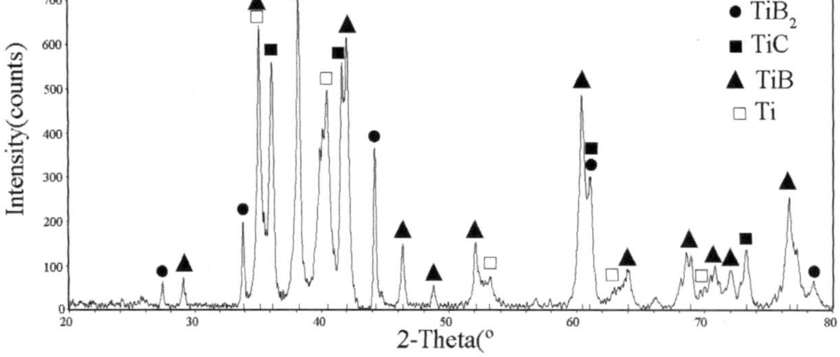

Figure 5. The XRD patterns of the fused joint area.

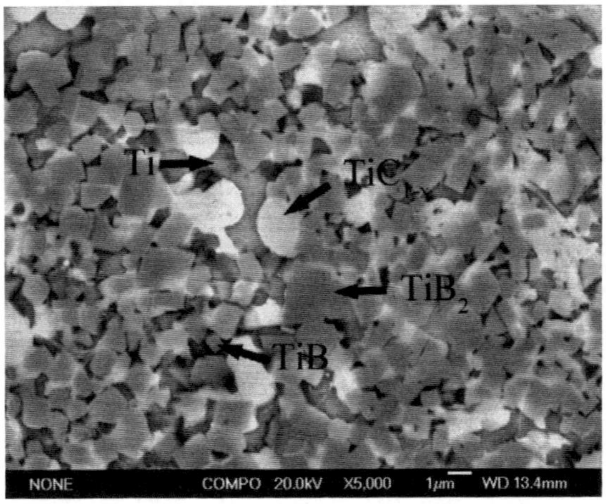

Figure 6. The FESEM image of fusion-induced joint between TiC-TiB_2 ceramic and Ti alloy.

Due to the instant release of high chemical energy of combustion system in high-gravity field, the surface of Ti alloy substrate at the bottom of the liquid TiC-TiB_2 is rapidly molten, while the introduction of high-gravity field makes the energy-release rate of chemical reaction increased, and combustion rate is so rapidly increased that the combustion characterization in high-gravity field is similar to one of thermal explosion, resulting in not only rapid escape of the gas in reactive blend during combustion reaction but also the formation of thermal vacuum circumstances around the products. Subsequently, C and B atoms in liquid TiC-TiB_2 ceramic rapidly diffuse into the molten Ti alloy substrate, while Ti atoms in the molten substrate of Ti alloy simultaneously diffuse into the nearby liquid TiC-TiB_2 ceramic. As a result, at initial stage of solidification a number of TiB_2 solids as the primary phases precipitate from TiC-TiB_2 liquid, followed by peritectic reaction (TiB_2 + Ti $\longrightarrow$ TiB) due to the presence of liquid Ti alloy, finally, a number of ultrafine TiB platelets (diameter of 0.5 to 1.0 μm and length of 0.5 to 1.5 μm) come to existence around the joint of ceramic to Ti alloy, resulting in the achievement of the ultrafine-grained microstructure in joint region and the ceramic nearby the joint, as shown in figure 4(b) and figure 6. Meanwhile, because of high diffusion rate of C relative to B in molten Ti alloy and the isotropy of TiC in crystallography, a number of Ti-enriched carbides develop at the joint area due to rapid growth of TiC crystals, moreover, some fine Ti-enriched carbides are also found at the area nearby the Ti solid alloy due to rapid atomic diffusion of C toward solid Ti substrate.

The high-magnification FESEM micrograph of the substrate and heat-affecting zone (HAZ) of Ti-6Al-4V alloy is shown in figure 7. The Ti-6Al-4V substrate consists of equiaxed α and intergranular β phases; however, the heat-affected zone of Ti-6Al-4V substrate nearby the joint area shows the presence of a number of needle-shaped grains, as shown in figure 4(d) and figure 7. XRD and EDS results further determined that those needle-shaped grains embedded in the Ti-6Al-4V substrate were in fact α'-Ti martensites. By considering the high heat conductivity of Ti alloy substrate, a part of α-Ti phases at Ti alloy substrate away from the joint area transform into β-Ti phases once the temperature at the area is somewhat higher than transformation temperature of $\alpha \rightarrow \beta$, as a result, some needle-shaped α'-Ti martensites inevitably come to existence in α-Ti matrix through martensitic transformation of β→α' due to rapid cooling of solid Ti alloy substrate, as shown in figure 7.

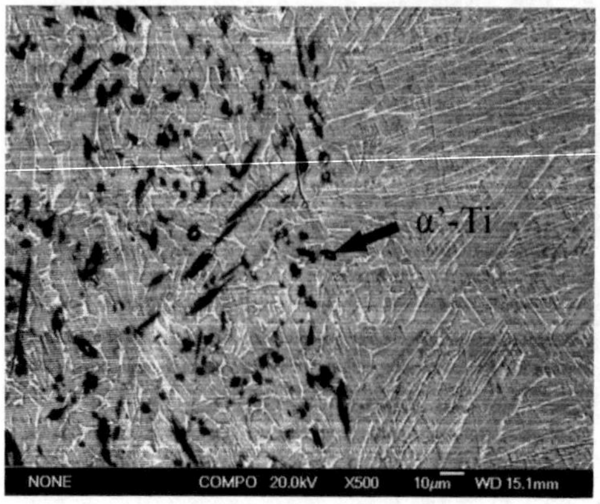

Figure 7. FESEM images of the HAZ.

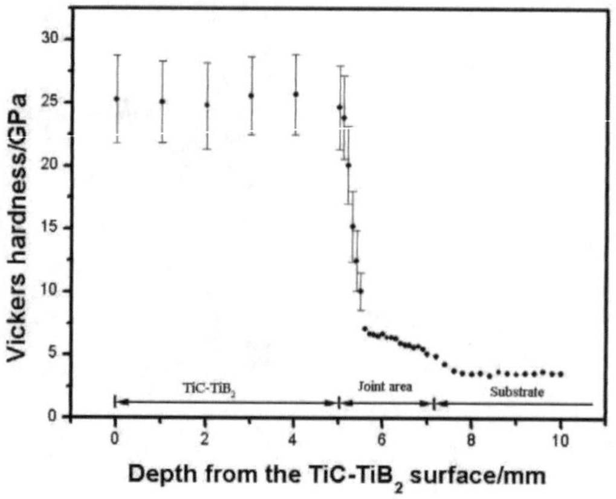

Figure 8. The dependence of the microhardness on the distance from ceramic to the Ti substrate.

The microhardness was measured from the ceramic to the substrate of Ti alloy, and the change in microhardness with the distance from the TiC-TiB_2 to the substrate was shown in figure 8. The maximum microhardness measured 25.1 ± 3.5 GPa at TiC-TiB_2, and high hardness of TiC-TiB_2 is considered to benefit from the achievement of fine-grained microstructure in near-full-density ceramic. The microhardness of the surface-alloyed substrate of Ti alloy measured 6.4 to 7.3 GPa, which was much lower than that of TiC-TiB_2 due to the presence of martensitic Ti, but presented a near-double increase over that that (about 3.2 to 3.5 GPa) of Ti alloy substrate, and the phenomenon is considered to be a results of the achievement of the surface-alloyed substrate of Ti alloy containing a number of hard TiC and TiB phases.

Meanwhile, the transitional change in microhardness was observed clearly from the ceramic to Ti alloy substrate, whereas shear strength of TiC-TiB_2 ceramic to Ti-6Al-4V substrate measured 450 ± 35 MPa. As discussed above, the ceramic on Ti alloy substrate is achieved by fusion bonding of liquid TiC-TiB_2 ceramic film and molten Ti-6Al-4V substrate to induce the surface alloying of Ti-6Al-4V substrate, thus, the joint of the ceramic to Ti alloy substrate is also achieved in continuously-graded composition by atomic interdiffusion between liquid film of TiC-TiB_2 and molten substrate of Ti alloy, resulting in the transition change in microhardness and high shear strength between the ceramic and the substrate of Ti alloy.

Results of the interlaminal shear strength tests showed that the fusion-induced combustion synthesis bonding in high-gravity field is a feasible and economical method to obtain joints of TiC-TiB_2 ceramic and Ti-6Al-4V alloy with useful strengths, comparable to some of those reported method for diffusion bonded specimens prepared at significantly higher temperature.

4. Conclusion

Based on preparation of high-performance solidified TiC-TiB_2 composite by utilizing combustion synthesis in high-gravity field, a novel available process for preparing TiC-TiB_2 ceramic on the substrate of Ti-6Al-4V alloy in continuously-graded composition has been achieved. The TiC-TiB_2 ceramic was composed of a number of fine TiB_2 platelets, irregular TiC grains and a few of Cr binder and Al_2O_3 inclusions, and physical, mechanical properties showed the density, relative density, microhardness and fracture toughness of TiC-TiB_2 ceramic reached 4.3 $g \cdot cm^{-3}$, 98.5%, 25.1 GPa and 14.5 ± 3.5 $MP \cdot m^{-0.5}$, respectively, and high fracture toughness of the ceramic benefited mainly from a coupled toughening mechanism of crack deflection, crack bridging and pull-out by fine TiB_2 platelets. XRD, FESEM and EDS results showed that there was a microstructure transformation between the ceramic and Ti alloy substrate, i.e. TiC-TiB_2 ceramic, surface-alloyed zone of Ti alloy, HAZ of Ti alloy and raw substrate of Ti-6Al-4V. Moreover, the high shear strength of 450 ± 25 MPa between TiC-TiB_2 ceramic and Ti-6Al-4V alloy was achieved.

5. Acknowledgement

This work is supported by National Natural Science Foundation of China (Grant No. 51072229).

References

[1] Locatelli M R, Dalgleish B J, Nakashima K, Tomsia A P and Glaeser A M 1997 *Ceram. Inter.* **123** 313
[2] Peteves S D, Paulasto M, Ceccone G and Stamos V 1998 *Acta mater.* **46** 2407
[3] Muraria A, Albrechtb H, Barzona A, Curiottoa S and Lottoa L 2003 *Vacuum* **68** 321
[4] Kakegawa K 1997 *Composites* B **28** 85
[5] Nishio S, Nakahira M, Abe T, Kawamura H, Sagawa S, Kawamura M, Yamazaki S 1996 *J. Nuc. Mater.* **233-237** 922-926
[6] Zhao Z, Zhang L, Song Y and Wang W 2009 *Scripta Mater.* **61** 281
[7] Huang X, Zhang L, Zhao Z and Yin C 2012 *Mater. Sci. Eng.* A **553** 105-111
[8] Vallauri D, Atlas I C and Adrian A 2008 *J. Eur. Ceram. Soc.* **28** 1697

Microstructure and properties of (Ti, Al) N coatings on WC-Co cemented carbide

Yudong Li, Zhen Liu, Jing Yang and Yigang Chen*

Department of Electronic Information Materials, School of Materials Science and Engineering, Shanghai University, 149 Yanchang Road, Shanghai 200072, China

E-mail: yigangchen@shu.edu.cn

Abstract. (Ti, Al)N films were grown on WC-Co cemented carbides by R.F. reactive sputtering using Ti/Al composite target. The crystalline structure, morphology and composition of (Ti, Al)N coatings were investigated by X-ray diffraction (XRD), optical microscope and X-ray energy dispersive spectroscope (EDS), respectively. The mechanical properties of (Ti, Al)N coatings with the variable Al content were investigated by applying nanoindentation. With the increase of the Al content, the crystal structure of (Ti, Al)N films changed from fcc to hexagonal structure. When the ratio of Ti to Al in (Ti, Al)N coatings was approximately 4:1, and the N content was 40.67 %, the nano-hardness of (Ti, Al)N coating was highly up to 28 GPa. By adjusting the Al content, the lowest value of friction coefficients was 0.1397.
Keywords: (Ti, Al)N; coating; nano-hardness; friction coefficient

1. Introduction

Ceramic hard coatings are commonly applied as protective materials on tools and mechanical components to enhance their performances in combination with extension of lifetime and decrease of wear. Transition metal nitride coatings are commonly used as protective coatings because of their high hardness, wear resistance, and chemical stability [1–7]. Among these materials, binary nitrides such as TiN are the most widely studied [8]. TiN films have wide ranging applications from erosion-resistant coatings for cutting tools [9,10] to diffusion barriers used in the construction of semiconductors devices [11,12]. One disadvantage of TiN in some applications is that it oxidizes easily in air at above 500°C. Recently, ternary nitride(Ti, Al)N was developed to provide better performance over TiN since the addition of Al in the form of solid solution increased hardness, decreased coefficient of friction and improved the oxidation resistance of the coating and cutting performance [13,14,15].

The purpose of this paper is to investigate the phase structure, morphology, composition, and the mechanical properties of the (Ti, Al)N films prepared by R.F. reactive sputtering, using composite targets, as a function of the ratio Ti to Al of the targets.

2. Experiment

WC-Co cemented carbide was used as the substrate whose surface was polished with SiC micro powder and diamond grinding cream. Then the substrate was ultrasonically cleaned in acetone, ethanol and deionized water in turn for two minutes per step. Finally, the substrates were dried with the high pure nitrogen.

(Ti, Al)N films were prepared by R.F. reactive sputtering using a series of Ti/Al composite targets (diameter of 118 mm, thickness of 5 mm and purity of 99.99%). The ratio of Ti to Al of the targets are 10:0, 9:1, 8:2, 7:3, 6:4 and 5:5, which were marked T1,T2,T3,T4,T5,T6, respectively.

Table 1 the conditions for (Ti, Al)N deposition process

Material	(Ti, Al)N
Target	Ti/Al
Substrate	WC-Co cemented
Substrate temperature	200°C
Power	350W
Gas contents	$Ar:N_2=1:2$
Sputtering pressure	0.6Pa
Deposition time	180min
Base pressure	5.0×10^{-3}Pa

Prior to the deposition, the chamber evacuated to a base pressure of 5×10^{-3}Pa, substrate heated to 200°C, then high purity argon was introduced into the chamber so that the pressure could be 0.4 Pa to clean the target and the chamber for 20 minutes by argon plasma. After cleaning, high purity argon (99.999%) and nitrogen (99.999%) gases were introduced. The (Ti, Al)N film was deposited for 180 minutes at a working pressure of 0.6Pa and the power of 350W. The deposition parameters are listed in Table 1. The crystalline structure, morphology and composition of (Ti, Al)N coatings were investigated by X-ray diffraction (XRD), optical microscope and X-ray energy dispersive spectroscope (EDS) using Cu Kα radiation at a condition of 40 kV and 100 mA. The nano-hardness and Young's modulus of the samples were determined by nanoindentor.

3. Results and discussion

3.1. Structure and surface morphology

X-ray diffraction investigations of coatings deposited onto cemented carbides using various targets are shown in Fig.1. Figure 1 shows that the (Ti, Al)N film was deposited with the TiN face-centered cubic (fcc) structure and AlN hexagonal structure. (111) plane is the close-packed plane of TiN with fcc structure. Growth along the close-packed plane has the lowest surface energy. So fcc structure was the main phase of the (Ti, Al)N film using T1 target. The introduction of the Al changes the crystal planes and also causes lattice distortion that induces distortion energy changing the preferred orientation. With the increase of the Al content, (111) orientation of (Ti, Al)N films is gradually weakened, while the (200) orientation have been enhanced [16, 17]. Hexagonal structure (002) appears using T3 target. With the Al content increasing, the (002) diffraction intensity was gradually increased, and (Ti, Al)N film structure changed from fcc to hexagonal structure, consistent with Ikeda and Satoh findings [18].

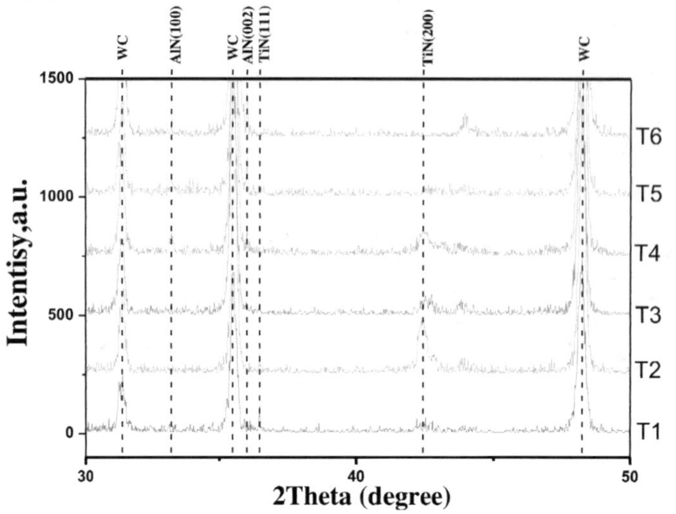

Fig 1.XRD patterns of the (Ti, Al) N films deposited with different targets

The orientation of the deposited film is the result of the interfacial energy, the strain energy and the surface dominant factors affect energy. When the Al content of target is low, fcc (111) plane of preferential growth and the film orientation at this time is dominated by the interfacial energy between the film and the substrate. (111) plane parallels to the substrates surface to reduce surface energy of the film. The higher the Al content is, the more serious the fcc phase lattice distortion is and the higher strain energy that produces. When the strain energy reaches a certain value, the lattice structure will shift to reduce the strain energy. Hexagonal phase appears to correspond to the transition point by using T4 target.

The morphology of the film has been observed by Metallurgical Microscope. Figure 2 presents the surface topography of the sample. There is no obvious defect on relatively flat surface of film from Figure A. The adhesion between the film and the substrate at the part edge of the sample is not good from Figure B which possibly is caused by defects and stress of the boundary.

3.2. element composition

Elements concentration (at. %) in (Ti, Al)N coatings as a function of the Al content of Ti/Al target taken by using EDS analysis is shown in Fig.3. The Al content increased slightly from 0 to 48 at. %; the Ti content decreased from 53 to 18 at. %; the N content gradually decreased from 47 to 33 at. %.The Al to Ti atom ratio in (Ti, Al)N films quickly increased from 0 to 2.7 with using T1 to T6 target. The Al content of the target increased from 0 to 50 at. %; meanwhile, the Ti content of the targets decreased from 100 to 50 at. %. Therefore the Al content of the (Ti, Al)N films gradually increased, otherwise the Ti content slowly decreased. While the proportion of Al and Ti of (Ti, Al)N film was not the same with the target. The Al content of(Ti, Al)N film was much higher than one of the target; on the contrary, the Ti was lower. According to the usual sputtering theory, the sputtering coefficient of Al is greater than Ti at the same condition. When the energy of the argon ions was 500eV, the sputtering coefficient ratio of the Al to Ti is 2.01 [19].

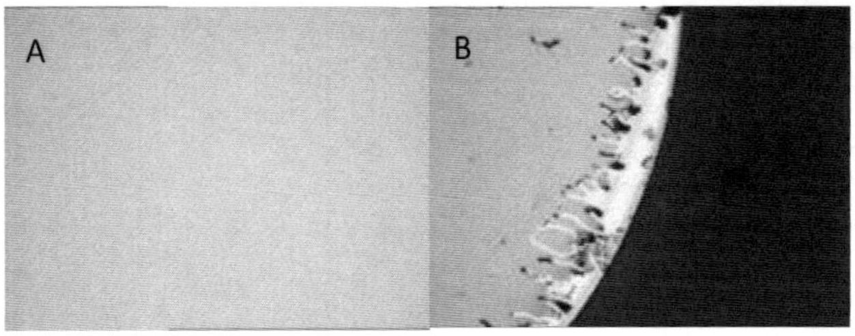

Fig 2. Optical microscopy images of sample surface topography 500 map

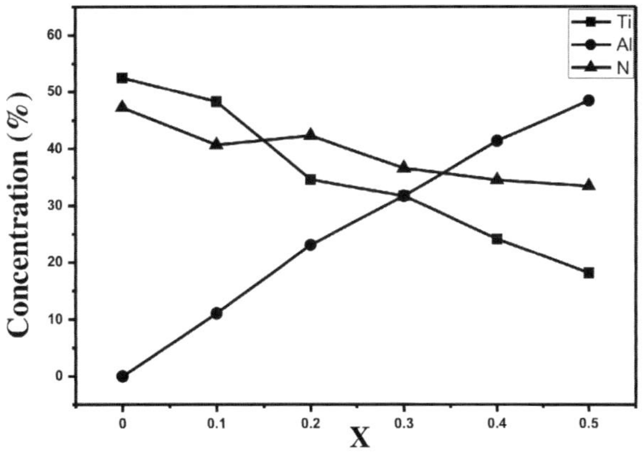

Fig 3. Elemental concentration (at. %) of (Ti, Al) N coatings
as a function of Al fraction x of Ti/Al target

In reactive sputtering, since the partial pressure of the reactive gas varied, it made the respective components of the composite target forming compounds. The formation heat of TiN (–305.6 kJ/mol) is lower than that of AlN (-241 kJ/mol) [20]. The N ions bombarded easily Ti on target surface with the low formation heat to form a thin nitride layer. Usually metallic state sputtering is greater than the non-metallic state sputtering. Therefore, in low reactive nitrogen, Ti and Al atoms sputtering from targets reacting with nitrogen generated (Ti, Al)N film. Another part cannot participate in the reaction existed in the form of metal amorphous of the (Ti, Al)N film.

3.3. Mechanical properties

The nano-hardness and Young's modulus of the (Ti, Al)N coatings deposited onto the WC-Co cemented substrates with different targets are shown in Fig. 4. The mechanical properties of (Ti, Al)N films were analyzed by Nanoindentation which is a very powerful technique [21-26]. In order to minimize the substrate effects, the maximum indentation force was limited to 5000μN so that the indentation depth was less than 10% of the film thickness. Three indentations at different places on the samples surface were performed and the values averaged were assigned as the nano-hardness and Young's modulus of the (Ti, Al)N films.

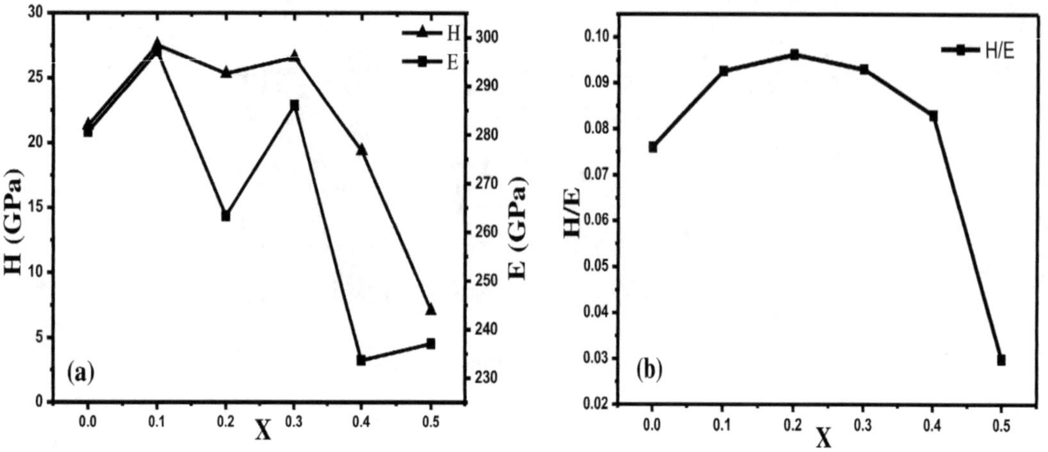

Fig 4. (a)Nano-hardness, Young's modulus, (b) H/E of (Ti, Al) N coatings as a function of Al fraction x of Ti/Al target

The maximum value of the nano-hardness occurred at using T1 target while the modulus at using T3 target. The maximum was approximately 28 GPa and 297 GPa, respectively. They both increased, firstly. With the further increased of the Al content of the targets, the nano-hardness values decreased. The value of Young's modulus was fluctuation, while also decreased in the whole. Vlasveld et al. [27] reported the highest nano-hardness of the (Ti, Al)N film was 34 Gpa using partially filtered arc technique. Zywitzki et al. [28] reported the highest Nano-hardness of (Ti, Al)N film was 38 Gpa deposited by reactive pulsed magnetron sputtering.

The hardness of (Ti, Al)N films strongly depends on the Al content. This may be ascribed to a solid-solution strengthening mechanism. Since the Al atomic radius (R_{Al} = 1.43 Å) is smaller than the Ti atomic radius (R_{Ti} = 1.46 Å), it is possible that the Al atom goes entirely into the fcc TiN cell, substituting part the Ti atom at a face-centered position , forming fcc (Ti, Al)N phase. With the A1 content increasing of (Ti, Al)N film, the vibration of each atom in the fcc cell changes by atom force. The position of each atom vibrating inwardly offset. That eventually leads to a change of the lattice constant to reduce the energy of the system. The more the Al content is, the smaller the lattice constant is. As described above, since the introduction of the Al atoms, the crystal structure was distorted, which resulted in the increase of the hardness of the film.

The nano-hardness value of Nitrides is relative directly with the key and the atom space, but also by the influence of the valence electron density in the unit cell (VEC) [29]. The VEC value is related to the composition of the coating and the vacancy concentration [30]. With the increase of the Al content in the coating, more and more Ti ions (Ti^{4+}) were replaced by Al ions (Al^{3+}). The value of VEC tended to decrease, which caused a higher hardness value. When the Al content exceeded a certain value, the hardness of the (Ti, Al)N coatings reduced for the crystal gradually disappearing and Al supersaturating.

The N content in (Ti, Al)N films was gradually decreased from 47% to 33%. In low nitrogen, the part Ti and Al atoms sputtering from targets cannot participate in the reaction existed in (Ti, Al)N film in the form of amorphous metal, which reduced the nano-hardness of the film. All these factors above could explain the reason of the hardness of (Ti, Al)N film changing.

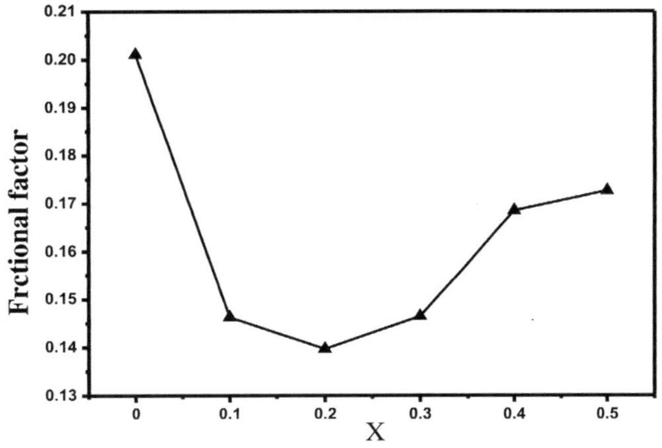

Fig.5 The friction coefficient of (Ti, Al) N coatings as a function of Al fraction x of Ti/Al target

The friction coefficient of (Ti, Al)N films deposited at various targets was illustrated in Fig.6. The lowest friction coefficient was 0.2011 at using T1 target. As introduction of the Al, friction coefficient decreased quickly to 0.1463. In other words, the friction coefficient of (Ti, Al)N was much lower than that of TiN. With further increase of the Al content, friction coefficient increased again. The reason of lower friction coefficient was probably ascribed to the better surface morphology of the (Ti, Al)N films. Wu et al. [31] reported the friction coefficient of the (Ti, Al)N films ranged from 0.2 to 0.3 by using unbalanced-magnetron sputter technique.

4. Conclusion

In this paper, the (Ti, Al)N films were deposited by R.F. reactive sputtering using Ti/Al composition target. The basic characteristics of the (Ti, Al)N films deposited under different targets were investigated systematically. With the increase of the Al content of the target, the crystal structure of the films changed from fcc to hexagonal structure. The Al content had great influence on the mechanical properties of the (Ti, Al)N films, such as hardness, friction coefficient. When the ratio of Ti and Al of the (Ti, Al)N coating was approximately 4:1 and the N content was 40.67 %, the nano-hardness of the (Ti, Al)N coating was highest. By adjusting the Al content, the highest values of the nano-hardness and friction coefficient could be 28 GPa and 0.1397, respectively.

References

[1] Sundgren J E, Johansson B O, Karlasson S E and Hentzell H T G 1983 *Thin Sol. Films* **105** 367-384

[2] Echigoya J, Liu Z T, Imamura A and Takatsu S 1991 *Thin Sol. Films* **198** 293-300
[3] He J W, Bai D D, Xu K W, Hu N S 1995 *Surf. Coat. Technol.* **74-75** 387-393
[4] Niederhofer A, Nesladek P, Mannling H D, Moto K, Veprek S and Jílek M 1999 *Surf. Coat. Technol.***120-121** 173-178
[5] Harris S G, Doyle E D, Vlasveld A C, Audy J and Quick D, 2003 *Wear* **254** 723-734
[6] Barshilia H C, Prakash M S, Jain A and Rajam K S 2005 *Vacuum* **77** 169-179
[7] Shi Y J, Long S Y, Fang L,Yang S C and Pan F 2008 *Appl. Surf. Sci.* **254** 5861-67
[8] Zhitomirsky V N, Grimberg I and Rapoport L 2000 *Surf. Coat. Technol.* **133-134** 114–120
[9] Auner G, Hsieh Y F and Padmanabhan K R 1983 *J. Vat. Sci. Technoi.* A **l** 275-278
[10] Sproul W D 1983 *Thin Sol. Films* **107** 141-147
[11] Hara T, Yamanoue A, Iio H, Inoue K, Wahiszu G and Nakamura S 1991 *Jpn. J. Appl. Phys.* **30** 1447-51
[12] Jin P and Maruno S 1992 *Jpn. J. Appl. Phys.* **31** 1446-52
[13] PalDey S and Deevi S C 2003 *Mater. Sci. Eng.* A **342** 58–79
[14] Knotek O, Bohmer M and Leyendecker T 1986 *J. Vat. Sci. Technol.* A **4** 2695-2700
[15] Mtinz W D 1986 *J. Vat. Sci. Technol.* A **4** 2717-25
[16] Yoon S Y, Kim J K and Kim K H 2002 *Surf. Coat. Technol.***161** 237-242
[17] LII D F 1998 *J Mater Sci* **33** 2137-45
[18] Ohnuma H, Nihira N, Mitsuo A, Toyoda K and Kubota K 2004 *Surf. Coat. Technol.* **177-17** 623-626
[19] Man B Y, Guzman L, Miotello A and Adami M 2004 *Surf. Coat. Technol.* **180/181** 9-14
[20] Mientus R and Ellmer K 1999 *Surf. Coat. Technol* **116-119** 1093-1101
[21] Jian S R, Chen G J, Jang Jason S C and Lai Y S 2010 *J. Alloy. Compd.***494** 219–222
[22] Jose F, Ramaseshan R, Dash S, Rajagopalan S and Tyagi A K 2009 *Appl. Ceram. Technol.* 243–248
[23] Yonenaga I, Shima T and Sluiter M 2002 *Jpn. J. Appl. Phys.***41** 4620–21
[24] Jian S R, Chen G J and Lin T C 2010 *Nanoscale Res Lett* **5** 935–940
[25] Li T S, Li H and Pan F 2001 *Surf. Coat. Technol.* **137** 225-229
[26] Phillips M A, Clemens B M and Nix W D 2003 *Acta Mater.***51** 3171-84
[27] Vlasveld A C, Harris S G and Doyle E D 2002 *Surf. Coat. Technol.***149** 217–224
[28] Zywitzki O, Klostermann H, Fietzke F and Modes T 2006 *Surf. Coat. Technol.* **200** 6522–26
[29] Roos J R, Celis J P, Vancoille E, Veltrop H, Boelens S, Jungblut F, Ebbrink J and Homberg H 1990 *Thin Sol. Films* **193-194** 547-556
[30] Konig U 1987 *Surf. Coat. Technol.* **33** 91-103
[31] Wu S K, Lin H C and Liu P L 2000 *Surf. Coat. Technol.* **124** 97–103

Fabrication and mechanical properties of a 14Cr-ODS steel

Helong Hu, Zhangjian Zhou*, Lu Liao, Man Wang, Shaofu Li

School of Materials Science and Engineering, University of Science and Technology Beijing, China 100083

* Corresponding Author: zhouzhj@mater.ustb.edu.cn

Abstract: In this paper, the fabrication method and the mechanical properties of a 14Cr-ODS steel are investigated. The pre-alloyed steel powders and oxide powders were mechanical alloyed after milling for 30 h under a rotation of 300 rpm. Y-Ti-O phase was obtained after annealing the MA powders at 1100 ℃. The tensile strength of the HIPed 14Cr-ODS steels at room temperature was about 1120 MPa. Microstructure observation showed that there were amount of nano dispersion particles in the base alloy which resulted in the high hardness and tensile strength.

1. Introduction

Based on safety and efficiency, the Generation IV forum has selected six innovative concepts for nuclear reactors. SCWR (Supercritical Water Reactor), which offered many advantages including the use of a single phase coolant with high enthalpy, the elimination of components and a much higher efficiency, was considered as a promising design and frequently investigated by researchers [1-3].

The design for the SCWR called for advanced structural materials. The main candidate structural materials for the SCWR are Ni-based alloys, ODS steels, ferritic/martensitic and austenitic steels. Ferritic steel is well known to have excellent irradiation resistance due to it's bcc structure [4-5]. However, its high temperature strength needs improvements. Nano oxide particle was added into the ferritic matrix to improve the high temperature strength and irradiation resistance properties, and ODS steels was developed [6].

The ODS steels is usually fabricated in a method of mechanical alloying due to its low cost and facile realization. Meanwhile, consolidation of MA powders by hot isostatic pressing (hipping) can avoid the anisotropy of the microstructure, compared with hot extrude. In this paper a 14Cr-ODS steels was fabricated, and the properties and the microstructure were investigated.

2. Experimental

Pre-alloyed 14Cr ferritic steel powders, Ti and Y_2O_3 were used as starting materials. The composition of the mixed powders was designed as Fe-14Cr-1W-0.2Si-0.5Ti-0.35Y_2O_3, as shown in table 1.

Table 1 Chemical composition of powders, in weight percent

Materials	C	Si	Cr	W	Ti	Y_2O_3	Fe
14CrNA	0.06	0.24	14.20	1.80	0.09	—	Bal.
14CrODS	0.06	0.24	14.20	1.80	0.50	0.35	Bal.

The starting materials were mechanical alloyed in a nitrogen atmosphere using a ball mill at 300 rpm for 30 h. The ball-to-powder mass ratio was 5:1. The mechanical alloyed powders were then consolidated by hot isostatic pressing at a temperature of 1150 ℃ under a pressure of 120 MPa for 3 h. The powders before and after milling were investigated using a scanning electron microscope (SEM, LEO-1450) and X-ray diffraction (XRD, D/MAX-RB). Microstructure and chemical composition were characterized by Transmission Electron Microscope (TEM, JEM-2010) with an energy dispersive spectroscope (EDS). Samples for TEM were prepared by mechanical polishing and electrochemical thinning using 7% perchloric acid + 93% ethanol as electrolyte. Tensile tests were carried out at room temperature and 700 ℃ with a strain rate of 5×10^{-2} s^{-1}.

3. Results and discussion

3.1 Precipitated phase

The contents of oxygen induced in the process of mechanical alloying played an important role in the formation of nano size particles. The size of the nano particle increased as the introduction of the excess oxygen. The nano particle related with the excess oxygen played a negative role in the densification and mechanical properties of the samples. The oxygen content of the pre-alloyed powders was about 0.069%. The oxygen content of the ODS powders was about 0.21%. The amount of oxygen induce in the ball milling process was about 0.067%.

Fig. 1 shows the XRD patterns of the pre-alloyed powders, mixed powders before and after mechanical alloying, as well as ODS powders after annealing. It is concluded from the figure that the main phase for the pre-alloyed powders is iron. There are Ti and Y_2O_3 peaks in addition to iron peak for the mixed powders. However, the Ti and Y_2O_3 peaks disappeared after ball milling for 30 h. Thus, ODS powders can be fabricated using mechanical alloying.

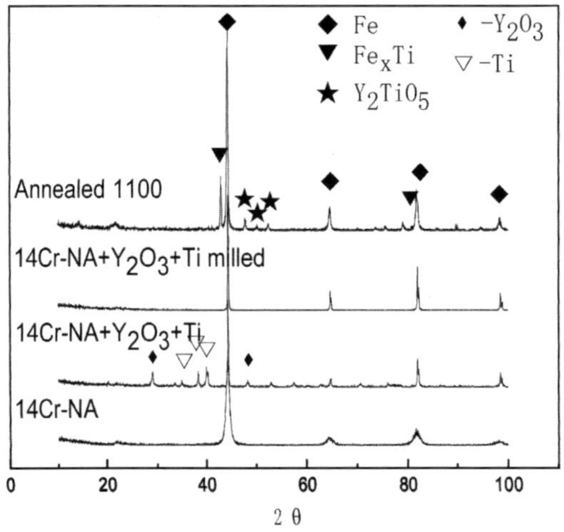

Fig. 1 XRD patterns for the powders

It can be concluded from patterns of the annealed samples that Y-Ti-O phase formed after high temperature annealing. This process can be illustrated as follows: Y_2O_3 which was added into the pre-alloyed powders decomposed to Y and O atoms. Y and O resolved in the base alloy. When the mechanical alloyed powders were annealed and consolidated at high temperature, Y-Ti-O particles formed with a smaller size, and the particles dispersed uniformly.

3.2 Tensile strength

Mechanical alloying is a nonequilibium process and makes pores forming between the powders. In this situation, the solubility of the gas in the mechanical alloyed powders is higher than that of ordinary powders. The gas in pores of the powders is hard to eliminate, while that between the powders are easy to eliminate at high temperature. Thus, the sample with high density consolidated at high temperature is possible to obtain. The theoretical density of 14Cr-ODS steel was about 7.77 kg/m^3, while the actual density was about 7.58 kg/m^3. Thus, the relative density of sample after consolidation was about 97.6%.

The operating condition in the reactor was very complex. It is unreasonable to acquire high strength or good ductility. Thus, it is rational to evaluate both strength and ductility at high temperature and room temperature. The tensile strength for the samples was tested at room temperature and 700 ℃.

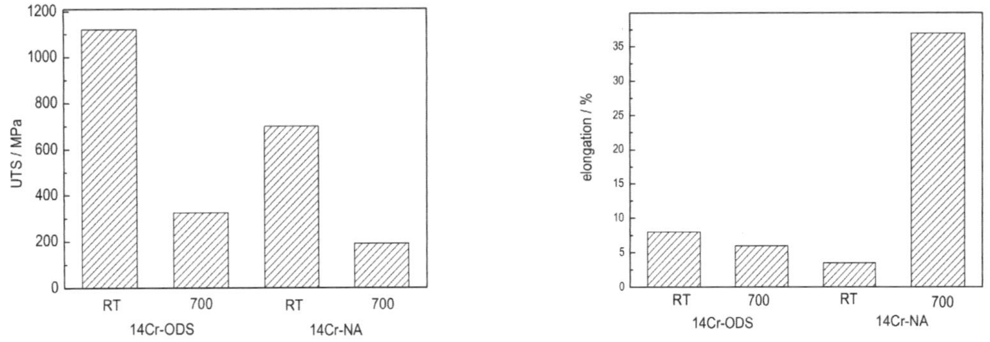

Fig. 2 Tensile properties for the samples of 14Cr-ODS and 14Cr-NA

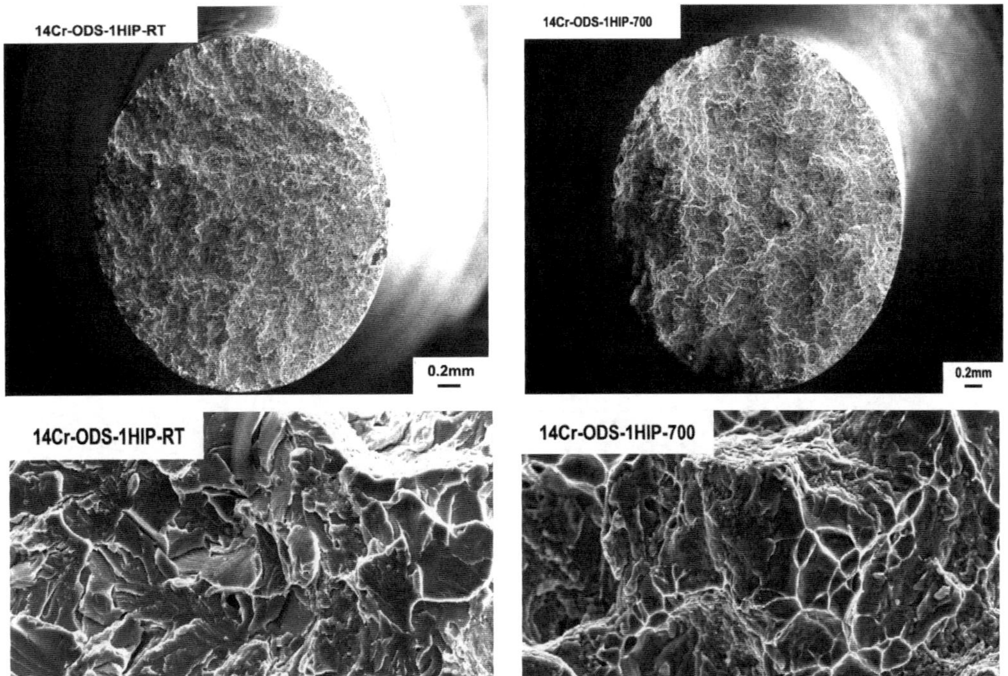

Fig.3 Fracture morphologies for the samples of room temperature and 700 ℃

The test results are shown in Fig. 2. UTS for the sample at room temperature and 700 ℃ are 1120 MPa and 325 MPa. UTS and yield strength decreased as the increasing temperature. At both temperatures, the UTS and yield strength of the ODS steel was higher than that of samples without Y_2O_3 addition. Especially at high temperature, UTS of the 14Cr-ODS steel was about twice of 14Cr-NA steel. It means that the addition of Y_2O_3 strengthens the ODS steels. Thus, the ODS steels exhibited excellent high temperature strength.

The fracture surface morphologies for the samples at high temperature and room temperature are shown in Fig. 3. There is almost no necking for both samples. Amount of cleavage surface was observed for the sample at room temperature, as well as river pattern and cleavage steps. Thus, it is typical brittle fracture for the sample at room temperature. However, there are dimple surface for the sample at high temperature. The sizes and distributions of the dimples are not evenly. This phenomenon is in related to the precipitation and nonmetallic inclusion. The hardness of the 14Cr-ODS steel and 14Cr-NA steel is shown in Table. The results show that the hardness of the 14Cr-ODS steel is higher than that of the 14Cr-NA steel. This is in accordance with that of the tensile strength.

Table 2 Hardness for the sample of 14Cr-ODS and 14Cr-NA

Samples	Hardness/HV
14Cr-NA	259.8
14Cr-ODS	373.7

3.3 Microstructure

The metallographic morphologies of the 14Cr-ODS steel and 14Cr-NA steel are shown in Fig. 4. A large number of boundaries were observed for the 14Cr-NA sample. The grain size was comparably with the size of the pre-alloyed powders. This means the grain boundary was the boundaries between the powders. The grain size for the samples of 14Cr-ODS steel is smaller than that of 14Cr-NA.

In order to investigate the strengthen mechanism of the nano particles, microstructure and the distribution of the nano particles were observed using TEM, which is shown in Fig. 5. It can be concluded from the figures that nano particles distributed evenly and the size of the particles was about several nano meters. Particles with a size of 20~30 nm was also observed. The chemical composition of the nano particles was indentified using EDS. Because the particle is too small, the chemical composition can only be indentified roughly. The results showed the particles might be nano Y-Ti-O particles.

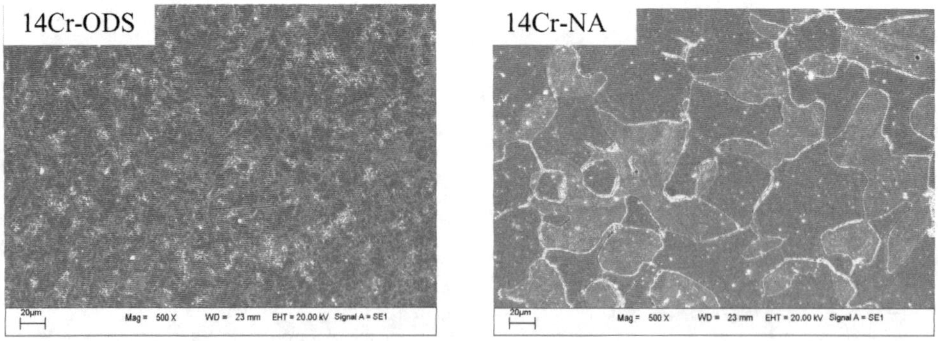

Fig.4 Metallographic morphologies for the samples 14Cr-ODS and 14Cr-NA

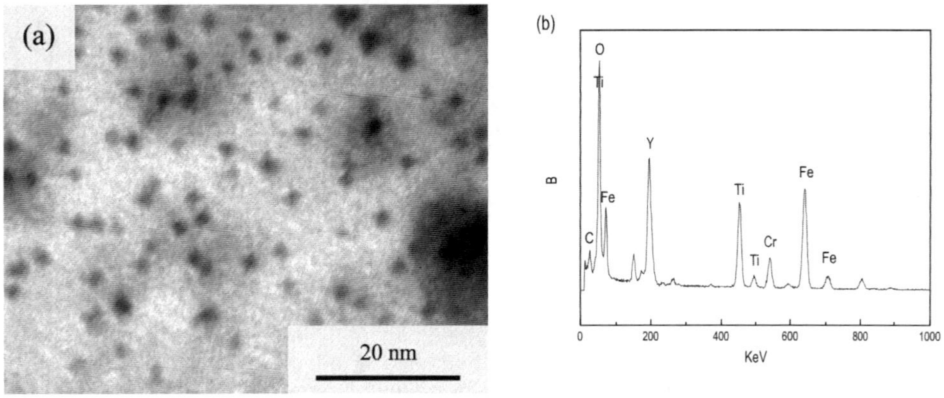

Fig. 5 Microstructure and nano particle photos for 14Cr-ODS

4. Conclusion

Homogeneous ODS powders can be fabricated after mechanical alloying for 30 h with a speed of 380 rpm. Y-Ti-O phase precipitated after annealing the MA powders at 1100 ℃ for 1 h.

The tensile strength of the 14Cr-ODS steel at room temperature and 700 ℃ was higher than those of 14Cr-NA steel, 1120 MPa and 325 MPa respectively.

Nano particles distributed evenly in the base alloy. EDS results show that the particles should be nano Y-Ti-O particles.

Reference

[1]. G.S. Was, S. Teysseyre, Challenges and recent progress in corrosion and stress corosion cracking of alloys for supercritical water reactor core components. 12th International Conference on Environmental Degradation of Materials in Nuclear Power System –Water reactors. Salt Lake city, United states. Aug.14-18, 2005.

[2]. G.S. Was, P. Ampornrat, G. Gupta, S. Teysseyre, E. A. West, T. R. Allen, K. Sridharan, L. Tan, Y. Chen, X. Ren, C. Pister. Corrosion and stress corrosion cracking in supercritical water, J. Nucl. Mater. 371 (2007) 176-201.

[3]. L. Tan, T.R. Allen, Y. Yang, Corrosion behavior of alloy 800H(Fe-21Cr-32Ni) in supercritical water, Corros. Sci. 53(2) (2011) 703-711.

[4]. S. Ukai, M. Harada, H. Okada, M. Inoue, S. Nomura, S.Shikakura, K. Asabe, T. Nishida, M. Fujiwara,. Alloying design of oxide dispersion strengthened ferritic steel for long life FBRs core materials, J. Nucl. Mater. 204 (1993) 65-73.

[5]. S. Ukai, M. Harada, H. Okada, M. Inoue, S. Nomura, S.Shikakura, T. Nishida, M. Fujiwara, K. sabe, Tube manufacturing and mechanical properties of oxide dispersion strengthened ferritic steel, J. Nucl. Mater. 204 (1993) 74-80..

[6]. T. Kaito, T. Narita, S. Ukai, High temperature oxidation behavior of ODS steels, J. Nucl. Mater. 329-333 (2004) 1388-1392.

Preparation and Oxidation of $ZrB_2/SiC/Zr_2Al_4C_5$ Multi-phase Ceramics with Spark Plasma Sintering

Qilong Guo[1], Junguo Li[1]*, Zhiyu Ma[2], Ye Nie[2], Qiang Shen[1], and Lianmeng Zhang[1]

[1]State Key Lab of Advanced Technology for Materials Synthesis and Processing, Wuhan University of Technology, Wuhan 430070, P. R. China

[2]Sinoma International Engineering CO.,Ltd.(Nanjing),106 Jiangjun avenue, Nanjing Jiangsu 211100, China

e-mail address: guoqilongfgm@whut.edu.cn

Abstract. The $ZrB_2/SiC/Zr_2Al_4C_5$ multi-phase ceramics were fabricated by spark plasma sintering (SPS) at 1800 °C for 3 min under 20 MPa in an vacuum. Oxidation behavior of multi-phase ceramics were investigated using thermo gravimetric analysis (TGA) from 20 °C to 1500 °C and muffle furnace in stagnant air at 1200 °C. Samples were analyzed after oxidation by X-ray diffraction (XRD), scanning electron microscopy (SEM) along with energy dispersive spectroscopy (EDS) to determine the reaction products and to observe the microstructure. The results showed that the aluminium borate and mullite crystallize on the surface in the samples oxidized. The effect of $Zr_2Al_4C_5$ content on the oxidation resistance of the ZrB_2 ceramics were discussed respectively, and oxidation mechanism was also analysised.

1. Introduction

Zirconium diboride (ZrB_2) as a member of ultrahigh-temperature ceramics (UHTCs) is currently expected to be a potential candidate material for aeropace applications, owing to its high melting point, high thermal and electrical conductivity, low theoretical density and good chemical inertness [1, 2]. Based on these properties, ZrB_2 ceramics are attracting more and more attention in ultra-high temperature applications where corrosion-wear-oxidation resistance is demanded [3].

However, oxidation resistance is a well-known bottleneck in the exploration of ZrB_2-based ceramics. Many studies reported that the addition of SiC could improve the oxidation resistance and fracture strength of the ZrB_2-based composites [4], and the oxidation of ZrB_2 can be improved by adding SiC to promote the formation of a silica-rich scale on the exposed surface. Other development had been also achieved on the oxidation behavior of ZrB_2-based ceramics with addition of $MoSi_2$, LaB_6, TaB_2 and $TaSi_2$ [5-7].

Our previous work has revealed [8] that addition of $Zr_2Al_4C_5$ was favorable to the densification and especially the toughness of ZrB_2-based ceramics. This paper focused on the oxidation behavior of $Zr_2Al_4C_5$-toughened ZrB_2 and ZrB_2-SiC ceramics using muffle furnace in air. By means of the investigation of microstructure and mass change after oxidation, effect of $Zr_2Al_4C_5$ content on the oxidation resistance of ZrB_2/SiC/$Zr_2Al_4C_5$ ceramics was newly explored.

2. Experimental procedures

Commercially available ZrB_2 powders (>99.5%, 2 μm), Zr powders (>99.9%, 10μm), Al powders (>99.99%, 8μm) and graphite powders (>99.9%, 1μm) powders were used as starting materials. $Zr_2Al_4C_5$ was prepared by the spark plasma sintering-reactive synthesis using Zr, Al, and C powders as sintering materials. The ZrB_2 and $Zr_2Al_4C_5$ powders were mixed in an agate mortar, and put into a graphite die with graphite foil and sintered using a SPS system.

The temperature was measured by an optical pyrometer focused on the surface of the graphite die. The samples were heated to 1800 °C at an average heating rate of 100 °C/min, and then the temperature holds constant for 3min. The sample was cooled naturally after the sintering period finished. The bulk density of the sintered products was determined by Archimedes' immersion method with water as the immersing medium. The phase compositions and microstructure were analyzed by XRD and SEM-EDS, respectively.

The oxidation test was carried out by two methods: TGA and Oxidation tests at a fixed temperature in air oven. Small specimens of either ZrB_2, ZrB_2/$Zr_2Al_4C_5$, and ZrB_2/SiC/ $Zr_2Al_4C_5$, having dimensions of approximately 0.5mm×2.0mm×1.5mm, were kept in alumina crucibles and then subjected to heating at the rate of 10 °C/min from 20 to 1500 °C in flowing air inside a thermogravimetric analyzer (TGA, Netzsch, STA449C, Germany). Oxidation tests were performed using an electronic furnace at 1200 °C for 30min in dry air. Specimens were heated to the designated temperature at a heating rate of 10 °C/min.

3. Results and discussion

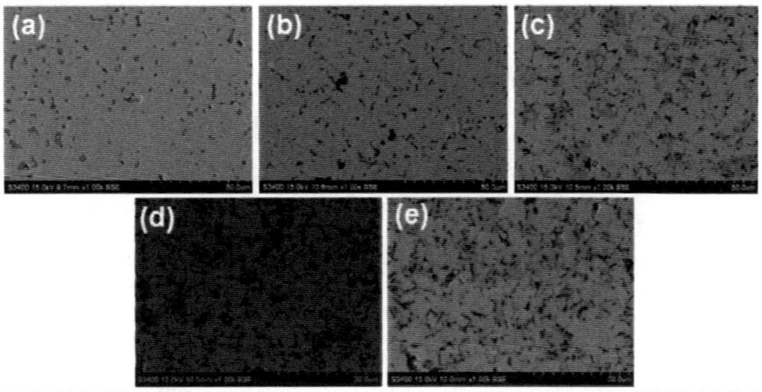

Figure 1. Backscattered electron images of ZrB_2/$Zr_2Al_4C_5$ composites surface (a) 0 vol.% (b) 10 vol.% (c) 20 vol.% (d) 30 vol.% and (e) 40 vol%.

Figure 1 shows the microstructure of sintered ZrB_2/$Zr_2Al_4C_5$ composites. The light phase was ZrB_2 whereas the gray phase consists of $Zr_2Al_4C_5$, the large columnar or plate-like gray $Zr_2Al_4C_5$ grains dispersed homogenously in all these composites. Near-theoretical density of composite was achieved by spark plasma sintering for ZrB_2/$Zr_2Al_4C_5$ composites with different content of $Zr_2Al_4C_5$. Our previous work has revealed [8] that addition of $Zr_2Al_4C_5$ can enhance the mechanical properties of ZrB_2-based ceramics.

Thermal gravimetric analysis studies on $ZrB_2/Zr_2Al_4C_5$ composites from 20 °C to 1500°C are presented in Figure 2. The weight did not change with the increased testing temperature for $ZrB_2/Zr_2Al_4C_5$ ceramics below 700 °C. The weight change of ZrB_2 ceramics occurred at 700 °C, such weight change indicated the presence of the oxidation of ZrB_2 according to reaction (1).

$$ZrB_2(s) + \frac{5}{2}O_2(g) = ZrO_2 + B_2O_3(l) \tag{1}$$

However, the weight of $ZrB_2/Zr_2Al_4C_5$ ceramics did not change, the inflection points of these ceramics at 1200 °C shown in Figure 2 imply that oxidation resistance have changed with the introduction of $Zr_2Al_4C_5$. The rapid increase in weight was observed in all the samples between 1200 and 1500 °C, which indicated that the rate of oxidation aggravated during the temperature. The overall weight gain observed in the ZrB_2 sample was larger than that of the $ZrB_2/Zr_2Al_4C_5$ samples at high temperature. In this case, the oxidation products were mainly B_2O_3, ZrO_2 and Al_2O_3, and a protective scale of glassy B_2O_3-Al_2O_3 phase formed, which was more viscous and stable [9].

In this study, the final increase in weight for the pure $Zr_2Al_4C_5$ and ZrB_2-30vol% $Zr_2Al_4C_5$ composites were 15.67% and 8.07%, respectively. It can be concluded that the introduction of $Zr_2Al_4C_5$ can improve oxidation property of ZrB_2 ceramic.

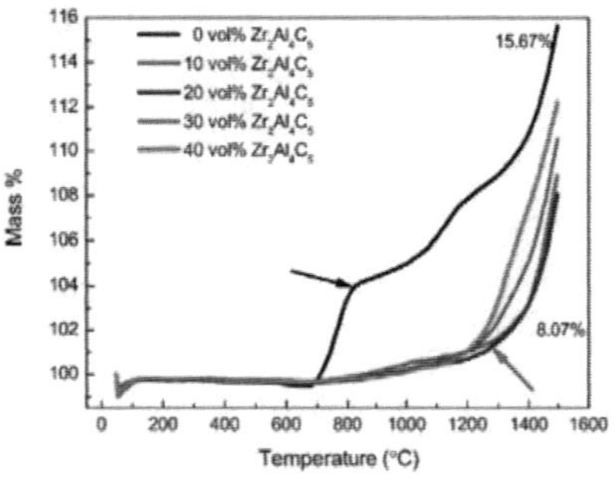

Figure 2. Thermal gravimetric analysis studies on $ZrB_2/Zr_2Al_4C_5$ composites from 20 to 1500 °C in air at the rate of 10 °C/min.

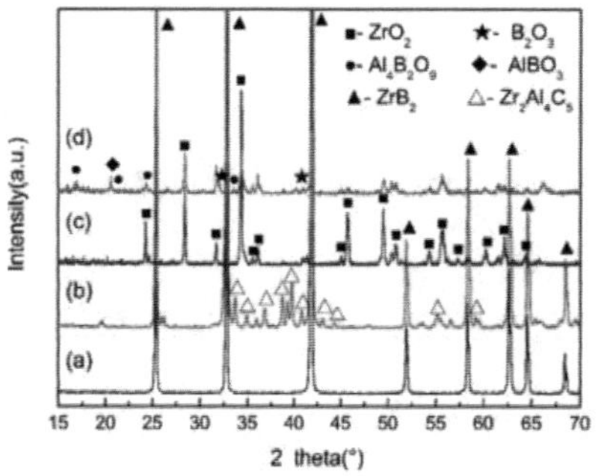

Figure 3. XRD pattern of top surface of the samples before and after exposure at 1200 °C for 30min: (a) and (c) ZrB_2, (b) and (d) ZrB_2/30vol% $Zr_2Al_4C_5$

Figure 3 shows the typical XRD patterns of ZrB_2 and ZrB_2-30vol%$Zr_2Al_4C_5$ before and after oxidized at 1200 °C. Before oxidation, ZrB_2 and $Zr_2Al_4C_5$ were detected for ZrB_2/ $Zr_2Al_4C_5$ composites. After oxidation, it was seen that the oxide layer on ZrB_2 contains significant ZrO_2 phases and a small amount of B_2O_3 phase. For the ZrB_2-30vol%$Zr_2Al_4C_5$ sample, the peaks of $(Al_2O_3)_2(B_2O_3)$, B_2O_3 and $AlBO_3$ were detected .

In addition, the intensity of ZrO_2 phase peaks decreased with the amount of $Zr_2Al_4C_5$ added, showing improved oxidation resistance, due to a continuous and/or partially amorphous B_2O_3-Al_2O_3 glassy scale was produced on the surface of samples. The formation of the scale, in particular continuous scale, inhibited the inward diffusion of oxygen in air, therefore improvement of oxidation resistance[10].

The SEM images depicting the top surface of the oxide scales formed on ZrB_2 and

ZrB_2-30vol% $Zr_2Al_4C_5$ ceramics subjected to exposure at 1200 °C for 30min are shown in Figure 4(a) and (b), respectively. Some pores were found on the surface of ZrB_2 ceramics, such pores appearing on the surface of ZrB_2 ceramic provided more paths for the inward flow of oxygen, and therefore led to the severe oxidation. EDS spectrum from bulk of the oxide scale formed on exposure at 1200 °C for 30min showed the peaks of O, Zr and B, which was suggestive of the presence of ZrO_2 and B_2O_3. The surface layers of the ZrB_2/30vol%$Zr_2Al_4C_5$ sample appeared compact and adherent, no macro-cracks or porous were observed. The bright phase was crystalline ZrO_2 and dark phase was glassy phase. EDS analysis showed that the nubbles were rich in Zr, Al, O and B (Fig. 7d), which should be a mixture of zirconium oxide and Al-B-O glassy phase. Based on EDS and XRD results, the oxidation products should be amorphous zirconium oxide and Al-B-O glassy phases.

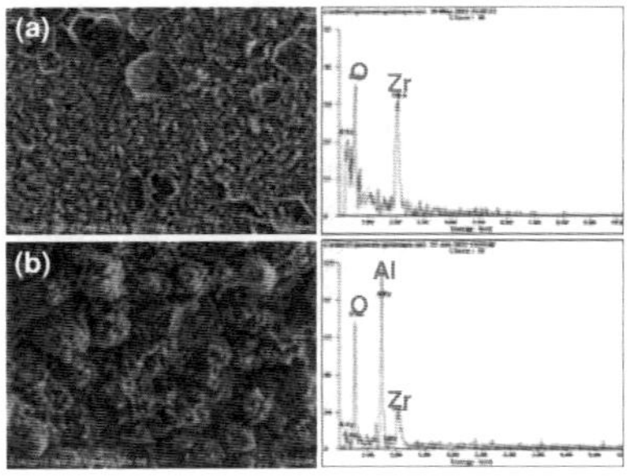

Figure 4. The surface morphologies of the samples after oxidation: (a) ZrB_2, (b) ZrB_2- 30vol% $Zr_2Al_4C_5$.

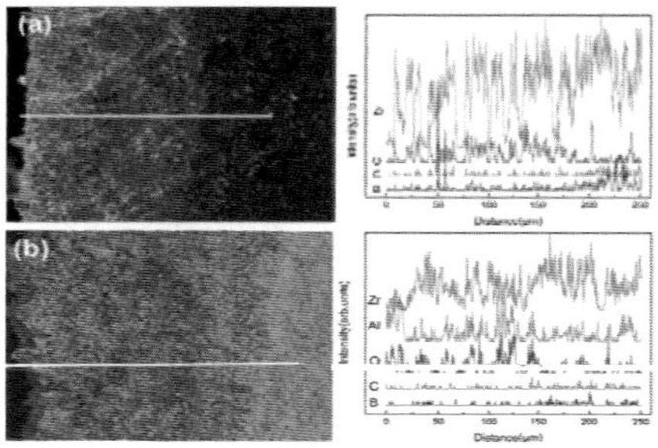

Figure 5. Cross-sectional morphologies and the corresponding EDS line scans of the samples after oxidation: (a) ZrB_2 and (b) ZrB_2-30vol%$Zr_2Al_4C_5$.

Figure 5(a) shows the cross-sectional morphology and corresponding EDS line scans of ZrB_2 oxidized at 1200 °C for 30min. The oxidized layer in the outer section was dominantly ZrO_2, and evident directional growth had been achieved for ZrO_2 grains. And obvious pores were further observed on the oxidized layer, which was favorable for the inward transport of oxygen. The dense oxide layer was about 200 μm thick. The cross-sectional morphology and corresponding EDS line scans of ZrB_2-30vol% $Zr_2A_4C_5$ exposed to the same oxidizing environment is given in Figure 5(b). The dark phase was Al-B-O glassy phases, and the thickness of the oxide scale was only 150 μm, which was less than that for ZrB_2 ceramics. Such layered structure along the cross-section could be testified by the line scanning with EDS in Figure 5. It could be found that the oxide scale had a monolayer structure mainly composed of dark and light grains, and was rich in Zr, Al, B and O. Based on EDS, XRD and Raman spectroscopy results, the oxidation products should be a mixture of ZrO_2, Al_2O_3 and Al-B-O glassy phase.

The oxidation tests these composites have been used to analyze the effect of $Zr_2Al_4C_5$ on the oxidation property of ZrB_2-SiC ceramics. Thermal gravimetric analysis studies on ZrB_2/SiC/ $Zr_2Al_4C_5$ composites from 20 °C to 1500 °C in the air are presented in Figure 6. The weight did not change with the increased testing temperature for ZrB_2/SiC/$Zr_2Al_4C_5$ ceramics below 1200 °C. The inflection points of these ceramics at 1400 °C shown in Figure 6 imply that oxidation resistance have changed with the introduction of SiC. The weight change of ZrB_2/SiC/$Zr_2Al_4C_5$ ceramics first increased and then decreased with increase in the amount of SiC. The rapid increase in weight was observed in all the samples between 1400 and

1500°C, which indicated that the rate of oxidation aggravated during the temperature.

In this study, the final increase in weight for the ZrB_2-30vol%$Zr_2Al_4C_5$ samples and ZrB_2-20vol.%SiC-30vol%$Zr_2Al_4C_5$ composites were 8.07% and 4.19%, respectively. It can be concluded that the introduction of SiC can improve oxidation property of ZrB_2/$Zr_2Al_4C_5$ ceramic. This was because that the oxidation products were mainly B_2O_3, ZrO_2 and Al_2O_3, and a protective scale of glassy B_2O_3-Al_2O_3 phase, Al_2O_3-SiO_2 phase, borosilicate glass and SiO_2 formed, which was more viscous and stable. Thus, oxidation resistance of ZrB_2/SiC/$Zr_2Al_4C_5$ ceramics was achieved by the formation of silica and borosilicate glass layer with oxygen permeability produced by oxidation of SiC, which provided an effective protection for the ZrB_2/SiC/$Zr_2Al_4C_5$ ceramics.

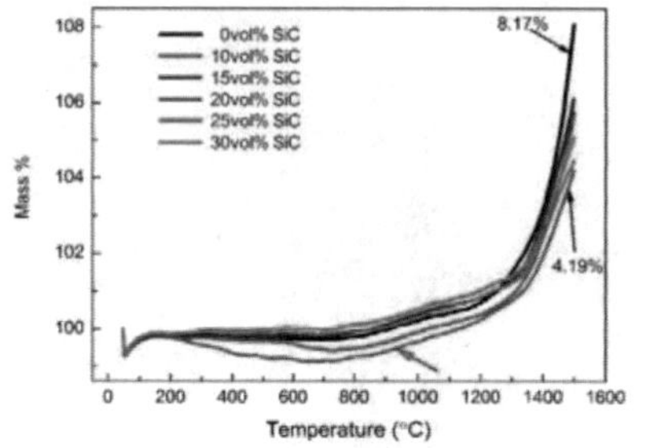

Figure 6. Thermal gravimetric analysis studies on ZrB_2/SiC/$Zr_2Al_4C_5$ composites from 20 to 1500 °C in air at the rate of 10 °C/min.

4. Conclusion

The ZrB_2/SiC/$Zr_2Al_4C_5$ multi-phase ceramics were fabricated by spark plasma sintering (SPS) at 1800 °C for 3 min under 20 MPa in an vacuum. Near-theoretical density of composite was achieved, and the introdution of $Zr_2Al_4C_5$ in restricting the coarsening of ZrB_2. Oxidation behavior of multi-phase ceramics were investigated using muffle furnace in stagnant air at 1200 °C and thermo gravimetric analysis (TGA) from 20 °C to 1500 °C. The results showed that oxidation protection had been attributed to the presence of a glassy Al-B-O phases, which acted as an oxygen diffusion barrier on the surface in the ZrB_2/$Zr_2Al_4C_5$ samples oxidized. At the some time, the introduction of SiC can improve oxidation property of ZrB_2/$Zr_2Al_4C_5$ ceramic, due to a fully dense protective layer with SiO_2-rich glassy scale on the exposed surface.

Acknowledgments

This work was supported by National Natural Science Foundation of China (51272190) and the 111 Project (B13035).

References

[1] Opeka M M, Talmy I G and Zaykoski J A 2004 J. Mater. Sci. **39** 5887
[2] Monteverde F, Guicciardi S and Bellosi A 2003 Mater. Sci. Eng. A. **346** 310
[3] Opeka M M, Talmy I G and Zaykoski 1999 J. Eur. Ceram. Soc. **19** 2405
[4] Hwang S S, Vasiliev A L and Padture N P 2007 Mater. Sci. Eng. A. **464** 216
[5] Sciti D, Brach M and Bellosi A 2005 Scr. Mater. **53** 1297
[6] Zhang X H, Hu P, Han J C, Xu L and Meng S H 2007 Scr. Mater. **57** 1036
[7] Peng F and Speyer R F 2008 J. Am. Ceram. Soc. **91** 1489
[8] Guo Q L, Li J G, Shen Q and Zhang L M 2012 Scr. Mater. **66** 296
[9] Huang X X, Wen G W, Cheng X M and Zhang B Y 2007 Corros. Sci. **49** 2059
[10] Chen G Q, Zhang R B, Zhang X H, Hu P and Han W B 2009 Scr. Mater. **61** 697

FGM 2012 IOP Publishing
Journal of Physics: Conference Series **419** (2013) 012031 doi:10.1088/1742-6596/419/1/012031

Fabrication and Characterization of Multiscale Graded SMAT-MAO Composite Coating Formed on the Surface of 2024 Al Alloy

Lei Wen[1], Yaming Wang[2], Ying Jin[1], Dongbai Sun[1]
[1] National Center for Materials Service Safety, University of Science and Technology Beijing, Beijing 100083, China
[2] Institute for Advanced Ceramics, Harbin Institute of Technology, Harbin 150001, China

Abstract. In order to improve the fatigue property of microarc oxidation (MAO) coated aluminum alloy, a new multiscale graded coating named SMAT-MAO graded coating, consisting of a bottom nanocrystalline layer covered by a top ceramic coating, was designed and fabricated on the surface of 2024 Al alloy by a duplex process with surface mechanical attrition treatment (SMAT) method prior to microarc oxidation process. Experimental results show that surface crystal size was refined to 52.8nm after SMAT for 15min. A 20μm thick nanocrystalline layer, with gradient changing crystal size from dozens of nm in the surface to 200~500nm deep into the substrate, was successfully obtained. The microarc oxidation ceramic coating with 5μm thickness grew by consuming part of the nanocrystalline layer. Since the local heat produced in the microarc discharge channels was not enough to lead to the growth of the substrate grain, the alloy substrate near to the coating/substrate interface still maintained a nano-grained size. Therefore SMAT-MAO coating, which consists of an outer ceramic coating and an inner nanocrystalline layer, was obtained.

1. Introduction

Microarc oxidation (MAO) technique is widely used to fabricate oxide ceramic coatings on Al [1-4], Mg [5], Ti [6, 7] and their alloys to enhance their corrosion- and wear-resistance properties, or to confer various other functional properties including anti-friction, thermal protection, optical and dielectric properties. However, the results that the MAO treated metal samples have a very low resistance to fatigue limited its applications [8-11].

The reduction of fatigue property can be ascribed to two possible factors: the one is micro defects in the coating, which result in an early fatigue crack initiation; the other one is tensile residual internal stress within the substrate adjacent to the coating, which not only leads an early fatigue crack initiation in the substrate, but also promotes the propagation of cracks. Thus, two ways can be considered to improve the fatigue property of MAO treated samples: Firstly, the structure defects of MAO coating, such as pre-existing cracks and micro pores, can be improved by adjusting experimental parameters. Furthermore, the tensile residual stress in the substrate can be offset by a cold-working pre-treatment. The combination of mechanical cold-working and microarc oxidation has been demonstrated to significantly improve thc fatigue resistance compared to the results of MAO coating alone. Asquith

[1] Corresponding author: Lei Wen, Tel.: +86-10-62321017-613; E-mail address: wenlei@ustb.edu.cn

et.al demonstrated that the loss in fatigue life can be recovered to a certain extent by the application of a suitable surface cold-working process prior to treatment [12]. In their study, shot-peening was employed as such a cold-working process and was demonstrated to recover fatigue life of MAO treated aluminum by approximately 85%.

Surface mechanical attrition treatment (SMAT) [13-19], producing a more severe cold-working compared with the conventional shot-peening, is a novel method which can not only introduce compressive residual stress into the surface layer, but also induce a grain refined gradient structure with nanometer to micro scale from the surface to the bulk substrate.

In our research, SMAT was used as the pretreated process of 2024 aluminum alloy, followed by the MAO treatment, in this way the specially modified graded structure (SMAT-MAO graded coating) with a refined grains bottom layer covered by a top conversion ceramic coating was fabricated. The composition, graded structure and morphologies of nanocrystalline layer and SMAT-MAO graded coating were investigated.

2. Experimental

2.1. Materials

The material used in the experiment is 2024 Al alloy with nominal composition 4.36% Cu, 1.49% Mg, 0.46% Mn, 0.25% Fe, 0.14% Si, 0.07% Zn, 0.01% Ti and balance Al. Tensile tests were conducted to ensure the mechanical properties of the alloy: yield strength 330 MPa, ultimate tensile strength 460 MPa and elongation 23%.

2.2. SMAT

The samples, 80 mm × 80 mm × 3 mm in dimensions, were ground with 400#, 800# and 1200# abrasive papers, ultrasonically washed with acetone and distilled water, and dried for SMAT. SMAT was performed in vacuum using a SNC-1 machine. A container of glass fiber reinforced plastic (GFRP) was placed in the steel chamber that was vibrated by a generator. ZrO2 balls with a diameter of 6mm, instead of steel balls used in the traditional SMAT, were placed at the bottom of GFRP container. This kind of set can effectively prevent the transfer of iron between shot and work-piece. The vibration frequency of the chamber is 50 Hz.

2.3. Coating preparation

The based electrolyte of alkaline silicate was prepared from the solution of Na_2SiO_3 (6.0 g/L), NaOH (1.2 g/L), $(NaPO_3)_6$ (35.0 g/L) and Na_2WO_4 (6.0 g/L) in distilled water. A 65 kW microarc oxidation device provides the voltage waveforms, and the main pulse parameters, such as pulse duration, voltage amplitude and duty cycle during both positive and negative biasing can be adjusted independently. The electrical parameters were fixed as follows: voltage 600 V, frequency 600 Hz, duty cycle 10.0%, and the treating time 10 min.

2.4. Characterization of nanocrystalline layer and graded coating

The microstructure of the sample surface treated by SMAT, MAO, and the combination of the two methods, was investigated by X-ray diffraction (XRD) with a Philips X'pert diffractometer using Cu Kα in the range of 30-100°, the measurements were performed with a continuous scanning mode at a rate of 4° /min with an incident angle of 4 °. The surface and cross-section morphologies of the coatings were observed by a Hitachi S-4700 scanning electron microscopy (SEM). A Philips CM-12 transmission electron microscopy (TEM) was used to examine the microstructure of the nanocrystalline layer and the crystal sizes in different depth. To prepare TEM samples, 0.2 mm thick slices were cut from the different depth of surface layer. After mechanical grinding to 40-50 μm, 3 mm-diameter discs were punched from the thin foils, which were then ion-milled with a small incident angle till perforation from single side only.

3. Results and discussion

3.1. Structure design of graded coating

The structure design and fabrication process of SMAT-MAO graded coating are illustrated in Fig. 1. Firstly, the substrate alloy was processed by SMAT, resulting in the fabrication of a nanocrystalline layer on the surface, secondly, the nanocrystalline surface layer sample with a gradient distribution of crystal size was immersed in the electrolyte solution to be oxidized by spark discharge during MAO process, finally a SMAT-MAO graded coating with a nanocrystalline bottom layer covered by a top conversion ceramic coating was fabricated on the surface of 2024 Al alloy.

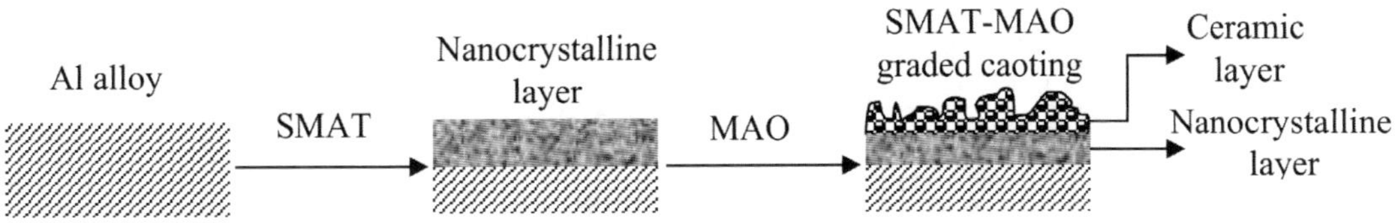

Figure 1. Structure design and fabrication process of SMAT-MAO graded coating formed on the surface of Al alloy

3.2. Microstructure of nanocrystalline layer

A surface layer with crystal size refined into nanometer level can be obtained on the surface of 2024 Al alloy after SMAT process. The XRD patterns of 2024 Al alloy samples before and after SMAT are shown in Fig. 2. It can be found that, compared with 2024 Al alloy, there is an evident broadening of the Bragg-diffraction peaks and a slight shift in the position of diffractions after SMAT. The Bragg-diffraction peak broadening in the surface layer may be attributed to grain refinement and the shift of the diffraction peaks is caused by the development of microstrain. During SMAT, a nanocrystalline surface layer forms due to large grain boundary misorientations, dislocation blocks and microbands [19]. Quantitative XRD measurements indicate that the average crystal size of treated surface layer was 52.8 nm after SMAT for 15 min.

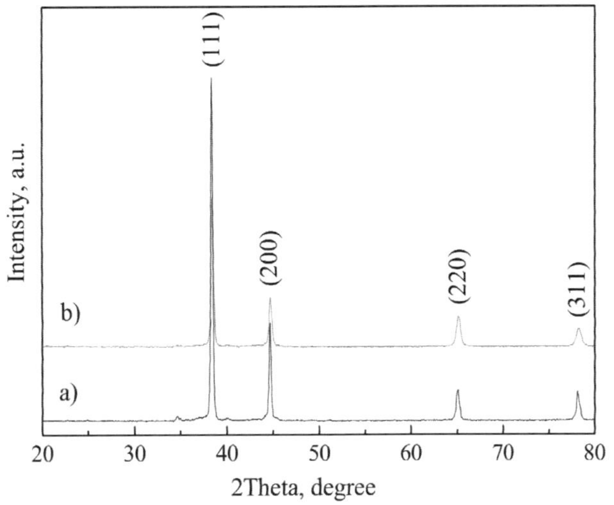

Figure 2 XRD patterns of 2024 aluminum alloy before and after SMAT
a) substrate alloy; b) SMAT

Bright-field TEM images at different depth from the top surface of 2024 Al alloy after SMAT for 15min are shown in Fig. 3a. It can be found that equiaxed shape grains of about dozens of nm are formed at the top surface of 2024 Al alloy after SMAT for 15 min. The crystal sizes at 10 μm and 20 μm depth from the sample surface increase to bellow 100 nm and several hundred nanometers, respectively, as shown in Fig. 3b and c. At about 30 μm depth from the top surface, the crystal size increases to a micron level.

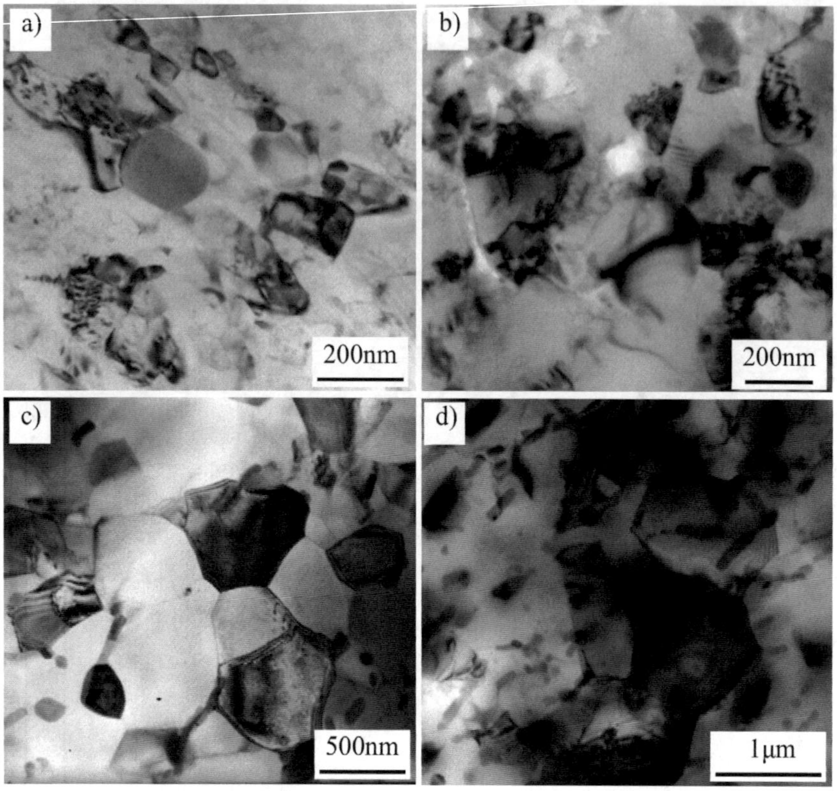

Figure 3. TEM micrographs of 2024 Al alloy after SMAT with ceramic balls:
a) surface; b) 10μm, c) 20μm and d) 30μm from the top surface

The hardness distribution along depth from the top surface of 2024 Al alloy after SMAT is shown in Fig. 4. It can be found that a maximum of hardness is at the top surface, and then the value decreased gradually along the depth. From the distribution of hardness, it can be concluded that the thickness of deformed layer caused by SMAT is less than 40μm. By making a combination analysis of Fig. 3 and 4, it can be deduced that, after SMAT, about a 20μm thick grain refined gradient structure layer, with a increasing from nanometer to micron meter scale grains along the surface to the bulk substrate was obtained at the surface of 2024 Al alloy sample.

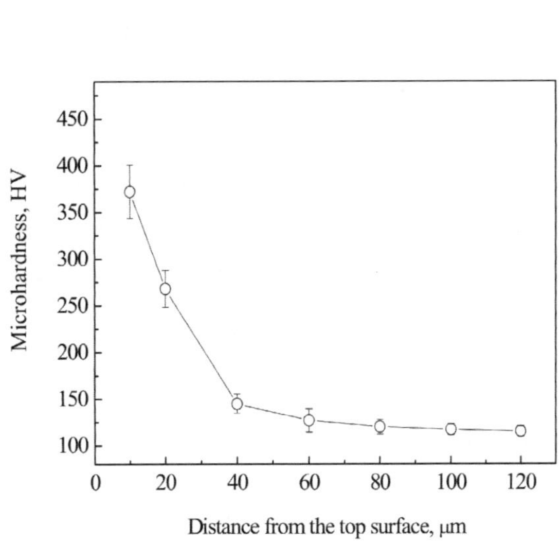

Figure 4. Microhardness distribution along the cross-section of 2024 Al alloy after SMAT 15min with ceramic balls

3.3. Characterization of SMAT-MAO graded coating

After SMAT, the nanocrystalline layer was processed by MAO to obtain a SMAT-MAO graded coating. The results of the XRD analysis performed on MAO coated and SMAT-MAO coated samples are presented in Fig. 5. It can be found that γ-Al_2O_3 phase is the main composition of both simple MAO and SMAT-MAO graded coatings. Furthermore, no obvious difference can be distinguished from the XRD patterns between MAO and SMAT-MAO coated Al alloy samples.

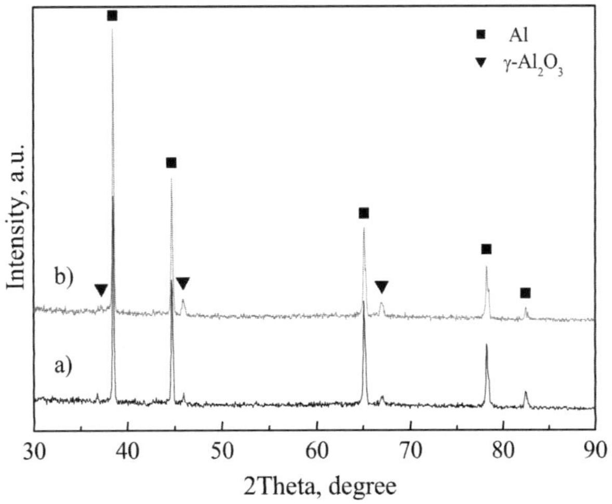

Figure 5. XRD patterns of coating: a) MAO coating; b) SMAT-MAO graded coating

Fig. 6 illustrates the surface and cross section morphologies of MAO coated and SMAT-MAO coated 2024 Al alloys. It can be seen that more micro pores can be found at the surface of SMAT-MAO graded coating compared with that of simple MAO one (Fig. 6a and b). This phenomenon may be due to the specially refined grain surface of the alloy after SMAT pre-treatment. After SMAT, there are much more defect sites at the nanocrystalline surface layer, at which it is easier to induce discharge during thc MAO process. Therefore, more discharge channels (micro pores at the surface) were formed at the surface of SMAT-MAO graded coating. The microarc oxidation ceramic coatings with

5μm thickness grew by consuming part of the nanocrystalline layer, as shown in Fig. 6d. Furthermore, the similar coating thickness and good adhesion between coating and substrate can be observed for both kinds of coating, as shown in Fig. 6c and d.

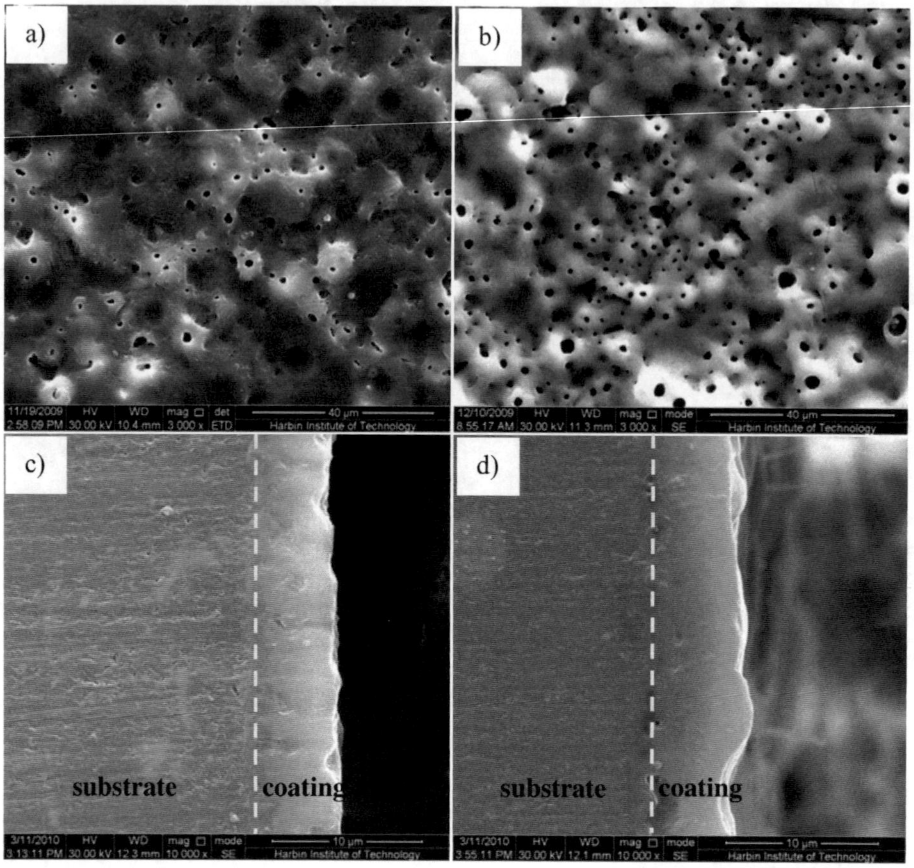

Figure 6. Surface and cross-section morphologies of coated 2024 Al alloy: Surface morphology of a) MAO coating; b) SMAT-MAO coating; Cross-section morphology of c) MAO coating; d) SMAT-MAO coating

In order to verify the gradient structure of SMAT-MAO graded coating, especially nanocrystalline layer under ceramic coating, TEM was employed to investigate the microstructure of 2024 Al alloy near the coating/substrate interface of SMAT-MAO graded coating, and the results were shown in Fig. 7. It can be found that a nanocrystalline structure characterized by equiaxed shape grains with crystal size bellow 100 nm can be observed in Fig. 7a. The corresponding selected area electron diffraction (SAED) pattern of A area in Fig. 7a, as shown in Fig. 7b, exhibits well-defined rings, illustrating the presence of fine grains with random orientation on the treated surface. Since the local heat produced in the microarc discharge channels was not enough to lead to the grain growth of the substrate, the alloy substrate near to the coating/substrate interface still maintained a nano-crystal size. Therefore SMAT-MAO graded coating, which consists of an outer ceramic coating and an inner nanocrystalline layer, was successfully fabricated on the surface of 2024 Al alloy.

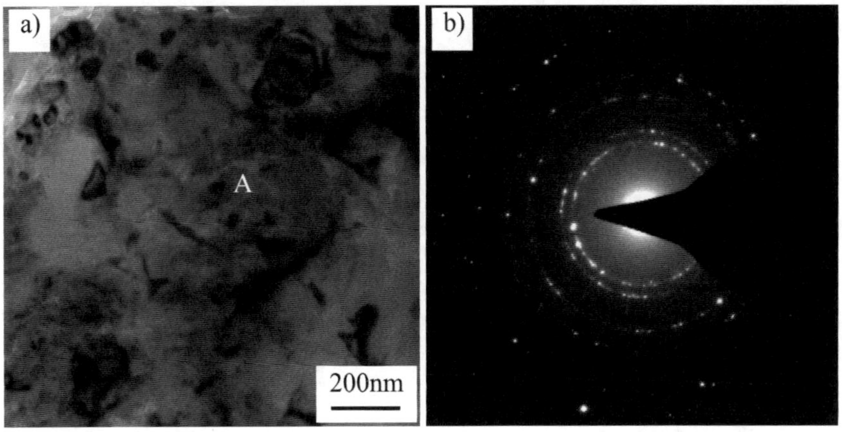

Figure 7. TEM images of 2024 Al alloy near the coating/substrate interface of SMAT-MAO graded coating
a) Bright-field TEM image; b) Corresponding SAED of A area in a)

4. Conlusions

A 20μm thick nanocrystalline layer, with gradient changing crystal size from dozens of nm in the surface to several hundred nanometers deep into the substrate, was obtained by SMAT process. SMAT-MAO graded coating, which consists of an outer ceramic coating and an inner nanocrystalline layer, was formed on the surface of 2024 Al alloy by consuming part of the nanocrystalline layer during MAO process. Since the local heat produced in the microarc discharge channels was not enough to lead to the grain growth of the substrate, the alloy substrate near to the coating/substrate interface still maintained a nano-crystal size.

Acknowledgements

The partial supports from the NSFC grant no. 51201008, the fundamental research funds for the central universities (FRF-TP-12-166A), the general (2011M500241) and special (2012T50044) financial grants from the china postdoctoral science foundation are gratefully acknowledged.

References

[1] Yerokhin A L, Voevodin A, Lyubimov V, Zabinski J and Donley M 1998 Surf Coat Technol. **110** 140
[2] Yerokhin A L, Lyubimov V V and Ashitkov R V 1998 Ceram Int. **24** 1
[3] Markov G A, Mironova M K, Potapova O G and Tatarchuk V V 1983 J Inorg Mater. **19** 1000
[4] Pushkarev O I and Sukhonos S I 2001 Refract Ind Ceram. **42** 15
[5] Timoshenko A V and Magurova Y V 2005 Surf. Coat. Technol. **199** 135
[6] Wang Y M, Lei T Q, Jiang B Q and Guo L X 2004 Appl Surf Sci. **233** 258
[7] Wang Y M, Jiang B Q, Lei T Q and Guo L X 2006 Surf. Coat. Technol. **200**
[8] Yerokhin A L, Shatrov A, Samsonov V, Shashkov P, Leyland A and Matthews A 2004 Surf Coat Technol, **182** 78
[9] Lonyuk B, Apachitei I, Duszczyk J 2007 Surf Coat Technol, **201** 8688
[10] Rajasekaran B, Rama S G S, Krishna L R, Joshi S V and Sundararajan G 2008 Surf Coat Technol, **202** 1462
[11] Rajasekaran B, Raman S G S, Joshi S V and Sundararajan G 2008 Int J Fatigue, **30** 1259
[12] Asquith D T, Yerokhin A L, Yates J R and Matthews A 2006 Thin Solid Films. **515** 1187
[13] Lu K and Lu J 1999 J. Mater. Sci. Technol. **15** 193
[14] Zhang H W, Hei Z K, Liu G, Lu J and Lu K 2003 Acta Mater. **51** 1871

[15] Liu G, Lu J and Lu K 2000 Mater. Sci. Eng. A **286** 91
[16] Zhang Y S, Han Z, Wang K and Lu K, Wear **260** 942
[17] Wu X, Tao N, Hong Y, Xu B, Lu J and Lu K 2002 Acta Mater. **50** 2075
[18] Tao N R, Wang Z B, Tong W P, Sui M L, Lu J and Lu K 2002 Acta Mater. **50** 4603
[19] Wei Y H, Liu B S, Hou L F, Xu B S and Liu G 2008 J. Alloys Compd. **452** 336

Fabrication of ATO/Graphene Multi-layered Transparent Conducting Thin Films

Na Li[a], Fei Chen[b,*], Qiang Shen[c], Chuanbin Wang[d], Lianmeng Zhang[e]
State Key Lab of Advanced Technology for Materials Synthesis and Processing, Wuhan University of Technology, Wuhan 430070, China

*Corresponding author Email: chenfei027@gmail.com

Abstract. A novel transparent conducting oxide based on the ATO/graphene multi-layered thin films has been developed to satisfy the application of transparent conductive electrode in solar cells. The ATO thin films are prepared by pulsed laser deposition method with high quality, namely the sheet resistance of 49.5 Ω/sq and average transmittance of 81.9 %. The prepared graphene sheet is well reduced and shows atomically thin, spotty distributed appearance on the top of the ATO thin films. The XRD and optical micrographs are used to confirm the successfully preparation of the ATO/graphene multi-layered thin films. The Hall measurements and UV-Vis spectrophotometer are conducted to evaluate the sheet resistance and optical transmittance of the innovative structure. It is found that graphene can improve the electrical properties of the ATO thin films with little influence on the optical transmittance.

1.Introduction

Indium doped tin oxide (ITO) is the dominant material used in transparent conducting oxides (TCOs) which have a wide range of applications as transparent electrodes[1-3]. In spite of the excellent optoelectronic properties, ITO suffers from the rare abundance and toxic sources making the application of ITO nonstainable. As an alternative, antimony doped tin oxide (ATO) have attracted vast interests due to prominent advantages as abundant, chemical stability, thermal stability, nontoxic, mechanical durability and especially for the low cost[4-6]. However, the electrical and optical properties of ATO are inferior to ITO, especially for the electrical properties.

Recent researches have focused on the development of thin layers of highly transparent conductive films. Graphene is an exciting material which has a high intrinsic mobility (200000 $cm^2v^{-1}s^{-1}$) and high optical transmittance (~97.7 %)[7-9]. However, the high cost for the preparation of the large area, high performance graphene restricts the application of graphene at present. Therefore, the concept of using the ATO/graphene multi-layered thin films is extraordinarily useful for a substitute of ITO to further improve the properties of transparent conducting oxide films. Moreover, the designable ATO/graphene multi-layered structures as metamaterials have the potential to enhance the optical absorption to improve the photoelectric conversion efficiency for thin film solar cell. Various methods to make graphene have been developed such as mechanical exfoliation from graphite, chemically converted graphene from solution-phase graphene oxide and chemical vapor deposition [7,10-11]. Chemical reduction method has attracted much attention due to the simple experimental condition, easily controlled system, high quality and yield of graphene.

In this paper, we prepare the ATO/graphene multi-layered thin films by spin-coating experimental setup using the graphene produced from the chemical reduction method. The structural, electrical and optical properties of the ATO/graphene multi-layered thin films have been investigated in details.

2. Experimental procedures

Graphite oxide (GO) was prepared from purified natural graphite (30 μm nominal particle size, Shanghai colloid chemical Inc.) by the modified Hummers method. The phase compositions were analyzed using X-ray diffraction (XRD, Rigaku Ultima III) with Cu Kα radiation. Surface morphology analysis was performed with a transmission electron microscope (JEM-2100F STEM/EDS, Japan) and optical microscope. Raman spectra were recorded from 200 to 2000 cm^{-1} on a INVIA Raman Microprobe (INVIA, England). The electrical resistivities were determined with a conventional four-probe instrument. The Hall measurements were made with Van der Pauw techniques. The optical transparency was performed with a Shimadzu UV-2550 spectrophotometer.

Graphite oxide (GO) was prepared from a modified Hummers method using 1.0 g of graphite powder (30 μm nominal particle size, Shanghai colloid chemical Inc.) dispersed in 23 mL of H_2SO_4 kept at 0 °C. Next, 0.5 g of $NaNO_3$ and 3 g of $KMnO_4$ were slowly added. After keeping the reaction for 3 h, the solution was transferred to a 35 °C water bath and stirred for 0.5 h. Then 140 mL of H_2O was added and the temperature was raised to 90 °C. The solution changed to a light brown color during the addition of 10 mL of 30% H_2O_2. Then the prepared GO dispersion was ultrasonicated until it became clear. $NaBH_4$ (200 mg) was added and the solution was heated in an oil bath at 95 °C for 5 h over which the reduced GO gradually precipitated out as a black solid.

The ATO/graphene multi-layered thin films were prepared by spin-coating experimental setup. The ATO thin films were grown on quartz glass substrate by pulsed laser deposition method at 550 °C in oxygen partial pressure of 8 Pa with Sb doping ratio of 12 at.%. Next, the graphene powder was dissolved in N,N-dimethyl formamide (DMF) and ethanol respectively at a concentration of 0.5 mg/L and the solution was spin coated onto the ATO thin films. The ATO/graphene multi-layered thin films were allowed spin at 3000 rpm and then dried at 100 °C inside a drying oven.

3. Results and discussion

The phase compositions and microstructure of the graphene oxide and graphene have been further investigated. The reduction of graphene oxide to graphene should result in structural change, especially the distance between the layers [12]. Fig. 1 shows the XRD patterns of natural graphite, prepared graphene oxide and graphene. It can be seen that the main peak of natural graphite appears at 2θ=26.5° corresponding to the (002) diffraction. The interlayer spacing between atomic planes of (002) diffraction is 0.34 nm [13]. However, the main peak becomes wider and shifts to lower angles of 2θ=10.1° corresponding to an interlayer spacing of 0.94 nm for graphene oxide [14]. The oxygen containing functional groups, such as C=O, C-OH and –COOH, introduced into the layers of graphite by the function of strong oxidizer result in the increase of the interlayer spacing. After the graphene oxide is reduced to graphene, it is observed that all the peaks are disappeared, indicating that the crystallinity is degraded and the graphene is efficiently exfoliated to sheet structure.

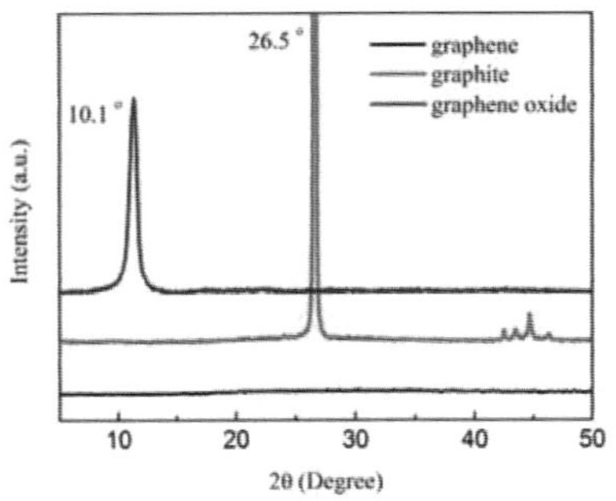

Fig. 1 The XRD patterns of natural graphite, prepared graphene oxide and graphene

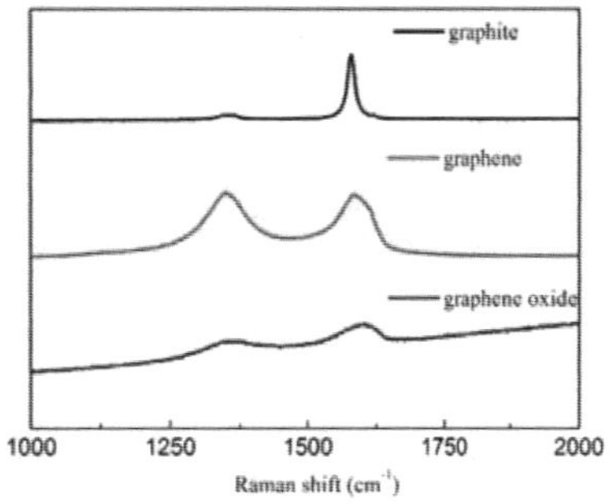

Fig. 2 Raman spectra of natural graphite, prepared graphene oxide and graphene

The chemical structures of natural graphite, prepared graphene oxide and graphene are further studied by Raman spectroscopy. As shown in Fig. 2, there exits two peaks for graphene oxide, which is G band locating at 1600 cm^{-1} and D band locating at 1349 cm^{-1}. After the reduction of graphene oxide, the peaks for G band and D band both shift to lower wave numbers. The G band moves to 1586 cm^{-1}, which equals the value of natural graphite (1586 cm^{-1}), indicating the reduction of the graphene oxide. However, a notable increase of the intensity of D band for graphene is observed which is compared with the natural graphite, reflecting the existence of large amounts of unrepaired defects on the graphene surface due to the removal of oxygen containing groups [15].

Optical microscope is used to analyze the surface morphology of the ATO/graphene multi-layered thin films. The obtained graphene are dissolved in DMF and ethanol respectively and the dispersion is spin coated onto the ATO thin films. Fig. 3(a-c) shows the optical micrographs of the ATO thin films, the ATO thin films coated with graphene dispersion in DMF and the ATO thin films coated with graphene dispersion in ethanol. The images illustrate that the graphene exhibits a black, spotty distributed appearance and the graphene sheets are scattered across the surface of the ATO thin films. It is observed that more graphene sheets are densely clustered for the ATO thin films coated with graphene dispersion in ethanol. Fig. 3(d) shows a TEM image of a partially folded ethanol-soluble graphene sheet. The graphene sheet appears transparent and is folded over on one edge with other graphene on its surface. The results reveal that the graphene is very thin with high optical transmittance and the obtained graphene is of the order of micrometres in size.

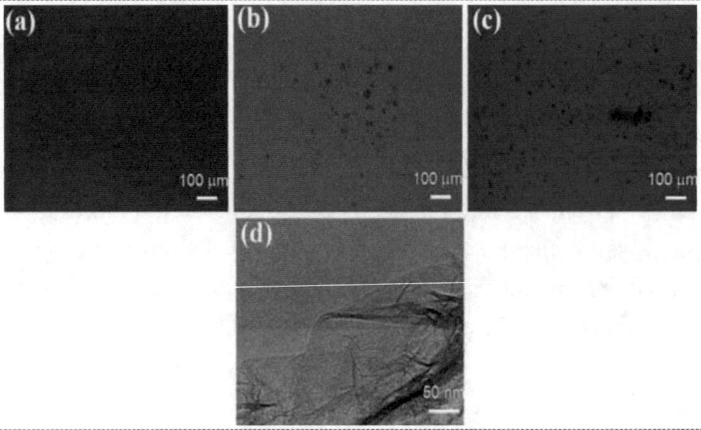

Fig. 3 Images of the ATO/graphene multi-layered thin films and grephane sheet: (a) optical micrographs of the ATO thin films; (b) optical micrographs of the ATO thin films coated with graphene dispersion in DMF; (c) optical micrographs of the ATO thin films coated with graphene dispersion in ethanol; (d) TEM image of a partially folded ethanol-soluble graphene sheet

Sheet resistance and optical transmittance are the main factors to evaluate the quality of transparent conducting oxide films. Table 1 compares the sheet resistance and average transmittance of the ATO thin films and the ATO/graphene multi-layered thin films. The average transmittance is calculated in the visible range of 300 nm～800 nm. As can be seen, the graphene can improve the electrical properties of the ATO thin films. Especially for the ATO thin films coated with graphene dispersion in ethanol, the sheet resistance is decreased by 15 %.

Table 1 Comparison of sheet resistance and average transmittance of the ATO thin films and the ATO/graphene multi-layered thin films

	Sheet resistance (Ω/sq)	Average transmittance (%)
ATO	49.5	81.9
ATO/graphene (DMF)	47.5	81.8
ATO/graphene (Ethanol)	40.6	80.5

The enhancement of the electrical properties for the ATO/graphene multi-layered thin films may result from the increase of electron moving in the whole system. However, spotty distributed graphene instead of the structure of the continuous large size will restrict the maximum electrical performance of the ATO/graphene multi-layered thin films. Nevertheless, the thin layer structure of the prepared graphene sheet can ensure less optical absorption which has a slight influence on the average transmittance of the ATO thin films. The optical transmittance is determined by the amount and dispersion of graphene across the surface of the ATO thin films. It is clearly that the ATO thin films coated with graphene dispersion in ethanol show lower average transmittance (80.5 %) due to the more densely clustered graphene sheet as shown in Fig. 3. The ATO/graphene multi-layered thin films provide a new method to further improve the electrical properties without the cost of transmittance for transparent conducting oxide films.

4. Conclusions

We fabricate the ATO/graphene multi-layered thin films by a spin-coating method. The graphene sheet is well reduced from the graphene oxide prepared by a modified Hummers method. The obtained graphene sheet shows good transmittance with thin layer structure. The graphene can improve the electrical properties of the ATO thin films. Especially for the ATO thin films coated with graphene dispersion in ethanol, the sheet resistance is decreased by 15 % from 49.5 Ω/sq to 40.6 Ω/sq. Moreover, the ATO/graphene multi-layered thin films exhibit atomically thin, spotty distributed appearance with less optical absorption to avoid the deterioration of transmittance. It provides a promising way to produce the transparent conducting oxide films with lower resistance, without lowering the transmittance.

Acknowledgements

This research was funded by the International Science & Technology Cooperation Program of China (No.2011DFA52650) and 111 Project (B13035).

References

[1] Kwark D J, Park M W. and Sung Y M 2008 Vaccum 83 113.
[2] Tak Y H, Kim K B and H.G. Park 2002 Thin Solid Films 411 12.
[3] Fortunato E, Ginley D and Hosono H 2007 MRS Bull 32 242.
[4] Montero J, Herrero J and Guillen C2010 Sol. Energ. Mat. Sol. C. 94 612.
[5] Ma J, Hao X T and Ma H L 2003 Appl. Surf. Sci. 214 208.
[6] Marcel C, Hedge M S and Rougier A 2001 Electrochim. Acta. 46 2097.
[7] Geim A K and Novoselov K S 2008 Solid State Commun. 146 351.
[8] Morozo S V, Novoselov K S and Katsnelson M I 2008 Phys. Rev. Lett. 100 016602.
[9] Tung V C, Allen M J and Yang Y 2009 Nature Nanotechnol. 4 25.
[10] Bae S, Kim H K and Lee Y B.2010 Nature Nanotechnol. 5 574.
[11] Chen W F and Yan L F 2010 Nanoscale. 2 559.
[12] Wang G X and Yang J Park J 2008 J. Phys. Chem. C. 211 8192.
[13] Boukhvalov D W and Katsnelson M I 2008 J. Am. Chem. Soc. 130 10697.
[14] Zhou Y, Bao Q Land Tang L A L 2009. Chem. Mater. 21 2950.

Study on the Preparation and Properties of Colored Iron Oxide Thin Films

Xianhui Zhao[1,2], Changhong Li[1], Qiuping Liu[1,2,4], Yandong Duan[2], Junjing He[1], Su Liu[3], Hai Wang[1], Song Liang[1]

[1] School of Mechanical, Electronic and Control Engineering, Beijing Jiaotong University, 100044, P. R. China
[2] Beijing National Laboratory for Molecular Sciences, Key Laboratory of Photochemistry, Institute of Chemistry, Chinese Academy of Sciences, Beijing 100190, P. R. China
[3] Planning Department, Beijing Oriental Hongsheng International Biotechnology Company, Beijing 100022, P. R. China
[4] Jiangxi university of science and technology, Nanchang 330013, P.R.China

E-mail:zxh860309@126.com

Abstract. Colored iron oxide thin films were prepared using Sol-gel technique. The raw materials were tetraethyl orthosilicate (TEOS), etoh ehanol (EtOH), iron nitrate, and de-ionized water. Various properties were measured and analysed, including the colour of thin films, surface topography, UV-Visible spectra, corrosion resistance and hydrophobicity. To understand how these properties influenced the structural and optical properties of Fe_2O_3 thin films, Scanning Electron Microscope (SEM), UV Spectrophotometer and other facilities were employed. Many parameters influence the performance of thin films, such as film layers, added H_2O content, and the amount of polydimethylsiloxane (PDMS). When the volume ratio of TEOS, EtOH and H_2O was 15: 13: 1, the quality of $Fe(NO_3)_3 \cdot 9H_2O$ was 6g, and pH value was 3, reddish and uniform Fe_2O_3 thin films with excellent properties were produced. Obtained thin films possessed corrosion resistance in hydrochloric acid with pH=1 and the absorption edge wavelength was ~350.2nm. Different H_2O contents could result in different morphologies of Fe_2O_3 nanoparticles. When 1.5 ml PDMS was added into the Sol, thin films possessed hydrophobility without dropping. Coating with different layers, thin films appeared different morphologies. Meanwhile, with the increment of film layers, the absorbance increased gradually.

1. Introduction

As nanosized magnetic metal, Fe reveals unique properties and it has a wide application in many fields, including thin film solar cells, data storage materials, magnetic resonance imaging and so on [1-2]. Silicon is a low cost and ideal material for modifying the surface performance. In the same time, the physical integrity of the underlying core nanocrystals still exists. Silicon coating of oxides can be easily performed using sol-gel technique and it binds to the silica through -OH surface groups [3-5]. The iron oxide is an important inorganic material. Nano-iron oxide has been paid attention by researchers because of its unique nature and application. It can be used for the preparation of catalyst, high-density magnetic recording materials, optical gas sensing materials, functional pigments and so forth. It has wide and potential application in the fields of acoustic, electronics, optics and thermo [6-7]. The performance of materials depends on their composition and microstructure. Therefore, iron oxide with different morphologies has become a hot research point.

PDMS/SiO_2 hybrid material is a hot research point in recent years. The formation of the PDMS/SiO_2 hybrid material could promote the molecule to produce conformational transformation easily, which could eliminate the stress in the gel conversion process. Meanwhile, it could inhibit the embrittlement of the material. PDMS/SiO_2 hybrid material possesses excellent optical and mechanical properties, superior high-temperature and aging resistance. Its outstanding feature is non-toxic [8-10]. Adjusting the PDMS content could control the hardness and elasticity, which could produce bioactive materials.

In 2010, Flora Heshmatpuor et al. [11] in Iran obtained a red and blue gel block by using sol-gel method. They added the inorganic salts respectively containing cobalt ions and iron ions into the solution. The effect of the heat treatment on the film color was discussed. The obtained thin film possessed best properties when the temperature was 600℃, including the uniform color, high transparency and so forth. Nowadays, this technique has been applied in optoelectronic products. For example, the anti-color LCD optical thin films come out in Japan. This product reveals different optical properties according to different colors. Recently, researchers in Tongji University prepared environmentally friendly colored glass with sol-gel technique. They joined one biological stain, such as methylene blue and phthalocyanine blue, as a raw material. On the one hand, obtained colored glass was uniform and transparent. On the other hand, the products could be recycled. However, there are some limitations for the biological stain as the raw material, including high pH value, superior solubility, strong absorption bands in the visible region and high melting point.

In this paper, sol-gel technique was utilized to prepare colored iron oxide thin films and the properties were analyzed. Transparent iron oxide thin film has a wide and significant application in our lives. Recently, the study of such films has become systematic and in-depth. Moreover, the research on iron oxide films develops quickly. Typically, deposited thin film is amorphous and it needs subsequent heat treatment for the crystallization. There exist many factors that affect the crystal structure of thin film, including the substrate material, the chemical properties of the solution, heat treatment conditions and so forth. Sol-gel method possesses many advantages. For example, the process is simple and the experiment cost is not high. And this technique could produce thin films with excellent properties. SEM, UV Spectrophotometer and other experimental apparatus were utilized to analyze the performance of obtained colored iron oxide thin films.

2. Experiments

Reddish iron oxide thin films were prepared by sol-gel method using TEOS and $Fe(NO_3)_3 \cdot 9H_2O$ as Si and Fe precursors respectively. At first, 15ml TEOS and 13ml EtOH were mixed together in a beaker at the temperature of 25℃. And about 5 drops of diluted hydrochloric acid were added in it until the pH value was regulated to 3. At the same time, 6g $Fe(NO_3)_3 \cdot 9H_2O$ and certain amount of de-ionized water were blended together in another beaker at room temperature. These two beakers were placed on the JB-3 Magnetism Stirrer for 15min. Then, the solution of the second beaker was added drop wise into the first beaker and the obtained mixed solution was stirred for 2h. After that, certain amount of PDMS was added into this mixed solution, which was then watered bath at 60℃ for 2h in an enclosure condition. Two hours later, the obtained Sol system was aged for 24h at room temperature. Then, the Sol was coated on the substrate by dip coating method. Finally, humid Gel was dried for 1h at the temperature of 100℃ and then sintered in the muffle at 450℃ for 30 min. Iron species, such as Fe_2O_3 nanoparticles, were successfully incorporated in the inner surfaces of mesoporous silica. And the crystallization of iron oxide particles inside the voids would reveal a better performance.

The morphology and size of the colored iron oxide thin films were determined by Field Emission Scanning Electron Microscopy (SEM) (Hitach, S-4300). To investigate the absorbance, a UV-spectrophotometer (U-3010) was employed. The absorption spectrum is between 300 and 800nm. And other properties of Fe_2O_3 thin films were analyzed by other facilities.

3. Results and discussion

3.1 The obtained colored iron oxide thin films

Fig. 1 shows the macroscopical image of obtained colored iron oxide thin films, which were prepared in a suitable sol viscosity by dip coating technique. The volume of TEOS, EtOH and H_2O was respectively 15 ml, 13 ml and 1 ml, the quality of $Fe(NO_3)_3 \cdot 9H_2O$ was 6g, and pH value was 3. From Fig. 1, it could be seen that the obtained thin films were transparent and uniform with reddish color. The sol viscosity plays a significant role in the forming of thin films. In the initial stage of sol aging, the sol viscosity was smaller than 2 cp. Under this situation, it was hard to form thin films on the glass slide. Meanwhile, the film thickness was so small that there existed apparent color interference fringes. The sol became larger than 5 cp in the later stage of sol aging, which would result in thin films with large thickness. This could produce apparent bubbles in the coating because the small bubbles on the surface of coating solution could not evaporate freely when the viscosity is too large.

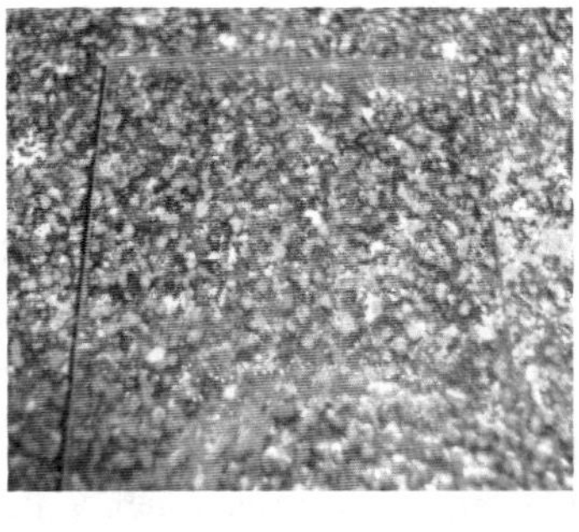

Fig. 1 The macroscopical image of obtained colored iron oxide thin films

3.2 The surface morphology of obtained colored iron oxide thin films

Figure 2 illustrates the FESEM images of colored iron oxide thin films produced by sol-gel method. As shown in Fig. 2, the surface was even and thin films were uniform. There was no crack and the bottom of thin films was connected together. What is more, the nanoparticles were distributed narrowly and uniformly, and the shape was regular. It could be found that the average nanoparticle size was ~ 55nm at the sintering temperature of 450℃. However, some island particles were distributed randomly on the surface of thin films. Through analysis, these particles are probably SiO_2 particles with small degree of polymerization. This phenomenon corresponds to the film formation process in the film nucleation growth theory, indicating that the film growth process on the substrate is essentially a heterogeneous nucleation and crystal growth process.

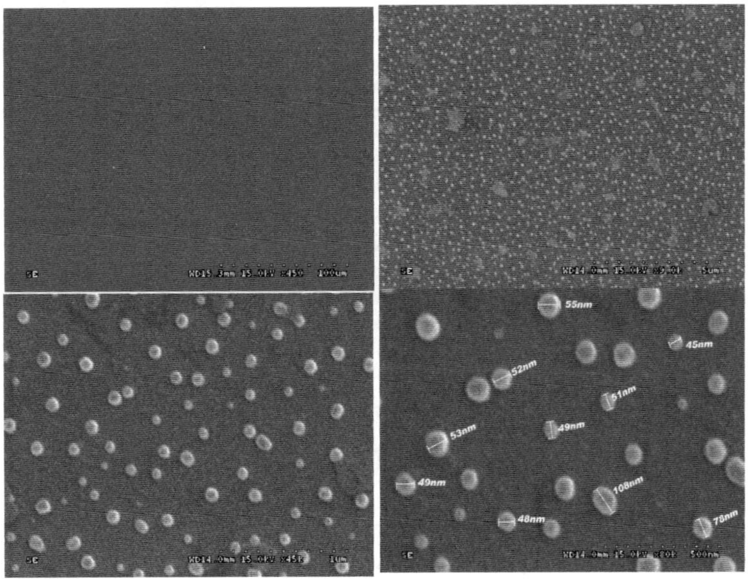

Fig. 2 FESEM images of colored iron oxide thin films

3.3 The effect of added H_2O content on the morphology of Fe_2O_3 nanoparticles

As shown in figure 3, there were three different morphologies of Fe_2O_3 nanoparticles. The morphology of Fe_2O_3 nanoparticles was spherical without adding de-ionized water. When 2 ml de-ionized water was added into the sol system, the shape of Fe_2O_3 nanoparticles became ellipsoidal and tetrahedral. However, the Fe_2O_3 nanoparticles were transferred to block-shaped ones with 6 ml H_2O. It was concluded the H_2O is a significant element for the preparation of colored iron oxide thin films, and adding varying amounts of de-ionized water could produce different morphologies of Fe_2O_3 nanoparticles. This could prove that de-ionized water has large influence on the anisotropic growth of Fe_2O_3 nanoparticles, which would cause different morphologies.

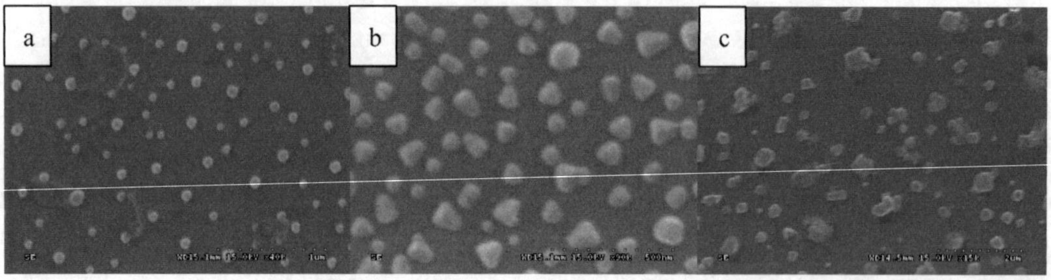

Fig. 3 Different morphologies of Fe_2O_3 nanoparticles with H_2O adding amount of a) 0 ml, b) 2 ml and c) 6 ml

3.4 The optical absorption of iron oxide thin films

Fe_2O_3/SiO_2 thin films are important silicon-based light-emitting materials because of the surface plasmon resonance of metal particles and local field enhancement effect. This kind of thin films possess excellent nonlinear optical properties in special wavelength range, including strong absorption, high nonlinear coefficient and fast response [12-13]. Study on the visible-near infrared light absorption of thin films, especially the effect of applied electric field on absorption, could help deepen the understanding of complex band structure of thin films, field-assisted photoemission mechanism and photoelectric conversion. When the beam shines in the composite thin films, energy is exchanged because of the interaction. Incident light in the thin films decays due to the optical absorption. Light intensity attenuation complies with the Bouguer-Lambert law [14-15]. The absorption spectrum between 300 and 800 nm was shown in figure 4 displayed the absorption edge wavelength of this sample, which was 350.2nm. This proves that obtained thin films possess excellent UV absorption performance.

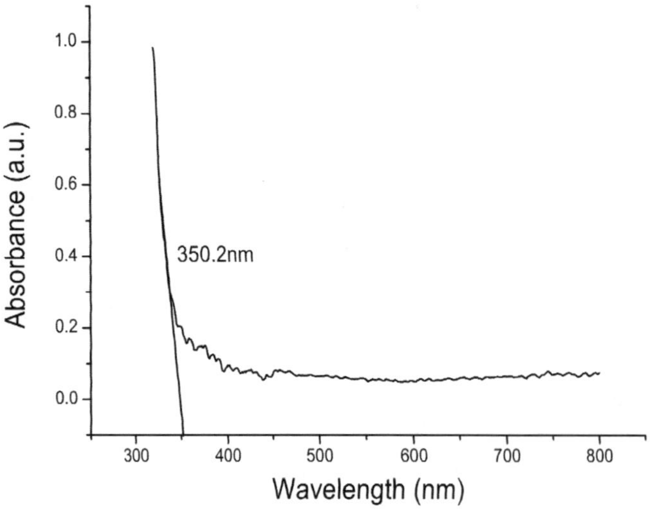

Fig. 4 Absorption spectra of colored iron oxide thin films

3.5 The hydrophobic property of thin films

The hydrophobic materials can be applied in the field of self-cleaning, micro-fluid system and many other fields. Non-wetting means the liquid surface adhesion on solid is smaller than the cohesion. In this case, liquid cannot spread on the solid surface and it tends to shrink into a spherical one on the contact surface [16-17].

Organic silicone polymers own -O-Si-O-Si- chain and $-CH_3$ side chain, which could provide with a hydrocarbon surface. This will form a layer of insoluble waterproofing thin films on this surface. The contact angle increases between water and substrate interface and it appears the "capillary effect", whose nature is to reduce the surface tension of thin films in order to prevent the water from inflowing into them. PDMS has a low surface energy and its $-CH_3$ is a common hydrophobic group. In this experiment, a simple method was made according to the principle of hydrophobic testing. The glass slide was placed on a horizontal table and 20 drops of de-ionized water were dropped wise on the central area of the surface of thin films added with certain amount of PDMS. The surface spread of water was recorded by a camera. Then the hydrophobic property of thin films was analyzed due to the size of the contact angle. As shown in figure, 20 drops of water were congregated together, showing a regular spherical. For thin films added with PDMS, the gravity of thin films to the water molecule is smaller than the gravitational force between water molecules. Then the molecular density of the adhesion layer is less than the density of water molecules, which results in the gravitation between the molecules within the adhesion layer. In this case, water shows the trend of contraction and there appears a non-wetting phenomenon. From figure 5, the contact angle was ~100°. Therefore, it is concluded that thin films added with suitable PDMS content possess hydrophobic properties.

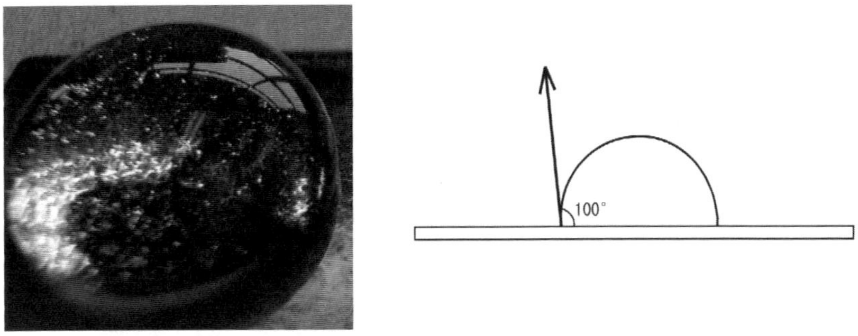

Fig. 5 The water on the surface of iron oxide thin films and the contact angle

3.6 The corrosion resistance of thin films

Figure 6 shows the corrosion situation of the film surface. Si-based coating possesses chemical stability and acid corrosion resistance. In the experiment, three substrates coating with thin films were soaked in the solutions with 40 ml volume, including hydrochloric acid with pH=1, ethanol, and NaOH solution with pH=10, which were shown in Fig. 6 respectively from left to right. After 40 days, they were fetched out and it was found from Fig. 6 that the glass substrate in the hydrochloric acid had no apparent change. In the ethanol, glass substrate had no any change. However, the surface of glass became pale with a lot of traces, and a part of the coatings dropped. It is concluded that iron oxide thin films possess corrosion resistance in hydrochloric acid with pH=1.

Fig.6 The corrosion situation of the film surface

3.7 The coating layers

3.7.1 The effect of coating layers on the morphology of thin films

Multi-layers of coating could improve the wear resistance of thin films. However, the viscosity becomes large when there are several layers of coating. Figure 7 shows the surface of thin films when there are several layers of coating. From Fig. 7 (a), the surface was even and there was nearly no crack. It could reduce reflection and had good transparency. The property of thin films is excellent. Fig. 7 (b) shows the surface of thin films when there were two layers of coating and there existed much regular crack. The property of thin films is good. Fig. 7 (c) illustrates the surface of thin films when there were three layers of coating and there existed much irregular crack. The property of thin films is not good. Figure 7 (d) shows the surface of thin films when there were four layers of coating. It was not even and there existed much irregular crack and the coatings appeared several layers. The property of thin films is not good.

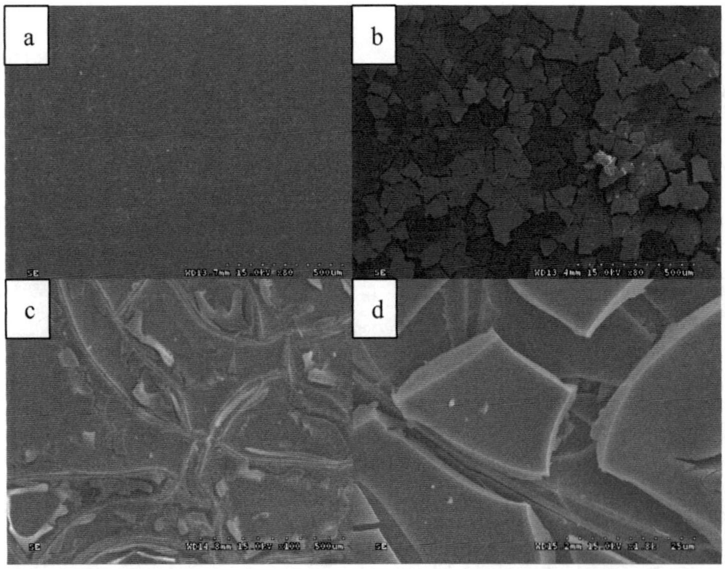

Fig. 7 The surface of colored iron oxide thin films

3.7.2 The effect of coating layers on the absorption of thin films

Figure 8 shows the UV-visible spectra of colored iron oxide thin films coated with different layers, including 4-layer film, 3-layer film, 2-layer film and 1-layer film. It can be seen that the absorbance value increased with the increase of the film layers. In the multi-layer structured thin films, the interface has some impact on the optical absorption of thin films. On the one hand, reflection happens on the film surface. On the other hand, the visible light would vibrate at the interface and it could cause interference if the film is not too thick. When the film layers increase, more reflection is produced on the surface of thin films. Thus, the light absorption rate increases along with it. Transparent metal-Au, Pt, Cu and Ag-thin films are not only excellent reflector of infrared light, but also reflector of visible light. The reason is the free carrier concentration of them is ~1020cm^{-1} [18-19]. The films should be so thin that they could increase the transparency of visible light region and maintain the strong reflectivity of infrared light region.

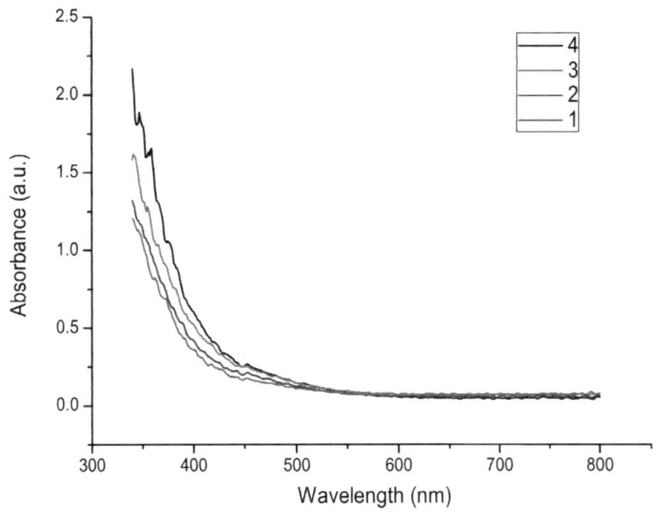

Fig. 8 The UV-visible spectra of colored iron oxide thin films coated with different layers of a) 4, b) 3, c) 2 and d) 1

4. Conclusions

Even and reddish iron oxide thin films with excellent performance are prepared when the volume ratio of TEOS, EtOH and de-ionized is 15: 13: 1. However, different H_2O amounts could lead to different morphology of iron oxide nanoparticles. The absorption edge wavelength of colored iron oxide thin films is 350.2nm and the absorbance increases with the increase of coating layers. The coating layer also influences the morphology of thin films. The band angle of Si-O bond in the structure of PDMS makes thin films possess superior performance. And obtained thin films possess corrosion resistance in hydrochloric acid with pH=1. When 1.5 ml PDMS is added into the Sol system, thin films possess hydrophobility.

References

[1] Cotica L F, Santos I A, Venet M, Garcia D, Eiras J A and Coelho A A 2008 *Solid State Communications.* **147** 123

[2] Huang J F 2005 *The principle and technology of Sol-Gel* (Beijing, Chemical Industry Press) pp

132-175 (in Chinese)

[3] Camponeschi E, Walker J, Garmestani H and Tannenbaum R 2008 *Journal of Non-Crystalline Solids.* **354** 4063

[4] Taboada E, Solanas R, Rodriguez E, Weissleder R and Roig A 2009 *Advanced Functional Materials.* **19** 2319

[5] Kijima N, Yoshinaga M, Awaka J and Akimoto J 2011 *Solid State Ionics.* **192** 293

[6] Duraes L, Oliveira O, Benedini L and Costa B 2011 *Journal of Physics and Chemistry of Solids.* **72** 678

[7] Predoi D, Crisan O, Jitianu A, Valsangiacom M C, Raileanu M, Crisan M and Zaharescu M 2007 *Thin Solid Films.* **515** 6319

[8] Sakurai S, Shimoyama J, Hashimoto K and Ohkoshi S 2008 *Chemical Physics Letters.* **458** 333

[9] Mazaleyrat F, Ammar M, LoBue M, Bonnet J P, Audebert P, Wang G Y, Champion Y, Hytch M and Snoeck E 2009 *Journal of Alloys and Compounds.* **483** 473

[10] Wang G J, Fang Y N, Kim P, Hayek A, Weatherspoon M R, Perry J W, Sandhage K H, Marder S R and Jones S C 2009 *Advanced Functional Materials.* **19** 2768

[11] Heshmatpuor F and Adelkhani H 2010 *Journal of Non-Crystalline Solids.* **357** 1409

[12] Borak B, Laskowski S, Heczko O, Aaltonen A, Baszczuk A, Jasiorski M, Soderberg O, Mazurek B, Oja M, Hannula S P and Maruszewski K 2007 *Journal of Sol-Gel Science and Technology.* **41** 185

[13] Khalil K M and Makhlouf S A 2008 Applied Surface Science. **254** 3767

[14] Battisha I K, Afify H H and Ibrahim M 2006 *Journal of Magnetism and Magnetic Materials.* **306** 211

[15] Li J H, Wei X T, Lin Y S and Su D 2008 *Journal of Membrane Science.* **312** 186

[16] Raap M B, Sanchez F H, Leyva A G, Japas M L, Cabanillas E and Troiani H 2007 *Physica B.* **398** 229

[17] Ortega D, Garitaonandia J S, Barrera-Solano C and Dominguez M 2008 *Sensors and Actuators A.* **142** 554

[18] Jitianu A, Raileanu M, Crisan M, Predoi D, Jitianu M, Stanciu L and Zaharescu M 2006 *Journal of Sol-Gel Science and Technology.* **40** 317

[19] Waseem M, Mustafa S, Naeem A, Koper G J and Din S 2010 *Journal of Non-Crystalline Solids.* **356** 2704

Journal of Physics: Conference Series **419** (2013) 012034 doi:10.1088/1742-6596/419/1/012034

Discrete Element Analysis of Sedimentary Body Density of Functionally Graded Materials derived from Particle Co-Sedimentation

Jing Ye, Guoqiang Luo, Anlin Yang, Qiang Shen, Lianmeng Zhang*
State Key Lab of Advanced Technology for Materials Synthesis andProcessing, Wuhan University of Technology, Wuhan 430070, P.R.China

Email: chenfei81@sohu.com

Abstract. In this paper, the particle co-sedimentation process is analyzed by discrete element analysis. The formation of deposits of continuous functionally graded materials derived from the particle co-sedimentation method is simulated. The porosity of the deposit is calculated and the variation of the porosity along the deposition direction is obtained. From the simulation, the method to increase the deposit density is firstly generated, which is helpful for preparation of functionally graded materials by particle co-sedimentation method.

1.Introduction

Functionally graded materials (FGM) is a kind of new functional composite materials of which main phase is made of two different materials. By designing the composition and structure of middle layer with continuous gradient variety, the properties of FGM vary gradiently along the design direction to meet the requirements of the special features of the material. Due to its widely applicable system, simple preparation and controllable size, co-precipitation process based on the continuous sedimentation of the particles in the liquid phase to obtain the continuous structural changes becomes the most prosperous method [1,2].

To obtain the FGM with efficiently gradient component, the size and distribution of particles is critical to the structure and density of precipitation system. In the previous co-precipitation method, the precipitation system is regarded as no voids (i.e. the accumulation ratio is 1.0) and a continuous, uniform structure, ignoring the effect of discrete particles, size distribution and shape difference on the formation of stacked system voids. In addition, theoretical analysis or calculation also ignores the effect of collision, friction and rolling (among the particles and between the particles and wall) on the porosity of the precipitated system in the liquid medium the precipitation process. Based on the discrete element method and EDEM discrete element [3], in this paper, the accumulation and precipitation process of the components have been analyze and simulated in the case of full consideration of factors to influence the porosity of stacked system.

Using the EDEM discrete element to simulate particles co-precipitation process is divided into two aspects: firstly, the simulation of particle co-precipitation process by visual presentation to clearly observe co-precipitation process; secondly, the simulation of particles accumulation by using the relationship between the particles of stacked system and the volume of voids to analyze the variation of the gap along the sedimentation direction.

1 Simulation of co-precipitation process

1.1 Component particle characteristics

We take the Al_2O_3-Cu ceramic-metal FGM prepared by the particle co-precipitation process as an example and the precipitation liquid medium as distilled water [2]. The component distribution of the precipitation systemis shown in Fig. 1. The precipitation liquid component parameter settings are shown in Table 1 and the component materials and contact characteristics are shown in Table 2， the gradient material component distribution function is a power function [4,5]:

$$C_A(x) = a \times \left(\frac{x}{l}\right)^p + b \,;\, C_B(x) = 1 - b - a \times \left(\frac{x}{l}\right)^p \qquad (1)$$

Where $C_A(x)$ and $C_B(x)$ represent the volume percentage content of the lower density component A and the higher density component B component at the point x of the sedimentary system; x is the position coordinate along the gradient in the thickness direction; l is the thickness of the graded layer; p is the component distribution index; a is a nonzero coefficient; b is a constant.

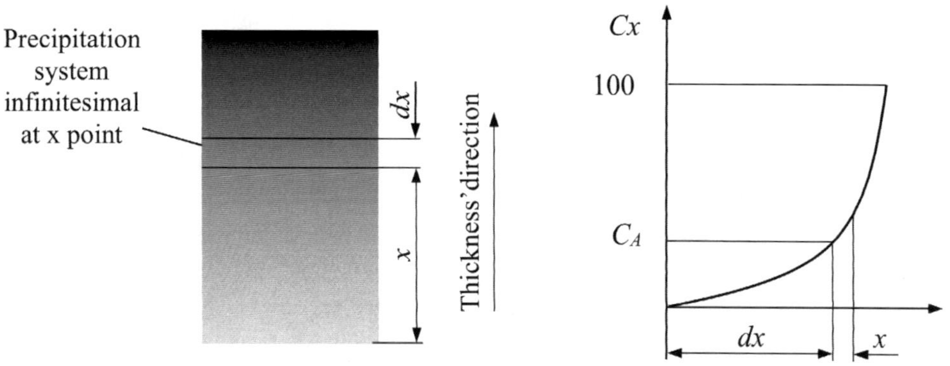

Fig. 1 The schematic graph of the component distribution of the precipitation system

Table 1 Theprecipitation liquid component parameter settings

Density of component A ρ_A/ kg·m^{-3}	Density of component B ρ_B/ kg·m^{-3}	Density of precipitation liquid ρ/ kg·m^{-3}	Viscosity of precipitation liquid μ/ Pa·s	Height of suspension H_1 / m	Height of clear liquid H_2 / m	a	b	p

8900	3970	1000	0.001003	0.4	0.8	1 0 2

Table 2 Component materials and contact characteristics

Material	Poisson's ratio	Sheer modulus (Pa)	Recoverycoefficient	Static friction coefficient	Rolling friction coefficient
Al_2O_3	0.22	9.06×10^{10}	-	-	-
Cu	0.33	3.9×10^{10}	-	-	-
Al_2O_3-Al_2O_3	-	-	0.45	0.45	0.01
Al_2O_3-Cu	-	-	0.48	0.40	0.01
Cu-Cu	-	-	0.50	0.17	0.01

1.2 Simulation parameters settings

Based on the EDEM (v2.4.1) software, the simulation parameters are set as follows:

① Defined the contact model: choosing Particle to Particle; setting custom particles in the liquid resistance in Particle Bodyforce.

② Defined the gravity and the material parameters and contact parameters of Al_2O_3 and Cu: wherein taking residual gravity of the particles.

③ Defined the original particles: the radius of the Al_2O_3 and Cu are 1μm multiples and thethe mass of particles, volume, and other parameters can be automatically calculated dependent on the previously defined density of material.

④Defined the geometric model: take the entities cylinder infinitesimal with a radius of 50 μm and a height of 500μm as a model.

⑤Setting the size distribution of Al_2O_3 and Cu: choose Lognormal distribution.

⑥Setting the particle factory: the particles are generated based on the total mass or the numbers and the way to fill the geometric model. To improve simulation efficiency, 300 particles are selected in simulation demo.

⑦Setting the calculation parameters: 30 % ideal time step are taken in the test of simulation results.

1.3 The result of precipitation simulation

After the simulation, the screenshots are used to clearly show the precipitation process (shown in Fig. 2). The Cu particles are marked as red color area and the Al_2O_3 particles are marked as green color area. The reasons for the long time calculation (for three days) as the follows:

First, the particle size is quite small. The small particles are only of a few micros which requires shorter time step in order to capture the changes in motion of the particles, resulting in the increase of iterative calculations and longer simulation time. The other reason is on the forces acting on particles. The precipitation of particles in a liquid mediumby residual gravity and resistance, wherein the resistance is programmed as Laminar flow state of motion resistance coefficientthrough API contributes to the slow simulation speed.

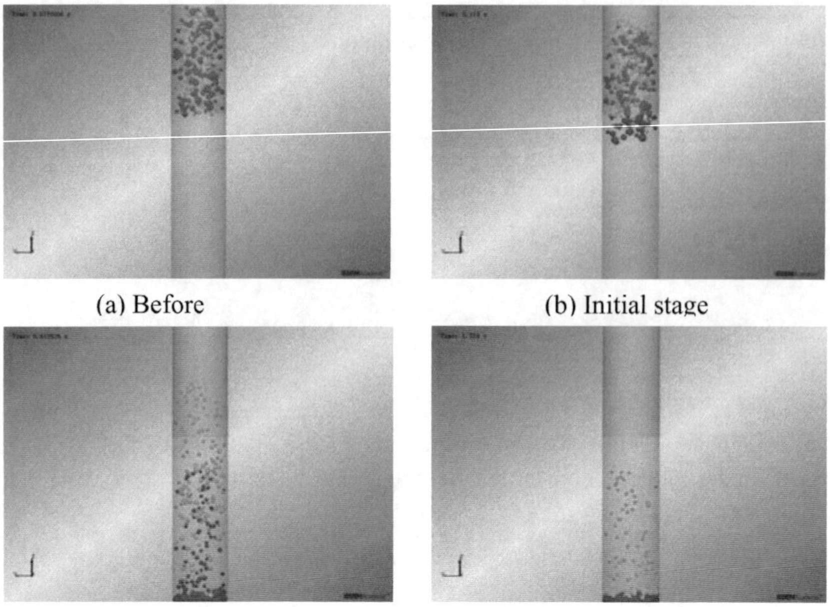

(a) Before (b) Initial stage

(c) Interim stage (d) Final stage

Fig. 2 Simulation of particle precipitation process

2 Simulation of accumulated precipitation structure

2.1 Particle size characteristics setting

Component A of raw material is the Al_2O_3-Cu system. The particles are generatedby the quality, wherein the total mass of the Cu is 5000g,$Al_2O_3$is 2300g. Set Al_2O_3 component granularity in line with the Gates-Gaudin-Schumann distribution equation [6]:

$$U_A(D) = 100\left(\frac{D}{50\times 10^{-6}}\right)^{0.35} (\%)\,; \qquad f_A(D) = U'_A(D) \tag{2}$$

At the time of t, the deposition amount of component A is that:

$$M_A(t) = m_A\left(\int_{D_t}^{D_{A\max}} f_A(D)dD - \int_{D_{t'}}^{D_t} \frac{H_2}{H_1} f_A(D)dD\right) + t\frac{dM_A(t)}{dt} \tag{3}$$

And at the time of t, the deposition amount of component B is that:

$$M_B(t) = m_B\left(1 - \left(1 + \frac{H_2}{H_1}\right)\int_{D_{B\min}}^{D_t} f_B(D)dD + \frac{H_2}{H_1}\int_{D_{B\min}}^{D_{t'}} f_B(D)dD\right) + t\frac{dM_B(t)}{dt} \tag{4}$$

Where $D_{t'} = \sqrt{\dfrac{18\mu H_2}{(\rho_A - \rho)gt}}$.

By numerical calculation, the undersize cumulative percentageof component B (copper powder Cu) is obtained. Part of the data is given in Table 3 due to the large number of numerical simulation.

Table 3 The cumulative percentage undersize of component B

Precipitation time t / s	$M_B(t)$ / g	Slope of the tangent	The undersize cumulative percentage / %	Particle size/ μm
…	…	…	…	…
1000	6.48762	1.28139E-06	32.87049	16.72
1100	6.83014	1.05052E-06	30.72643	15.94
1200	7.13312	8.78255E-07	27.82187	15.26
1300	7.40376	7.46010E-07	25.38739	14.66
1400	7.64755	6.42090E-07	23.74932	14.13
1500	7.86871	5.58835E-07	21.82987	13.65
1600	8.07058	4.91020E-07	20.63998	13.22
1700	8.25585	4.35010E-07	19.10149	12.82
1800	8.42669	3.88175E-07	18.03379	12.46
1900	8.58490	3.48580E-07	16.73271	12.13
2000	8.73196	3.14795E-07	15.91682	11.82
2100	8.86913	2.85725E-07	14.82111	11.54
2200	8.99748	2.60520E-07	14.04059	11.27
2300	9.11789	2.38515E-07	13.16125	11.03
2400	9.23117	2.19185E-07	12.47745	10.79

In order to improve the simulation efficiency and easy to observe, all micron particles are amplified $\times$ 10^3 times to the millimeter, and this treatment does not affect the analysis of the precipitation porosity.

2.2 Simulation results of the precipitation system

Afterthe simulation of the precipitationsystem, the precipitation results are shown in Fig. 3. The precipitationsystem data is obtained at the same time as shown in Table 4, and the porosity of the precipitationsystem is analyzed. Precipitationsystem in the height direction in order to investigate the relationship between porosity in the deposition height, the average is divided into 10 units, respectively, the porosity of each unit is obtained, and can be obtained by the porosity of Fig. 4 with the variation in the height relationship.

Fig. 3 the results of simulation of precipitation system

Table 4 the porosity of precipitation system

No.	The volume of cell / $\times 10^{-4}m^3$	Height / m	Porosity / %
1	1.82206	0.01275	9.02004
2	1.83403	0.03825	8.42528
3	1.83015	0.06375	8.61930
4	1.83037	0.08925	8.60788
5	1.83727	0.11475	8.26374
6	1.84613	0.14025	7.82096
7	1.84114	0.16575	8.07026
8	1.84578	0.19125	7.83895
9	1.84430	0.21675	7.91213
10	1.84391	0.24225	7.93232

Note: the apparent volume of unit is $2.00277 \times 10^{-4}m^3$

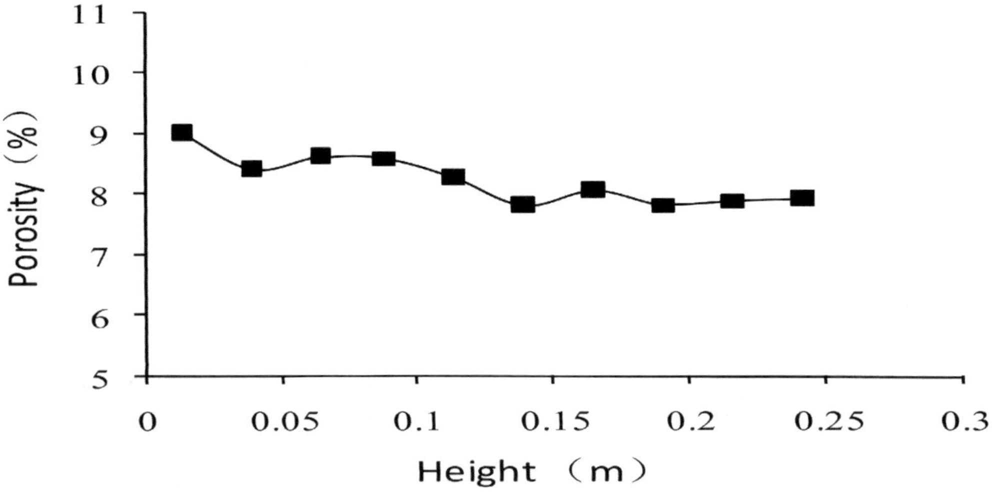

Fig. 4 the variation for the porosity of precipitation system

As shown in Fig. 4, the bottom of the precipitation system mainly consists of large particles and the porosity of stacked system is relatively large and the density is low. However, the top of the sedimentary body mainly consists of small particles and the porosity of stacked system tends to decrease and the density increases. The fluctuation of the porosity in this rangereflectsthe effects of two component particles on stacked structure in the size and distribution of particles.

4 Conclusions

(1) Using the EDEM discrete element to analyze a continuous gradient material precipitation process by the co-precipitation particles prepared by the particle co-precipitation method. The effect of collision, friction and rolling (among the particles and between the particles and wall) will be considered to simulate the accumulation and precipitation process of the components, contributing to more accurate and reliable design of continuous gradient material.

(2) Under the condition of particle size distribution of raw materials corresponding with the dense accumulation, the particle size distribution value of another group is obtained by Component distribution function and the precipitation process associated numerical solution.The variation of porosity of precipitation system along precipitation height and the dense precipitation with accumulation rate of 90.98% to 92.18% are attained by the EDEM discrete element simulation.

(3) According to the simulation, if the particle distribution of two components is set as Rosin-Rammler distribution equation, the porosity of precipitation system reaches 48.14% ~ 39.15% when that the uniformity coefficient is set to be 2 and 2.5 makes it significantly deviation from the dense packing conditions (0.5 ~ 0.7). This indicates that the density of precipitation system can increase more obviously by the dense-graded structure of the particle distribution than that the particle distribution is not controlled and graded.

Acknowledgments

The authors gratefully acknowledge the sponsorship from the National Natural Science Foundation of China (No. 51202175), the Nature Science Foundation of Hubei Province (No. 2011CDB245) for the financial support and 111 Project (B13035).

References

[1] Kawasaki A and Watanabe R 1997 Bulletin of the Japan Institute of Metals. **36** 55.
[2] Lee W Y, Stinton D P and Berndt C C 1996 J Am Ceram Soc. **79** 3003.
[3]Tong X F, Zhang L M and Yuan R Z 1994 Journal of the Chinese Ceramic Society **22** 44.
[4] Shen Q, Zhang L M and Yuan R Z 1997 Journal of the Chinese Ceramic Society **25** 406.
[5] Kiebback B, Neubrand A and Riedel H 2003 Mater. Sci. Eng. A. **362** 81.
[6] Cundall P A and Strack O D L 1979 Geotechnique **29** 47.
[7] Matuttis H G, Luding S and Herrmann H J 2000 Powder Technol. **109** 278.
[8] Zhou Y C, Xu B H, Yu A B and Zulli P 2002 Powder Technol. **125** 45.
[9] Khane V, Mueller G E and Al-Dahhan M H, Continuation of experts in discrete element Modeling(EDEM) validation study for packed bed structural properties. AIChE-2011 Annual Meeting, Conference Proceedings, Minneapolis, November 7-12, 2011
[10] Yang Z M and Zhang L M 2002 Journal of the Chinese Ceramic Society **05** 17
[11] Yang Z M, Zhang L M and Zhen Q 2001 J. Mater. Sci. Technol. **17** 275.
[12] Jullien R,Pavlocith A and Meakin P 1992 J Phys A: Math Gen. **25** 4103.
[13] Visscher W M and Bolterli M 1972 Nature **239** 504.

[14] Aparicio N D and Cocks C F 1995 Acta Metall Mater. **43** 3875.
[15] Latham J P, Lu Y and Munjiza A 2001 Geotechnique **51** 871.
[16] Liu H B 1991 Journal of the Chinese Ceramic Society **19** 164.
[17] Dinger D R and Funk J E A, Paper Presented at Fine Particle Society. 13th Annual Meeting, USA, April 12-14, 1982.
[18] Funk J E and Dinger D R 1988 Amer. Ceram. Soc. Bull. **67** 890.

Structure and nanomechanical properties of $Al_{1-x}Sc_xN$ thin films

Jing Yang, Yudong Li, Miaomiao Cao and Yigang Chen*

Department of Electronic Information Materials, School of Materials Science and Engineering, Shanghai University, 149 Yanchang Road, Shanghai 200072, China

E-mail: yigangchen@shu.edu.cn

Abstract. In this study, c-axis oriented AlN and $Al_{1-x}Sc_xN$ films have been successfully grown on Si (100) and quartz glass by DC magnetron reactive sputtering method. The crystalline structure, optical properties and nanomechanical properties of AlN thin films are investigated by X-ray diffraction (XRD), Raman spectroscopy and nanoindentation techniques, respectively. The XRD patterns show that the crystal structure of the $Al_{1-x}Sc_xN$ films was (002) orientation. The frequency of the E_2 (high) mode observed in the $Al_{1-x}Sc_xN$ films shows higher red shift compared to that observed in AlN film. The nanoindentation hardness and elastic results of $Al_{1-x}Sc_xN$ films were 16 GPa and 190 GPa compared to that of 11.2 GPa and 110.4 GPa for AlN film.

Keywords: Thin film; Scandium; Aluminum nitride; Nanomechanical properties

1. Introduction

Wide-bandgap III–V nitride semiconductors have received much attention due to their importance in both scientific researches and technological applications [1-6]. Among the group III nitrides, AlN exhibits the largest direct band gap (6.2 eV) [3, 4], a high thermal conductivity (285 W/mK), and a high melting point (3237 K) [7, 8]. In recent years, attention has also been paid to rare-earth (RE)-doped IIIA-nitride semiconductor thin films [11], for example, scandium doping in AlN. Scandium is a RE element no.21 in the periodic system. The first principle calculation indicated that it is possible to fabricate Sc-IIIA-N nitrides [9, 10]. Akiyama [6] has reported that $Al_{1-x}Sc_xN$ films with a Sc concentration of 43% exhibited a four times larger piezoelectric response than pure AlN films, which is the largest piezoelectric response among the known tetrahedrally bounded semiconductors. In a recent paper, we show the structure and optical properties of the $Al_{1-x}Sc_xN$ films. In this work, we investigate the mechanical properties of the $Al_{1-x}Sc_xN$ films.

The film have been prepared by various methods including radio frequency (RF) magnetron sputtering, CVD, and reactive molecular beam epitaxy [12-14]. Among them, sputtering has been the most frequently used technique to obtain the thin films. It has advantages over conventionally high temperature techniques used for thin film deposition since its simplicity, low cost, and the ability to obtain films with the properties required for many applications [15-19]. So in this work c-axis oriented AlN thin films had been successfully grown on Si(100) substrates by DC magnetron reactive sputtering method. By Sc-doping in AlN thin films, $Al_{1-x}Sc_xN$ alloy phase is formed. Then we investigate the effect of Sc contents on the mechanical properties related to the microstructure of $Al_{1-x}Sc_xN$ thin films. The structure and optical properties were characterized carefully by X-ray diffraction (XRD) and Raman spectroscopy, respectively. The mechanical properties were examined by nanoindentation measurements.

2. Experiment

$Al_{1-x}Sc_xN$ thin films were prepared on Si(100) substrates by DC magnetron reactive sputtering method. The growth conditions of the films are listed in Table 1. The aluminum sputtering targets were 60mm in diameter and 99.999% of purity. In this work, we grew $Al_{1-x}Sc_xN$ thin films with different Sc concentrations by changing the number of Sc tips which were set on the Al target. The sputtering chamber was evacuated to a pressure below 3.5×10^{-4} Pa and then high-purity argon (99.999%) and nitrogen (99.999%) gases were introduced. The growth pressure was 0.7 Pa.

Table 1. Sputtering conditions of $Al_{1-x}Sc_xN$

Material	$\mathbf{Al_{1-x}Sc_xN}$
Target	Al
Substrate	Si (100)
Substrate temperature (℃)	Room temperature
Power (W)	280
Gas contents	$Ar:N_2$=1:1
Sputtering time	1 hour
Sputtering pressure (Pa)	0.7

3. Results and discussion

The crystal structure and orientation of the films are determined by XRD. The film thickness of $Al_{1-x}Sc_xN$ films is approximately about 700nm. The Sc concentration was measured by energy dispersive X-ray spectroscopy (EDS). The XRD patterns of $Al_{1-x}Sc_xN$ films are shown in Fig. 1(a). From Fig. 1(a) it can be found that when the films are doped with Sc, the diffraction intensity of (002) peak remarkably decreases and the peak shifts to the low angle side. Generally, the AlN has the wurtzite structure while the ScN has the rock-salt (nonpolar) [21-24]. When Sc was doped into AlN films, the Sc atoms were substituted on the Al sub-lattice. It is probably that the crystal structure of $Al_{1-x}Sc_xN$ films is a hexagonal intermediate phase between wurtzite AlN and rocksalt ScN.

The Raman spectra were recorded for $Al_{1-x}Sc_xN$ films with different Sc contents. The intense band at 520 cm^{-1} is the O (Γ) phonon from the silicon substrate. The Raman peaks of

$Al_{1-x}Sc_xN$ films at 750-800 cm^{-1} can be observed from Fig. 1(b) but they are very weak. Compared to AlN E_2 (high) Raman peak, the Raman peak of $Al_{1-x}Sc_xN$ films is higher red

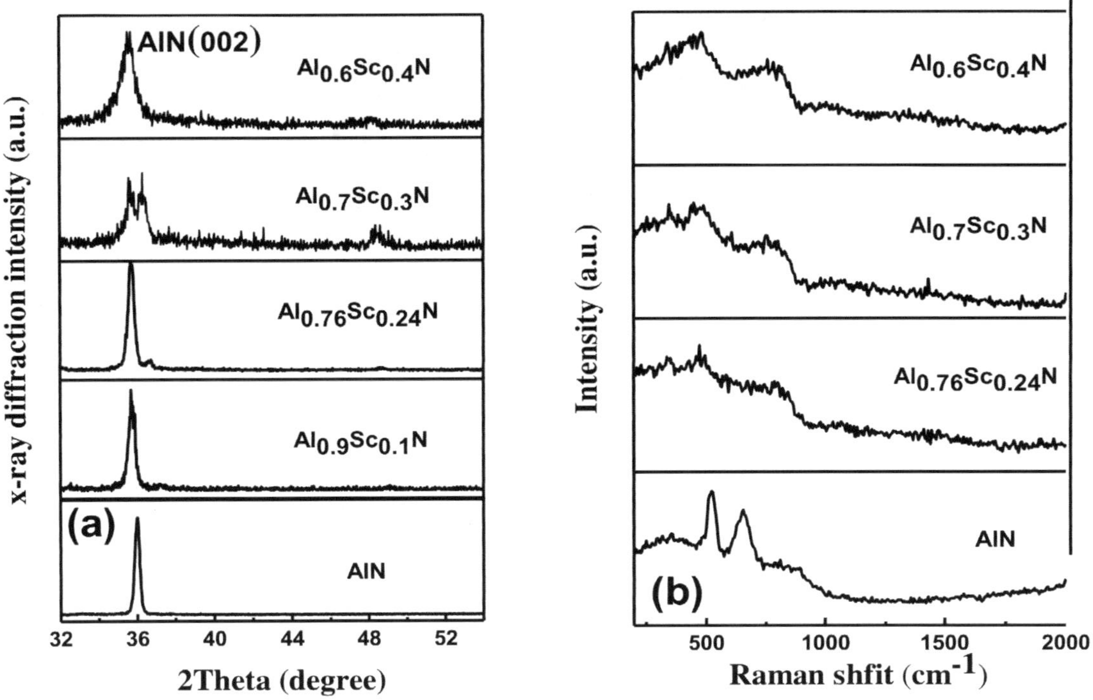

Figure 1. (a) XRD patterns of $Al_{1-x}Sc_xN$ films with different Sc mole fractions; (b) Raman spectra of $Al_{1-x}Sc_xN$ film.

shift. With the reference to the results of our XRD, the observed peak shifting can be attributed to the change of the crystal structure.

It was reported that a red shift was observed in Raman spectra with the decrease of grain size [25]. In our cases, when AlN films are doped with Sc, the grain size decrease. These are consistent with the results of XRD. The E_2 (high) mode is usually being used to analyze the stress state information in the sample due to its high sensitivity to stress which affects the E_2 phonon frequency [26]. An increase in the E_2 phonon frequency indicates compressive stress, whereas a decrease points to tensile stress. In Fig. 3, the Raman peak of $Al_{1-x}Sc_xN$ films shift to low wave number was also found with the increasing of Sc concentration, which indicates the compressive stress in $Al_{1-x}Sc_xN$ films.

The mechanical properties of thin films were analyzed by Nanoindentation which is a very powerful technique [7,27-31]. The load-displacement response obtained by nanoindentation may be regarded as a film 'fingerprint' and contains a lot of information about the elastic behavior and deformation mechanism which the system is undergoing. In order to minimize the substrate effect, the maximum indentation force was limited to 2000μN so that the indentation depth was less than 10% of the film thickness. Fig. 2 gives a typical loading unloading curve of $Al_{1-x}Sc_xN$ films. The initial loading segment contained an elastic-plastic displacement while the unloading process released the elastic energy [28-30].

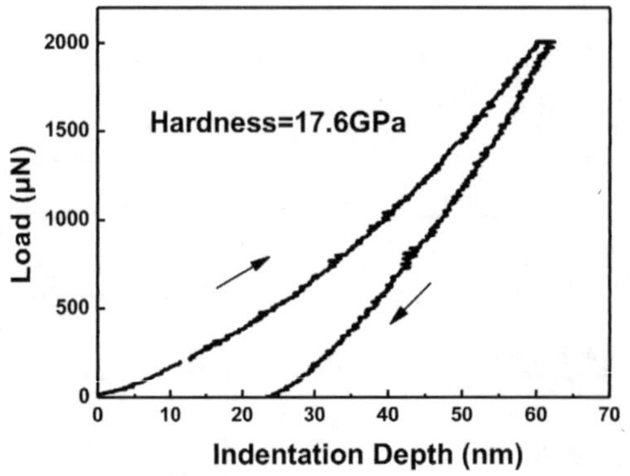

Figure 2. Typical load vs. displacement curve for an $Al_{1-x}Sc_xN$ film at 2000μN load.

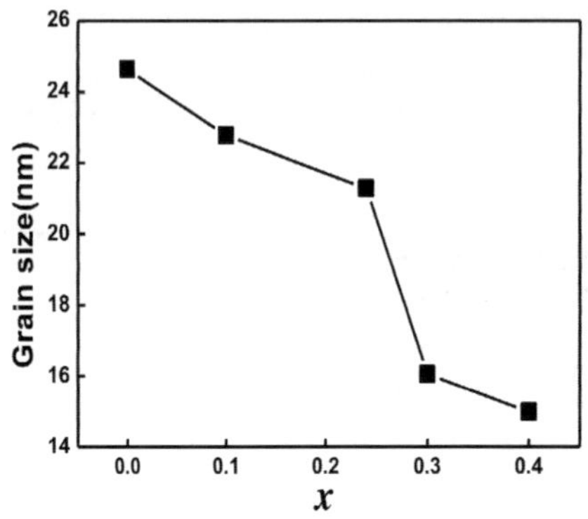

Figure 3. The grain size of $Al_{1-x}Sc_xN$ calculated using Scherrer's equation.

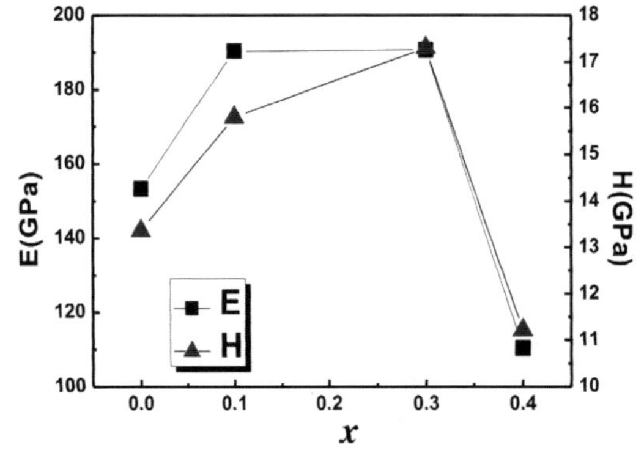

Figure 4. Hardness (H) and elastic modulus (E) of $Al_{1-x}Sc_xN$ films with different Sc mole fractions.

Fig. 4 shows the hardness and elastic modulus of $Al_{1-x}Sc_xN$ films with different Sc contents. For the AlN film, the mean values of hardness and elastic modulus obtained from three independent measurements were 13.4 and 153.3 GPa, respectively. The hardness and elastic modulus of the films increase with increasing Sc contents, and reach a maximum value for the film with 30 at.% Sc (hardness = 17.3 GPa and elastic modulus = 190.7 GPa). With further increase in the Sc contents, both hardness and elastic modulus values decrease drastically. Based on the results obtained and the above discussion, the crystal structure of $Al_{1-x}Sc_xN$ film is a hexagonal intermediate phase between wurtzite AlN and rocksalt ScN. Thus, wurtzite AlN surrounded by a matrix of rocksalt ScN will assist the relaxation of stress and restrict the generation of dislocations and defects in the film. From the results of XRD the grain size of non-doped AlN film was calculated by using the Scherrer formula [32] to be about 24.6 nm. The grain size of the $Al_{1-x}Sc_xN$ films (show in Fig. 3) was in the range of 14-23 nm. Then with the Sc concentration increases, the grain size of $Al_{1-x}Sc_xN$ films decreases. It is reasonable to consider that Sc may cause the Sc–N phases forming and enwrapping the AlN grains, which restrict the growth of AlN grains. These as-formed new phases play a role as a barrier to dislocation propagation, which reduces grain boundary sliding under stress, and thus hardness is improved.

4. Summary

In this study, $Al_{1-x}Sc_xN$ thin films were prepared on both Si (100) and quartz glass substrates by DC magnetron reactive sputtering method. All films show (002) preferred orientation. The nanoindentation hardness and elastic results of $Al_{1-x}Sc_xN$ films were 16 GPa and 190 GPa compared to that of 11.2 GPa and 110.4 GPa for AlN film, and reach a maximum value for the film with 30 at.% Sc. With further increase in the Sc contents, both hardness and elastic modulus values decrease drastically. From the results, we consider that the hardness enhancement appears in $Al_{1-x}Sc_xN$ films with a proper Sc contents.

5. Acknowledgments

The authors thank for the financial support of Shanghai Eastern-scholar program and Shanghai Pujiang Program, China.

References

[1] Vergara L, Olivares J, Iborra E, Clement M, Sanz-Hervás A, Sangrador J 2006 *Thin Solid Films* **515** 1814 – 1818

[2] Kar J.P, Bose G, Tuli S 2005 *Materials Science in Semiconductor Processing* **8** 646–651

[3] Liu B, Gao J, Wu K.M, Liu C 2009 *Solid State Communications* **149** 715–717

[4] Onuma T, Chichibu S. F, Sota T, Asai K, Sumiya S, Shibata T and Tanaka M 2002 *Appl. Phys. Lett* **81** 652-654

[5] Kuokstis E, Zhang J, Fareed Q, Yang J. W, Simin G, Asif Khan M, Gaska R, Shur M, Rojo C and Schowalter L 2002 *Appl. Phys. Lett* **81** 2755-2757

[6] Kamohara Toshihiro, Akiyama Morito, and Kuwano Noriyuki 2008 *Appl. Phys. Lett* **92** 093506

[7] Jian Sheng-Rui, Chen Guo-Ju, Jang Jason S.-C, Lai Yi-Shao 2010 *J. Alloy. Compd* **494** 219–222

[8] Slack G.A, Tanzilli R.A, Pohl R.O and Vandersande J.W 1987 *J. Phys. Chem. Solids* **48** 641-647

[9] Takeuchi Noboru 2002 *Phys Rev B* **65** 045204

[10] Luis Mancera, Rodríguez Jairo A, and Takeuchi Noboru 2004 *Phys. stat. sol. (b)* **241** 2424 – 2428.

[11] Kenyon A.J 2002 *Progress in Quantum Electronics* **26** 225–284
[12] Chen Chi-San, Hwang Bing-Hwai, Lu Hong-Yang and Hsu Tzu-Chien 2002 *Appl. Phys* **35** 2608-2613
[13] Oliveira I.C, Grigorov K.G, Maciel H.S, Massi M, Otani C 2004 Vacuum 75 331–338
[14] Yoshida, S 1979 *J. Vac. Sci. Technol* **16** （4） 990-993.
[15] Auger M.A, Vazquez L, Jergel M, Sanchez O, Albella J.M 2004 *Surf. Coat. Technol* **180 –181** 140–144
[16] Kusaka K, Taniguchi D, Hanabusa T, Tominaga K 2000 *Vacuum* **59** 806-813
[17] Mahmood A, Machorro R, Muhl S, Heiras J, Castillon F.F, Farias M.H, Andrade E 2003 *Diam. Relat. Mater* **12** 1315–1321
[18] Dimitrova V, Manova D, Paskova T, Uzunov Tz, Ivanov N, Dechev D 1998 *Vacuum* **51** 161-164
[19] Manova D, Dimitrova V, Karpuzov D, Yankov R. 1999 *Vacuum* **52** 301-305
[20] Jang Kyungsoo, Lee Kwangsoo, Kim Junsik, Hwang Sunghyun, Lee Jeongin, Dhungel Suresh Kumar, Jung Sungwook, Y Junsin 2006 *Materials Science in Semiconductor Processing* **9** 1137-1141
[21] Akiyama Morito, Kamohara Toshihiro, Kano Kazuhiko, Teshigahara Akihiko, Takeuchi Yukihiro and Kawahara Nobuaki 2009 *Adv. Mater* **21** 593–596
[22] Oussaifi Y, Fredj A. B, Debbichi M, Bouarissa N and Said M 2007 *Semicond. Sci. Technol* **22** 641-646
[23] Paudel T. R and Lambrecht W. R. L 2009 *Phys. Rev. B* **79** 085205
[24] Gall D, Petrov I, Madsen L. D, Sundgren J. E and Greene J. E 1998 J. Appl. Phys **84** 6034-6041
[25] Zhao Qing, Zhang Hongzhou, Xu Xiangyu, Wang Zhe, Xu Jun and Yu Dapeng 2005 *Appl. Phys. Lett* **86** 193101
[26] Kuball M 2001 *Surf Interface Anal* **31** 987–999
[27] Feby Jose, Rajagopalan Ramaseshan, Sitaram Dash, Subbiah Rajagopalan, and Ashok Kumar Tyagi 2009 *Appl. Ceram. Technol* **6** [2] 243–248
[28] Yonenaga Ichiro, Shima Toshiyuki and Sluiter Marcel H. F. 2002 *Jpn. J. Appl. Phys* **41** 4620–4621
[29] Jian Sheng-Rui, Chen Guo-Ju, Lin Ting-Chun 2010 *Nanoscale Res Lett* **5** 935–940
[30] Li T.S, Li H, Pan F 2001 *Surface and Coatings Technology* **137** 225-229
[31] Phillips M.A, Clemens B.M, Nix W.D 2003 *Acta Materialia* **51** 3171-3184
[32] Patterson A. L 1939 *Phys. Rev* **56** 978-982

Microstructure and mechanical properties of 16 Cr-ODS ferritic steel for advanced nuclear energy system

Shaofu Li , Zhangjian Zhou*, Man Wang, Helong Hu, Lei Zou, Guangming Zhang, Liwei Zhang

School of Material Science and Engineering, University of Science and Technology Beijing, Beijing 100083

*** Corresponding author, E-mail:zhouzhj@mater.ustb.edu.cn**

Abstract: A 16Cr-0.5Ti-1W-0.35Y_2O_3 oxide dispersion strengthened (ODS) ferritic steel was fabricated by mechanically alloying and hot isostatic pressing (HIP). Subsequent thermo-mechanical treatments were performed to improve the microstructure homogeneous and service properties of the HIPed 16Cr-ODS steel. Nano-oxide particles were observed by TEM, which can be identified to be (Y, Ti) complex oxide by EDS and SAED. The mechanical property was measured by tensile test, and the oxidation behaviour of the ODS steel was performed at high temperature in a muffle.

Keywords: 16Cr-ODS ferritic steel; Mechanical alloying(MA); Hot isostatic pressing(HIP); Micro-structure; Mechanical properties

1. Introduction

The candidate cladding materials of Generation IV nuclear energy systems must have a high resistance to neutron irradiation embrittlement and void swelling as well as a good performance of mechanical properties at elevated temperatures. In addition, a good corrosion resistance of the claddings in the relevant environments has been certainly required for practical long term operation of the advanced fission reactors [1].

Oxide dispersion strengthened (ODS) ferritic steel which contains a high number density of small oxide particles has been considered as one of the most promising structural materials for advanced nuclear systems [2-5]. Due to the unique microstructure, the ODS ferritic steels show both excellent high-temperature tensile properties, creep resistance, and promising irradiation resistance [3].

For fission application, the Cr content in ODS steels is usually in the range of 9-14% [6-7]. However, the low-Cr-ODS ferritic/martensitic steels are not suitable for super critical water reactor (SCWR) owing to an insufficient corrosion resistance of the materials. Corrosion resistance of iron based alloys is influenced by chromium. The Cr content can be balanced between a merit of corrosion resistance and a demerit of aging embrittlement with maintaining strength at elevated temperatures [8].

In this work, a 16Cr-0.5Ti-1W-0.35Y_2O_3-ODS ferritic steel was fabricated by high-energy mechanical ball milling and HIP. The microstructure was observed by TEM. Mechanical properties and Oxidation resistance were tested and evaluated.

2. Experiment

2.1 Sample preparation

The nominal composition of the ODS ferritic steel is listed in Table 1.The average size of nitrogen-gas-atomized metallic powders and nano-Y_2O_3 were 45 μm and 30 nm respectively. The pre-alloyed powders were mechanical alloyed in a high-energy planetary ball mill under a pure argon atmosphere. The weight ratio of grinding media to material was 5:1 and rotation speed of 380 revolutions per minute (RPM) were conducted with milling time up to 30 h. In the next step, the milled powders were degassed and sealed in a stainless-steel container in vacuum so that it was suitable to be consolidated by HIP. The process of HIP had two stages: the first step was 1100℃ for 2h and the second one was 1150◦C for 1h. Both of these two stages, the pressure was around 200 MPa. Then the as-HIPed specimen would be given hot working treatment, which includes forging and rolling. The effects were reducing the defects in sample and Ar bubbles which resulted from progress of mechanical alloying [9]. The process of forging was1150◦C with a forging ratio of 3:1. Additional hot rolling was performed three times with a reduction ratio of about 20% each stage. Then microstructure observation can be obtained and service properties test would be performed.

Table 1 Nominal composition of ODS pre-alloy powders

element	Cr	W	Ti	Y_2O_3	Fe
wt%	16.0-16.5	1.0-1.5	0.5	0.35	Bal.

2.1 Microstructure observation

The disk shaped specimens with a diameter of 3 mm and a thickness of 0.25 mm, were cut and their surfaces were polished by mechanical polishing and electropolishing. Final thinning of the foils was performed using a standard twin-jet electropolishing technique in an electrolyte (90vol.% acetic acid+10vol.% perchloric acid) at 30V at room temperature. Microstructure and chemical composition were identified by Transmission Electron Microscope (TEM,JEM-2010) with an Energy Dispersive Spectroscope (EDS). Combining with selected area electron diffraction (SAED), the software Gatan Digital Micrograph and database Pcpdfwin were used to simulate electron diffraction patterns in order to identify the crystal structures of nano-oxide particles in the 16Cr-ODS steel.

2.2 Mechanical properties test

Tensile specimens with a gauge length of 15mm were made along the deforming direction and tensile tests were carried out at 23 ℃ and 700 ℃ with a strain rate of $5\times10^{-2}s^{-1}$. The tensile fracture morphologies were investigated by Scanning Electron Microscope (SEM, LEO-1450). The specimens for ultrasonic non-destructive test were cut into size of 10mm×10mm×10mm. And the instrument model is MH-6.

2.3 Oxidation resistance test

Specimens were cut from as-HIPed product with size of 10mm×10mm×10mm for the oxidation resistance testing. There were two stages in this test: at first, put the sample into the muffle furnace at 700℃ for 100h in atmosphere, and calculated oxidation weight gain rate of the post-oxidation sample. Then put the same sample into the muffle furnace at 1000℃ for 2h in atmosphere, and calculated the oxidation weight gain rate again, then observed the oxidation layer by SEM.

3. Results and discussion

3.1 Mechanical alloying effect

Fig 1 shows the comparison between the SEM images of original gas atomized powders and MA powders. In the aspect of raw atomized powders, the shape was easily specified for regular sphere, and the particle sizes were ranging from 10 μm to 30 μm. During the process of MA, the powders were forced to collide with others and with much larger and more hardened balls contained in a mill. The collisions were very energetic, involved large contact pressures and led eventually to the formation of an intimate solid solution. Refractory yttrium oxides can also be finely dispersed into the mechanically alloyed powders in order to obtain dispersion strengthening. Due to the high energy ball milling role, the accumulated strain and work hardening led to increasing brittleness and deformation happened [10]. Thus the shape of particles became irregular and the size distribution was not very uniform, and the mean particle size was about 50 μm.

Excess oxygen (total oxygen concentration except for oxygen concentration in Y_2O_3), resulted from the process of high energy ball mill, was harmful to the microstructure and mechanical properties of ODS Ferritic steel, which can be estimated by the method named inert gas pulse infrared thermal conductivity (ASTME1019-2005). The excess oxygen in this MA powders is only 0.066%, which shows that the parameter control of MA is reasonable.

Fig.1 Mechanical alloying effect on original atomized powder
(a) Original atomized powder (b) Mechanical alloying powder

3.2 Microstructure observation

Typical morphologies of dispersed particles in the specimens of as-HIP-16Cr-ODS ferritic steel are shown in Fig.2, there are many small grains with sizes ranging from a few nanometers up to less than 30 nm in the grains. The small particles distribute well and uniformly, as shown in Fig. 2(a). EDS analysis showed that the chemical composition of the dispersed particles contains Fe, Cr, Y, Ti, and O element. Because the dispersed oxide particle with a diameter is smaller than the thickness of the foil having several tens nm thick, the EDS data of oxides includes both the matrix and oxide [11]. However, this result of EDS inhibits what main composition the fine particle contains.

Recent studies demonstrated that Y-Ti complex-oxide cores can form in Ti-contained ODS steels such as MA957 steel, and Y_2O_3 cores can form in ODS steels with no additions of Al and Ti [12, 13], . Fig.2(b) shows the selected area electron diffraction pattern (SADP) of the fine particle indicated by the red arrow in Fig.2(a), which was observed from the direction zone axis $<11\bar{1}>$. By means of calculating the crystal lattice plane spacing and crystal plane angle, the complex oxide particles were charactered to be cubic structure $Y_2Ti_2O_7$, which was supposed to enhance the properties of materials.

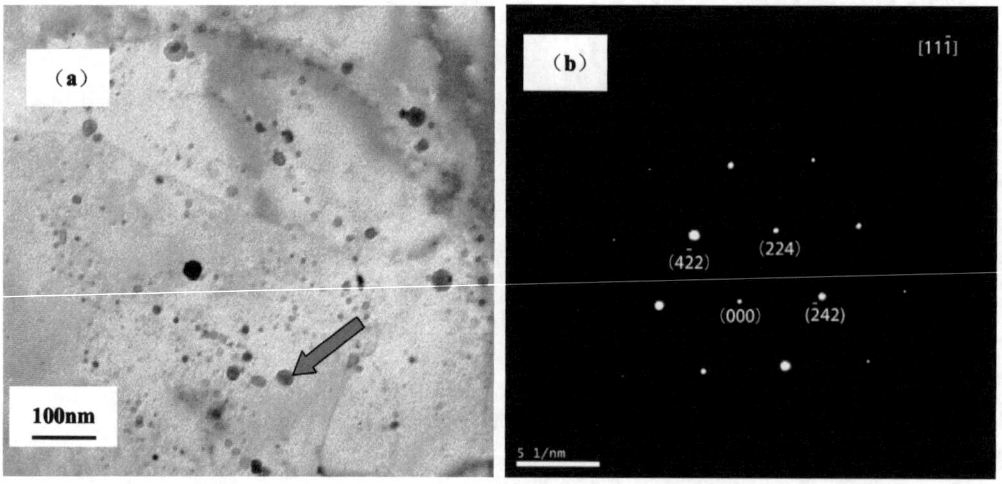

Fig.2 Observation and analysis of precipitate by TEM and SADP
(a) Morphology of microscopic particle (b) SADP analysis of diffraction of particle

3.3 Mechanical properties test

Table.2 shows the tensile properties of 16Cr-ODS Ferritic steels at room temperature and high temperature. The results demonstrate that the as-HIPed 16Cr-ODS ferritic steel exhibited relatively qualified at room temperature, the ultimate tensile strength (UTS) reached about 767MPa, and the ductility represented excellent, the elongation and percentage reduction of area(R.A) reached 17.00% and 57.0% respectively. By means of hot plastic deformations process, both the UTS and elongation have been improved in different degree. Compared with the as-HIPed sample, the UTS enhanced 22.30% and the elongation also enhanced 31.76%. Plastic deformations can improve the microstructure of materials by applying force to specimens and reduce the porosity [10]. At high temperature, the UTS of material will decrease significantly, and the ductility will obtain a observably improvement.

Take conventional ODS-EUROFER for a comparison [3], at room temperature, the UTS and R.A of ODS-EUROFER are 1085Mpa and 11.7% respectively. The UTS of 16Cr-ODS ferritic steel is inferior to ODS-EUROFER slightly. However, the elongation of this specimen is almost double times to ODS-EUROFER. This result demonstrated that 16Cr-ODS Ferritic steel possess reasonable ultimate tensile strength and excellent ductility. Fig.3 shows the fracture morphology of 16Cr-ODS ferritic steel, high density of deep dimples can be observed, which represent excellent ductility ,and it is difficult to find any cleavage cracks due to the very small grains in this specimen.

Though the ultrasonic waves non-destructive test, Young modulus and shear elasticity of specimen can be measured, they were 211.47Gpa and 82.4Gpa, compare with conventional commercial steel, the range of Young modulus is 190Gpa to 200Gpa, the 16Cr-ODS Ferritic fabricated in this experiment exhibited a better pressure resistance and hardness.

Table 2 Tensile properties of 16Cr-ODS ferritic steels

samples	temperature ℃	UTS MPa	$Rp_{0.2}$ MPa	elongation %	R.A %
16Cr (HIP)	23	767	594	17.00	57.0
16Cr(HIP+foring+Rolling)	23	938	476	22.40	53.0
16Cr (HIP)	700	189	164	45.50	74.0

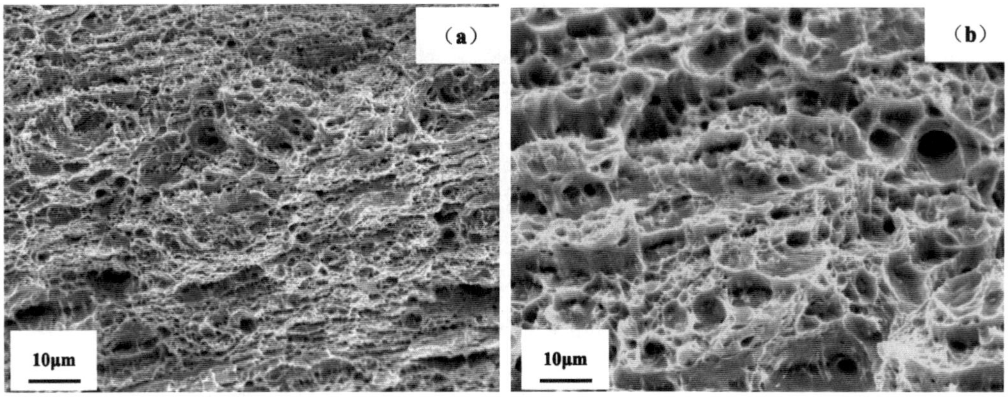

Fig.3 SEM image of the fracture surface of sample after tensile strength test at RT.
(a) Low magnification (b) High magnification

3.4 Oxidation resistance test

This experiment was divided into 2 stages. The high temperature oxidation behavior of 16Cr-ODS ferritic steel was performed in a muffle furnace from 50h at 800 ℃ to 2h at 1000 ℃ in dry air atmosphere. The two stages oxidation weight gain rates were 0.002445% and 0.1934% respectively. The results showed a excellent oxidation resistance.

Fig.4 shows the cross-sectional morphology of alloys after oxidation test. The thickness of oxidation layer was about 2μm. The connection along the oxidation layer/alloy substrate interface was compact. There was no cracks and voids in the contact areas. EDS analysis indicated that the oxide layer was mainly Cr-enriched oxide layer. The superior oxidation resistance of ODS ferritic steels was attributed to the early formation of a protective α-Cr_2O_3 layer on their surface, which is stabilized by the finely dispersed Y-content oxide precipitates. It may contribute to scale formation by promoting a fine grain size in the matrix that increases the diffusion of Cr to the scale-metal interface.

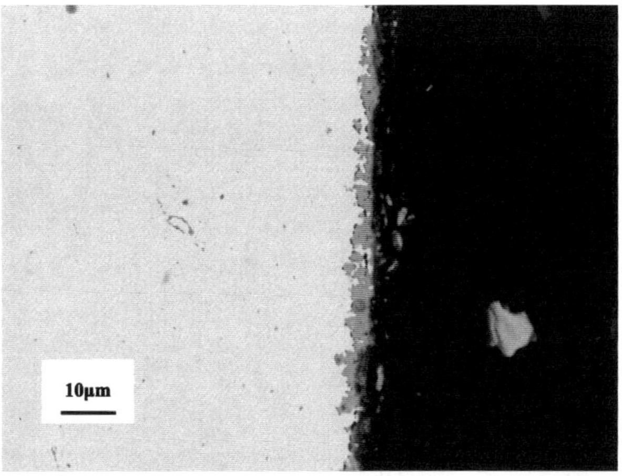

Fig.4 SEM image of cross section morphology of oxidation layer

4. Conclusions

16Cr-0.5Ti-1W-0.35Y_2O_3 ODS ferritic steel was fabricated by MA and HIP. Subsequent thermo-mechanical treatments were carried out to make the microstructure and service properties better. Though microstructure observation and mechanical properties tests, the main results are summarized as follows:

(1) A high number density of nano-scale particles can be observed by TEM in specimen, by the means of EDS analysis and SADP, the fine particles can be identified to be cubic structure $Y_2Ti_2O_7$, which was supposed to enhance the properties of materials.

(2) The results of tensile strength tests of 16Cr-ODS ferritic steel show that the ultimate tensile strength reached nearly 767MPa at room temperature and the elongation reached 17%. At 700℃, the UTS of specimen still maintained around 189MPa. It represented that 16Cr-ODS Ferritic steel possess reasonable UTS and excellent ductility. Hot working treatments, such as forging and rolling, can improve the UTS and ductility significantly of material. The result of ultrasonic waves non-destructive test demonstrated 16Cr-ODS Ferritic fabricated in this experiment exhibited a better pressure resistance and hardness.

(3) 16Cr-ODS ferritic steel also possess excellent oxidation resistance. After exposured in air at 800℃ for 50 h and then 1000℃ for 2h. The thickness of oxidation layer was only about 2 μm. The oxidation layer was dense and showed a good adherence to the matrix. Fine Y-content oxide precipitates and the Cr content in ODS Ferritic steel controlled in a appropriate range was helpful to oxidation resistance of material.

Acknowledgments

The authors would like to express their thanks for the financial support of National Basic Research Program of China under Grant No. 2007CB209801, and thanks for the support of IAEA Coordinated Research Project (CRP code: T11006) under research con- tract No. 16763.

References

1, A. Kimura, R.Kasada, N. Iwata, H.Kishimoto 2011 Development of Al added high-Cr ODS steels for fuel cladding of next generation nuclear systems *Journal of Nuclear Materials* **417 176-179**

2, Sanghoon Noh, R.Kasada, A.Kimura 2011 Microstructure and mechanical properties of friction stir processed ODS ferritic steels *Journal of Nuclear Materials* **417 245-248**

3, D.A.McClintock, M.A.Sokolov, D.T.Hoelzer 2009 Mechanical properties of irradiated ODS-EUROFER and nano-cluster strengthened 14YWT *Journal of Nuclear Materials* **392 353-359**

4, S.Ukai, T.Okuda, M.Fujiwara, T.Kobayashi 2009 Evaluation of threshold stress of the MA957 ODS ferrtic alloy *Journal of Nuclear Materials* **511-514**

5, K.Asano, Y.Kohno, A.Kohyama, T.Suzuki and H.Kusanagi 1988 Microstructural evolution of an oxide dispersion strengthened steel under charged particle irradiation *Journal of Nuclear Materials* **928-934**

6, M.K.Miller, D.T.Hoelzer, E.A.Kenik, K.F.Russell 2005 Stability of ferritic MA/ODS alloys at high temperatures *Intermetallics* **13 387-392**

7, M.J.Alinger, G.R.Odette, D.T.Hoelzer 2009 On the role of alloy composition and processing parameters in nanocluster formation and dispersion strengthening in nanostructure ferritic alloys *Acta Mater* **57 392-406**

8, A.Kimura, R.Kasada, N.Iwata, H.Kishimoto 2011 Development of Al added high-Cr ODS steels for fuel cladding of next generation nuclear systems *Journal of Nuclear Materials* **417 176-179**

9, M.Klimiankou, R.Lindau, A.Moslang 2004 TEM characterization of structure and composition of nanosized ODS particles in reduced activation Ferritic-martensitic steels *Journal of Nuclear Materials* **329-333**

10, Man Wang, Zhangjian Zhou, Hongying Sun, Helong Hu and ShaofuLi 2013 Microstructural observation and tensile properties of ODS-304 austenitic steel *Materials Science and Engineering A* **559 287-292**

11, S.Ukai, M.Harada, H.Okada, M.Inoue, S.Nomura 1993 Alloying design of oxide dispersion strengthened ferritic steel for long life FBRs core materials *Journal of Nuclear Materials* **204 65-73**

12, H.Kishimoto, Kasada O.Hashitomi and A.Kimura 2009 Stability of Y-Ti complex oxides in Fe-16Cr-0.1Ti ODS ferritic steelbefore and after heavy-ion irradiation *Journal of Nuclear Materials* **386-388 533-536**

13, Ming Li, Zhangjian Zhou, Pei He, Lu Liao, Yingli Xu and Changchun Ge 2010 Microstructure and mechanical property of 12Cr oxide dispersion strengthened ferritic steel for fusion application *Fusion Engineering and Design* **85 1573-1576**

14, Luke L.Hsiung, Michael J.Fluss, A.Kimura 2010 Structure of oxide nanoparticles in Fe-16Cr MA/ODS ferritic steel *Materials Letters* **64 1782-1785**

Effect of strain rate on the plastic strain gradient beneath the deformed surface of iron

Xianguang Zhang[1], Kiyotaka Matsuura[2], Munekazu Ohno[2]
[1] Graduate Student, Graduate school of Engineering, Hokkaido University, Kita 13 Nishi8, Kita-ku, Sapporo, Hokkaido 060-8628, Japan
[2] Division of Materials Science and Engineering, Faculty of Engineering, Hokkaido University, Kita 13 Nishi 8, Kita-ku, Sapporo, Hokkaido 060-8628, Japan

E-mail: matsuura@eng.hokudai.ac.jp

Abstract. Plastic strain gradients along the depth direction of a locally deformed surface of pure iron have been studied under different levels of strain and strain rates. Deformation at high strain rates was carried out by using a ball-dropping test, while deformation at low strain rates was performed by a ball-pressing test on a computer controlled servohydraulic machine. Evaluation of the local plastic strain was conducted by using the electron backscatter diffraction (EBSD) analysis. Experimental evidences indicate that high strains with high strain rates lead to large gradients of the plastic strain distribution profile and also a localized severe deformation near the surface. Furthermore, deformation at high strain rates is favourable to impose plastic deformation at the selected area near the surface.

1. Introduction

It is widely known that most of plastic deformation is performed from the surface of materials. The degree of local plastic strain gradient is crucial in determining the mechanical and microstructural features of the work-piece surface, because plastic deformation strongly affects grain structure development in recrystallization and phase transformation during heat treatment processes. Furthermore, most of material failures originate from the exterior layers of the work piece (fatigue, fretting corrosion, corrosion, wear, etc.). These phenomena are all extremely sensitive to the structure and properties of the surface of the material [1-3]. Therefore, good understanding of the local plastic strain gradient near the surface of the work-piece processed by surface local plastic deformation (SLPD) is crucial.

The plastic deformation behavior of materials in principle depends on many intrinsic and extrinsic factors such as structure and stacking fault energy of the material, the intensity of strain and strain rates, deformation temperature, and so on[4]. It is well known that the strain and strain rate play a dominant role in determining the degree of plastic deformation in work-piece. Therefore, it is important to investigate the influences of strain and strain rate on the local plastic strain gradient in the depth direction of the component.

The present study aims to understand the effects of strain and strain rate on the magnitude, gradient of local plastic strain distribution in the depth direction of SLPD-processed work-piece. EBSD analysis is performed to provide quantitative description of the local plastic strain. As will be shown in the following sections, the present study confirms that the gradient of local plastic strain strongly depend on strain rate. High strain rates lead to plastic strain distribution profiles of large gradients with high peaks not at the surface but slightly beneath the surface.

2. Experimental procedure

The material used in the present study is commercial pure Fe with a purity of 99.95 mass%. The as-received pure Fe bars were annealed at 1273 K for 60 min to diminish the effect of mechanical processing and then slowly cooled to room temperature to obtain homogeneous ferrite grains. All specimens were taken from the same lot in each case. The high strain rate deformation was carried out by ball-dropping test, in which a steel weight with a spherical steel ball attached on its bottom was dropped from a height of 2 or 0.5 m onto a 30 mm in diameter and 10 mm in thickness pure Fe plate with a flat surface. The ball had a diameter of 8 mm and the weights of 1.5 to 5 kg. To make the experimental parameters less complicated, single impact was performed for each case. The corresponding low strain rates tests were performed by ball-pressing test which was carried out on a computer controlled servohydraulic machine. All the tests were carried out at room temperature in air. The strain rates used in this study are summarized in table 1.

Table 1. Strain rate used in this study

Deformation form	Ball velocity, m/s	Strain rate , s^{-1}
BD, H=2 m	6.26	6.83×10^{2}
BD, H=0.5 m	3.13	3.07×10^{2}
BP, Case I	8.33×10^{-5}	1.26×10^{-2}
BP, Case II	1.67×10^{-6}	2.35×10^{-4}

After deformation, longitudinal mid-plane sections were prepared for EBSD measurement. The surface was polished up to 100 nm alumina polishing slurry followed by colloidal silica in order to achieve relatively flat surfaces free from damage. The preparation is a necessary step to obtain good diffraction patterns of the surface. The EBSD system can obtain a crystal orientation map automatically by scanning a square area on the surface of the specimen, which is tilted 70° from the horizontal. The step size was 3 μm in all cases. Furthermore, angle misorientations bigger than 5° were considered grain boundaries.

3. Results and discussion

3.1. Relation between misorientation and plastic strain

Fig. 1 shows the crystal orientation maps of the samples with crater depth (CD) of 1.1, 0.41, and 0.0 mm at strain rate of 1.26×10^{-2} s^{-1} within 1.0 mm beneath the treated surface.

Figure 1. Orientation map of pure iron: (a) CD=0.0 mm, (b) CD=0.41 mm, (c) CD=1.1 mm

The plastically deformed maps show a varying orientation within individual grains. By comparison, the orientation of undeformed material is almost uniform within each grain. The range of crystal orientations within individual grains becomes more notable as the crater depth increases. This indicates that the plastic deformation caused the rotation of crystal orientation inside a grain and the local misorientation increases as the degree of plastic deformation increase. This shows good agreement with [5].The process used in the present study to estimate the local plastic strain by local misorientation method, which quantifies the variation in crystal orientation of each point from a reference orientation that is evaluated as an average of the crystal orientation distribution. Subsequently, the misorientation data obtained were averaged by the following equation:

$$M_{\text{ave}} = \frac{\sum_{k=1}^{n_g} \left(\sum_{i=1}^{n_k} f(p_r, i) \right)}{\sum_{k=1}^{n_g} n_k} \tag{1}$$

where n_g is the number of grains and p_r denotes the reference point with the average orientation of the grain k. $f(p_r,i)$ denotes the misorientation between average orientation and other points i within the grain, and n_k is the number of points included in the grain k.

3.2. Average local misorientation profile

Fig. 2 shows the average local misorientation (ALM) profile as a function of strain rate and the position measured from the bottom of the crater for the specimen deformed with 1.1 mm depth crater.

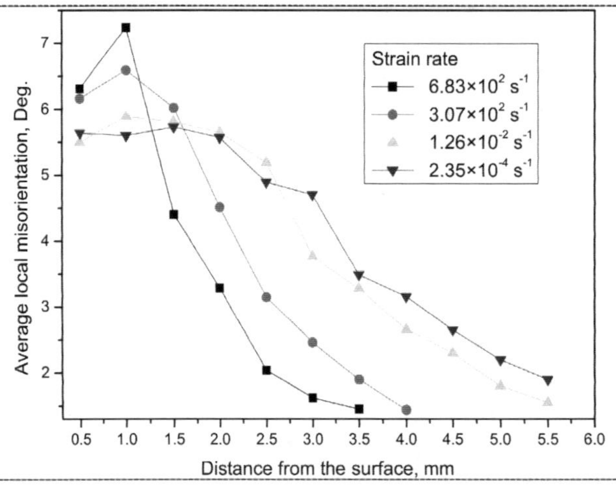

Figure 2. Local plastic strain profile versus distance from the crater bottom.

It can be seen that the ALM profiles are quite different between high and low strain rates, which indicates that strain rate plays a key role in controlling strain profile beneath the treated surface under the same crater depth. Furthermore, the profile shown in Fig. 1 reveals that the maximum ALM thus the plastic strain is not right at the treated surface but immediately beneath the surface, as for strain rates higher than 2.35×10^{-4} s^{-1}. This result is consistent with the famous Hertz contact analysis [6], which predicts that the maximum shear stress and thus the plastic strain is at a small depth of the contact surface. Similar result was reported by Dai et al [7]. Noteworthy, the ALM near the surface (within 1.0 mm beneath the contact surface) obtained from high strain rates is always larger than that at low strain rates even though deformed with the same crater depth, which is in good accordance with the previous studies in [8-10]. And this indicates that high strain rates lead to higher level near-surface local plastic strain.

3.3. Local plastic strain gradient

The local plastic strain gradient K is defined as the slope of the ALM curve. The maximum and minimum ALM values were used to calculate K. The strain gradient is plotted against strain rate and crater depth in Fig. 3. It is evident that the strain gradients obtained at high strain rates are significantly different from that at low strain rates. High strain rates produced much larger strain gradient than that at low strain rates regardless of crater depth. This indicates that strain rate plays a key role in determining the local plastic strain gradient. And high strain rate is favourable to impose plastic deformation on the selected area near the surface.

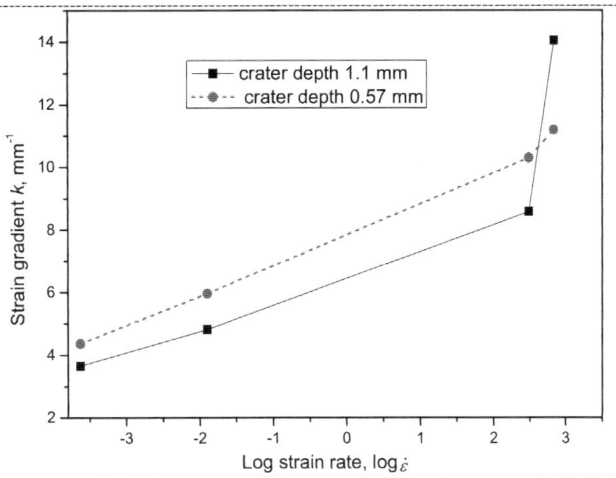

Figure 3. The influence of macroscopic deformation on the strain gradient at various level of strain rate

4. Conclusions

The pure iron has been plastically deformed under various macroscopic strains and strain rates. The results show that the strain rate plays a key role in controlling the local plastic strain gradient beneath the treated surface. The high strain and high strain rate lead to larger strain gradient and higher near-surface local plastic strain level. Correspondingly, low strain rate and high degree deformation are favourable to produce uniform near-surface deformation.

References

[1] Tao N R, Wang Z B, Tong W P, Sui M L, Lu J, and Lu K 2002 Acta Mater. 50 4603–16
[2] Wu X, Tao N, Hong Y, Liu G, Xu B, Lu J, Lu K 2005 Acta Mater. 53 681 - 691
[3] Zhu K Y, Vassel A, Brisset F, Lu K, Lu J 2004 Acta Mater. 52 4101–10
[4] Lu K, Lu J 2004 Mater. Sci. Eng. A 375–377 38–45
[5] Masayuki K, Angus J. Wilkinson, John M. Titchmarsh 2006 Acta Mater. 54 539 - 548
[6] Timoshenko S. and Goodier J N 1961 *Theory of Elasticity*, (New York: McGraw-Hill), pp 372-381
[7] Dai K, Shaw L 2007 Mater. Sci. Eng. A 463 46-53
[8] Lee WS, Lam HF. 1996 J. Mater. Proc. Technol. 57:233
[9] Staudhammer K P and Murr L E 1980 Mater. Sci. Eng. 44 97-113
[10] Chiem C Y and Duffy J 1983 Mater. Sci. Eng. 57 233-247

Thermal-mechanical analysis on W/CuCrZr plasma facing component with functionally graded material interlayer

D H Zhu[1, 3], J L Chen[1], Z J Zhou[2] and R Yan[1]

[1] Institute of plasma physics, Chinese academy of sciences, Hefei 230031, P. R. China

[2] University of science and technology Beijing, Beijing 100083, P. R. China

E-mail: dhzhu@ipp.ac.cn

Abstract. Thermal-mechanical analysis is performed on W/CuCrZr plasma facing component with functionally graded material (FGM) interlayer under the steady-state heat load of 5 MW/m^2 using a finite element method (ANSYS code). The influences of the FGM interlayer on the stress, strain and temperature distribution in the whole component is evaluated and compared to the joint without an interlayer and with OFHC interlayer. It is shown that the inserting of W/CuCrZr FGM interlayer between W and CuCrZr effectively alleviates the thermal stress, whose reduction can reach up to 70% when the W/CuCrZr FGM interlayer thickness exceeds 6 mm. Meanwhile, the optimization of the structure and component distribution for W/CuCrZr FGM interlayer was also carried out. The thermal stress tends to saturate with the layer number exceeding 4. For 6 mm 4-layered FGM interlayer, the preferred component distribution exponent (p) is 1.8.

1. Introduction

Plasma facing component in fusion devices, especially for the divertor target, always faces the extreme conditions with the high heat load in the range 5 - 20 MW/m^2 [1, 2]. It consists of a plasma facing and a heat sink material, fulfilling the different functions that require different material properties. Tungsten (W) is a promising plasma facing material and is widely used in present tokamaks (EAST, JET, ASDEX, etc) attributed to its high melting point, low tritium retention, low sputtering yield and good thermo-mechanical properties [3 - 6]. Furthermore, W has been selected as the armor material in the activated phase of ITER and currently the most suitable candidate for the first wall in DEMO and future fusion reactors [7, 8]. Meanwhile, copper alloy, i.e. CuCrZr is proposed as the heat sink material because of the high thermal conductivity and other acceptable properties [9]. However, due to the large differences in the coefficient of thermal expansion (CTE) and elastic modulus, the joining of W and CuCrZr results in a high thermal stress concentration at the interface when a W/CuCrZr component is exposed to a quasi-stationary heat load. This stress concentration may lead to the failures in forms of cracking and detachment reducing the lifetime of the components.

In the past decades, a large amount effort has been performed to mitigate the thermal stress when joining of W on CuCrZr to form high heat load component. From the engineering point of view, the solution is to use brush-like or lamellar type of W armor design. The advantage of the brush structure

[3] Corresponding author: Zhu Dahuan; Tel: +86-551-5593282; Fax: +86-551-5591310.

is that the single armor elements are free to expand under the heat load deposition, reducing the thermal stress [10]. On the other hand, the insert of an interlayer has also been demonstrated as an effective method to join W armor and CuCrZr heat sink. The interlayer materials are often typically ductile in order to accommodate the dissimilar properties, and it is also preferable that the CTE of an interlayer material is intermediate between W and CuCrZr. If the intermediate interlayer is graded, the stress mitigation would be better than that of a single homogenous interlayer. So W/CuCrZr functionally graded material (FGM) interlayer is the most suitable interlayer when joint W armor material on CuCrZr heat sink material [11].

In this paper, thermal-mechanical simulation was performed on a divertor scaled W/CuCrZr component with FGM interlayer using ANSYS code based on finite element method. The effect of the W/CuCrZr FGM interlayer on the temperature, stress and strain distribution in the entire component is examined and compared to the joint without an interlayer and with oxygen free high conductivity copper (OFHC) interlayer. The interlayer thickness is of special concern in our analysis. In addition, the optimization of the structure and component distribution of W/CuCrZr FGM interlayer was also carried out.

2. Model and materials

To perform the thermal and mechanical simulation a commercial finite element code (ANSYS 10.0) was used. The 2D axis-symmetry geometry analyzed is a divertor scaled model, i.e. a flat-type actively cooled W/CuCrZr mock-up with FGM interlayer as depicted in figure 1. The model contains three pieces of different materials. At the top, the W armor faces to the plasma, providing the protection from the plasma irradiation. At the bottom, the CuCrZr heat sink material with water cooling tube serves to transport heat rapidly, providing the actively cooling of the component. Between the W armor and CuCrZr heat sink, the W/CuCrZr FGM interlayer, i.e. graded layers, is used to join the dissimilar materials. The height of the bottom CuCrZr heat sink and the top W armor are 20 and 2 mm, respectively, while the thickness of the W/CuCrZr FGM interlayer changes from 0 to 6 mm to investigate the effect of interlayer thickness on the stress, strain and temperature distribution. The length of the W/CuCrZr FGM mock-up is 30 mm. The diameter and thickness of water cooling tube are 10 mm and 1mm, respectively, and the distance between the bottom interface and center tube wall is 8 mm. In our simulation, the model element was described by plane 55 for thermal analysis and plane 42 for structure analysis. The analytical model was meshed for finite element model with mapped and free meshing for different pieces using the quadrilateral-shaped elements. Fine meshing through the thickness was performed to ensure the saturated results. The meshing size near the interfaces was refined to an ultra-fine size to 0.05 mm because this area was under very high stress concentration.

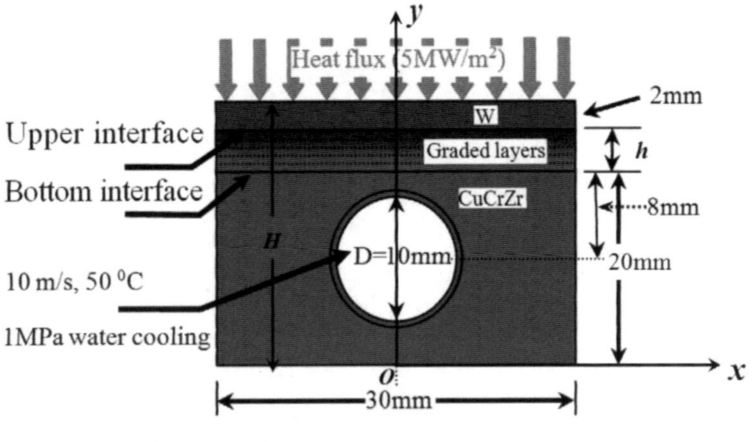

Figure 1. Schematic of cross section view of flat-type actively cooled W/CuCrZr mock-up with FGM interlayer.

Journal of Physics: Conference Series **419** (2013) 012038 doi:10.1088/1742-6596/419/1/012038

The thermophysical and mechanical temperature-dependent properties of the W and CuCrZr for finite element analysis (FEA) are listed in **Table 1**, referencing from the similar work [12, 13]. The spatial composition of the CuCrZr varies discontinuously and step-wisely from the heat sink substrate to armor material and obeys an exponential function distribution as follows

$$V_{CuCrZr}=\begin{cases} 1 & (y \le 20) \\ (\frac{y\text{-}20}{h})^{p} & (20 \le y \le H\text{-}2) \\ 0 & (y \ge H\text{-}2) \end{cases} \quad (1)$$

where the V_{CuCrZr} is the volume fraction distribution of CuCrZr through *y* axis, the p is the component distribution exponent to characterize the volume distribution of CuCrZr in the W/CuCrZr FGM interlayer, h and H are the thickness of graded interlayer and the whole component, respectively. The Kerner rules are used to calculate the properties of W/CuCrZr mixed composite layers [14].

In the process of numerical simulation, a homogenous steady-state heat load of 5 MW/m^2 was deposited on the top of W armor surface, while a water velocity of 10 m/s, an average temperature of 50 °C and a water pressure of 1 MPa relevant to actual tokamak cooling condition were applied for active cooling [15]. To simplification, 1/2 part of the axisymmetric model is used, and the isotopic behavior of materials as well as perfect thermal contact at interface were assumed. During thermal stress simulation, the stress free temperature of 50 °C was used as the reference temperature. The center line, i.e. the symmetry axis was to be constrained at horizontal direction, while the origin point was pinned to be restricted at vertical direction in order to avoid the movement of the model. All other edges were free so that the expansion of the model was performed during the heat loading. The maximum von Mises stress was used as the first criterion to evaluate the effect of the W/CuCrZr FGM interlayer to joint W on CuCrZr as compared to the joint with OFHC interlayer and without an interlayer. The auxiliary criterions are the changes of maximum von Mises plastic strain and maximum surface temperature.

Table 1. Temperature-dependent properties of the W and CuCrZr for FEA.

Materials	Temperature (°C)	Thermal conductivity (Wm^{-1}K^{-1})	Coefficient of thermal expansion (10^{-6}K^{-1})	Elastic module (GPa)	Yield strength (Mpa)	Tangent modulus (GPa)
	20	379	15.7	128	293	0.9
CuCrZr	250	355	17.3	118	257	0.7
	500	352	18.6	102	194	0.6
	20	173	3.93	398	1360	1.3
W	500	133	4.21	390	854	1.0
	1000	110	4.51	368	465	0.8

3. Result and discussion

3.1. Temperature distribution

The temperature gradient distribution in the vertical direction on the W/CuCrZr mock-up with different interlayer architectures is obtained by applying the heat load at W armor surface and removing heat at the cooling tube in CuCrZr heat sink material. As expected, the maximum surface temperature occurs at the edge corner of W armor surface far from the cooling tube. Figure 2 shows the temperature distribution through the edge line with different architectures, including W/CuCrZr FGM interlayer, OFHC interlayer and sharp joint. The maximum surface temperature is 342 °C for the sharp joint geometry under the analysis condition. For 1 mm OFHC interlayer and W/CuCrZr FGM interlayer with liner distribution, the surface temperature is 351 and 361°C, respectively. The lower temperature rise for OFHC interlayer as compared to that for FGM interlayer was attributed to its excellent thermal conductivity. If the FGM interlayer thickness increases to 6 mm, the maximum surface temperature reaches to 473°C, which is acceptable for W operation. The thick FGM interlayer leads to the long heat transfer distance and thus the more thermal heat accumulates at surface, causing high surface temperature. So, the surface temperature rise caused by inserting FGM interlayer should be given more concern during operation. It is interest that the temperature distribution shows no significant difference in CuCrZr heat sink with different interlayer architectures, even the vacuum plasma spraying (VPS)-W coating with low thermal conductivity as the armor material. That is to say, the inserting of several millimeters W/CuCrZr FGM interlayer causes negligible influence of the temperature distribution in CuCrZr heat sink substrate. Moreover, the reasonable design of the structure for CuCrZr heat sink as well as the cooling system helps to control the temperature in its operation range.

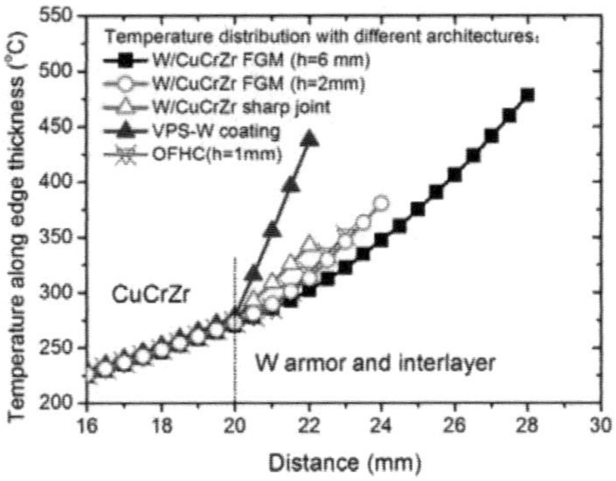

Figure 2. Temperature distributes through thickness with different interlayer architectures.

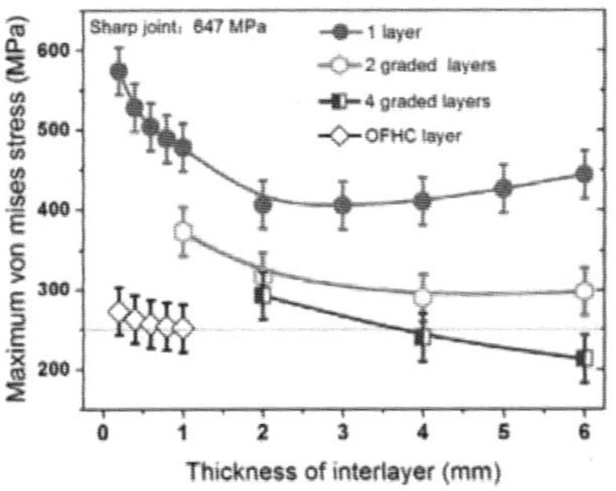

Figure 3. The peak von Mises stress changes through interlayer (graded layers and OFHC layer) thickness with different interlayer architectures.

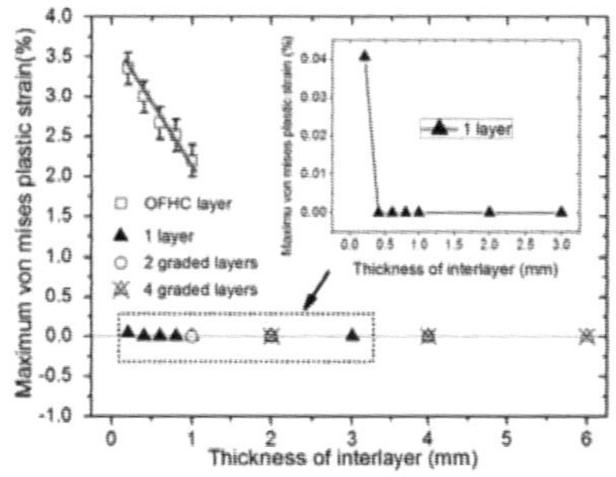

Figure 4. The peak von Mises plastic strain changes through interlayer (graded layers and OFHC layer) thickness with different interlayer architectures.

3.2. Stress and strain distribution

For the sharp joint, as expected in the general viewpoint, the peak von Mises stress is found in the singular corner site in the W layer, while the peak von Mises plastic strain is located in in the singular corner site in CuCrZr substrate. For different interlayer architectures, the comparison of the maximum von Mises stress and maximum von Mises plastic strain through interlayer thickness is plotted in figure 3 and 4, respectively. Without an interlayer, the maximum von mises stress and maximum von mises plastic strain are 647 Mpa and 0.7308%. However, when several millimeters W/CuCrZr FGM interlayer is inserted into the joint, the reduction of peak von Mises stress is significant, which can reach up to 70% as compared to that in a sharp joint, illustrating the feasibility of W/CuCrZr FGM interlayer to alleviate the thermal stress, which is largely related to the FGM interlayer thickness. In general, the maximum von mises stress decreases with increasing the interlayer thickness, because the thicker transition distance provide the large stress relaxation. However, for 1-layered W/CuCrZr interlayer, it is found that the maximum von mises stress increases with its thickness exceeding 3 mm. That is not surprise, because the higher thickness results in the high temperature gradient, especially the high temperature at upper interface, resulting in the large thermal expansion. It is concluded that the interlayer poses a critical thickness for minimal thermal stress, above which the thermal stress increases. If the 4 mm 4-layered W/CuCrZr FGM interlayer is used, the decreasing effect of the maximum von mises stress is the same as compared the OFHC interlayer. Moreover, if the 4-layered FGM interlayer thickness increases to 6 mm, the maximum von mises stress decreased to 213 MPa, which is 15% lower than that using 1 mm OFHC interlayer.

As reported in the reference [11], for OFHC interlayer, the reduction of the thermal stress is accompanied with significant plastic strain, and that is why the maximum thermal stress changes little with increasing the interlayer thickness. The large von Mises plastic strain is found in OFHC interlayer as high as 2%, which is 3 times than that in a sharp joint. The plastic strain may further be amplified during many cycled operation, reducing the component lifetime. However, for the joint using W/CuCrZr FGM interlayer, the plastic strain disappeared with FGM interlayer thickness exceeding 0.4 mm. According to the stress mitigation effect and the plastic strain, the W/CuCrZr FGM interlayer seems to be more preferred than OFHC interlayer.

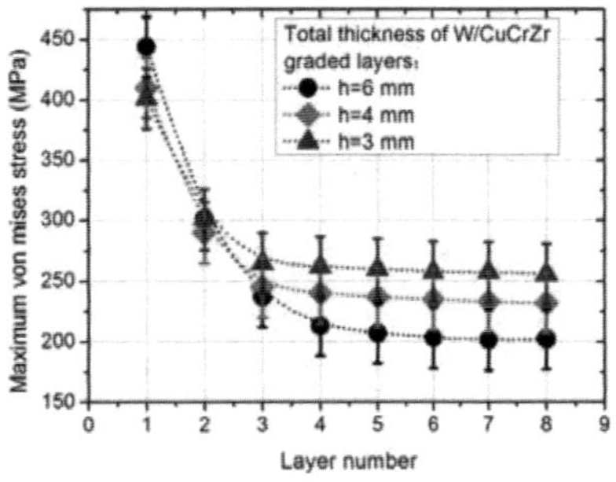

Figure 5. The change of the maximum thermal stress with increasing of graded layer number.

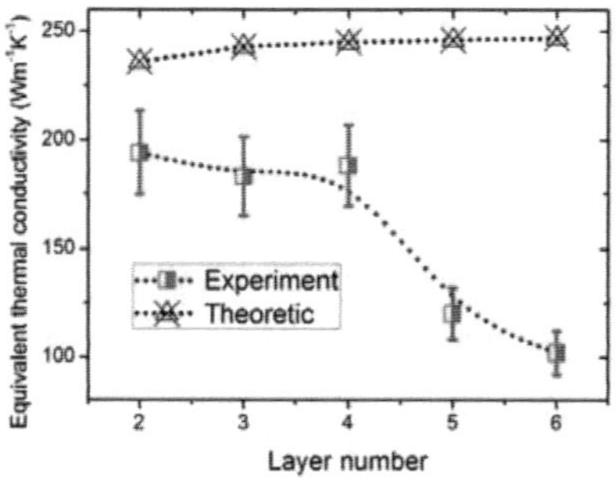

Figure 6. The change of equivalent thermal conductivity of W/CuCrZr FGM with increasing of graded layer number.

3.3. Optimization of structure and component distribution

Figure 5 shows the change of the maximum von Mises stress with increasing of graded layer number with different interlayer thickness of 3, 4 and 6 mm. The total thickness of graded layers was fixed as constant, and the homogenous thickness of each layer was assumed. The maximum von Mises stress decreases with increasing layer number. For single layer, the maximum von Mises stress is as high as 444 Mpa, however, for 4-graded layers with liner distribution, the maximum von Mises stress decreases to 200 MPa, demonstrating that the graded effect on stress alleviation is significant. The maximum von Mises stress decreases rapidly with increasing the layer number below 4, and then saturates after the layer number exceeding 4, suggesting that the graded layer number should not be less than 4 to most relieve the thermal stress. However, continually increasing numbers of total layers may lead the difficulties of materials processing, enhancing the interface heat resistance which should be of special concern. For example, figure 6 shows the equivalent thermal conductivity of the W/CuCrZr FGM with increasing layer numbers, in which the W/CuCrZr FGM was sintered by resistance sintering under ultra-high pressure method. The theoretic thermal conductivity almost keeps invariant with increasing of layer number, however, the experiment curve shows a sharp decrease when the layer number exceeds 4 due to the interfaces thermal resistance, illustrating that in actual application, the choosing of the promising graded layers number should balance the thermal stress alleviation and heat transfer ability.

Figure 7 shows the variation of the maximum thermal stress through the exponent distribution (p). For 4-graded layers with total thickness of 6 mm, the maximum thermal stress decreases rapidly with p below 1.2, then seems to saturate with p between 1.2 and 2, and finally increases with p exceeding 2. Such evolution is valid because the FGM interlayer properties diverge from CuCrZr heat sink and close to W armor with increasing of p. Moreover, it is possible to predict that such distribution of maximum thermal stress is very agreement with the results, if the exponent distribution is infinitesimal or infinite (i.e. graded layers degenerate to pure tungsten or CuCrZr material). In addition, it should also point out that the maximum thermal stress is found in the surface W layer near the upper interface with p below 1.2, while in the heat sink material at the bottom interface with p exceeding 1.2. For 4 and 6 mm graded layers, the promising p for minimal maximum thermal stress are 1.8 and 2, respectively. The promising p for minimal maximum thermal stress is not uncertain and depends on the structure of whole component. In a word, the promising p is greater than 1, illustrating the FGM interlayer should be rich in W. Meanwhile, figure 8 shows the temperature evolution with increasing of p. From the curve, we found that the temperature at bottom interface almost keeps invariant, which is consistent with the above discussion in section 3.1. Both the temperature at surface and upper interface increases with increasing p. But the temperature rise seems to be not substantial, which can be ignored. The temperature change is below 30 °C with p in the range 1.2 to 2.

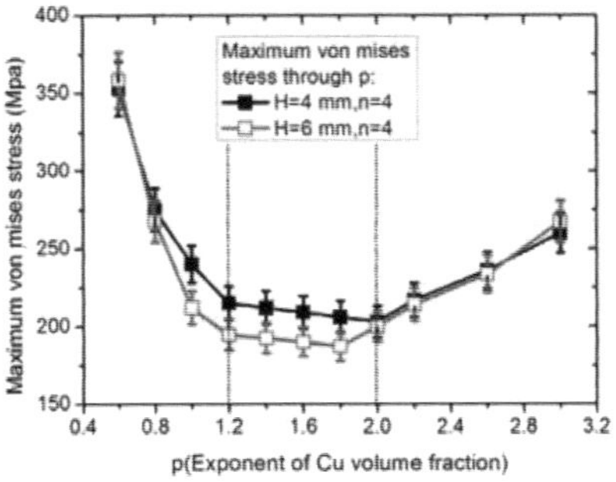

Figure 7. The maximum thermal stress changes through the component distribution exponent (p) with different architectures.

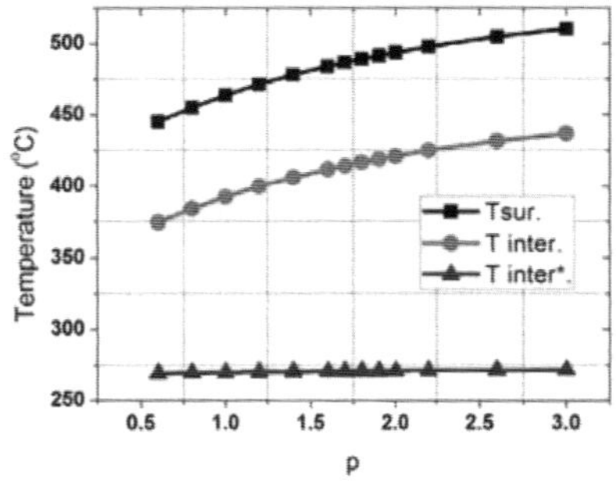

Figure 8. The change of surface temperature (T sur.) and interface temperature (T inter. is the temperature at upper interface, and T inter*. is the temperature at bottom interface)

4. Conclusion

Thermal mechanical analysis was carried out on W/CuCrZr component with FGM interlayer under the steady state heat load of 5 MW/m^2 using a finite element method. The effect of the FGM interlayer on the stress, strain and temperature distribution is evaluated and compared to the joint without interlayer and with OFHC interlayer. The using of W/CuCrZr FGM interlayer with several millimeters thickness can effectively alleviate the stress and decrease the plastic strain in the entire component. The reduction of stress can reach up to 70% with the disappearance of the plastic strain when 6 mm W/CuCrZr FGM interlayer is used. The thermal stress alleviation is largely related to the interlayer thickness. Meanwhile, the use of the thicker W/CuCrZr FGM interlayer causes acceptable temperature rise at W armor surface and negligible influence of the temperature distribution in CuCrZr heat sink substrate.

The optimization of the component distribution and structure for W/CuCrZr FGM mock-up was also carried out. The maximum thermal stress changes rapidly with increasing the graded layer number initially and then saturates after the graded layer number exceeding 4. The variation of p not only changes the value of thermal stress, but also the thermal stress distribution. The promising p for minimal maximum thermal stress is not uncertain and depends on the structure of whole component. The change of p in the range 1.2 to 2 leads a negligible surface temperature rise. For W/CuCrZr plasma facing component with 6 mm 4-layered FGM interlayer, the preferred component distribution exponent (p) is 1.8.

Acknowledgments
This work was subsidized by the Key Project of Chinese Academy of Sciences (grant No. KJCX2-YW-N35) and the National Natural Science Foundation of China (No. 11175205)

Reference

[1] Shimomura Y 2007 *J. Nucl. Mater.* **363 - 365** 467
[2] Janeschitz G 2001 *J. Nucl. Mater.* **290 - 293** 1
[3] Luo G N, Zhang X D, Yao D M and Gong X Z 2007 *Phys. Scr.* T **128** 1
[4] Brezinsek S, Fundamenski, W Eich T and Coad J P 2011 *J. Nucl. Mater.* **415** 936
[5] Neu R, Dux R, Geier A, Gruber O and Kallenbach A 2003 *Fus. Eng. Des.* **65** 367
[6] Philipps V 2011 *J. Nucl. Mater.* **415** 1
[7] Cambi G, Cepraga D G and Petrizzi L 2009 Fus. Eng. Des. **84** 530
[8] Raffray A R, Malang S and Wang X 2000 *Fus. Eng. Des.* **84** 1553
[9] Park J Y, Jung Y J and Byung-K 2011 *J. Nucl. Mater.* **417** 916
[10] Barabash V, kAiba M, Cardella A and Mazul I 2000 *J. Nucl. Mater.* **283-287** 1248
[11] Chapa J and Reimanis I 2002 *J. Nucl. Mater.* **303** 131
[12] Weber T and Aktaa J 2011 *Fus. Eng. Des.* **86** 220
[13] Zhang P B, Zhao J J, Qin J and Wen B 2011 *Fus. Eng. Des.* **86** 45
[14] Gasik M M 1998 *Compu. Mater. Sci.* **13** 42
[15] Chong F L, Chen J L and Li J G 2008 *Mater. Des.* **29** 1675

Development of Thermal Spraying and Coating Techniques by Using Thixotropic Slurries Including Metals and Ceramics Particles

S Kirihara[1], Y Itakura[2] and S Tasaki[1]

[1] Joining and Welding Research Instite, Osaka University, Osaka 567-0047, Japan
[2] Graduate School of Engineering, Osaka University, Osaka 565-0871, Japan

E-mail: kirihara@jwri.osaka-u.ac.jp

Abstract. Thermal nanoparticles coating and microlines patterning were newly developed as novel technologies to fabricate fine ceramics layers and geometrical intermetallics patterns for mechanical properties modulations of practical alloys substrates. Nanometer sized alumina particles were dispersed into acrylic liquid resins, and the obtained slurries were sputtered by using compressed air jet. The slurry mists could blow into the arc plasma with argon gas spraying. On stainless steels substrates, the fine surface layers with high wear resistance were formed. In cross sectional microstructures of the coated layers, micromater sized cracks or pores were not observed. Subsequently, pure aluminum particles were dispersed into photo solidified acrylic resins, and the slurry was spread on the stainless steel substrates by using a mechanical knife blade. On the substrates, microline patterns with self similar fractal structures were drawn and fixed by using scanning of an ultra violet laser beam. The patterned pure metal particles were heated by the argon arc plasma spray assisting, and the intermetallics or alloys phases with high hardness were created through reaction diffusions. Microstructures in the coated layers and the patterned lines were observed by using a scanning electron microscopy.

1. Introduction

Thermal nanoparticles spraying had newly developed as a novel coating technique to create fine ceramic layers without structural defects on practical alloys substrates [1,2]. The nanometer sized alumina ceramic particles were dispersed into liquid resins. Fluid viscosity of the mixt pastes were measured and optimized to realize homogenized particles dispersions and smooth slurry flows. The obtained thixotropic slurry of rheological flow was sputtered by using compressed air jet, and the slurry mist blew into the arc plasma with an argon gas spraying. On a stainless steel substrate, the ceramic layers were formed. The formation of micro cracks and pores in the coated layers will be discussed by scanning electron microscope observations. Subsequently, thermal microline patterning to realize metal phases drawing on the alloys substrates had been developed [3,4]. Micrometer sized pure aluminum particles were dispersed into the photo solidified liquid resins. The slurry pastes were spread on the stainless steel substrate, and micro patterns were drawn and fixed by using an ultra violet laser scanning. The patterned pure metal particles were heated by the argon arc plasma spraying, and the intermetallics and alloys phases with high hardness were created through reaction diffusions. Microstructures and composite distributions in the vicinity of formed alloy and metal interfaces were observed and analyzed by using an electron microscope. Load dispersion abilities of the network were evaluated by using conventional mechanical tests and compared with simulated and visualized profiles by using a numerical analysis simulation. The modulation effects of stress distributions on the material surfaces by design and fabrication of these geometric fractal patterns will be discussed.

2. Experimental Procedure

The nanometer and micrometer sized alumina and aluminum particles of 200 nm and 10 μm in average diameters were dispersed into the acrylic liquid resins at 40 % and 60 % in volume fractions, respectively. These obtained slurries were mixed in airtight containers through planetary configuration of rotation and revolution movements. Fluid characteristics of the slurry pasts were evaluated by using a viscosity and viscoelasticity measuring instrument to realize thixotropic fluid flows. The thixotropic slurry was injected into a gas flow chamber perpendicularly from a syringe nozzle of 40 mm in inner diameter, and the fluid viscosity was decreased by shear stress loading as schematically illustrated in Fig. 1. The slurry mists were sprayed toward a glass plate to measure and observe droplet diameters and spattering patterns. The micro sized mists were formed through the compressed air jet of 2 atm in gas pressure, and introduced into the into the arc plasma with an argon gas spray of 50 slpm in flow late through the stainless still guide nozzle of 4 mm in caliber size. Figure 2 shows the mist supplying equipment mounted on the plasma spraying apparatus. The alumina coated layer was formed on the SUS-304 stainless steel substrate of 50×50×1 mm in size placed at 140 mm in distance from the plasma gun. Crystal phase of the coated layer was analyzed by X-ray diffraction, and cross sectional micro-structures and compositional distributions were observed by using an optical and scanning electron microscopy. Subsequently, self similar patterns of Hilbert curve with stage numbers 1, 2, 3 and 4 of the fractal line structures were designed by using a computer graphic application, and these graphic images were converted into the numerical data sets by computer software. These patterns of 25×25 mm in whole size were composed of arranged lines of 400 μm in width. These graphic models were transferred into the processing apparatus as operating data sets. The pure aluminum particles were patterned on the SUS-304 stainless steel substrate by a stereolithographic pattering system as schematically illustrated in Fig. 3. The mixed resin paste with metal particles was spread with 100 μm in layer thickness on the substrate of 30×30×2 mm in size by a mechanically moved knife edge. An ultraviolet laser beam of 355 nm in wavelength and 100 μm in beam spot was scanned on the resin surface. The solid pattern was obtained by light induced photo polymerization. Figure 4 shows the appearance of stereolithographic system. After the removing uncured resin by ultrasonic cleaning in ethanol solvent, the patterned sample was heated above the reaction temperature by using the argon gas arc plasma spraying. The microstructures and composite distributions were observed by using the scanning electron microscopy and energy dispersive X-ray spectroscopy. The stress distributions on the patterned samples were simulated by using the finite element method of the numerical simulation.

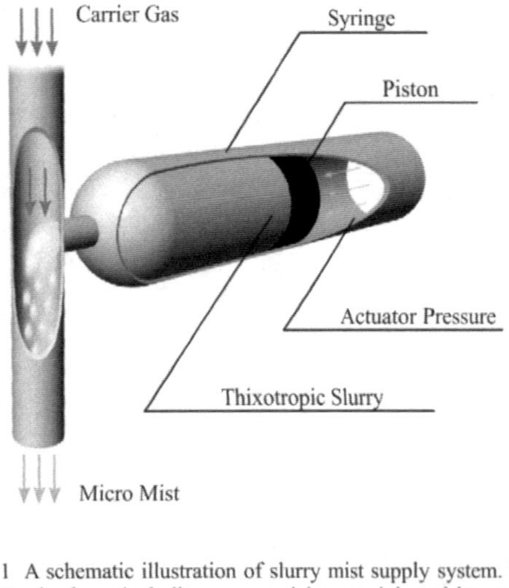

Fig. 1 A schematic illustration of slurry mist supply system. The thixotropic slurry including nanoparticles was injected into a gas flow chamber from an actuator moved syringe.

Fig. 2 An appearance photograph of the slurry mists supplying system mounted on a plasma torch. The slurry syringe piston is moved by a mechanical actuator in micromater order accuracy.

Journal of Physics: Conference Series **419** (2013) 012039 doi:10.1088/1742-6596/419/1/012039

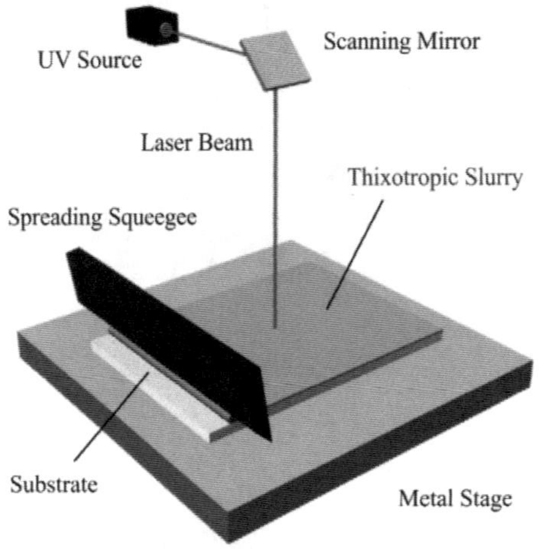

Fig. 3 A schematic illustrated stereolithograpy system using laser scanning processes. The thixotropic slurry can be spread smoothly on a metal substrate by using a mechanically movedsqueegee.

Fig. 4 The appearance photograph of stereolithograpy equipment with the slurry supplying syringe. An ultra violet laser beam was irradiated from the top position of this apparatus for a metal stage.

3. Results and Discussions

The acrylic liquid resin with the alumina nanoparticles of 200 nm in diameter at high volume contents exhibited the thixotropic characteristics of a non-Newtonian fluid. The formed slurry with the particles dispersion above 42 % in volume contents could not flow smoothly. For the stable materials supplying and the continuous mist creation, the maximum volume content of the nanoparticles can be optimized at 40 %. In this slurry fluid, the nanoparticles are considered theoretically to spatially disperse without coagglutinations. The obtained slurry blew toward the glass plate to observe and measure the mist droplets shapes and sizes. These droplet diameters became finer as shown in Fig. 5, and the variations could be reduced by decreasing the slurry supply rate. The minimum diameter size was 50 μm. Figure 4 shows the fine alumina coated layer formed on the stainless steel substrate. The ceramic layers of 50 μm in thickness were formed at 300 gpm in supply rate. The structural defects of cracks or pores were not observed through microscopic observations. The coated ceramics layer and alloy substrate were joined successfully without the defects of voids or exfoliations in the interfaces. Moreover, the carbon element contaminations in the created alumina layers derived from the acrylic liquid resins were not detected though the X-ray diffraction analysis. Subsequently, the iron aluminide micro pattern with the fractal structure of Hilbert curve was formed successfully on the stainless steel substrate. Figure 4 shows the formed fractal polyline of the number 3 in fractal stage. The micrometer order geometric structure was composed of fine intermetallics lines of 450 μm in width. The part accuracy of these microline patterns were estimated as 10 % approximately. The iron aluminide composite was formed widely comparing with the designed line width, though the reaction diffusion between the aluminum and stainless steel substrates. The defects of crack or pores could not observed in the cross sectional microstructure of the formed intermetallics. And the intermetallic layer and the alloy substrate were joined successfully without void formations through the reaction diffusion. The composite phase could be identified as Al-rich Fe_3Al. The stress distributions on the patterned surfaces were visualized for the Hilbert curve of stage number 3 through the numerical simulation. The required mechanical properties of Young's modulus were defined along the compositional analysis and the phase identifications. The stress intensities concentrate into the vicinity of fixed edge and are distributed along the patterned lines and the corner points with the higher hardness. The fractal patterns can include the more numbers of sides and nodes in limited aria comparing with the periodic arrangements of polygon figures.

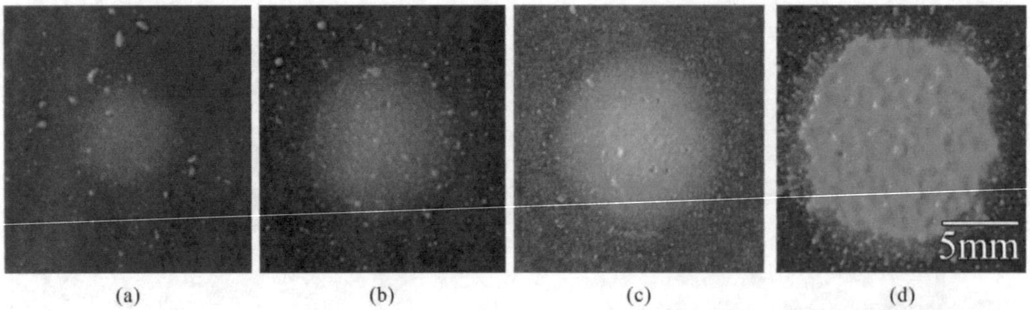

Fig. 5 Micro mist spraying patterns formed by blowing of the thixotropic resin slurry including the alumina nanoparticles. The slurry supplying rate was varied systematically to create the fine mist droplet. The mist droplet sizes in these spraying patterns (a), (b), (c) and (d) are measured by using a digital optical microscope as about 50, 100, 200 and 500 μm, respectively.

Fig. 6 A microstructure of an alumina coated layer formed on a stainless steel substrate. A side vew of the coated sample was observed to show the thermal sedimention of nanoparticles.

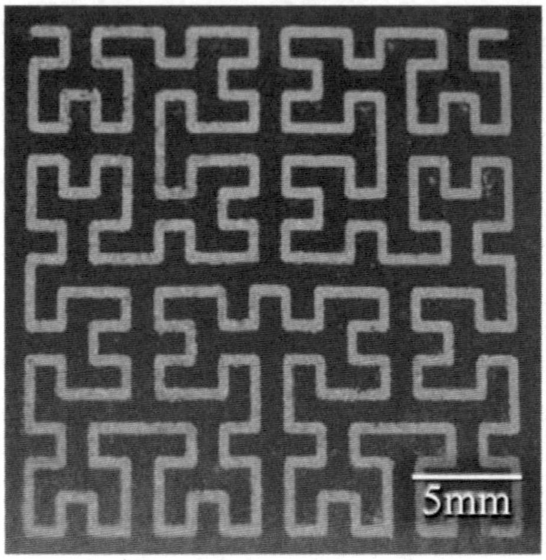

Fig. 7 An iron aluminide micro pattern with Hilbert curve of a fractal structure formed on the stainless steel substrate through reaction diffusion assisted by the plasma spray irradiation.

4. Conclusion

Fine alumina coated layers and aluminide composite lines could be created successfully on stainless steel substrates by thermal nanoparticles spraying and microlines patterning to improve thermal and mechanical properties on components surfaces. As raw materials using these techniques, thixotropic slurries including nanometer and micrometer sized particles were formulate systematically. The slurry handling methods are key techniques to realize the nanoparticles arrangements create useful functional surfaces on various components. In the near future, the investigated coating and patterning techniques will become the candidate for efficient processes to realize effective thermal barrier coating without structural defects and mechanical properties improvement on various mechanical components

References

[1] S. Kirihara: Thermal Spraying Technology, 30 (2010) 44-50.
[2] S. Kirihara: Journal of Japan Welding Society, 80 (2011) 6-9.
[3] Y. Uehara, S. Kirihara: Journal of Smart Prosessing, 1 (2012) 186-189.
[4] Y. Uehara, S. Kirihara: Joural of Functionally Graded Materials, 22 (2012) 36-42.

Study on Cd vacancy in CdZnTe Crystal by Positron Annihilation Technology

Weiwei Liu, Jiahua Min*, Xiaoyan Liang,Jijun Zhang, Xiaoxiang Sun,Linjun Wang,An Ran,Bangjiao Ye

Department of Electronic Information Materials, School of Materials Science and Engineering, Shanghai University, 149 Yanchang Road, Shanghai 200072, China

E-mail:minjh@shu.edu.cn

Abstract. Cd vacancies in cadmium zinc telluride(CdZnTe) crystals have an important effect on the crystal properties.In this paper, position distribution and concentration change of Cd vacancy in CdZnTe crystal grown by the temperature gradient solution growth (TGSG) were investigated by positron annihilation technology(PAT) , which was based on the potential energy distribution and probability density of the positron in the crystal. The results showed that, the density of Cd vacancy increased obviously from the first-to-freeze to stable growth of the ingots, while decreased along the radial direction of the ingots.

Keywords: CdZnTe; Cd vacancy; positron annihilation Technology (PAT)

1. Introduction

Owing to its large average atomic number, wide band gap, high resistivity, and good electron transport properties, Cadmium Zinc Telluride(CdZnTe) has become one of the most exploited materials for x-ray and gamma-ray radiation detection at room temperature, which has broad prospects in medicine, space science, airport, port security, nuclear waste monitoring and other applications of nuclear technology[1-3].

High resistivity, good integrity of CdZnTe monocrystal is essential for high-performance detectors. However, the relatively high equilibrium vapor pressure of Cd easily leads the lack of Cd in CdZnTe single crystals, therefore imperfections such ac Cd vacancies, Te inclusions, and Te antisite are included in the process of crystal growth[4,5]. The above problems would affect the stoichiometric ratio of CdZnTe monocrystalline and cause crystal component segregation, resulting in many defects in the crystal as-grown, which is detrimental to the detection performance of the device, and thus the application of CdZnTe monocrystal.

Now there are a lot of methods to analyze the defects of CdZnTe single crystal, such as photoluminescence (PL), electron paramagnetic resonance (EPR), electron microscopy (SEM and TEM) and so on. However, these methods are difficult to obtain the information about the microscopic structure on the atomic scale. In the past 20 years, positron annihilation technology has been widely used to study neutral vacancy-type and negatively charged vacancy-type defects in the solid. As the neutral vacancy and negatively charged vacancy will become the captured trap of positive electrons, so that the positive electrons which have been injected in will be captured by the neutral vacancy and negatively charged vacancy after thermalization and diffusion. Then the positive electrons will annihilate and release two γ photons. The characteristics of defects in the solid can be analyzed by the information carried by the γ photons.

As we all know that the rate of positron annihilation is proportional to the electron concentration of the place where the annihilation appears. There are different positron lifetime according to the annihilation appears at the different types of vacancies[6]. The experimental data of positron annihilation can not only give the relationship between lifetime and temperature, but also identify the charge state and the ionization energy of the defect.

The CdZnTe crystal in this research was prepared through the temperature gradient solution growth method (TGSG) in the lower growth temperature. Furthermore, the radial and longitudinal distribution of Cd vacancies can be analyzed by the positron annihilation technology to obtain the properties of defects.

2. Experiment

2.1 Experimental samples

Test samples are prepared with the CdZnTe crystal made through the method of temperature gradient of the solution growth (TGSG), 14 groups of wafers were cut from the same ingot, and sequentially each sample was made of the adjacent two wafers with the size of 10mm×10mm×1mm on average. In order to remove the surface oxide layer and the stress layer, thus reduce the interference during the measurement process, the wafers were mechanically and chemically polished with Al_2O_3 powder and bromine methanol of 2 to 3%.

Table 1. The relative position parameter of the sample on radial direction test
(r=0: the center of ingot)

sample	logitudinal direction (mm)	radial direction (mm)	sample	logitudinal direction (mm)	radial direction (mm)	sample	logitudinal direction (mm)	radial direction (mm)
a1#	Z=10	r= -10	b1#	Z=20	r= -10	c1#	Z=30	r= -10
a2#	Z=10	r= 0	b2#	Z=20	r= 0	c2#	Z=30	r= 0
a3#	Z=10	r= 10	b3#	Z=20	r= 10	c3#	Z=30	r= 10

Table 2. The relative position parameter of the sample on logitudinal direction test

sample	4#	5#	6#	7#
logitudinal direction(mm)	Z=5	Z=15	Z=25	Z=35

2.2 positron annihilation lifetime test

The Positron Annihilation test was carried out using ORTEC positron spectrometer , with ^{22}Na as the positron source. The instrument has a FWHM of 210ps, and its total number of each lifetime spectrum is 10^6.

3. Results and Discussion

When the point defects in the solid are capturing the positron, the main process is 2γ annihilation. For the two-state capture model, computer fitting can be used on positron annihilation lifetime spectroscopy curve decomposition.We have used General POSITRONFIT fitting procedure, obtained two life values τ_1, τ_2 and the intensities were I_1, I_2, respectively. Therefor the trapping model average life τm and capture rate κ of the two-state positron was as follows:

$$\begin{aligned} & I_1+I_2=1 \\ & \tau_m=\tau_1 I_1+\tau_2 I_2 \\ & \kappa=I_2(\tau_1^{-1}-\tau_2^{-1}) \end{aligned} \quad (1)$$

Table 3. The positron annihilation lifetime(PAL)of CdZnTe crystal in the radial direction

	a1#	a2#	a3#	b1#	b2#	b3#	c1#	c2#	c3#
τ1/ps	214.5	215.2	214.6	215.1	216.3	215.3	216.4	217.4	216.5
τ2/ps	353.7	353.6	353.9	354.5	354.4	354.6	355.6	355.1	360.3
I1/%	80	81.4	79.7	81.4	81.6	81.1	82.4	83.4	82.5
I2/%	20	18.6	20.3	18.6	18.4	18.9	17.6	16.6	17.5

Table 4. The positron annihilation lifetime(PAL)of CdZnTe crystal in the logitudinal direction

	τ_1/ps	τ_2/ps	I_1/%	I_2/%	τ_m/ps	$\kappa/s^{-1}\times10^8$
4#	212.3	353.6	79.4	20.6	233.1	2.77
5#	215	354.9	82.3	17.7	239.8	3.24
6#	215.6	356.1	81.7	18.3	248.12	4.25
7#	217.5	367.2	76.8	23.2	254.8	4.89

Positrons generally annihilate freely in the idealized crystal lattices. If vacancies or micro-structural defects (such as vacancy clusters and dislocations) present in the medium, positrons can be easily captured by defect, forming captured states. Usually species of point defects including vacancies, interstitial atoms, and substitutional defects may exist in CdZnTe crystals.

As a general rule, positrons can be captured by negative ions, neutral impurity and dislocations, forming weak bound states in CdZnTe crystals as well. But these defects can be ignored, because they have little effect on positron lifetime. Additionally, interstitial atoms, substitutional defects and positron can also trap positrons shallowly. The positron annihilation lifetime in these states is similar to the bulk material[7], which have a relatively small influence on positron lifetime. So for CdZnTe single crystal, the vacancy is the main capture potential for positrons. Te vacancies in CdZnTe crystals do not influence the lifetime[8], because they are positively charged and can not capture positron.

However, Cd vacancies are negatively charged and can be the trapping center of the positive electrons, so that they will have the greatest impact on the lifetime of positron. Since the Zn occupy and randomly distribute at the position of the Cadmium which forms negatively charged cationic vacancies, so that Zn in CdZnTe single crystal is just one kind of substitution of Cd. In a word, all cationic vacancies are naturally equivalent.

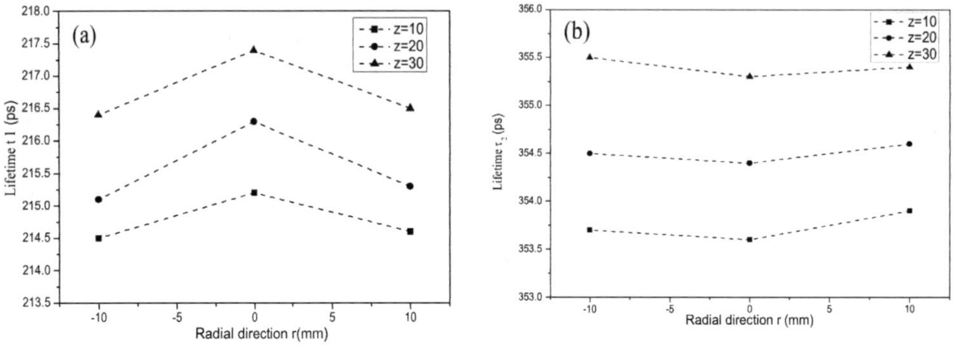

Figure 1. The relation between the positron annihilation lifetime τ_1(a) ,τ_2 (b)and the radial direction parameters

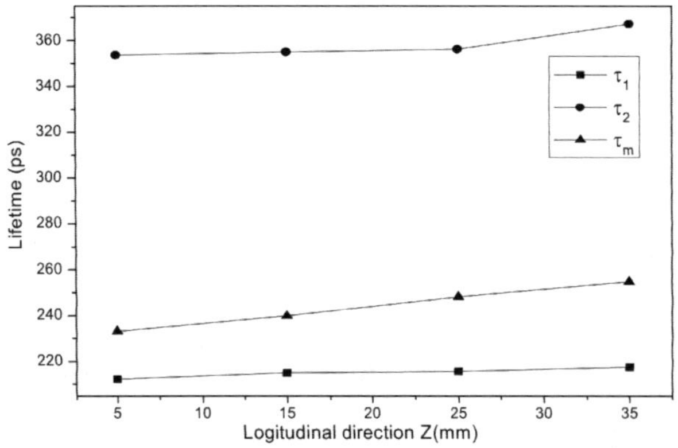

Figure 2. The relation between the positron annihilation lifetime τ_1、 τ_2, the average lifetime and the logitudinal direction parameters

In Fig. 1, it can be seen that in the radial direction, compared with the parameter τ_1, the comprehensive lifetime of the free state of positron and the captured annihilation of single vacancy decreased, both the parameter τ_2 which is the long lifetime value of comprehensive lifetime related to the vacancy cluster and the defects of microstructure and the average lifetime of τ_m[9]related to the sizes of defects and concentration have no dramatical changes.

Due to the change of the content of Zn component[10,11], the Cd vacancies in the center are more than that on the edge. In order to prove this relationship, scanning electron microscopy spectrometer(SEM) is used to test three samples CZT1,CZT2 and CZT3, which were at different positions along the growth axis direction in the same ingot, so that the content of Zn components can be measured evenly spaced along the radial direction and the results are shown in Fig. 3 (r=0: the center of ingot).

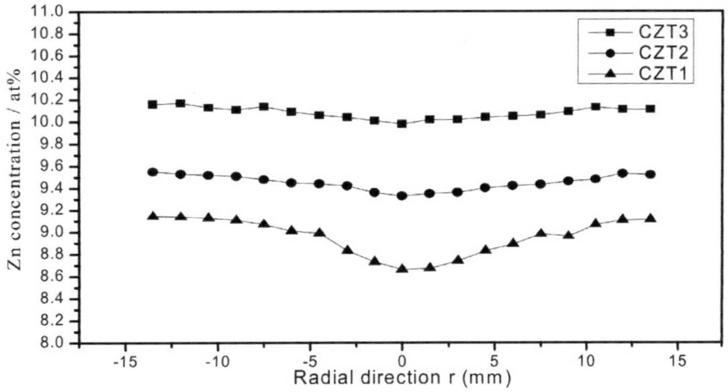

Figure 3. Zn component distributions in radial and axial of the crystals

As shown in Fig. 3, the content of Zn component increases along the radial axis in the crystal, but the lifetime τ_1 shows a contrary tendency. Therefore it is found that the concentration of Cd vacancy will decrease with the increasing of Zn component. In Fig.3, it is clear that τ_1 gradually increases in the direction of the crystal growth, meanwhile, τ_2, τ_m and κ also increase. All that is resulted from the increasing of single vacancies along the growth direction part of them aggregate or coalesce and become bigger, so the relative vacancy cluster increase which is not only affected by the change Zn component. At the same time, since the liquid-solid interface is concave at the end of crystal growth which can easily promote the production of dislocation, the crystal quality declines. It is one of the impact factors which cause the enlargement of lifetime.

From Table 3 and 4, τ_1 always fluctuate around 216ps, so it can be concluded that positron free state annihilation life of CdZnTe crystal grown by the temperature gradient solution growth (TGSG) is about 216ps.

4. Summary

In this paper, the concentration and distribution of Cd vacancies in CdZnTe crystal grown by the TGSG were investigated by PAL, which was based on the potential energy distribution and probability density of the positron of Positron in the crystal.The results showed that Cd vacancies had an increasing trend when the crystal ingot grew from initial growing region to stable growing region, while the content of Cd vacancies had a decreasing trend along the radial direction of the crystal (from center to edge). The lifetime of positron free state annihilation in CdZnTe crystal is about 216ps.

5. Acknowledgments

This work is supported by National Nature Science Foundation(No.11275122), Science and Technology Commission of Shanghai (No.11530500200) and Innovation Program of Shanghai Municipal Education Commission (No.12ZZ096).Thanks to the help from School of Physical Sciences ,University of Science and Technology of China.

References

[1] Shin Hwa-Yuh , Sun Cherng-Yuan 1998 *J. Cryst. Growth.* **186** 67-78.

[2] Limousin O 2003 *Nucl. Instrument. Methods Phys. Res.* A **504** 24-37.

[3] Czock K.H, Arlt R 2001 *Nucl. Instrument. Methods Phys. Res.* A **458** 175-182

[4] Turjanska L, Hoschl P, Belas E, Grill R, Franc J, Moravec P 2001 *Nucl. Instrument. Methods Phys. Res.* A **458** 90-95

[5] Sang Wenbin, Qian Yongbiao, Shi Weiming 2000 *Journal of Crystal Growth.* **214/215** 30-34.

[6] Krause-Rehberg R, Leipner H.S 1999 *Positron Annihilation in Semiconductors, Springer, Berlin.*

[7] Dlubek G 1986 *J.Phys.C:Solide State Phys.* **19** 331 344.

[8] Gely Sykes C 1991 *Solid State Commun.* 80:79.

[9] Hautojarvi P 1995 *Mater Sci Forum.* **47** 175~ 178.

[10] Vydyanath H.R. and Ellsworth J 1999 *J.Mater .Sci Lett.* **18** 695-697.

[11] Vydyanath H.R, Ellsworth J. A, Fisher R. F 1993 *J. Electron. Mater.* **29** 1067-1071.

FGM 2012 IOP Publishing
Journal of Physics: Conference Series **419** (2013) 012041 doi:10.1088/1742-6596/419/1/012041

Thermal residual stresses in SiC particle reinforced aluminium composites: a study by the Taylor-based nonlocal theory of plasticity

D F Cao[1], X Liu[2], H Mei[2], Z T Lei[3], and L S Liu[1,4]
[1]State Key Laboratory of Materials Synthesis and Processing, Wuhan University of Technology, Wuhan 430070, P. R. China
[2]Department of Engineering Structure and Mechanics, Wuhan University of Technology, Wuhan 430070, P. R. China
[3]Sheng Li oil field Sheng Li engineering construction group.Co.LT, Dongying 257000, P. R. China

E-mail:Liulish@whut.edu.cn

Abstract. Strain gradient theories have been considered to be an effective means for capturing the size effects on ceramic particle-reinforced metal-matrix composites (MMCp), but the results predicted by them are significantly lower than the corresponding experimental data. One reason might be, that the thermal residual stresses caused by thermal contraction mismatch between the metal matrix and the ceramic particle were neglected in the numerical models. By incorporating the Taylor-based nonlocal theory (TNT) of plasticity, the finite element method (FEM) is applied in the present research, to investigate the effect of thermal residual stresses on the yield stress and average axial stresses of the aluminium matrix reinforced by silicon carbide particles (SiCp/Al). The elements of the matrix have been implemented in the ABAQUS finite element code through its USER-ELEMENT (UEL) interface for TNT plasticity. Some comparisons with the associated literature demonstrate that the numerical model with the thermal residual stresses is more in agreement with the experimental results.

1. Introduction

Ceramic particle-reinforced metal-matrix composites (MMCp) display strong size effect when the characteristic length scale associated with non-uniform plastic deformation is on the order of microns. The most famous case which has described these multiple-scale characteristics of MMCp, is the Lloyd's experiment[1] in 1994. He investigated the particle size effect for A356 aluminum alloy reinforced by 15 vol.% SiC particles with two different particle sizes, either 7.5 or 16μm diameter particles. The research showed that small silicon carbide particles with 7.5μm diameter gave significantly higher plastic work hardening than large particles with 16μm diameter.

Various continuum micromechanics methods, such as the modified shear-lag theory[2], Eshelby equivalent inclusion methods[3] and finite element method (FEM)[4-6], failed to predict the influence of particle size on the overall deformation behavior for MMCp, since all the classical plasticity theories do not inherently include an intrinsic length scale[7]. In the past several decades, some plastic

[4]Corresponding author, Tel.: +86 27 87651820; Fax: +86 27 87860863
E-mail: liulish@whut.edu.cn (L S Liu)

strain gradient theories have been considered to be an effective means for capturing the size effects in numerical computations, such as phenomenological theories[8, 9], MSG[7] (mechanism-based strain gradient plasticity theory), TNT[10, 11] (Taylor-based nonlocal theory of plasticity), and so on. Strain gradient plasticity theory postulates that the stress at a given point depends not only on the current values and possibly also the previous history of deformation and temperature, but also the neighbourhood of that point. Quite dissimilar to the phenomenological strain gradient plasticity theories[8, 9] , TNT and MSG plasticity have clear physical mechanism, which can be applied to void growth, cavitation instabilities and particle-reinforced composites.

Xue et.al[12] used MSG plasticity to investigate the particle size effect and found good agreements with the experiments of aluminum matrix reinforced by silicon carbide particles (SiCp/Al). However, the stress–strain curves predicted by MSG plasticity are still lower than those reported in Lloyd's experiments[1]. That may be because they have not accounted for other influencing factors in their numerical model such as the effect of thermal residual stress, etc. In addition, Gao and Huang[10] also used TNT plasticity to model the size dependence of MMCp by constructing a spherically symmetric model, which was simplistic and several approximations were made in order to obtain an analytical solution. Yan et al.[6] modified the constitutive relation of metal, with introducing Gao's analytical solution of TNT to describe the flow behavior of the composite matrix. However, all their work did not take into account the effect of residual stresses, which were caused by differences in the coefficient of thermal expansion between the matrix and the ceramic particles.

In the study, by incorporating the TNT plasticity, FEM is used to investigate the effect of thermal residual stresses on the mechanical behavior of the SiCp/Al composites under the quasi-static uniaxial compressive loads. The elements of the matrix have been implemented in the ABAQUS finite element code through its USER-ELEMENT (UEL) interface for TNT plasticity. In the simulation, the average axial stress of composite matrix follows Taylor hardening law and the strain gradient at each Gaussian point can be obtained in a form of nonlocal integral. The numerical results will be compared with Lloyd's experiment results[1].

2. Numerical model

2.1. TNT plasticity

TNT plasticity is based on a multi-scale framework linking the microscale notion of statistically stored and geometrically necessary dislocations(GNDs) to the mesoscale notion of plastic strain and strain gradient. The constitutive equations contain neither higher-order stresses nor strain gradients. As a nonlocal integral of strains, strain gradients are introduced in the form of nonlocal variables. The basic principles in the nonlocal constitutive framework can be written as follows: (1) The flow stress obeys the Taylor hardening relations; (2) The GND density is calculated as nonlocal variables expressed in terms of a weighted average of plastic strain; (3) The essential structure of classical plasticity is preserved. In TNT plasticity, the flow stress σ_{flow} is obtained from the Taylor dislocation model as in equation (1).

$$\sigma_{flow} = M\alpha\mu b\sqrt{\rho} = M\alpha\mu b\sqrt{\rho_G + \rho_S} = \sigma_{ref}\sqrt{f^2(\varepsilon) + l\eta} \tag{1}$$

Where ρ is the total dislocation density, ρ_S is the statistically stored dislocation (SSD) density which accumulates by trapping each other in a random way, ρ_G is the GND density, which is required for compatible deformation of various parts of the non-uniformly deformed material, M is the Taylor factor which acts as an isotropic interpretation of the crystal line anisotropy at the continuum level, α is an empirical material constant in the Taylor dislocation model ranging between 0.1 and 0.5, μ is the shear modulus, b is the magnitude of Burgers vector, σ_{ref} is a reference stress in uniaxial tension, the function $f(\varepsilon)$ is the non-dimensional function determined from the uniaxial stress-strain curve, η is the effective strain gradient, and l is identified as the intrinsic material length introduced by Fleck and Hutchinson[8] as in equation (2).

$$l = 18\alpha^2(\mu / \sigma_{ref})^2 b \tag{2}$$

In equation (1), the effective strain gradient η is obtained from the deviatoric strain gradient tensor η'_{ijk} by

$$\eta = \sqrt{\eta'_{ijk}\eta'_{ijk} / 4} \tag{3}$$

where

$$\eta'_{ijk} = \varepsilon_{ik,j} + \varepsilon_{jk,i} - \varepsilon_{ij,k} - \frac{1}{4}(\delta_{ik}\varepsilon_{pp,j} + \delta_{jk}\varepsilon_{pp,i}) \tag{4}$$

The gradient term, $\varepsilon_{ij,k}$ in equation (4) is calculated as nonlocal variables expressed in terms of the weighted average of strains. An eight-node isoparameter element, has been used to evaluate strain, strain gradient and stress. There are two different levels of Gaussian integration in the finite element analysis for TNT plasticity. One is at the element level, which is the same as the classical plasticity theory. Another is at the mesoscale cell level, which is special for the TNT plasticity. There is a square (for a two-dimensional problem) mesoscale cell surrounding each element-level Gaussian integration point. The stresses and effective strain gradients of each element-level Gaussian integration point can be expressed in terms of the nodal displacements via Gaussian integration in the mesoscale cell.

In the numerical model, it is considered that α is equal to 0.5 and l is equal to 3.028 μm.

2.2. Unite cell model

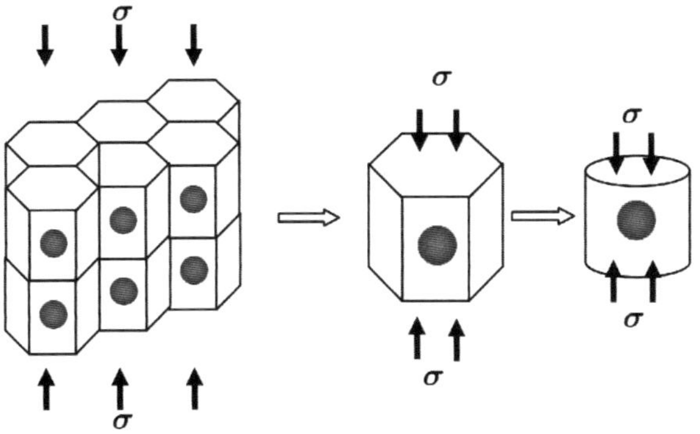

Figure 1. A schematic diagram of the unit cell model for a finite metallic matrix containing a single ceramic particle.

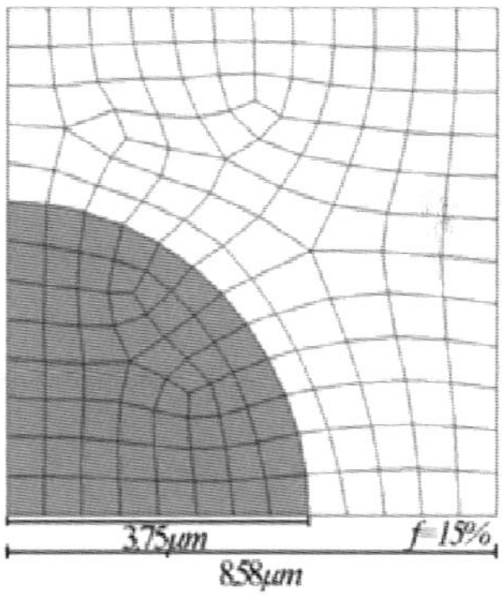

Figure 2. A finite element model with the particle diameter d of 7.5μm and volume fraction of 15%.

A two-dimensional axial symmetry unite cell model[13] was used to investigate the effect of thermal residual stresses on the mechanical characteristics in SiCp/Al composites. Figure 1 is a schematic diagram of the unit cell model for a finite metallic matrix containing a single spherical ceramic particle. Figure 2 is a finite element model with the particle diameter d of 7.5μm and the particles' volume fraction of 15%.

The elements of the matrix have been implemented in the ABAQUS finite element code through its UEL interface for TNT plasticity. These elements, which are similar to the conventional eight-node isoparametric elements, also have eight nodes and nine Gauss integration points. The bonding between the particles and the matrix is assumed to be perfect in the analysis.

2.3 Thermal residual stress and material parameters

The thermal contraction mismatch between the reinforcement and the matrix results in a residual stress state, when the composite is cooled from the processing temperature to room temperature. The processing causing thermal residual stresses are simulated by cooling the model composite from the solutionizing temperature of 500°C, where the composite is assumed to be free of internal stresses, to room temperature 20°C, at a constant cooling rate of 10°C·s^{-1}. The thermal deformations only generate linear strains, while the shear strains are zero. The linear strain by thermal expansion is $\alpha \cdot \Delta T$ at arbitrary integration point, when the temperature variation is ΔT, where α is the thermal expansion coefficient(CTE).

In the axisymmetric model, the thermal strain caused by the temperature variation in the form of tensor can be written as:

$$\varepsilon_{ij}^{0} = \alpha \Delta T \delta_{ij} \tag{5}$$

In the classical thermal elastic-plastic method, the consistency condition is used to derive a point-wise relation between the stress rate and strain rate,

$$\dot{\boldsymbol{\sigma}} = \boldsymbol{D}^{ep}(\boldsymbol{\sigma}) : \dot{\epsilon} \tag{6}$$

where $\epsilon_{ij} = \varepsilon_{ij} - \varepsilon_{ij}^{0}$, and $\boldsymbol{D}^{ep}(\boldsymbol{\sigma})$ is the local elastic-plastic stiffness tensor. In contrast, the TNT flow theory has a nonlocal relation between the stress rate and strain rate,

$$\dot{\sigma} = G(\epsilon, \dot{\epsilon}, \langle \epsilon \rangle, \langle \dot{\epsilon} \rangle) \tag{7}$$

where $\langle \cdots \rangle$ denotes integral average over a small representative volume.

The flow stress accounting for the effect of GNDs and thermal residual stresses can be given as

$$\sigma_{flow} = \sigma_{ref} \sqrt{f^{2}(\epsilon) + l\eta} \tag{8}$$

where the effective strain gradient η is described as the equation (3). The effective strain ϵ in equation (8) is determined in terms of the deviatoric tensor of strain $\epsilon' = (\varepsilon'_{ij} - \varepsilon'_{0ij})$ as

$$\epsilon = \sqrt{1/2 \epsilon'_{ij} \epsilon'_{ij}} \tag{9}$$

The effective strain gradient η, which is dissimilar to TNT plasticity without thermal residual stresses, is also the function of ϵ'.

The yield stress of Al matrix dependent of temperature is defined by the experimental result of LEVY[14]. The modulus of elasticity, which is also dependent of temperature, refers to the experimental data of Lloyd[1].

The SiC particle is assumed to be linearly elastic and isotropic. The matrix at room temperature is characterized by using a piecewise elastic-power law hardening relation, to model the uniaxial stress–strain behavior for unreinforced Al alloy. Their material parameters are shown in table 1.

Table 1 The material parameters of SiC and Al alloy at room temperature

	Young's modulus (GPa)	Poisson's ratio	Initial yield stress (MPa)	Power exponent	CTE ($10^{-6}\times K^{-1}$)
SiC	427	0.17	-	-	4
Al alloy	76	0.33	208	0.136	24

3. Result and discussion

3.1 Effects of thermal residual stresses on average axial stresses

The stress–strain curves without thermal residual stresses, predicted by TNT plasticity and classical elastic-plastic theory and Lloyd's experimental data[1], are shown in figure 3. The average axial strain in the cell is evaluated by the ratio of the uniform displacement on the top surface dividing the half-cell height. The average axial stress in the cell is determined by the total axial force multiplying the cell cross section area, where the total axial force is obtained from the nodal force of all nodes on the top surface.

The average axial stress predicted by the classical elastic-plastic theory without thermal residual stresses is lower than that predicted by TNT theory, since the classical elastic-plastic theory does not take into account the strain gradient caused by GNDs. But the flow stress predicted by TNT theory is still below Lloyd's experimental data [1]. Compared with MSG theory, the stress–strain curve without thermal residual stresses predicted by TNT theory for particle diameter size of 7.5μm quantitatively agrees with that done by Xue[12]. It means that some other factors, which were not taken into account in our current model, influence the predicted results. In fact, there are many factors which may lead to the deviation from the experimental data, such as thermal residual stresses, particle shape, and particle distribution, etc. In this section, we will discuss the effect of thermal residual stresses on the average axial stress of SiCp/Al composites.

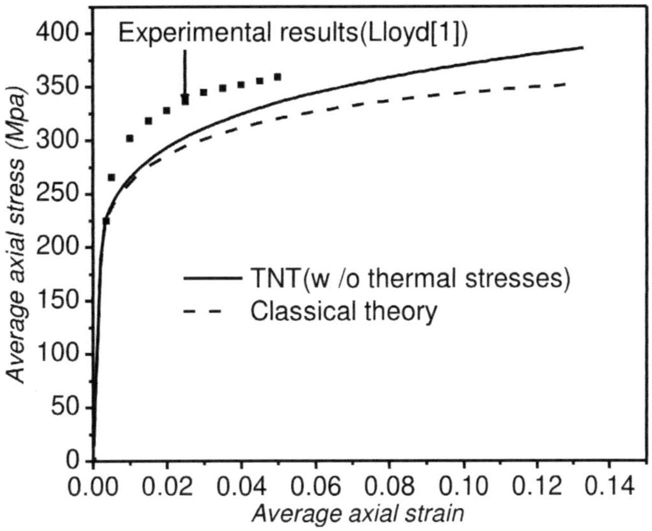

Figure 3. The stress–strain curves without thermal residual stresses predicted by TNT plasticity and classical elastic-plastic theory and Lloyd's experimental data.

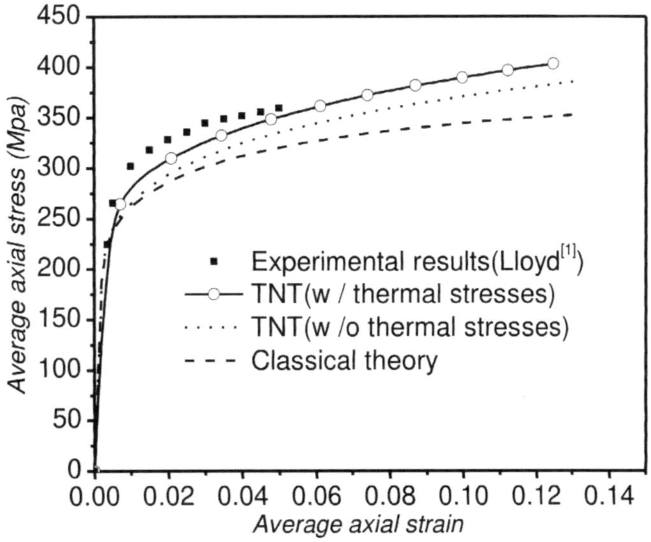

Figure 4. Comparisons of compressive stress-strain curves with and without thermal residual stresses (including Lloyd's experimental data and the curve predicted by classical theory).

Figure 4 shows comparisons of compressive stress-strain curves with and without thermal residual stresses (including Lloyd's experimental data and the curve predicted by classical theory). The curve with the existence of thermal residual stresses, shows lower slopes at the early stage of deformation, due to the lower apparent moduli (arising from prior plastic deformation). It explains that the presence of residual stresses enhances the initial strain hardening. The curve then crosses the corresponding curve without thermal residual stresses, and results in higher values of the average axial stress thereafter. The differences in average axial stresses beyond the crossover point persist throughout the plastic deformation stage. Compared with Lloyd's experimental data[1], a deviation from the experimental data still exists, but the amplitude of average axial stress is closer to the experimental curve. Both strain hardening and strain gradient due to the thermal residual stresses may lead to the additional plastic hardening. But we think that the strain hardening, rather than strain gradient, provides a major contribution. The effects of strain hardening caused by thermal stresses[14, 15] have been investigated in the past decades, thus we will discuss effects on thermal residual stresses of the effective strain gradient in next section.

3.2 Effects of thermal residual stresses on equivalent strain gradient

For deeper insight into the effects of thermal residual stresses of the equivalent strain gradient during subsequent mechanical loading, the average equivalent strain gradient with and without thermal residual stresses are compared, as shown in figure 5. The average equivalent stain gradient η_{av} in the compressive direction of the composite are calculated by using an area average approach as follows,

$$\eta_{\text{av}} = \frac{1}{S}\int_s \eta dS = \frac{1}{S}\sum_{j=1}^{m}\left(\sum_{i=1}^{n}\eta^{ji}\left|J\right|^{ji} H^{ji}\right) \qquad (n = 1, 2 \cdots 9) \tag{10}$$

where η_{av} is the average equivalent stain gradient, S is the total area, n is the total number of Gauss integral points of each element, m is the total number of elements, η^{ji} is the equivalent stain gradient at the Gauss integral point, and H^{ji} and $\left|J\right|^{ji}$ are the Gaussian integral weight coefficient and Jacobian determinant at the integral point, respectively.

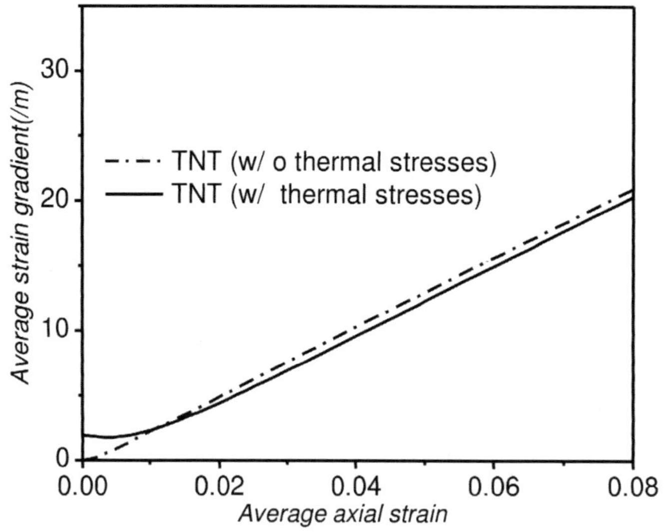

Figure 5. Comparison of the average equivalent strain gradient with and without thermal residual stresses

We can find that the curve with the existence of thermal residual stresses has a nonzero value at the onset of subsequent mechanical loading after cooling. The curve with the existence of thermal residual stresses shows a higher level of η_{av}, and then crosses the corresponding curve without thermal residual stresses at average axial strain of 0.01. This means that overall yielding takes place at a higher average axial stress value. After the axial strain of 0.01, η_{av} without the existence of thermal residual stresses is slightly greater than that with the existence of thermal residual stresses. This reveals that thermal residual stresses have been redistributed and prior strain by thermal residual stresses offsets the subsequent increase of η_{av} to a certain extent. In a word, the thermal residual stresses improve the overall initial yield stress of composites, but offer a less accumulation of strain gradient beyond the crossover point.

The contours of η_{av} predicted by TNT plasticity with and without thermal residual stresses at the average axial strain of 0.02 are shown in figure 6. It can be seen that both distribution characteristics are similar, except that the area of high strain gradient region is slightly greater for the case without the thermal residual stresses. This contour also confirms the conclusion of figure 5.

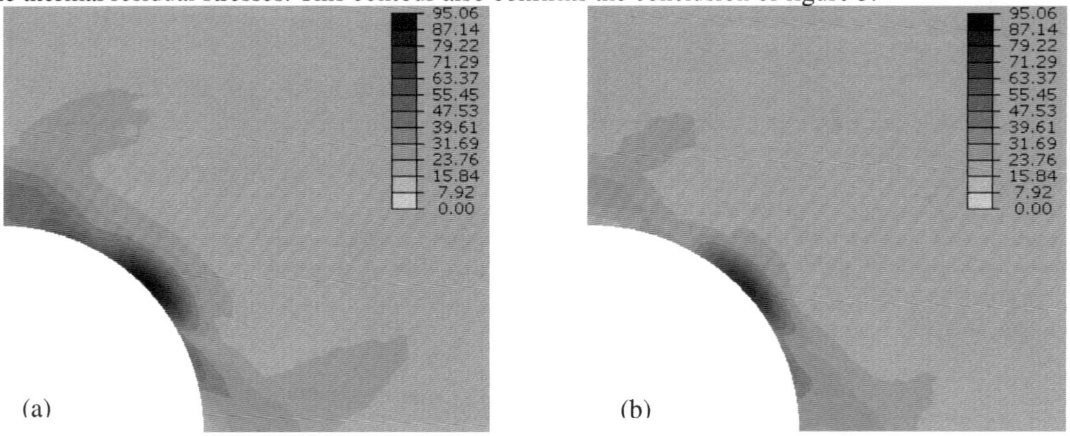

Figure 6. The contour of average equivalent strain gradient predicted by TNT plasticity at the average axial strain of 0.02: (a) without thermal residual stresses; (b) with thermal residual stresses.

4. Summary

In this investigation, a two-dimensional axial symmetry model based on the TNT plasticity is applied to investigate the effects of thermal residual stresses on the strain gradient in SiC particle reinforced aluminium composites. We have obtained the following conclusions:

(1) The numerical model with the thermal residual stresses is more in agreement with the experimental results and the prior thermal residual stresses can improve the initial yield stress and flow stress of SiCp/Al composites.

(2) The strengthening caused by thermal residual stresses mainly attributes to strain hardening of metal-matrix, not to strain gradient.

(3) The contours of the average strain gradient with and without thermal residual stresses have similar distribution characteristics, except that the area of high strain gradient region of the matrix is slightly higher for the case without the thermal residual stresses.

Acknowledgements

The work was supported by Fundamental Research Funds for the Central Universities(2011-IV-124). References

References

[1] Lloyd DJ 1994 *Int. Materials Reviews* **34** 1
[2] Nardone VC, Prewo KM 1986 *Scr. Metall.* 43
[3] Mori T, Tanaka K 1973 *Acta Metall.* **21** 571
[4] McHugh PE, Asaro RJ, Shih CF 1993 *Acta Metall. Mater.* **41** 1489
[5] Bruzzi MS, McHugh PE, O'Rourke F, Linder T 2001 *Int. J. Plast.* **17** 565
[6] Yan YW, Geng L, Li AB 2007 *Mater. Sci. Eng., A* **448** 315
[7] Gao H, Huang Y, Nix WD, Hutchinson JW 1999 *J. Mech. Phys. Solids* **47** 1239
[8] Fleck NA, Hutchinson JW 2001 *J. Mech. Phys. Solids* **49** 2245
[9] Fleck NA, Hutchinson JW 1993 *J. Mech. Phys. Solids* **41** 1825

[10] Gao H, Huang Y 2001 *Int. J. Solids Struct.* **38** 2615
[11] Guo Y, Huang Y, Gao H, Zhuang Z, Hwang KC 2001 *Int. J. Solids Struct.* **38** 7447
[12] Xue Z, Huang Y, Li M 2002 *Acta Mater.* **50** 149
[13] Bao G, Hutchinson JW, McMeeking RM 1991 *Acta Metall. Mater.* **39** 1871
[14] Levy A, Papazian JM 1991 *Acta Metall. Mater.* **39** 2255
[15] Shen YL, Finot M, Needleman A, Suresh S 1995 *Acta Metall. Mater.* **43** 1701

FGM 2012 IOP Publishing
Journal of Physics: Conference Series **419** (2013) 012042 doi:10.1088/1742-6596/419/1/012042

Effect of particle shapes on effective strain gradient of SiC particle reinforced aluminum composites

X Liu[1], D F Cao[2], H Mei[2], Z T Lei[3], and L S Liu [2,4]
[1]Department of Engineering Structure and Mechanics, Wuhan University of Technology, Wuhan 430070, P. R. China
[2]State Key Laboratory of Materials Synthesis and Processing, Wuhan University of Technology, Wuhan 430070, P. R. China
[3]Sheng Li oil field Sheng Li engineering construction group.Co.LT, Dongying 257000, P. R. China

E-mail:Liulish@whut.edu.cn

Abstract. The stress increments depend not only on the plastic strain but also on the gradient of plastic strain, when the characteristic length scale associated with non-uniform plastic deformation is on the order of microns. In the present research, the Taylor-based nonlocal theory of plasticity (TNT plasticity), with considering both geometrically necessary dislocations and statistically stored dislocations, is applied to investigated the effect of particle shapes on the strain gradient and mechanical properties of SiC particle reinforced aluminum composites(SiC/Al composites). Based on this theory, a two-dimensional axial symmetry cell model is built in the ABAQUS finite element code through its USER-ELEMENT (UEL) interface. Some comparisons with the classical plastic theory demonstrate that the effective stress predicted by TNT plasticity is obviously higher than that predicted by classical plastic theory. The results also demonstrate that the irregular particles cause higher effective gradient strain which is attributed to the fact that angular shape particles give more geometrically.

1. Introduction

In recent years, the ceramic particle-reinforced metal-matrix composites (MMCp) have offered designers much advantage on applications under high temperature, for their good strength, good structural rigidity, dimensional stability, light weight and low thermal expansion. Lots of work has also improved the mechanical properties of MMCp, and the fundamental relationships between microstructure and mechanical properties have become the focus of researches. There are many microstructure characteristics influencing the mechanical properties of MMCp, such as particles' volume fraction[1], particles' shape[1-3], particles' size[4], particles' topology dis tribution[5], and relative particle size[6], etc.

There are some reports about particle shape in the open literature. Christman and et al.[1] predicted the stress-strain behavior for whisker aspect ratios of 4 and 5, and cell aspect ratios of 3, 5, and 6, respectively. It's showed that an increase in the cell aspect ratio results in a decrease in the predicted

[4]Corresponding author, Tel.: +86 27 87651820; Fax: +86 27 87860863
E-mail: liulish@whut.edu.cn (L S Liu)

elastic modulus. Weissenbek and et al. [3] have compared different unit cell models (including plane stress, plane strain, axisymmetric, and three-dimensional models), each with different particle shapes (spherical, cubic, and cylindrical). They concluded that reasonable overall mechanical properties (in tension and compression) could be predicted by using the axisymmetric model, and it's also pointed out that spherical inclusions give rise to the softest response. In view of the irregular shapes of the particulate typically used in real particle-reinforced MMCp, thus spherical inclusions may not be the best choice for unit cell models of real materials. In an excellent comparison of experiment and modeling, Song and et al.[4] performed uniaxial tension tests on particulate-reinforced 6061-Al matrix composites reinforced by spherical particles and by angular particles, and also performed FEM calculations based on an axisymmetric unit cell model. Considering that the particles in the majority of particle-reinforced MMCp are angular in shape, unit cell models that use cylindrical particles may be more appropriate for modeling the overall mechanical properties.

The above discussion was implemented in the framework of classical plastic theory, which only takes into account the strain hardening resulted from the statistically stored dislocations (SSDs). When the characteristic length scale associated with non-uniform plastic deformation is on the order of microns, the stress at a given point depends on not only the current values and possibly the previous history of deformation and temperature, but also the neighborhood of the point. Lloyd[4] investigated the particle size effect for A356 aluminum alloy reinforced by 15 vol.% SiC particles with two different diameter sizes, 7.5 and 16μm. The research showed that the small silicon carbide particles with 7.5μm diameter gave significantly higher plastic work hardening than the large particles with 16μm diameter. The classical plastic theory possesses no internal length scale and cannot explain the size dependence of plasticity observed in experiments, which is reflected by the collective behavior of a vast number of SSDs.

In the study, we have investigated the effect of particle shape on the mechanical property of SiC particle reinforced aluminum composites(SiC/Al composites) by applying the Taylor-based nonlocal theory (TNT) of plasticity [7, 8].The effective strain gradients of matrix in SiC/Al composites with different particle shapes have been discussed by a numerical analysis. In our analysis, an axisymmetric unit cell with the SiC particle size of 13.5μm diameter and particles' fraction of 20% has been built to investigate the plastic hardening work of MMCp caused by effective strain gradient.

2. TNT plasticity and finite element model

2.1. TNT plasticity

TNT plasticity(developed by Gao and Huang[7]) is based on a multi-scale framework connecting the micro-scale notion of statistically stored and geometrically necessary dislocations to the meso-scale notion of plastic strain and strain gradient. The constitutive equations contain no higher-order stresses nor strain gradients. As a nonlocal integration of strains, strain gradients are introduced in the form of nonlocal variables. The basic principles in the nonlocal constitutive framework can be written as follows: (1) The flow stress obeys the Taylor hardening relations; (2) The density of geometrically necessary dislocations is calculated as nonlocal variables expressed in terms of a weighted average of plastic strain; (3) The essential structure of classical plasticity is preserved. In TNT plasticity, the flow tress σ_{flow} is obtained from the Taylor dislocation model as in equation (1).

$$\sigma_{flow} = M\alpha\mu b\sqrt{\rho} = M\alpha\mu b\sqrt{\rho_G + \rho_S} = \sigma_{ref}\sqrt{f^2(\varepsilon\ ) + l\eta} \tag{1}$$

The dislocation density ρ is composed of the density ρ_S for statistically stored dislocations (SSDs), which accumulate by trapping each other in a random way, and the density ρ_G for geometrically necessary dislocations (GNDs), which are required for compatible deformation of various parts of the non-uniformly deformed material. Where M is the Taylor factor which acts as an isotropic interpretation of the crystal line anisotropy at the continuum level, α is an empirical material constant in the Taylor dislocation model ranging between 0.1 and 0.5, μ is the shear modulus, and b

is the magnitude of Burgers vector, σ_{ref} is a reference stress in uniaxial tension, the function $f(\varepsilon)$ is a non-dimensional function determined from the uniaxial stress-strain curve, η is the effective strain gradient, and l is identified as the intrinsic material length introduced by Fleck and Hutchinson[9] as in equation (2).

$$l = 18\alpha^2(\mu / \sigma_{\text{ref}})^2 b \tag{2}$$

In equation (1), the effective strain gradient η is obtained from the deviatoric strain gradient tensor η'_{ijk} by

$$\eta = \sqrt{\eta'_{ijk}\eta'_{ijk} / 4} \tag{3}$$

where

$$\eta'_{ijk} = \varepsilon_{ik,j} + \varepsilon_{jk,i} - \varepsilon_{ij,k} - \frac{1}{4}(\delta_{ik}\varepsilon_{pp,j} + \delta_{jk}\varepsilon_{pp,i}) \tag{4}$$

The gradient term, $\varepsilon_{ij,k}$ in equation (4) is calculated as nonlocal variables expressed in terms of a weighted average of strains. An eight-node isoparameter element, have been used to evaluate strains, strain gradients and stresses. The isoparametric transformation is shown as figure 1. There are two different levels of Gaussian integration in the finite element analysis for TNT plasticity. The first is at the element level, which is the same as the classical plasticity theory. The second is at the mesoscale cell level, which is special for TNT plasticity. There is a square (for a two-dimensional problem) mesoscale cell surrounding each element-level Gaussian integration point. The stresses and effective strain gradients of each element-level Gaussian integration point can be expressed in terms of the nodal displacements via Gaussian integration in the mesoscale cell.

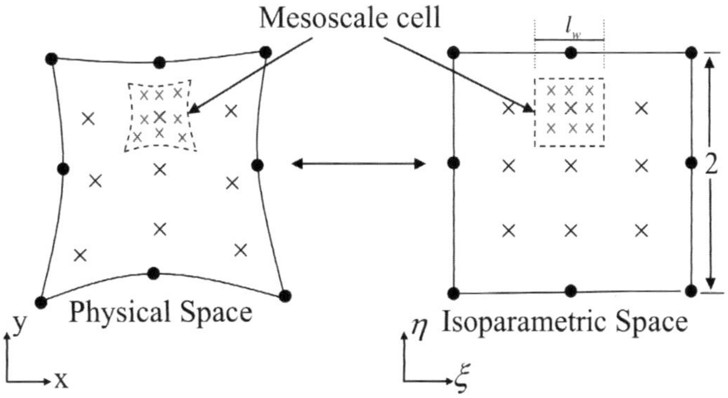

Figure 1. A schematic diagram of the isoparametric transformation.

Consider the Taylor expansion of a strain component ε_{ij} in the neighborhood of point $\boldsymbol{x}$

$$\varepsilon_{ij}(\boldsymbol{x}+\boldsymbol{x}) = \varepsilon_{ij}(\boldsymbol{x}) + \varepsilon_{ij,m}\xi_m + o(|\boldsymbol{x}|^2) \tag{5}$$

where $\boldsymbol{x}$ denotes local coordinates centered at $\boldsymbol{x}$. Integrating this equation with $\boldsymbol{x}_k$ over a small representative volume V_{cell} containing $\boldsymbol{x}$, we find

$$\int_{Vcell}\varepsilon_{ij}(\boldsymbol{x}+\boldsymbol{x})\xi_k = \varepsilon_{ij}(\boldsymbol{x})\int_{Vcell}\xi_k dV + \varepsilon_{ij,m}\int_{Vcell}\xi_m\xi_k dV \tag{6}$$

where the characteristic size l_w of V_{cell} is assumed to be sufficiently small such that terms of higher order are negligible. Therefore, the gradient term $\varepsilon_{ij,k}$ can be expressed in terms of strain,

$$\varepsilon_{ij,k} = \int_{Vcell}[\varepsilon_{ij}(\boldsymbol{x}+\boldsymbol{x}) - \varepsilon_{ij}(\boldsymbol{x})]\xi_k dV(\int_{Vcell}\xi_m\xi_k dV)^{-1} \tag{7}$$

where $(\int_{Vcell} \xi_m \xi_k dV)^{-1}$ is the inverse of $\int_{Vcell} \xi_m \xi_k dV$.

2.2. Finite element model

We adopt a unit cell model to investigate the effect of particle shape on the quasi-static mechanical behavior of the SiCp/Al composites. Both particles and matrix are meshed by the conventional eight-node isoparametric elements. The elements of the matrix have been implemented in the ABAQUS finite element code through its USER-ELEMENT (UEL) interface for TNT plasticity. Those elements, which are similar to the conventional eight-node isoparametric elements, also have eight nodes and nine Gaussian integration points. The bonding between the particles and the matrix is assumed to be perfect in the analysis. The computational meso-mechanics models for various particle shapes are shown in figure 2.

The SiC particle is assumed to be linearly elastic and isotropic, with Young's modulus E_{sic} =427GPa and Poisson's ratio v_{sic} =0.17. The matrix is characterized by TNT plasticity in section *2.1* by using a piecewise elastic-power law hardening relation to model the uniaxial stress–strain behavior for unreinforced A359 Al alloy as in equation (8). Where E_{Al} =76GPa is the aluminum Young's modulus, σ_{y0} =208MPa and ε_{y0} are the initial yield stress and yield strain, n=0.136 is the power exponent, and Poisson's ratio of aluminum is v_{Al} =0.33.

$$\sigma = \begin{cases} E_{Al}\varepsilon & if\, \varepsilon > \varepsilon_{y0} \\ K \cdot \varepsilon^n \quad (K = E_{Al}^n \sigma_{y0}^{1-n}) & if\, \varepsilon > \varepsilon_{y0} \end{cases} \tag{8}$$

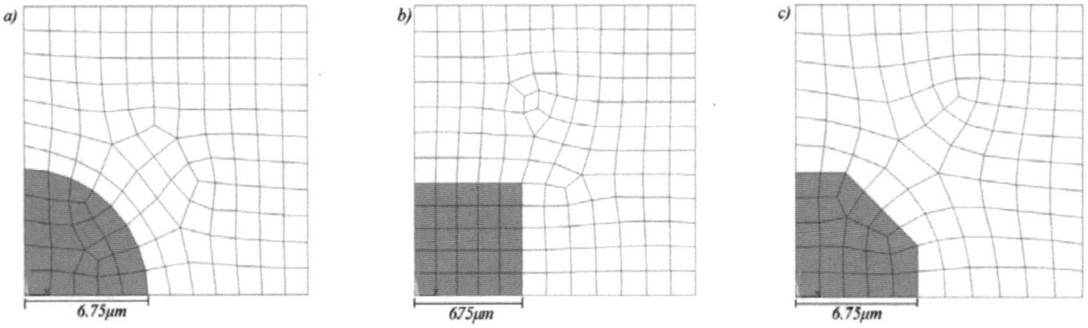

Figure 2. Computational meso-mechanics models for various particle shapes: (a) *circle*; (b) *square*; (c) *octagon*.

3. Results and discussion

3.1. Particle shape effect of the reinforcement

The stress–strain curves for various particle shapes predicted by TNT theory and classical elastic-plastic theory are shown in figure 3. The equivalent axial stain in the cell is evaluated by the ratio of the uniform displacement u_0 on the top surface to the half-cell height *h*. The equivalent axial stress in the cell is determined by the total axial force and the cell cross section area, where the total axial force is obtained from the nodal force of all nodes on the top surface. Some conclusions can be given as follows: (1) Both TNT plasticity and classical plasticity show that the strengthening effect of square is better than that of octagon and circle, and the effect of octagon is slightly stronger than that of circle, which is consistent with the other numerical research[10]. (2) The flow stresses of all particle shapes predicted by TNT plasticity are significantly higher than those predicted by the classical elastic-plastic theory, which is consistent with the result predicted by mechanism-based strain gradient plasticity[11].

The contours of the equivalent stresses predicted by TNT for various particle shapes are in figure 4. The particle volume fraction is 20%. It's known by observing the von mises stress contours that there is obvious stress concentration at sharp corners of square model, whose area of high stress region (the stress range from 0 to 420 MPa for all contours) is also obviously lager than that of circle and octagon.

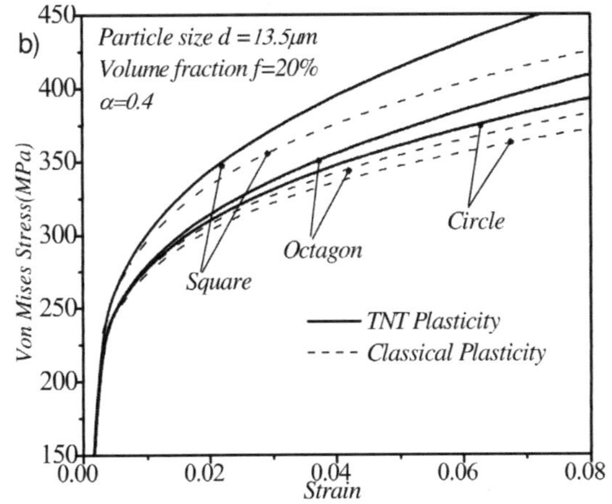

Figure 3. The equivalent stress–strain curves predicted by TNT theory for various particle shapes.

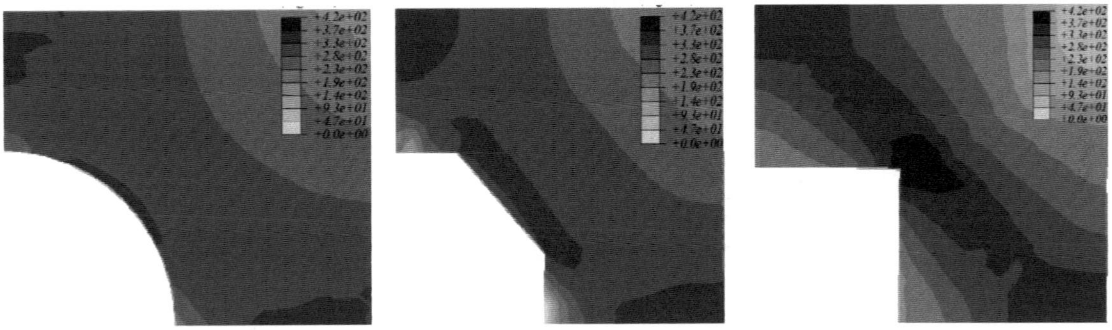

Figure 4. The contours of von mises stresses predicted by TNT for various particle shapes: (a) *circle*; (b) *square*; (c) *octagon*.

3.2. Effect of particle shape on the effect strain gradient

The overall equivalent strain gradient η_{av} in the compressive direction of the composite are calculated by using a area average approach as follows,

$$\eta_{av} = \frac{1}{S}\int_S \eta dS = \frac{1}{S}\sum_{j=1}^{m}(\sum_{i=1}^{n}\eta^{ji}|J|^{ji}H^{ji}) \qquad (m, n=1, 2, 3, \ldots) \qquad (9)$$

where η_{av} is the average equivalent stain gradient, S is the total area, n is the total number of Gaussian integration points of each element, m is the total number of elements, σ^{ji} and η^{ji} are the equivalent stress and the average equivalent stain gradient at the Gaussian integration point, H^{ji} and $|J|^{ji}$ are the Gaussian integrations weight coefficient and Jacobian determinant at the integration point, respectively. The average strain gradient versus strain curve predicted by TNT for various particle shapes is presented in figure 5. From the curves, it can be found that the average equivalent stain gradient for square shape is higher than that for circle and octagon. In accordance with the Taylor hardening law in equation (1), higher equivalent strain gradient contributes to higher flow stress of SiCp/Al composites for square shape, compared with circle and octagon model.

The contours of the equivalent strain gradients predicted by TNT plasticty for various particle shapes are presented in figure 6. It can be seen that the distribution of effective strain gradient in matrix is highly inhomogeneous due to the mismatch of elastic module between particles and matrix.

The largest strain gradient of the matrix is located at the region near particles. There is an obvious concentration phenomenon at sharp corners of square model, and the area of the high strain gradient region is also larger than that of octagon and circle. It is attributed to the fact that square shape particles give more geometrically necessary dislocations on the macroscopic plastic work hardening of alloys or metal–matrix composites.

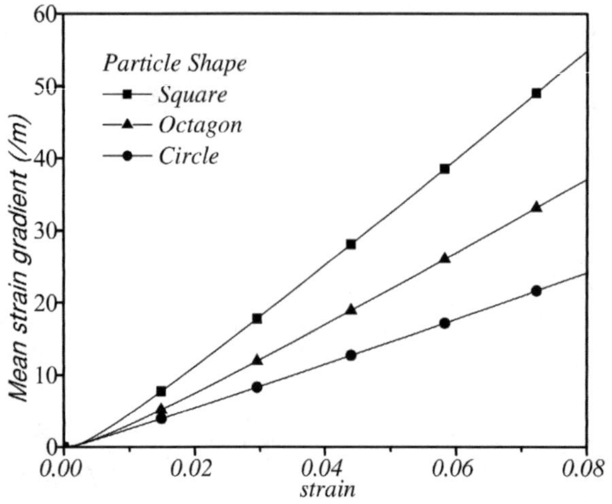

Figure 5. The equivalent strain gradient versus strain curve predicted by TNT for various particle shapes.

It's noteworthy that the present study does not account for cracking and debonding between the particles and matrix. The high level of stress concentration often means that the fracture of particles is more likely to happen. Generally speaking, broken particles are unable to afford tension and shear effectively, and the macroscopic mechanical properties of SiCp/Al composites will decrease with the increase of particles' fracture. Therefore, although the strengthening effect of square inclusions is better than that of circle and octagon inclusion before particles' fracture happens, the square particle may not be the best inclusion.

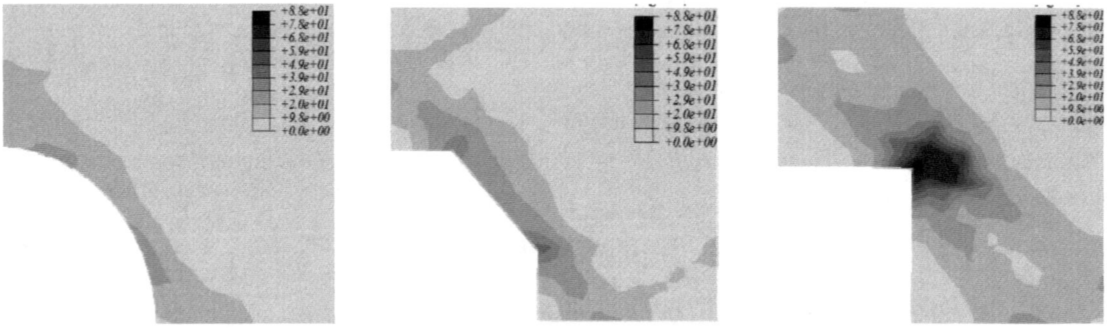

Figure 6. The contours of equivalent strain gradients predicted by TNT for various particle shapes: (a) *circle*; (b) *square*; (c) *octagon*.

4. Summary

In this investigation, a two-dimensional axial symmetry model based on the TNT plasticity is applied to investigate the effect of particle shape on the plastic hardening work of SiC particle reinforced aluminum matrix composites. We have reached the following conclusions:

(1) With the predicted numerical results of various particles shapes, it's observed that the strengthening effect of square model is more effective than circle and octagon (whether TNT plasticity or classical theory).

(2) The effective stresses predicted by TNT plasticity for various particle shapes are significantly higher than those predicted by classical theory, since TNT plasticity includes the influence of strain gradient on plastic hardening work of Al matrix.

(3) The effective strain gradient for square shape (predicted by TNT plasticity) is significantly higher than that for circle and octagon shape. It is attributed to the fact that the sharp corners of square give more geometrically necessary dislocations, which cause higher effective strain gradient.

Acknowledgements
The work was supported by Fundamental Research Funds for the Central Universities (2011-IV-124).

References

[1] Christman T, Needleman A, Suresh S 1989 An experimental and numerical study of deformation in metal-ceramic composites *Acta Metallurgica* **37** 3029
[2] Song S, Shi N, Gray G, Roberts J 1996 Reinforcement shape effects on the fracture behavior and ductility of particulate-reinforced 6061-Al matrix composites *Metallurgical and Materials Transactions A* **27** 3739
[3] Weissenbek E, Böhm HJ, Rammerstorfer FG 1994 Micromechanical investigations of arrangement effects in particle reinforced metal matrix composites *Computational Materials Science* **3** 263
[4] Lloyd DJ 1994 Particle reinforced aluminium and magnesium *Int. Materials Reviews* **34** 1
[5] Tszeng TC 1998 The effects of particle clustering on the mechanical behavior of particle reinforced composites *Composites Part B: Engineering* **29** 299
[6] Prasad VVB, Bhat BVR, Mahajan YR, Ramakrishnan P 2002 Structure–property correlation in discontinuously reinforced aluminium matrix composites as a function of relative particle size ratio *Materials Science and Engineering: A* **337** 179
[7] Gao H, Huang Y 2001 Taylor-based nonlocal theory of plasticity *International Journal of Solids and Structures* **38** 2615
[8] Guo Y, Huang Y, Gao H, Zhuang Z, Hwang KC 2001 Taylor-based nonlocal theory of plasticity: numerical studies of the micro-indentation experiments and crack tip fields *International Journal of Solids and Structures* **38** 7447
[9] Fleck NA, Hutchinson JW 2001 A reformulation of strain gradient plasticity *Journal of the Mechanics and Physics of Solids* **49** 2245
[10] Shen YL, Finot M, Needleman A, Suresh S 1995 Effective plastic response of two-phase composites *Acta Metallurgica et Materialia* **43** 1701
[11] Gao H, Huang Y, Nix WD, Hutchinson JW 1999 Mechanism-based strain gradient plasticity-- I. Theory *Journal of the Mechanics and Physics of Solids* **47** 1239

The Application of Ti/Al_2O_3 Composite in W-Mo-Al System Wave Impedance Graded Flier-plate

Z WANG[1], S FENG[1], G P SHI[1], Q SHEN[2] and L M ZHANG [2]

[1]School of Materials Science and Engineering, University of Jinan, Jinan 250022, China
[2]State Key Laboratory of Advanced Technology for Materials Synthesis and Processing, Wuhan University of Technology, Wuhan 430070, China

E-mail: wangzhi@ujn.edu.cn

Abstract: Ti/Al_2O_3 composite of different components were prepared by Spark Plasma Sintering technique, mechanical and acoustical properties of Ti/Al_2O_3 composite were analyzed, weldability of Ti/Al_2O_3 composite with Mo and Al was also studied. The results indicate that 38.5vol%Ti/Al_2O_3 composite has good mechanical performance with relative density >99.20%, bending strength 897MPa, fracture toughness 17.38 $MPa{\cdot}m^{1/2}$ and micro-hardness 17.13GPa. The wave impedance of the composite was between which of Mo and Al, and the 38.5vol% Ti/Al_2O_3 welded well with Mo and Ti. These demonstrate that 38.5vol%Ti/Al_2O_3 composite could serve as one of interlayer materials in W-Mo-Al system graded flier-plate.

1 Introduction

In research fields of astrophysics, geophysics, solid-state physics and inertial confinement nuclear fusion etc, it is necessary to study the dynamic behavior and response characteristics of materials at extremely high pressure (1-2TPa)[1,2]. An important technique, in which a flier-plate is driven to a high velocity to impact on a flat-plate target by two-stage light gas gun[3,4], is widely used for research on dynamic high-pressure physics. So, it is important to design the structure of the flier-plate besides optimizing the super high-speed launch devices. The study shows that wave impedance graded flier-plate, compared with traditional homogeneous flier-plate, can load a nearly isentropic compression to target at a lower temperature when the target is impacted on[5]. Various systems of graded flier-plate were reported recently years[6-10]. However, the study on W-Mo-Ti-Al system flier-plate is difficult because of the bad welding property and dramatic reactions between titanium and aluminum [11], which affect quasi-isentropic compression effect and structure stability of composite both during the impact time and high-speed flying time. So, one proper substituted interlayer material is very important. Ti/Al_2O_3 composite is considered one of candidates of graded flier-plate interlayer materials due to good mechanical performance and intermediate density between Mo and Al. This article will demonstrate the feasibility of the Ti/Al_2O_3 composite as the interlayer material of W-Mo-Ti/Al_2O_3-Al system wave impedance graded flier-plate. In this experiment, Ti/Al_2O_3 composites were prepared by Spark Plasma Sintering (SPS) technique.

2 Experimental

2.1. Preparation of samples

High purity alumina, titanium and niobium powders were raw materials. The mean particle sizes of the three powders were 1.5μm, 7.76μm and 5.7μm respectively. Different proportions of powders were mixed uniformly as design in Table 1, and then poured them into a graphite mold. The sintering was performed in a spark plasma sintering device (SPS-1050, Sumitomo Coal Mining Co. Ltd. Japan) at 1300℃ for 10min with heating rate of 200 ℃/min, the applied mechanical pressure was 30MPa. The sizes of samples were Ø32×5mm.

Table 1. Component of Samples/vol%

Sample NO.	Ti	Nb	Al_2O_3
T2	18.5	1.5	80
T4	38.5	1.5	60
T6	58.5	2.0	40
T8	78.5	2.0	20

2.2. Mechanical property testing

After being polished using 120μm, 80μm, 20μm SiC abrasive and 10μm, 1.5μm diamond paste in turn, samples were incised bars of 26mm×3mm×4mm and 26mm×2mm×4mm (with incision of 0.2mm width and 1.8mm length). The final densities of samples were determined by the Archimedes method using distilled water. Three-point bending strength and fracture toughness were measured at room temperature using a universal testing machine (INSTRON 5569, USA) with the pressure head down speed of 0.5mm/min and 0.05mm/min, and the supporting span is 20mm. The micro-hardness was measured using a digital micro-hardness tester (HDX-1000) at a load of 1Kg and indention time of 20s. The Scanning Electron Microscope (SEM, HITACHI S-2500) with Energy Disperse Spectroscopy (EDS, LINKISIS-300) was used to observe the micro structures of sample sections.

2.3. Acoustical properties testing and welding experiment

Ultrasonic pulse-echo method was adopted to measure the shear wave velocity and vertical wave velocity of the composite, the instrument used in this method were Ultrasonic Signal Receiver (Panametrics5072PR) and Oscilloscope (TDS2022). Wave impedance and elasticity mechanical were calculated from hear wave velocity and vertical wave velocity using the equations followed[12]:

$$C_b^2 = C_l^2 - \frac{4}{3}C_t^2 \tag{1}$$

$$Z = \rho \cdot C_b \tag{2}$$

Where C_b, Z, ρ, C_l and C_t were volume velocity, wave impedance, density shear wave velocity and vertical wave velocity respectively.

$$K = \rho C_b^2 \tag{3}$$

$$G = \rho C_t^2 \tag{4}$$

$$E = \rho C_t^2 \frac{3C_l^2 - 4C_t^2}{C_l^2 - C_t^2} \tag{5}$$

$$\nu = \frac{C_l^2 - 2C_t^2}{2\left(C_l^2 - C_t^2\right)} \tag{6}$$

Where K is bulk modulus, G is shear modulus, E is Young modulus and ν is Poisson's ratio.

Diffusion welding method was adopted to test the weldability of Ti/Al_2O_3 composite with Mo and Al as technological parameters listed in Table 2. And SEM was used to observe the microstructure of weld joints.

Table 2. Parameters for diffusion welding

Welding joint	Temperature/℃	pressure/MPa	t/min
Mo-T4	1300	5	30
Al-T4	600	5	30

3 Results and discussion

3.1 Mechanical Properties of Ti/Al_2O_3 Composite

Table 3 shows that with Ti content increasing, relative densities of Ti/Al_2O_3 composite increase firstly, then decline slightly, but are all larger than 99.10%; bending stress, fracture toughness and micro-hardness also show trend of rise firstly then decrease with Ti content increasing. This dues to that with the metal phase increasing, the combination way of metal phase and ceramic phase and the modification mechanism of second phase to matrix performance change.

Table 3. Mechanical properties of Samples

Sample NO.	Relative density/%	Bending stress/MPa	Fracture toughness/MPa·m$^{1/2}$	Micro hardness/GPa
T2	>99.10	893±60.74	10.94±0.80	19.85±1.64
T4	>99.20	897±49.32	17.38±1.07	17.13±0.82
T6	>99.20	429±37.11	9.62±0.68	12.02±0.77
T8	>99.15	274±22.06	5.89±0.54	11.14±0.79

When the Ti content is much higher, it is metal matrix composite. The dispersed Al_2O_3 particles change the flow stress behavior and limit the deformation of matrix, which make the cracks formation around alumina grain boundaries. These cracks reduce fracture toughness and bending stress of composite. When the Al_2O_3 phase is major phase, Al_2O_3 particles are coated by metal thin layer and the interface reaction also improves interfacial energy[13], these make alumina particles connected tightly and transgranular fracture failure appear. Figure 1 shows cross section SEM photos of T2 and T4, where the light part is Al_2O_3 and the grey is titanium. The photos show that grains grow finely and combine compactly. And from the grain fracture and draw traces on the cross section, the mixed intergranular and transgranular fracture mode can be determined. The transformation of load phase from ceramic to metal with the Ti content increase leads the micro-hardness decline.

3.2 Acoustical and welding Properties of Ti/Al_2O_3 Composite

Shear wave velocity and vertical wave velocity tested by ultrasonic pulse-echo method and parameters of acoustics and elasticity mechanics calculated are all listed in Table 4.

The data demonstrates that not only the wave impedance value of Ti/Al_2O_3 composite is in between the higher value of Mo and the lower value of Al, but the mechanical performance of composite is good enough to keep the stability of flier-plate.

From the analysis above, T4 sample is selected to weld with Mo and Al because its good properties and proper wave impedance value. The SEM photo in Fig. 1 shows the welding interface of Mo-T4 is smooth, and the corresponding EDS photos show one thick transition layer in interface area.

Table 4. Parameters of acoustics and elasticity mechanics

	Mo[10]	T2	T4	T6	T8	Al[14]
ρ/g·cm^{-3}	—	4.074	4.086	4.278	4.423	—
Ct/m·s^{-1}	—	5670	5203	4740	4384	3133
Cl/m·s^{-1}	—	9856	9480	8421	7767	6500
Cb/m·s^{-1}	5160	7367	7333	6400	5891	5400
Z/10^{6}kg·m^{-2}s^{-1}	52.74	30.22	30.90	27.65	26.08	14.58
E/GPa	325	330	293	246	216	71
G/GPa	260	132	114	97	85	27
K/GPa	261	223	227	177	154	79
v	0.293	0.253	0.285	0.268	0.266	0.348

SEM and EDS photos of T4-Al are shown in Figure 3. The interface is also flat, but the T4 and Al cannot distinguish obviously from SEM photo. The result from two reasons: one is alumina brightener insert in the surface, the other is a few aluminum is oxidized to alumina during polishing; the latter reason is also certified from corresponding EDS photo where aluminum area is detected higher oxygen concentration.

Figure 1. Cross section SEM photos of T2 (a)and T4 (b)

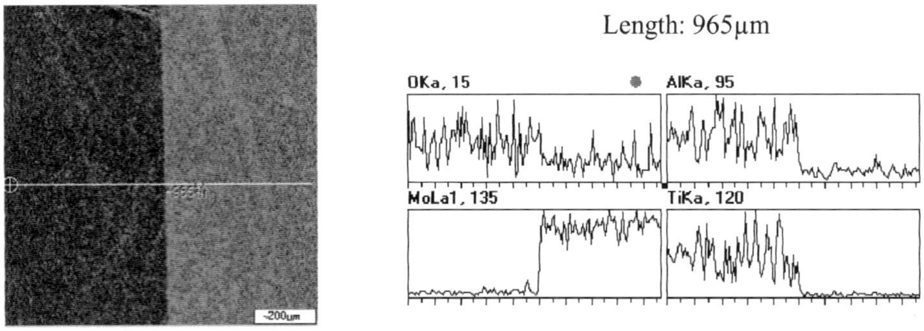

Figure 2. Microstructure of Mo-T4 welding joint and the corresponding line distributions of elements Mo, Al, Ti and O

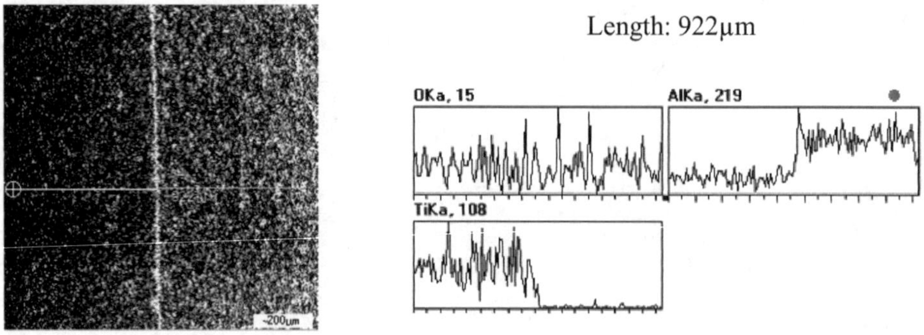

Figure 3. Microstructure of Al-T4 welding joint and the corresponding line distributions of elements Al, Ti and O

4 Conclusion

38.5vol%Ti/Al_2O_3 composite has good performance with relative density 99.23%, bending strength 897MPa, fracture toughness 17.38 MPa·$m^{1/2}$ and micro-hardness 17.13GPa and its wave impedance value is in between which of Mo and Al. The Ti/Al_2O_3 composite satisfies the requested properties of mechanics and wave impedance.

This component composite can be bonded well with Mo and Al. The welded Mo-T4-Al laminate material has good parallelism and the interfaces of Mo-T4 and Al-T4 are fine. So the 38.5vol%Ti/Al_2O_3 composite is suitable to W-Mo-Al system wave impedance graded flier-plate as interlayer material.

Acknowledgement
This work was sponsored by Natural Science Foundation of China (No.51142010)

References

[1] Li B S, Kung J and Liebermann R C 2004 *Phys. Earth. Planet. In* **143** 559-74
[2] Huang H J, Shen Q, Luo G Q and Zhang L M 2007 Chin. Phys. Soc. 56 1538-41
[3] Masahide K, Atsushi T, Kumi N, Shirou K, Haruhisa M and Yukihito K 2010 Mater.Sci. Eng. B 173148-154
[4] Yokoo M, Kawai N, Nakamura K G, Kondo K 2008 Int. J. Impact. Eng 35 1878-83
[5] Wang C B, Shen Q and Zhang L M 2002 J. Wuhan University of Technology- Mater. Sci. Ed. 17 32-35
[6] Shen Q, Zhang LM, Tan H, Jing F Q 2004 J. Wuhan University of Technology- Mater. Sci. Ed. 19 56-80
[7] Wang C B, Zhang L M and Shen Q 2003 Mater. Sci. Forum 423 77-80.
[8] Bai JS, Tang M, Luo GQ, Yu J D, Yuan S, Dai C D, Wu Q and Tan H 2011 *Phy. Mech. Astr.* **54** 1710-16
[9] Deng J X, Bai C, Zhang L M and Yuan R Z 2005 *J. Wuhan Uni. Sci.Tech*(Natural Science Edition) **28** 228-230
[10] Shen Q, Zhang L M and Xiong P H 2000 Chin. Sci. Bull. **45** 878-81
[11] Chen Y H, Ni Q, Ke L M 2012 *Trans. Nonferrous Met. Soc. China* **22** 299-304
[12] Hua JS, Jing FQ, Tan H and Zhao Z D 1988 *Chin. J. High Press. Phy.***12** 150-155
[13] Wang Z, Xu K, Chen Y M and Shen Q 2004 Chin. J.Mater. Sci.Eng **22** 799-801
[14] Wang C B 2003 *A Flier-plate Material with Graded Impedance Used in Dynamic High-pressure Physics* (Wuhan: Wuhan University of Technology)

Numerical Simulation of the Dynamic Performance of the Ceramic Material Affected by Different Strain Rate and Porosity

Zhen Wang[1],L S Liu [2], H Mei[1], X Lai[1],P C Zhai[2],D F Cao[2]

[1]*Department of Engineering Structure and Mechanics, Wuhan University of Technology,*
Wuhan 430070, China,
[2]*State Key Laboratory of Advanced Technology for Materials Synthesis and Processing, Wuhan University of Technology, Wuhan 430070, China.*

E-mail: liulish@whut.edu.cn

Abstract. Ceramic materials are frequently used in protective armor applications for its low-density, high elastic modulus and high strength. It may be subject to different ballistic impacts in many situations, thus many studies have been carried out to explore the approach to improve the mechanical properties of the ceramic material. However, the materials manufactured in real world are full of defects, which would involve in variable fractures or damage. Therefore, the defects should be taken into account while the simulations are performed. In this paper, the dynamic properties of ceramic materials (Al_2O_3) affected by different strain rate (500-5000) and porosity (below 5%) are investigated. Foremost, the effect of strain rate was studied by using different load velocities. Then, compression simulations are performed by setting different porosities and random distribution of pores size and location in ceramic materials. Crack extensions and failure modes are observed to describe the dynamic mechanical behavior.

1. Introduction

A number of expressions for the effects of different strain rate and porosity on the mechanical performance of ceramic have been proposed by various investigators. Many of these articles already discussed the influence of porosity and pore size on the compressive strength or fracture properties of porous ceramic materials[1-2], which have been widely used in bone replacement, but barely studied the mechanical properties of a high-density aluminum oxide ceramic (porosity below 5%). This is because all the required factors (e.g., random pore sizes, geometrical shapes and distributions)are not clearly studied in practical ceramic bodies. And also the porosity are not so easy to control when prepare test pieces before experiment[3]. Herein, a numerical simulation was carried out to do the research. The mechanical properties of ceramic with low porosity was studied, especially the dynamic performance.

Several semi-empirical functions have been used to describe the relationship between fracture strength and porosity of ceramic[4-7].Meanwhile, it has been frequently studied the effect of porosity on Young's modulus. DEAN[8]and SPRIGGS[9]et al. have proposed several empirical relations to fit Young's modulus and porosity.

The effects of strain rate on the mechanical properties of ceramics have been observed in many studies[9-10].Considering the very comprehensible degree to which both porosity and strain rate influence the dynamic performance of ceramic, there is a remarkable lack of accurate information in the literatures. Furthermore, almost none of the available information deals with the combined effect of these factors on dynamic properties.

In the present study, a numerical simulation has been made (1) to study the dynamic properties of high-density aluminum oxide ceramic in different strain rate and porosity, (2) to observe the situation of damage accumulation and crack propagation, and (3) to provide reference for further engineering application in engineering and other sectors.

2. Numerical simulations

Specimens of alumina ceramic belong to non-strictly axial symmetry. Since pores are separated in specimens randomly, which have small size and few quantities, we reckon that specimens are homogenous, and distributions of stress and strain are axisymmetric. Hence an axisymmetric model was taken to do the research.

Simulations were carried out by different models with porosity of 0, 1%, 3%, 5%and strain rate of 500, 1000, 3000and 5000 separately. Dynamic mechanical properties of alumina ceramic are studied using models referred before.

2.1. Model and Mesh

Models was constructed by a software ANSYS. Since the pores size is small, if the original model was used in simulation, it generates huge mesh quantity, which means large-scale simulations. Hence the simulation used a tiny cell which is cut out from ceramic panel and has a size of 0.5cm×0.5cm.Pores are all in the form of circle with diameters ranging from 10nm to 200μm[11] randomly. The porosities are 0%, 1%, 3% and 5% respectively as shown in figure1(a), (b),(c) and(d).

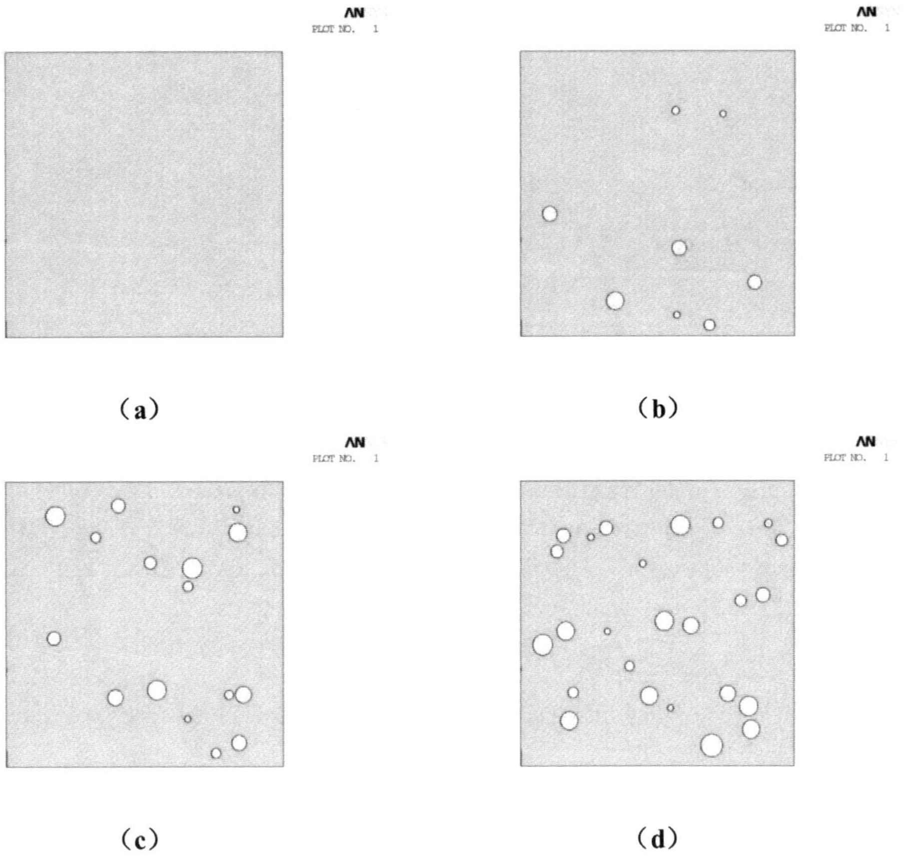

Figure 1. Models of different porosities

The geometries of computational models are regular and the mesh size is set to 0.002cm, hence the model with no porosity contains 126002 nodes and 62500 elements.

2.2. Constitutive models

Ceramic is a kind of typical brittle material. When ceramic is under impact load, the damage appears in the place of micro-cracks. With the development of cracks, material yields or breaks eventually.

One of the most widely used constitutive model for simulating the response of ceramic materials is the JH-2 model which is also used in simulation and engineering design of glass and cement. This constitutive model does account for strain rate effects and shows the relationship among the pressure, strength and damage of material which can be obtained from figure 3.

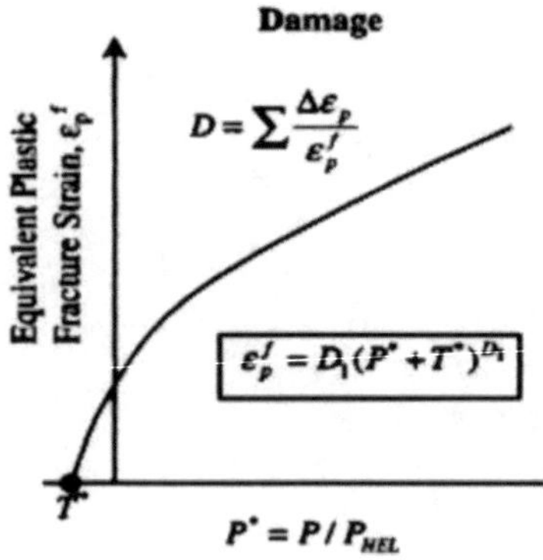

Figure2. Description of the JH-2 ceramic model[12]

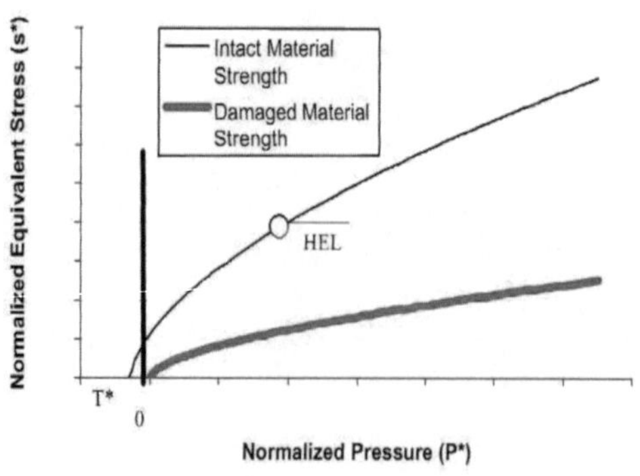

Figure3. Strength versus pressure for intact and damaged material[13]

JH-2 model include a damage parameter that can describe the material from intact to yield which can be used to simulate damage. As in figure 2, damage is controlled by parameter D ranging from 0 to 1. The material is intact when D equals 0, while it is totally damaged, D was set to 1.The parameters of ceramics available for this simulation are cited from other literature [13].

2.3.Loading methods and boundary conditions

In order to achieve loading at different strain rates, the loading methods and boundary conditions are set as follows:

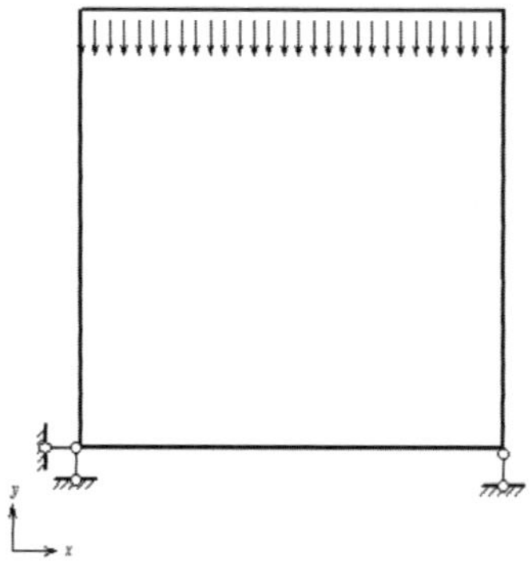

Figure4. Load and boundary conditions

The interfacial effect of models under loading is ignored. The bottom is constrained in displacement to fix the model, while the top is loaded by different speed of displacement to simulate impact loading. The left and right sides are set as nonreflecting boundary conditions.

3. Results and discussion

It is evident that the strain rate and porosity do have some influence on the performance of ceramic material due to the result analyzed below.

3.1. the effect of porosity

Since the effect of porosity was discussed, here we take a fixed strain rate (500) as an example. The fracture strengths are 8.84GPa, 3.63GPa, 2.74GPa and 2.32GPa with the increase of porosity, respectively for 0%, 1%, 3% and 5% when strain rate is set to 500, shown in figure 5(a).

The strength decreased with the porosity increased, which is in good accordance with the literature. The strength of ceramic for porosity 0% (8.84GPa) is much higher than others, mainly because porosity 0% means 'ideal material'—there are no pros in the ceramic body, which is impractical to obtain that kind of material during production process. In a word, it's only an ideal value. That's the reason why we never got such a high strength through experiments.

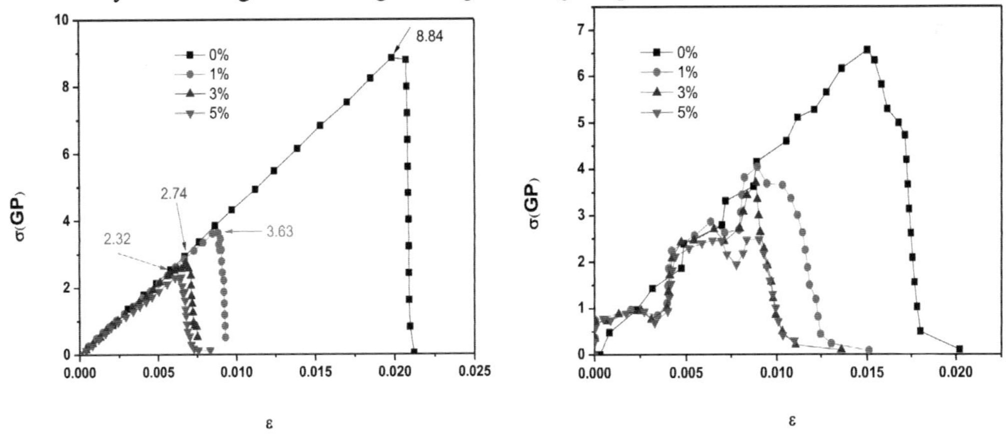

(**a**) strain rate 500 (**b**) strain rate 5000

Figure5. Stress-strain curve of different porosities under different strain rates

when strain rate is set to 5000 as in figure5(b), the issue was turned into a high strain rate situation which should take impact effect in to consideration. Since the shock wave prorogation is complex, transmission and reflection in material caused fluctuation in the curve. So the stress-strain curve of figure5(a) is smoother than that of (b).

Table 1. Performance of ceramic under strain rate 500

Porosity	E（GPa）	ΔE	σ（GPa）	Δσ
0%	442.1	/	8.84	/
1%	423.7	18.5	3.63	5.21
3%	415.6	8.1	2.74	0.89
5%	377.8	37.8	2.32	0.42

Table 1 shows material properties for ceramic at strain rate 500. Two fitting functions can be obtained from the data listed.

$$E=441.9e^{-2.88P} \quad (1)$$

$$\Delta\sigma=0.003P^{-1.57} \quad (2)$$

Those two expressions are in good accordance with the result of KNUDSEN [9] (in form of $E=E_0e^{-bP}$, where E is the Elastic Modulus with porosity, E_0 is the Elastic Modulus of non-porous ceramic, b is a fitting parameter, P is porosity) and Dunn[6] (in form of $\Delta\sigma = aP^b$, where $\Delta\sigma$ is stress difference, P is porosity, and $a > 0 > b$) under lower porosity. Meanwhile, the value of E_0 obtained by fitting is 441.9GPa, which is in good accordance with the result 442.1GPa gained by this simulation. But the fitting function of stress difference $\Delta\sigma$ is not so agreed with the results Dunn obtained. That's because there are only three data used to fit the function which brings the unusual margins of error. Generally, the strength has decreased 5.21GPa since the porosity increased from 0% to 1%; 0.89GPa since the porosity increased from 1% to 3%; and 0.42GPa since the porosity increased from 3% to 5%. That means the decrease of strength slows down with the increase of porosity.

The Elastic Modulus and failure strength calculated from the simulation decreased with porosity increased when strain rate is 500, which is in good accordance with the literature[9], and validate the simulation method at the same time.

3.2. The effect of strain rate

The computational results from the comparison analysis shows the stress-strain curves of ceramic model under different strain rate for porosity 0%, 1%, 3% and 5% respectively.

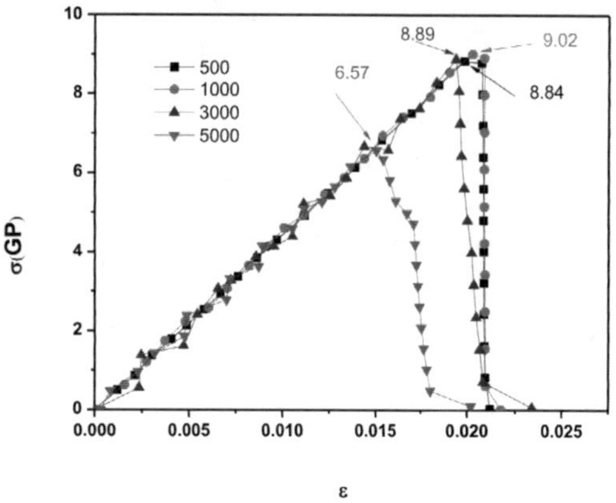

(**a**) porosity 0%

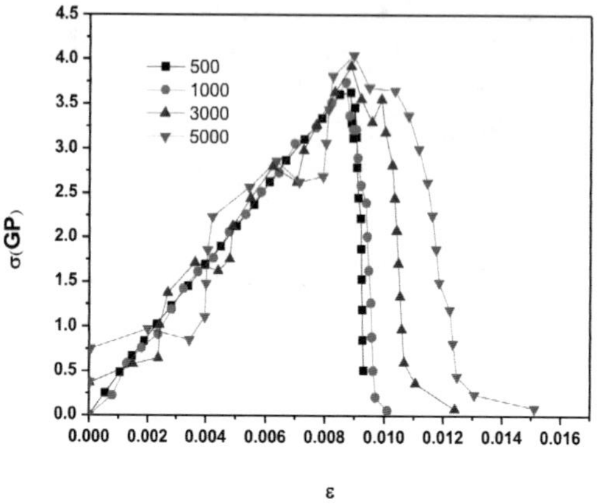

(**b**) porosity 1%

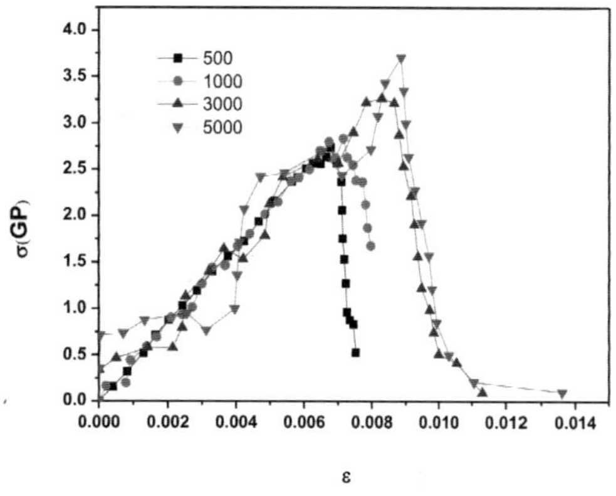

(**c**) porosity 3%

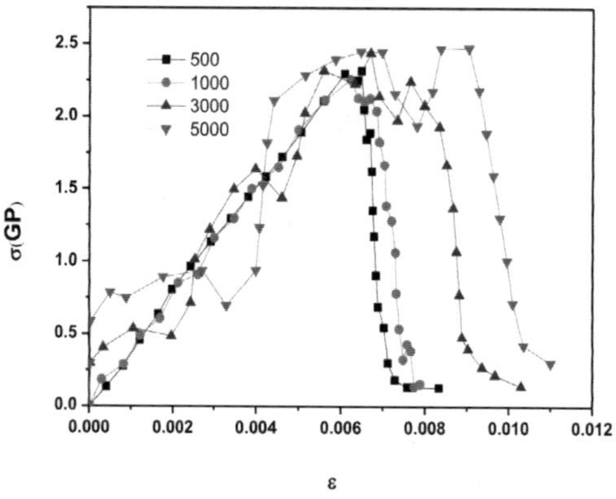

(**d**) porosity 5%

Figure6. Stress-strain curves of different strain rates under different porosities

Discrepancy can be easily observed from the figures. When porosity was set to 0% , shown in figure 6(a), the fracture strengths of ceramic are 8.84GPa, 9.02GPa, 8.89GPa and 6.57GPa with the corresponding strain rate 500, 1000, 3000 and 5000 respectively. The 'ideal ceramic' has lower fracture strength when strain rate reaches 5000 which reveals the impact performance at high strain rate are unsatisfactory. When loading at low strain rate, Stress is well-distributed in material, so the fracture strength shows the load bearing capacity of the material; but when loading at high strain rate, damage initially emerged at the load end, thus the current material strength represents the load bearing capacity after local fracture, not what the material really has.

Models with pores have the similar conclusion: larger strength comes with higher strain rate, which lead to a better impact performance than ceramic with no pores.

3.3. Damage accumulation and fracture

The colored areas in figure 7 represent the damaged area, and when the color turns into red, that means this part of material is entirely ineffective.

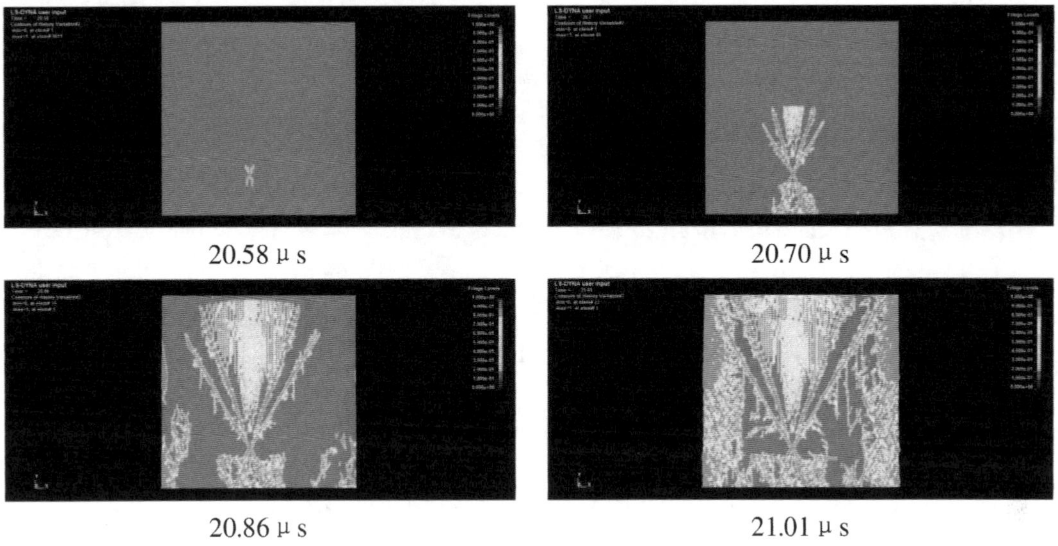

Figure7. Damage contour of ceramic when porosity sets 0%

Under compressive loading, damage begins to accumulate when the deviator stress exceeds a critical value. It can be seen clearly in figure 7, the damage generate initially at the direction of $\pm 45^{\circ}$ plane which shear failure occurred due to the shear stress reaches its limit. As we all know, the strength of compressive is greater than shear for ceramic materials, so the results displayed that the fracture of ceramic material did agree with the maximum shear stress criterion.

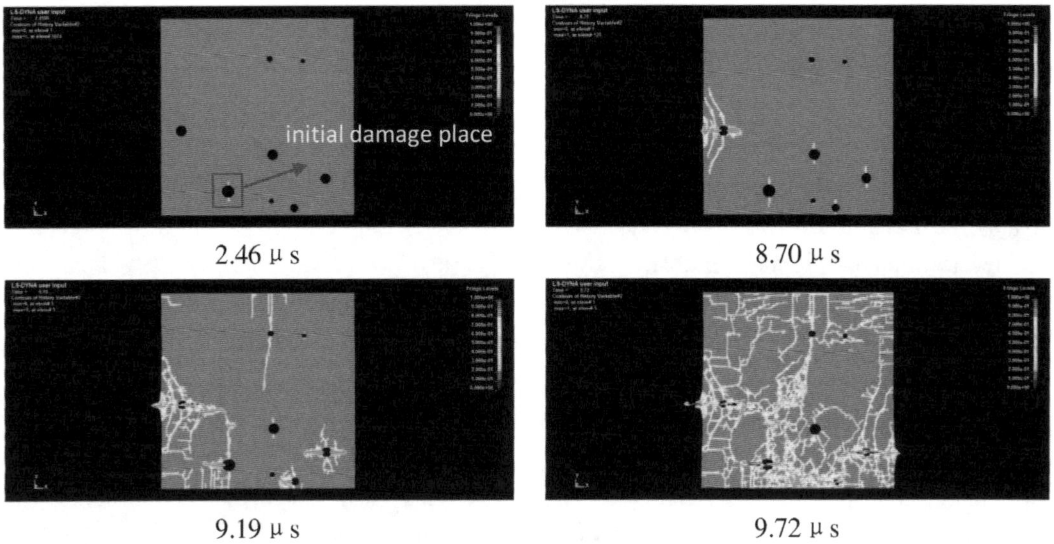

Figure8. Damage contour of ceramic when porosity sets 1%

when pores exist in ceramic material, take porosity 1% as an example, figure 8 shows the damage propagation and fracture situation: in normal conditions, damage generate initially at the place of more dense pores, larger pore size or the vicinity of the boundary which form stress concentration easily along with the impact loading. Thus it can be seen instantaneous damage occurred at a larger pore size close to the bottom surface, after that failure emerged near the left boundary due to the damage accumulation, ultimately an increment in damage leads to material bulking.

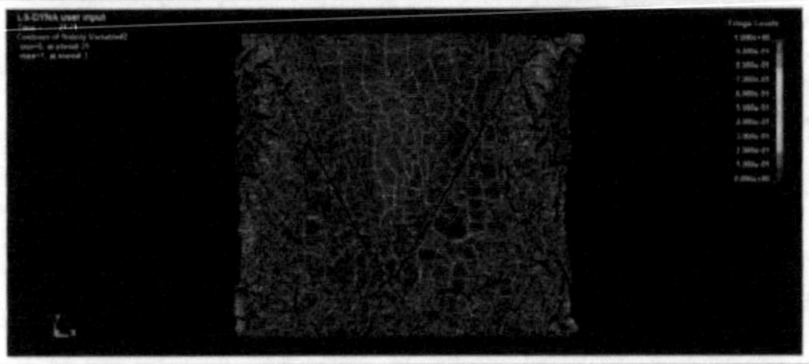

(a) porosity 0%

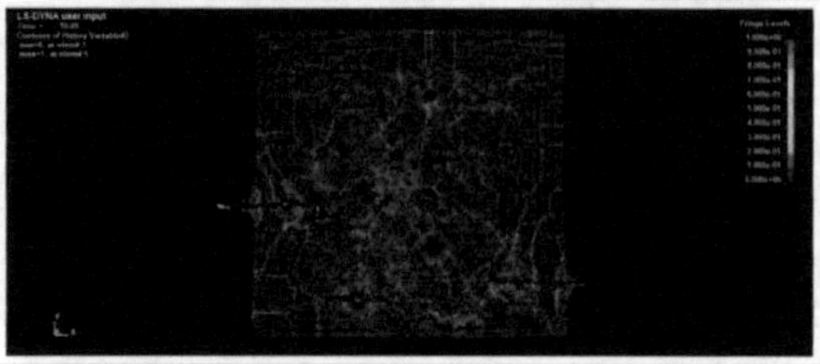

(b) porosity 1%

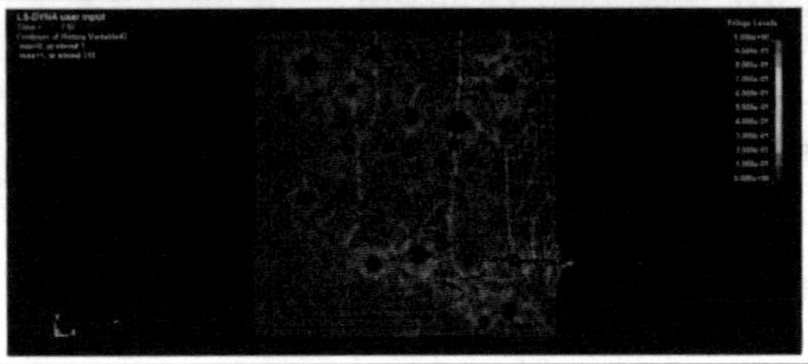

(c) porosity 3%

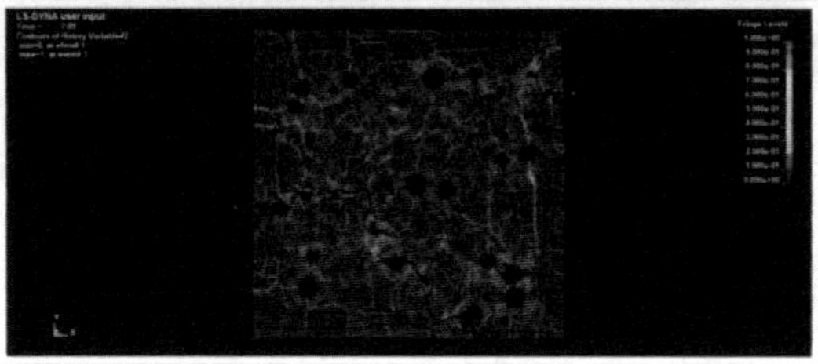

(d) porosity 5%

Figure9. Final fracture configuration under strain rate1000

Figure9 shows the fracture of ceramic without pores (0%) were controlled by shear stress when compression. But the ceramic with pores fractured in other way-varies from different distribution, size and shape of pores.

4.Conclusions

It should be emphasized that the present results refer to specific ideal material, and should be expected to predict the 'dynamic property' at given porosity and strain rate in ceramics. The results presented here should not be expected to apply directly to substantially different types of porosity.

Meanwhile, fitted curves of E-porosity and fracture strength-porosity show great acceptance with literature. With the increase of porosity, elastic modulus and strength of ceramic both decreased. And the decrease of strength slows down with the increase of porosity.

High-density alumina ceramic (porosity below 5%) is analyzed among different porosities. The fracture strength increased as the strain rate raised from 500 to 5000 when pros exit in ceramic material, but it went to the opposite direction when there are no pros in ceramic body during simulation. From that it can be concluded the dynamic property of ceramic with pros (porosity below 5%) performed better than without it. But it has been noted that the strain rate effects are typically secondary compared to the porosity effects.

Lastly, the fracture pattern of ceramic without pores is totally different from that with pores: the fracture of ceramic without pores acts as normal brittle material—shear stress reaches its limit; the process of damage is different from ceramic with pores, which involved local damage coursed by stress concentration. With the accumulation of damage in the formation of micro-cracks, micro-crack propagation and eventually cross throughout the entire material, then came the material failure.

Reference:

[1] Rice R W 1996 *J Mater Sci.* **31**1969-83
[2] Michael AM, James YYand Carlos GL 2001 *J Am Ceram Soc.***84**2594-602
[3] DEANM L 1997*J Mater Sci Mater Med.***8**227-32
[4] David E D,Lester J L and Robert EJ 1973 *J. Geophys. Res.***78**2403-17
[5] KNUDSENFP 1959*J Am Ceram Soc.***42**376-87
[6] Thomas JA, Gust W Hand and Royce E B 1968*J. Appl. Phys.* **39** 4610-6
[7] Wagh A S Singh J P andPoeppel R B1993 *J Mater Sci.* **28**3589-93
[8] DEAN E A andLOPEZ J A 1983*J Am Ceram Soc.***66**366-70
[9] SPRIGGS R M 1961 *J Am Ceram Soc.* **44**628-9
[10] Weitao YAlaTabie 2002Strain Rate Dependent Micro-Mechanical Composite Material Model for Finite Element Impact Simulation.*7th Int. LS-DYNA Users Conf.*(Detroit:Materials)p 16-5
[11] Le Huec J C, SchaeverbekeT andClement D 1995*Biomaterials.* **16**113-8
[12] Holmquist J, Templeton W and Bishnoi D 2001*Int J Impact Eng.* **25**211-31
[13] Khahn Bui Duane S. Cronin 2003Implementation and Validation of the Johnson-Holmquist Ceramic Meterial in Ls-Dyna*4th European LS-DYNA Users Conf.*(Ulm:Materials) p 53
[14] KNUDSEN F P 1962*J Am Ceram Soc.* **45**94-5

Multiscale Numerical Simulation of the Shaped Charge Jet Generated from Tungsten-Copper Powder Liner

Liu Jintao[1, 2], Cai Hongnian[1, 2], Wang Fuchi[1, 2], Fan Qunbo[1, 2*],

[1] School of Materials Science and Engineering, Beijing Institute of Technology, China,

[2] National Key Laboratory of Science and Technology on Materials under Shock and Impact, China

*Corresponding author Email: fanqunbo@bit.edu.cn

Abstract. Formation process of the shaped charge jet of W-Cu powder liner was simulated with smoothed particle hydrodynamics (SPH) method of LS-DYNA software. With the digital image process technique and macro-micro coupling method, a multiscale finite element model was established, and the high speed deformation process of the microstructure driven by explosive detonation in the liner of shaped charge was successfully simulated. The Cu phases were susceptible to serious deformation while the tungsten phase has less deformation. Besides, the temperature field of the microstructure during the shaped charge deforming was calculated, and a discussion of the deformation mechanism of the liner was given. The methods proposed in this paper would be of help in microstructure design of shaped charge materials.

1. Introduction

The linear shaped charge is a kind of shaped charge warhead[1]. After the initiation of detonation, the shaped charge liner collapse, then a high speed jet is formed and stretched in a short time[2]. The Shaped charge jets with high energy and good direction has a strong cutting penetration effect and it is widely applied to the defense industry and civil engineering.[3-6] The shaped charge liner is the carrier of energy, and the property of liner materials acts an important role on the energy accumulation of shaped charge. Therefore, studies on shaped charge material receive great attention in recent years[7-13]. Fe, Cu, Ta, U, W-Cu alloy are all good candidate materials. Especially, the W-Cu material has a high density characteristics and a good ductility capacity[14-15]. Compared with the traditional Cu materials, the tungsten Cu shaped charge liner has more advantages in higher initial density, very short stand off and deep penetration, so it has been extensively investigated. Jamet[15] put forwarded that W alloy with high density and high dynamic ductility can replace the Cu to become a good liner material. Seong[13] studied the penetration performance of forging tungsten Cu shaped charge liner. Jaekowski[16], Lichtenberger[17],et al discussed the influence of processing technology, grain size and purity on W - Cu material jet properties; Zhang[18] found that the density uniformity of the shaped charge liner and tungsten particle size were important factors in affecting penetration depth. Most of these studies based on the experiment result. However, because of the large-deformation dynamics and the complex multi-materials interaction over very a short time[19], some experimental phenomena and results are hard to be observed. Therefore, numerical simulation has become necessary to study the behaviors of the shaped charge jets. Maysele[20] studied the effect of liner porosity on jet properties by using a 2-D hydrodynamics code. Li[21] took advantage of Lagrange function and adaptive mash method to model the process of jet formation. Huerta et al[22] used an analytical code to investigate the penetration properties of the conical shaped charge and compared it with experimental results. Lei [23]simulated the formation process of the shaped charge jet of tungsten-Cu powder liner by multi-material ALE algorithm. Qiang [24]developed a multi-phase SPH

(MSPH) method to improve the numerical simulation of shaped charge jet. Though some positive achievements have been made on understanding the action of shaped charge jet formation, almost all the previous numerical simulations were for the macro jet formation process, and the microstructure evolution were not taken into account.

In the current study, formation process of the shaped charge jet of W-Cu liner was simulated by SPH method in LS-dyna, firstly. Subsequently, by extracting the load information of the local large deformation region and acting it on a microstructure finite element model, the microstructure evolution at the early formation stage of the jet was successfully captured.

2. Numerical Simulation Model

2.1. Characteristics of the Linear Shaped Charge SPH Model

The geometrical model for a linear shaped charge consists of the liner and explosive, as illustrated in Fig1. Because of the symmetrical structure of the linear shaped charge, the model is established using 2D plane strain approximation, which is composed of 8367 SPH elements. As shown in figure 1, the wall thickness is 3mm; the angle of the liner is 80° ; the length of the liner is 30mm; and the charging height is 32 mm. The material used for the explosive is 8701 explosive. The material model "High Explosive Burn" with an equation of state "JWL" is used for the explosive. The material parameters for the explosive are listed in table 1[25]. The material model for the W-Cu alloy is *MAT_STEINBERG with an equation of state *EOS_GRUNEISEN in LS-DYNA. The material parameters for the liner are listed in table 2[22].

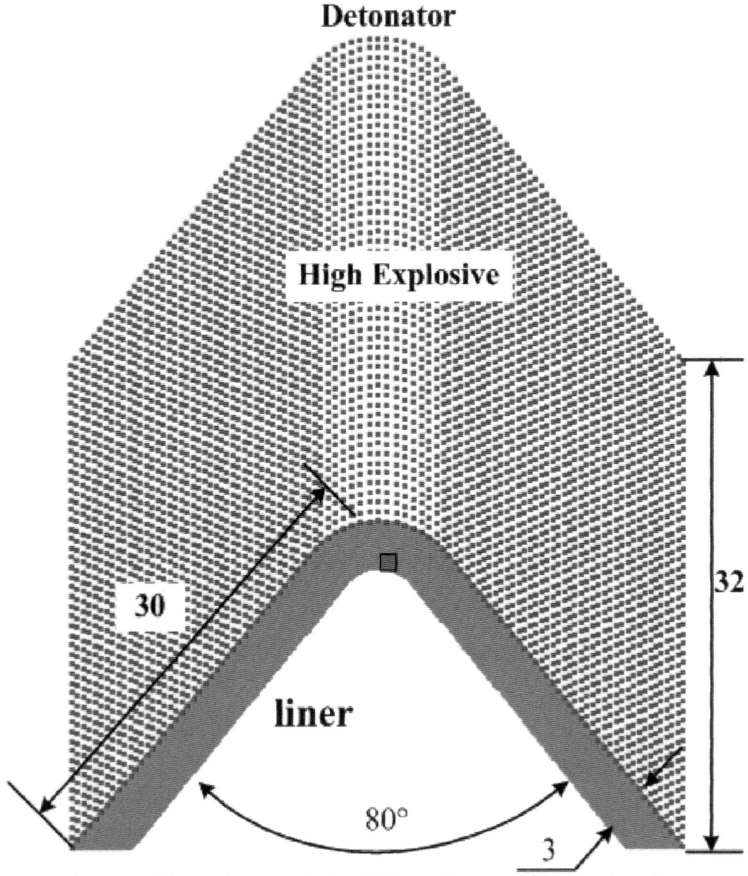

Figure 1 Numerical model of Liner Shaped Charge (mm)

Table 1 The JWL state equation parameters of explosive products.

ρ /(g•cm^{-3})	D/(km • s^{-1})	P_{CJ}/GPa	e/(kg • cm^{-3})	A/GPa	B/GPa	R_1	R_2	ω	V_0
1.7	8.1	29.5	8.35	581.2	6.8	4.10	1.10	0.34	1.0

Table 2 The constitutive model parameters of W-Cu material.

ρ /(g•cm^{-3})	G_0/GPa	b/s^2 • kg$^{-2/3}$	b/s^2 • kg$^{-2/3}$	h	f	a	T_{m0}/K	γ$_0$	σ$_0$/GPa
18.98	84.4	4.739	4.739	8.056	0.001	1.5	1710	2.42	0.12

2.2. Finite Element Model of the W-Cu Alloys Microstructure

In order to study the response of the microstructure of W-Cu shaped charge liner on experiencing detonation loading, a finite element model based on the actual microstructure of W-Cu alloys needs to be established. By using the Microstructure Oriented Finite Element Software (MOF) designed in our group [26], a finite element model consistent with the real microstructure of W-Cu alloys is generated, as shown in figure 2. This finite element model is composed of 34080 shell finite elements, and the dimensions are 258μm×193μm.

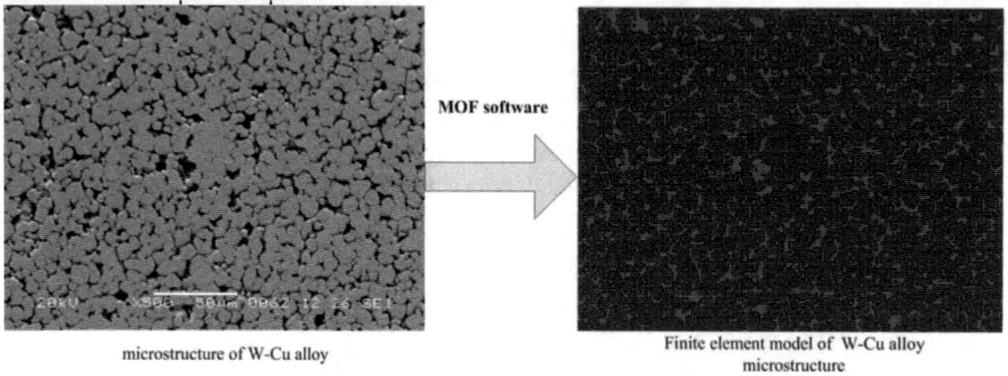

Figure 2 The establishment of the microstructure finite element model

In this model, the *MAT_STEINBERG is chosen to model the W and Cu materials. The parameters are given in table 3 and table 4 , respectively.

Table 3 The constitutive model parameters of W material.

ρ /(g•cm^{-3})	G_0/GPa	b/s^2 • kg$^{-2/3}$	b/s^2 • kg$^{-2/3}$	h	f	a	T_{m0}/K	γ$_0$	σ$_0$/GPa
19.22	160	4.70	4.70	8.06	0.001	1.838	3560	2.02	0.12

Table 4 The constitutive model parameters of Cu material.

ρ /(g•cm^{-3})	G_0/GPa	b/s^2 • kg$^{-2/3}$	b/s^2 • kg$^{-2/3}$	h	f	a	T_{m0}/K	γ$_0$	σ$_0$/GPa
8.93	47.7	2.83	2.83	3.77	0.001	0.635	1790	2.02	0.12

2.3. Coupling of the macro and micro numerical models

Based on the simulation of the jet formation, the severe deformation process of the liner is observable, and the load conditions of each region can be obtained accordingly. Extract the load conditions of the region which we concern, e.g. the small red square position in figure 1, and load it into the finite element model of the microstructure, the response of the microstructure of this region under the corresponding loads can be tracked. The coupling method of micro-macro numerical simulation is illustrated in figure 3.

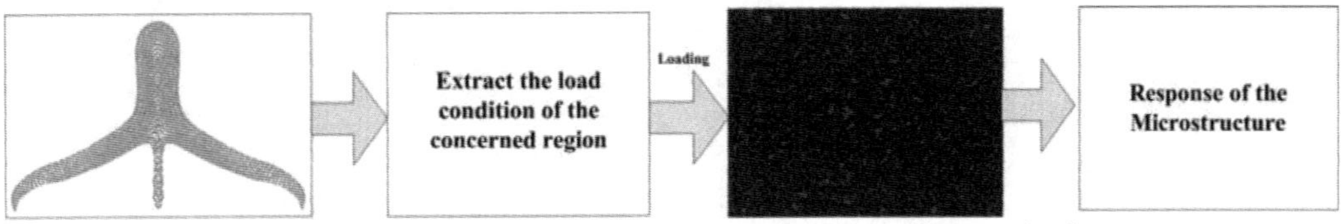

Figure 3 The sketch of the macro-micro coupling method

2.4. Heating Mechanism of the Liner

When the explosive charge is initiated, the detonation wave spreads, and the liner material is accelerated and collapsed under the high pressure. At the same time, the temperature of the liner is increased significantly under the action of the shock waves and the severe plastic deformation.

The temperature increase Ts caused by shock waves is calculated as:[27]

$$T_s = \exp\left[\left(\frac{\gamma_0}{V_0}\right)(V_0 - V)\right] \times \left(\frac{1}{2C_v}\int_{V_0}^{V}\left[p(V) + (V_{00} - V)\frac{dp(V)}{dV}\right] \times \exp\left[\left(\frac{\gamma_0}{V_0}\right)(V - V_0)\right] dV + T_0\right) \quad (1)$$

where V_{00} and V_0 are volumes of porous materials and dense materials. p and γ_0 are the pressure in the liner and Gruneisen coefficient of the material, respectively. C_v is the heat capacity at constant volume and T_0 is the room temperature.

When pressure unloading, the residual temperature can be obtained form the isentropic adiabatic equation:

$$T_r = T_s \exp\left[\left(\frac{\gamma_0}{V_0}\right)(V_s - V_r)\right] \quad (2)$$

T_r is the residual temperature after the shock waves. V_s and V_r are liner volumes under the situation of shocks and after shocks.

The increasing temperature of liner ΔT caused by shock wave is obtained: $\Delta T_1 = T_r - T_0$

The temperature rise ΔT_2 due to plastic deformation can be calculated as:[28]

$$\Delta T_2 = \frac{\eta}{\rho C_p}\int_0^{\varepsilon} \sigma d\varepsilon \quad (3)$$

Where ρ and C_p are the density and heat capacity at constant pressure of the material; σ and ε are stress and strain of the liner; η is the thermal transformation efficiency.

The general temperature rising is: $\Delta T = \Delta T_1 + \Delta T_2$.

3. Numerical Simulation of Shaped Charge Jet Formation

Upon initiation of the explosive, the detonation wave spreads, and the liner material is accelerated and collapsed under the high pressure. The simulation results shown below demonstrate the progress of jet forming process.

Figure 1(a) shows the initial state of the liner. The explosive is initiated, but the shock wave has no interact with the liner. The liner deformation state at t = 10μs is presented in figure 4(b). The collapse of the liner material on the centerline generates a jet where the jet tip can travel in excess of 1800m/s. As seen in figure 4(c), the Jet and slug are generated at t=20μs, and the further jet evolution is illustrated in figure 4(d).

With the continuous flow of the jet, the jet tip accelerates up to 1870m/s. Because of the different speeds of the slug and jet tip, the jet is stretched. It can also be found that the mass of the jets primarily comes from the bottom of the liner, and the part at the top of liner becomes the slug.

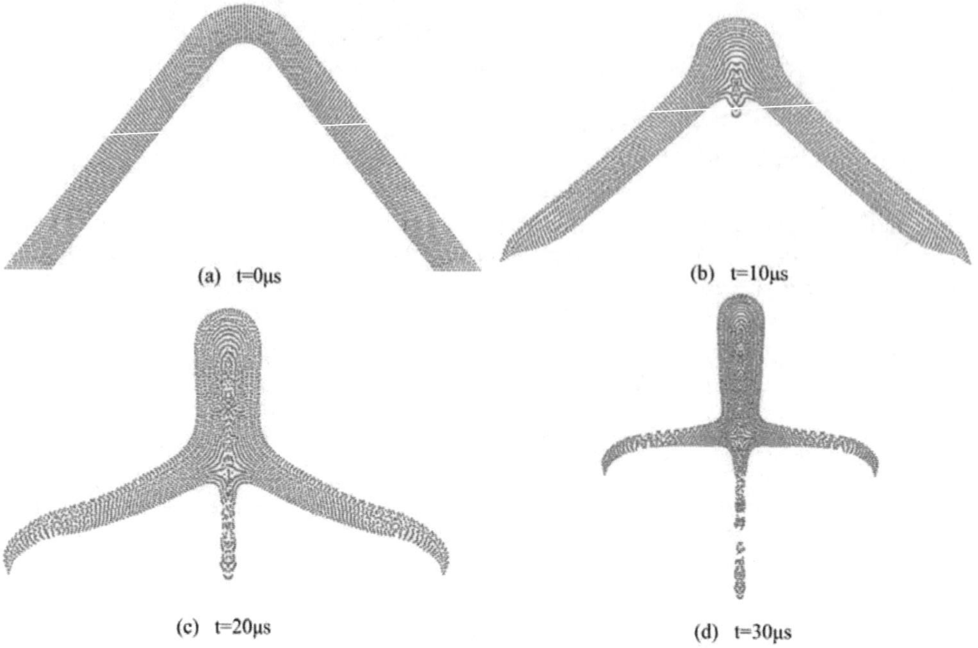

Figure 4 .Shaped charge jet forming and stretching

4. Numerical Simulation for Microstructure Evolution of W-Cu Alloys

4.1. Numerical simulation of the deformation of W-Cu alloys

By using the method described in section 2.3, the deformation behavior of the W-Cu microstructure can be tracked.

The microstructure deformation at t=6.8μs is presented in figure 5. A serious deformation is found in the finite element model of the W-Cu microstructure under the detonation load. It is clear that there are slip bands on Cu phase, along the boundaries of W phase and W/Cu interfaces. It might be noted that the bands appear on Cu phase but not reach the fracture limit, which is believed to be responsible for its excellent plasticity. On the other hand, there is little plastic deformation in W phase due to its poor ductility. A comparison between the experimental result and the numerical result is shown in Fig. 5, it can be seen that the simulation result agrees well with the corresponding experimental result.

The distribution of the effective stress and effective plastic strain of the microstructure finite element model at t=6.8 μ s are shown in figure 6. According to the contour shown in figure 6（a）, the stress value at the W phase is about 1GPa, which is much higher than that of 0.2~0.3GPa in Cu phases. The maximum stress appears at the Cu/W interfaces, up to 1.19GPa. During deformation, the stress in Cu phase is higher than its yield strength 57MPa, while the stress in W phase is far below than its yield strength 1200MPa, which means that plastic deformation occurs in Cu phase primarily. As shown in figure 6 (b), the strain value in the Cu phase is much higher than that of W phase, and the maximum strain value reaches 10.

Figure 5 Comparison of experimental and simulation results

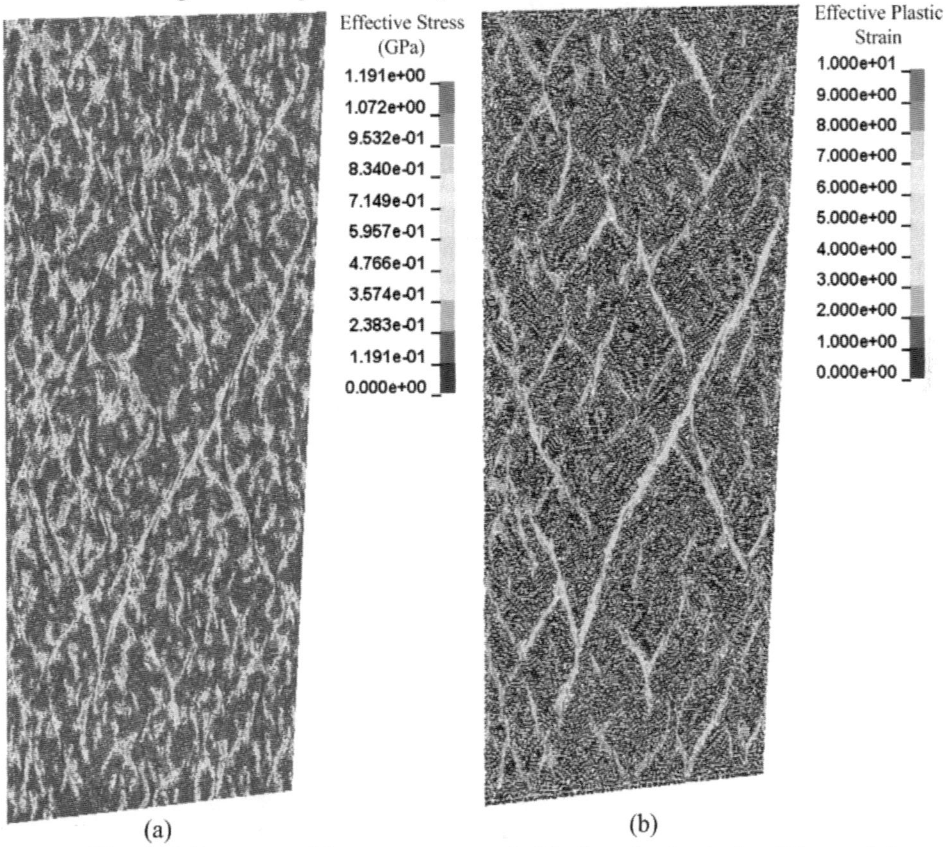

(a) (b)

Figure 6 Contours of the v-m stress and effective plastic strain of the model

4.2. Heating effect of W-Cu alloys

The temperature contours due to the heating effect of shock wave and plastic deformation work is shown in figure 7. As shown in figure 7 , the temperature at the W skeleton is about 500~600℃ while the Cu phase temperature can reach 400 ~ 550 ℃ .The W generates more energy when plastic deformation occurs for its high strength. Besides, the heat capacity and thermal conductivity of W are 136 J/(kg •K)and 174 W/(m •K) while that of the Cu are 385 J/(kg •K) and 403 W/(m• K) respectively. Because of the high plastic deformation energy with the low heat capacity and thermal conductivity, the temperature of W phase is higher than that of Cu. But the temperature of W is much lower than its melting point of 3387℃ while the Cu phase has reached 0.4 times of the melting point, 1083℃, i.e. the dynamic recrystallizational temperature. Under the high-speed service conditions, the refined recrystallizational Cu grains make great contribution to the superplasticity conditions during jet formation.

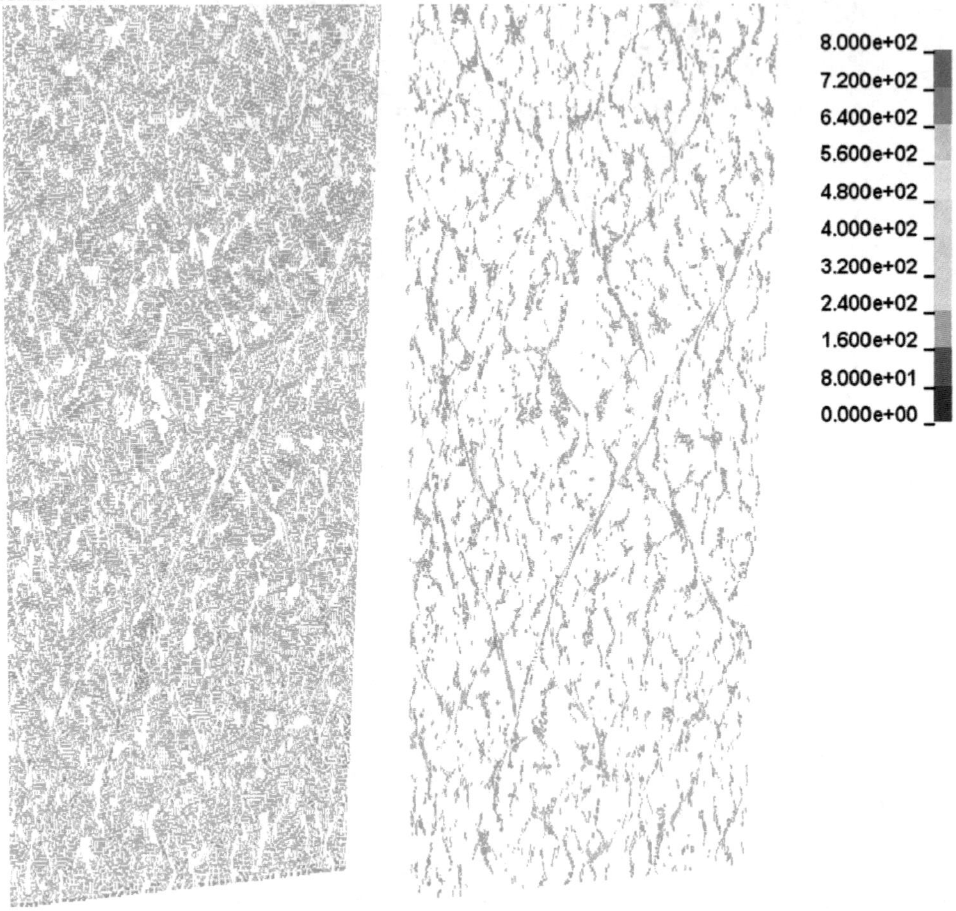

(a) (b)

Figure 7 Contours of the temperature of the W-Cu microstructure model

5. Conclusions

1. In this paper, a 2D shaped charge jet model is established using SPH elements. The formation process of the shaped charge jets is simulated, which is in good agreement with the experimental results.

2. Multiscale numerical simulation by Macro-micro coupling is accomplished by extracting the load of a micro region onto the microstructure model. Deformation of the microstructure of the W-Cu

liner is successfully captured. It is found that the slip band appears on the Cu phase while the W phase changes little.

3. The heating effect of the W-Cu microstructure, due to shock wave and plastic deformation work, is studied, indicating thatthe temperature of Cu is higher than 0.4 times of its melting point, thus meeting the recrystalline conditions and making great contribution to its superplasticity conditions during the jet formation.

6. Acknowledgements

The authors would like to acknowledge Professor Li shukui and Dr. Liu jinxu for helping in this research. This research is also supported by Dr. Guo wenqi for providing W-Cu microstructure pictures.

References

[1] Yun S R and Zhao H Y 2005 *Explosion Mechanics* Beijing:National Defence Industry Press 171
[2] Liu Z H and Gou R J 2008 *MechManageDev.* **23** 19
[3] Johnson J L and German R M 1993 *The International Journal of Powder Metallurgy* **4** 201-209
[4] Wang W F 1997 *Powder Metallurgy*, **40** 295
[5] Raghu T,Sundaresan R and Ramakrishnan P 2001 *Mechanical Alloying Materials Science and Engineering A* **304** 438
[6] Gao W and Zhang C X 2006 *Advanced Materials Industry* **2** 57
[7] Richard F,Jeffrey K and William N 2004 *The proceedings of the 21th international symposium on ballistics* 721
[8] Song S C, Gao P and Cai H N 2003 *Applied Mathematics and Mechanics* **24** 1075
[9] Jia W M, Shi H G and Zhang Q X 2006 *Rare Metal Materials and Engineering* **29** 33
[10] Jia W M, Zhang X Q and Bai Z G 2007 *Rare Metal Materials and Engineering* **36** 1512
[11] Zhao T, Luo H and Jia W M 2007 *Ordnance Material Science & Engineering* **30** 78
[12] Held M 2005 *20th Int Symp on Ballistics* Orlando 425
[13] Lee S, Kim E P and Kim Y 2005 *22 th International Symposium on Ballstics*
[14] Bransky I, Faibish E and Miller S 1986 *Proceedings of 9th International Symposium on Ballistics* England **Part 11** 237
[15] Jamet F and Lichtenberger A 1986 *Proceedings of 9th International Symposium on Ballistics* England **Part 11**,233
[16] Jaekowski and Wlodarczyk E 2006 J. *Mat. Pro. Tec* **171** 21
[17] Lichtenberger A 1997 4th International Conference on Tungsten Refractory Metals and A1loys 66
[18] Zhang Q X and Yao D 2009 *RareMetal Materials and Engineering* **38** 527
[19] Guo W, Li S K and Wang F C 2009 *ScriptaMater* **60** 329
[20] Mayseless M, Harvey W B and Hetz A 1999 Shock Compression of Condensed Matter 505 367
[21] Li Y C, Yu T F, Xu Q J, Xia W G and Tang Y 2002 *Jurnal of PLA University of Science and Techology* **3** 71
[22] Huerta M, Vigil M G 2006 *International Journal of Impact Engineering* **32** 1201
[23] Lei F B and Cao D L 2010 *ChineseJournal of Explosives&Propellants* **33** 112
[24] Qiang H F, Wang K R and Gao W R 2008 *Trans. Tianjin Univ* **14** 495
[25] Chen H 2007 The Spread of Shock Wave in Copper and Its Influence to the Microstructure Nanjing University of Science and Technology 26
[26] Shen W, Wang F C, Fan Q B, Ma Z and Yang X W 2011 Surface&Coating Technology **205** 2964
[27] Tao G, Chen H and Shen C Q 2008 *Explosion and Shock Waves* **28** 336
[28] Bai Y and Dodd B 1992 *Adiabatic shear localization* Now York:Pergamon Press

A Study on Propagation Characteristic of One-dimensional Stress Wave in Functionally Graded Armor Composites

S Y Yang[1], X Liu[1], D F Cao[2], H Mei[1], Z T Lei[3], and L S Liu[2,4]
[1]Department of Engineering Structure and Mechanics, Wuhan University of Technology, Wuhan 430070, P. R. China
[2]State Key Laboratory of Materials Synthesis and Processing, Wuhan University of Technology, Wuhan 430070, P. R. China
[3]Sheng Li oil field Sheng Li engineering construction group.Co.LT, Dongying 257000, P. R. China

E-mail:Liulish@whut.edu.cn

Abstract. The development of Functionally Graded Materials (FGM) for energy-absorbing applications requires understanding of stress wave propagation in these structures in order to optimize their resistance to failure. One-dimensional stress wave in FGM composites under elastic and plastic wave loading have been investigated. The stress distributions through the thickness and stress status have been analyzed and some comparisons have been done with the materials of sharp interfaces (two-layered material). The results demonstrate that the gradient structure design greatly decreases the severity of the stress concentrations at the interfaces and there are no clear differences in stress distribution in FGM composites under elastic and plastic wave loading.

1. Introduction

Early composite targets were made by simply bonding a ceramic tile to a metal backing plate. The underlying idea is to use the hard ceramic layer to defeat and erode the projectile, and to use the backing plate to absorb the impact energy and increase the fracture resistance of the target. The full potential of ceramic hard facing layers has not been achieved because the interface between ceramic and metal has an unfavorable impedance mismatch that may even induce tensile failure. At present, functionally graded materials (FGM) was introduced to armor structure design. The advantage of using FGM is their superior resistance to interfacial failure. Analytical and computational studies of the evolution of stresses and displacements in FGM show that the utilization and optimization of structures and geometry of a graded interface between two dissimilar layers can reduce stresses significantly. The emergence of functionally graded materials (FGM) provides a possible avenue for solving this problem since an FGM layer between the ceramic and metal can make the impedance change continuous rather than abrupt. FGM materials could, therefore, reduce the intensity of reflected

[4]Corresponding author, Tel.: +86 27 87651820; Fax: +86 27 87860863
E-mail: liulish@whut.edu.cn (L S Liu)

stress waves, so the tensile stresses might be reduced or even removed entirely.

Some simplified analytical models [1-3] and numerical simulations [4] have been developed and used for the analysis and optimization of materials consisting of layers of different materials. But there is as yet no consistent method available to evaluate and compare composites with different layering configurations and varying layer properties. It is difficult to use case-by-case simulations to gain a systematic understanding of the ballistic response of composite structures. Without such a systematic understanding, development and optimization of composite structures often becomes a trial-and-error process. In this paper, we want to investigate one-dimensional stress wave in FGM composites under elastic and plastic wave loading and analyze the stress distribution through the thickness.

So this paper tries to demonstrate and understand the propagation characteristics of elastic and plastic stress wave in FGM. The numerical simulation has been implemented in accordance with standard split Hopkinson pressure bar (SHPB). The striker bar impact the incident bar at velocities of 15 and 25 m/s, respectively, which lead to the elastic and plastic stress wave in the specimen. A notional FGM of aluminum nitride (AlN) and aluminum (Al) was modeled as a series of discrete (bonded) layers, with adjusted material parameters to approximate a gradient structure. Some comparisons have been done with the materials of sharp interfaces (bi-layered specimens)

2. Modeling approach

2.1 Computed models

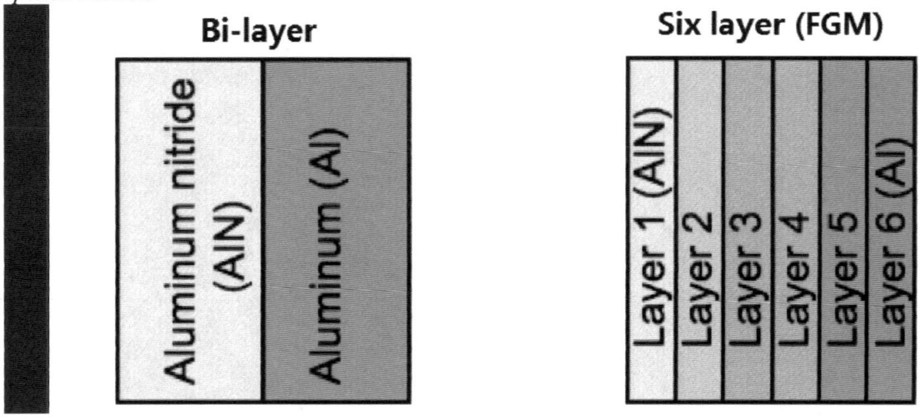

Figure 1. Initial geometry for the two (standard) and six (gradient) layered specimens

The most difficult aspect of modeling a FGM is how to account for the changing material behavior going from the top layer to the bottom layer. The design goal of a FGM is to produce a specimen that has a gradual change in material behavior, typically going from a strong, brittle response on the top, to a soft ductile response on the bottom. For this work, two specimens are investigated, a two-layered specimen and a six-layer specimen with a linear function as shown in figure 1. The two-layer specimen is a traditional configuration consisting of a hard, brittle top layer (aluminum nitride) backed by a soft, ductile layer (aluminum). The six-layer specimen is treated as a system of discrete layers, which transit from the top layer (AlN) to the bottom layer (Al) in a systematic manner. The JH-2 model is used to describe the top three layers (layers 1-3) and the JC model is used for the bottom three layers (layers 4-6). The distribution characteristics of ceramic volume fraction for the six layer specimen are shown in figure 2.

In the numerical model, the interfaces between all layers are regarded as perfect interfaces, in which shear slip is ignored. The height and diameter of all specimens are 2cm and 6mm, respectively.

2.2 Material parameters

The modeling approach used to model a graded material incorporates the use of two models. The JH-2

[5] model is used for the rock-like materials. The Johnson-cook(JC) [6]strength and fracture models are used for the metal. The JH-2 model is used to represent the more brittle layers. The JC models are used to represent the more ductile layers. In our work, the JH-2 model is used to describe the left three layers (layers 1-3), and the JC models are used for the right three layers (layers 4-6). The general idea is to modify the properties of each layer such that they transit from the top layer (AlN) to the bottom layer (Al) in a systematic manner. The constants are listed in table 1 where the AlN is described by layer 1 and the Al is described by layer 6. The six-layer (gradient) specimen is an attempt at modeling a FGM specimen by gradually changing the material properties for each layer. Similarly, the strength and fracture characteristics could also be transited from a strong and brittle behavior for layer 1 to a weak and ductile behavior for layer 6.

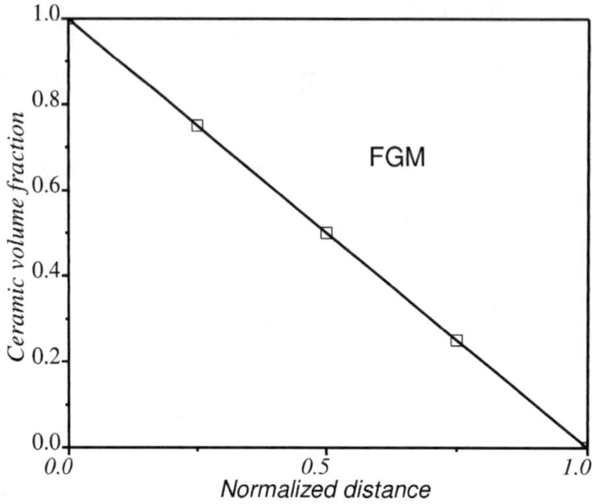

Figure 2. Distribution characteristics of ceramic volume fraction for the six-layer specimen

Table 2. Material model constants used for the two and six layer specimens

Constant	Layer1	Layer2	Layer3	Layer4	Layer5	Layer6
Density (Kg/m^3)	3226	3134	3043	2951	2860	2768
Bulk Modulus (GPa)	201	176	151	127	102	77
Shear Modulus (GPa)	127	107	87	66	46	26
K2 (GPa)	260	234	207	181	154	128
K3 (GPa)	0	25	50	75	100	125
HEL(GPa)	6.41608	5.33188	4.3959			
P_{HEL}	3.54462	2.99539	2.52756			
Equivalent Stress (GPa)	4.3072	3.50475	2.8025			
Tensile Strength (GPa), T	0.50	0.75	1.00			
Strength Constant (GPa), A	0.96246	0.93725	0.90785			
Fractured Constant B (GPa)	0.20	0.16	0.12			
Fractured Exponent M	0.21	0.21	0.21			
D1	0.16	0.56	0.63			
D2	1.00	1.26	1.47			
Specific Heat (J/Kg °C)	735	763	791	820	848	876

3. Computed result and discussion

3.1. Propagation characteristic of elastic stress wave

Computations were done to determine the propagation characteristic of elastic stress wave, while the striker bar impact at the velocity of 15m/s. Figure 4 shows the pressure-time histories from the bi-layered specimen. Numerical sample points in the specimen are shown in figure 3. Because the bi-layered (ceramic/metal) specimen has a sharp interface which is more likely to lead to tensile stresses, it is necessary to observe the load characteristic of the bi-layered specimen at first. From figure 4, some phenomenon can be observed: (1) the pressures value of all numerical sample-points are still equal to zero after unloading finish, which indicates that materials do not exhibit yield behavior; (2) The lowest pressure is still greater than zero, which indicates that there is no negative pressure zone when the stress waves go through the specimens.

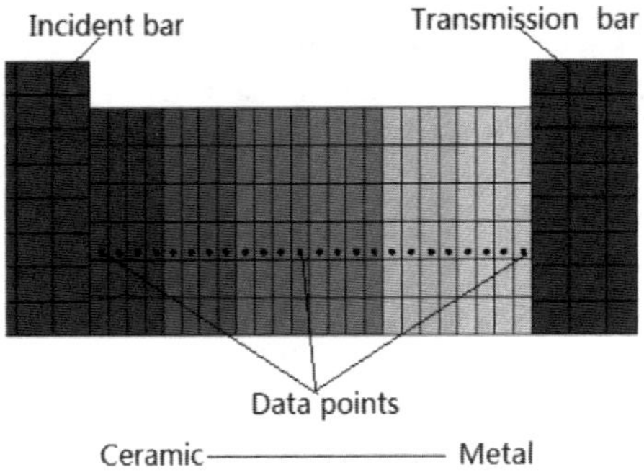

Figure 3. Numerical sample-point in the specimen

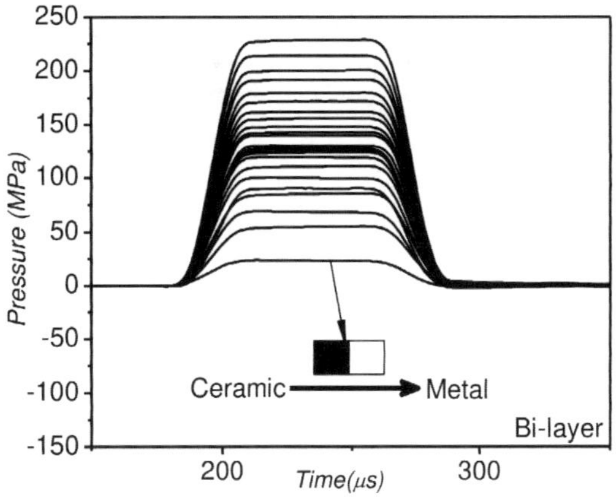

Figure 4. The pressure-time histories for all numerical sample points of bi-layered specimen.

Figure 5 shows the effective stress distribution through the thickness of bi-layered and FGM specimens. Stress concentration is found at the ceramic/aluminum interface in the bi-layered specimen and the maximum stress value locates at the ceramic layer near interface. In FGM specimen, there is no obvious stress concentration, and the maximum stress value locates at the middle of the metal-like components. The results reveal that the gradient structure design greatly decreases the severity of the stress gradient at the interfaces during the elastic deformation.

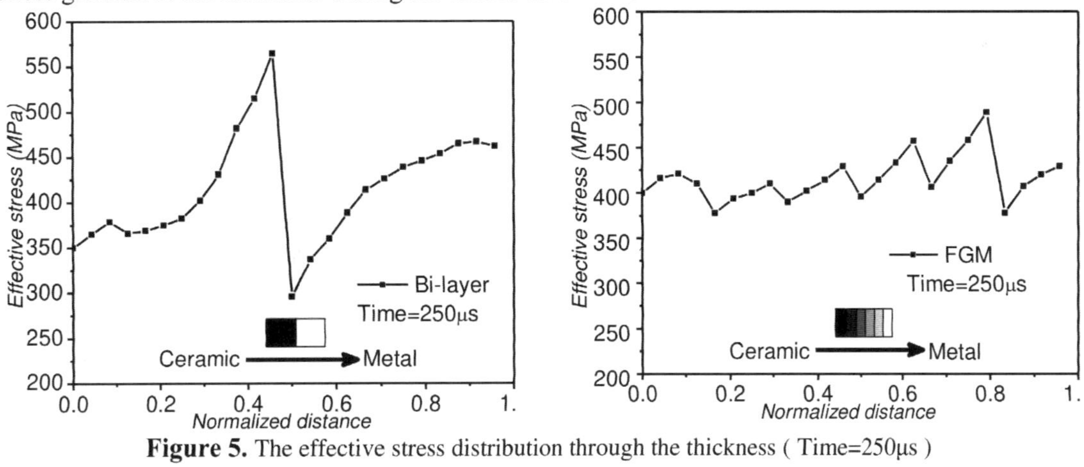

Figure 5. The effective stress distribution through the thickness (Time=250μs)

3.2 Propagation characteristic of plastic stress wave

Computations were done to determine propagation characteristic of plastic stress wave, while the striker bar impact at the velocity of 25m/s. Figure 6 shows the pressure-time histories from the bi-layered and FGM specimens. Numerical sample-points in the specimen are the same to that of the figure 3. We can find both bi-layered and FGM specimens have experienced plastic deformation (some pressure values still retain negative after the stress waves go through the specimens). During the stress wave loading, a negative pressure zone at the interface between ceramic and metal have appeared in the bi-layered specimen, while the FGM specimen still retains compressive status.

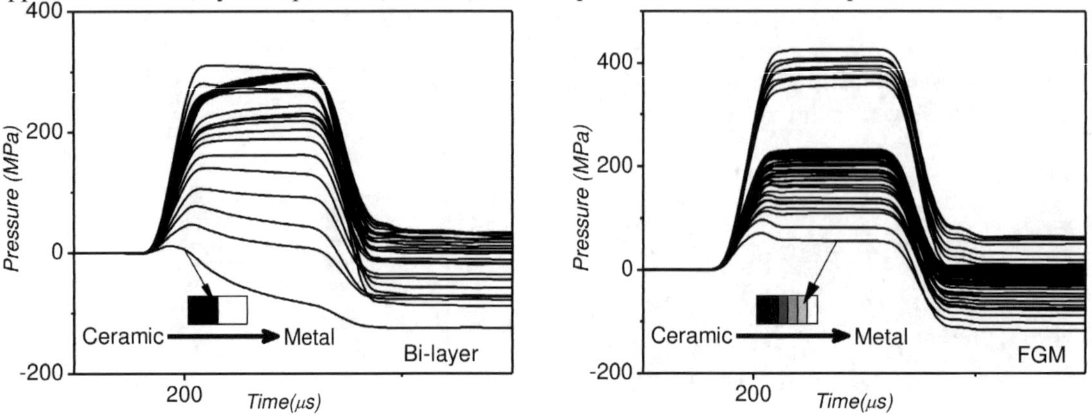

Figure 6. The pressure-time histories for all numerical sample points of bi-layered and FGM specimen.

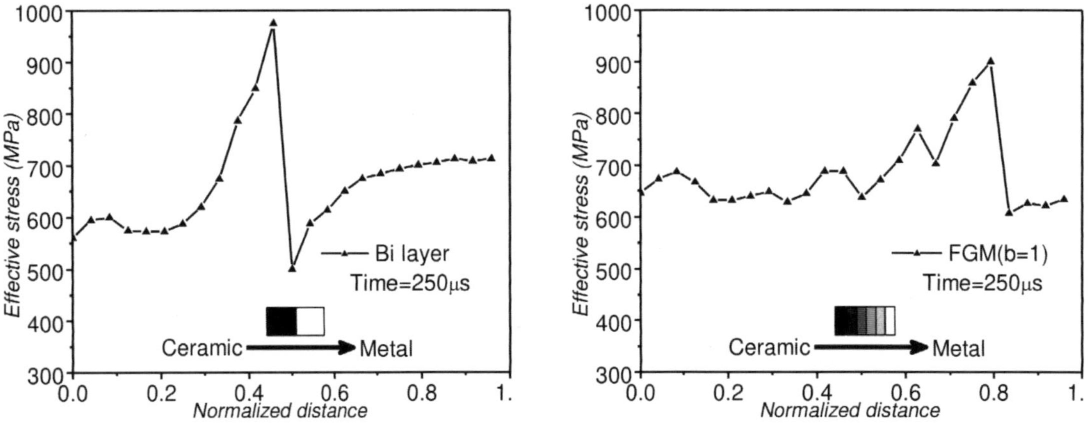

Figure 7. Comparisons of the effective stress distribution through the thickness (Time=250μs)

Figure 7 shows comparisons of the effective stress distribution through the thickness at 250μs (at the higher striker bar velocities). When compared with the low striker bar velocities, no obvious difference of distribution characteristics can be found except higher stress values. The region of high stress concentration still appears at the ceramic layer near ceramic/aluminum interface in the bi-layered specimen. In FGM specimen, the high stress still located at the middle of the metal-like components but the stress gradient become severe.

The literature [7] also compared the elastic and plastic deformation of FGM. The results show plastic deformation of the materials can significantly affect their response to wave propagation. In particular, a major effect of plastic deformation is to lead to a more homogeneous stress distribution within the components. In our work, the similar mechanical phenomenon has not been found.

4. Summary

A notional FGM of aluminum nitride and aluminum was modeled as a series of discrete (bonded) layers, with adjusted material parameters to approximate a gradient structure. The Johnson-Holmquist ceramic model was used for the AlN, and the Johnson-Cook metal model was used for the aluminum, with the computations performed using the LS-DYNA code. We have investigated the propagation characteristic of one-dimensional stress wave with the low and high striker bar velocities.

Comparing with bi-layered specimens, we obtain the following conclusions.

(1) The bi-layered specimen has obvious stress concentrations at the ceramic layer near the interface between the ceramic and metal (whether elastic or plastic wave). The gradient structure design greatly decreases the severity of the stress concentration at the interface.

(2) There is no negative pressure zone in both bi-layered and FGM specimens under the elastic loading. But under the plastic loading, the bi-layered specimen has a negative pressure zone at the interface between ceramic and metal.

(3) Compared with the low striker bar velocities, there is no obvious distinction in the distribution characteristics in the higher striker bar velocity, except higher stress values.

Acknowledgements
The work was supported by Fundamental Research Funds for the Central Universities (2011-IV-124).

References
[1] Hazell PJ, Fellows NA, Hetherington JG 1998 *Int. J. Impact Eng.* 21 589
[2] Zaera R, Sánchez-Gálvez V 1998 *Int. J. Impact Eng.* 21 133
[3] Bruck HA 2000 *Int. J. Solids Struct.* 37 6383
[4] Li Y, Ramesh KT, Chin ESC 2001 *Int. J. Solids Struct.* 38 6045
[5] G.R. Johnson TJHaSRB 2003 *J. Appl. Phys.* 94 1639
[6] G. R. Johnson WHC 1983 *7th International Symposium on Ballistics* 541
[7] Tasdemirci A, Hall IW 2007 *Int. J. Impact Eng.* 34 1797

Synthesis and characterization of Dopamine graft compound N-methacryloyl 3,4-dihydroxyl-phenylamine

Xiong Xiong[1,2],**Qu Shu-xin**[2],**Liu Yu-mei**[2]

[1]Southwest Jiaotong University-emei, Emei 614202; [2]School of Material Science and Engineering, Key Lab of Advanced Technologies of Materials, Ministry of Education, Southwest Jiaotong University, Cheng du 610031

Email address:qushuxin@swjtu.edu.cn

Abstract. In order to obtain adhesive biomaterials inspired by mussels, the intermediate derivatives of dopamine, N-methacryloyl 3,4-dihydroxyl-phenylamine (dopamine methacrylamide DMA), was synthesized by grafting methacrylate anhydride to dopamine. The structure of the compound was confirmed by fourier transform infrared spectroscopy and nuclear magnetic resonance. The thermal stability of DMA was also characterized by thermo gravimetric analysis and differential scanning calorimeters techniques. The surface morphology of DMA crystal was analysed by scanning electron microscope analyses. The present result showed that the synthesis of new monomers was successfully fulfilled and the new compounds retain the hydroxyl functional groups. The surface morphologies and thermal stability of DMA crystal were also altered by grafting reaction.

1. Introduction

Adhesive biomaterials have many potential medical applications, such as wet tissue adhesives in minimally invasive surgery, and as vehicle for drug delivery to mucosal surfaces. [1] Clinically successful adhesive biomaterials should have the following primary attributes: (1) the ability to rapidly solidify or polymerize *in situ* from a fluid precursor; (2) the ability to form strong and durable bonds to tissue surfaces even in the biological environment; and (3) biocompatibility. [1,2] However, the current medical adhesives seldom meets these requirements due to lack of adhesion strength, in particular, in the presence of biological fluids, sensitization and allergic response, and inflammation. [2] Therefore, it is necessary to develop the safe and effectively curable adhesive biomaterials on wet tissue surface with minimal inflammatory insult.

Marine and freshwater mussels are attracted many attentions due to their remarkable adhesive capabilities to natural or manmade surfaces in the wet environment. [3] Mussels secrete specialized adhesive proteins containing a high content of L-3,4-dihydroxyphenylalanine(DOPA), which residue is thought to be responsible for adhesion.[4] Proteins incorporating the DOPA functionality contribute to strong adhesive structures, allowing marine organisms to fix their bodies to various types of surfaces, such as polymers, ceramics, and metals, even in aqueous conditions. Under oxidizing conditions, the covalently cross-linking occurs between DOPA molecules or between DOPA molecule and biologically relevant nucleophiles, such as primary amines and thiols, by reversal of the Michael addition. [5-8] Lee et. al. demonstrated that the adhesive strength of DOPA was declined compared to that of the unoxidized DOPA, in particular when it adhered to metallic surface, by the atomic force microscope(AFM) experiments. [9] The oxidized forms of DOPA that result from such reactions are believed to be less adhesive than unoxidized DOPA. [10] Therefore, several groups have reported the synthesis and characterization of DOPA-containing peptides and polymers to avoid the oxidization of DOPA and to maintain its adhesive. [11-13] However, most previous efforts incorporating DOPA into polymer have been hypothesized to reduce the adhesive properties of DOPA and require reagents that are harmful to biological tissues. Both adhesive and polymerize in situ of DOPA are depended on the two hydroxyl in the benzene. DOPA containing proteins and polymers exhibit better adhesion to both metallic and mucosal surfaces when DOPA is not oxidized. [11, 14] Hence it is necessary that two hydroxyl on the DOPA are preserved when the solidification of the liquid adhesive to improve adhesive strength.

The aim of this research was to graft methacrylate anhydride with DOPA in order to develop materials for the adhesive biomaterials. Monomers of N-methacryloyl 3,4-dihydroxyl phenylamine was synthesized and these monomers were confirmed by fourier transform infrared spectroscopy (FT-IR), nuclear magnetic resonance(NMR) techniques. The thermal stability of DMA monomer was characterized by thermo gravimetric analysis (TG) and differential scanning calorimetry (DSC). The morphologies of DMA crystal were observed by Scanning Electron Microscopy (SEM).

2. Experimental

2.1. Materials

3,4-dihydroxyphenethylamine hydrochloride (dopamine-HCl) were purchased from Sigma (Steinem, Germany); methacrylate anhydride were purchased from Aladdin(Shang Hai,China); sodium borate, sodium bicarbonate, sodium hydroxide, hydrochloric acid, tetrahydrofuran, ethylacetate and hexane were purchased from Kelong(Cheng Du, China).

2.2. Synthesis of DMA

20g of $Na_2B_4O_7$ and 8g of $NaHCO_3$ were dissolved in 200mL of deionized water and bubbled with N_2 for 20 min. 10g of dopamine-HCl (52.8 mmol) was then added followed by the dropwise addition of 9.4 mL of methacrylate anhydride (58.1 mmol) in 50 mL of tetrahydrofuran, during which the pH of solution was kept above 8 with addition of 1mol/L NaOH as necessary. The reaction mixture was stirred 12h at room temperature with N_2 bubbling. At this time, a white slurry-like solution had formed and was then washed twice with 50 mL of ethyl acetate. The resulting solid in the solution was vacuum filtered and the obtained aqueous solution was acidified to pH 2 with 6M of HCl solution. The organic layer of the solution was extracted three times from the acidified aqueous solution with 50 ml of ethyl acetate. The extracted clear brown organic layer in the ethyl acetate was dried over $MgSO_4$. The solution volume was reduced to 25 mL with a rotary evaporator. The obtained solution was added to 250 mL of hexane with vigorous stirring to precipitate a brownish solid and then the formed suspension was refrigerated to maximize crystal formation size. To purify, the resulting light brown solid was dissolved in 20 ml of ethyl acetate and precipitated in 300 ml of hexane. The final solid powder was dried in a vacuum overnight. The synthetic route of the DMA monomer and structure of the product were shown in Figure1 and Figure 2.

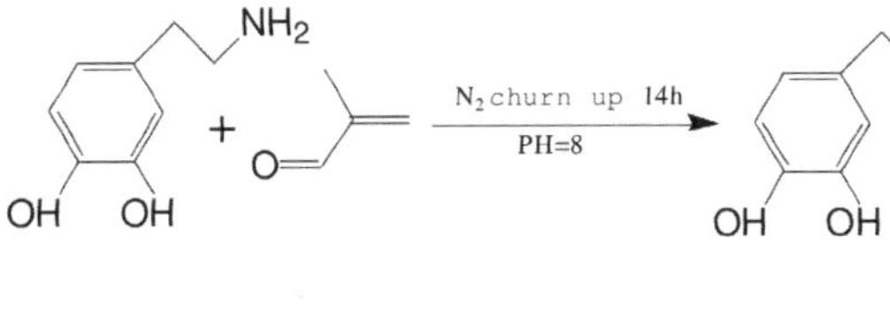

Figure 1. Synthetic route of DMA

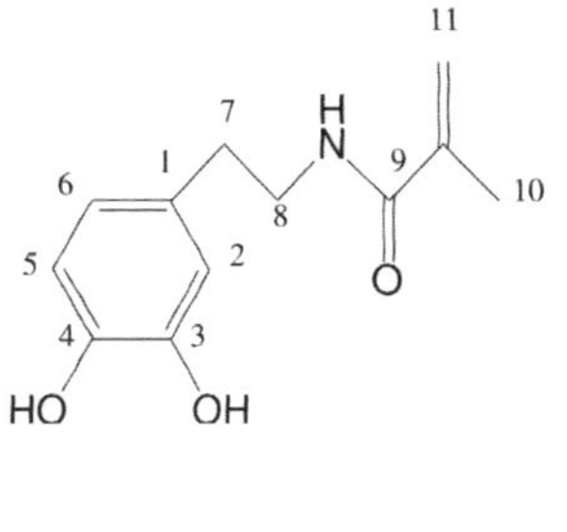

Figure 2. Structure of DMA

2.3. FT-IR spectroscopy

FT-IR spectroscopy (5700, Instron, USA) was performed to analysis DMA composition. Small amount of DMA was ground together with KBr and then pressed into pellets (2mg/300mg KBr) for FT-IR with a resolution of 4.00 cm^{-1}. Infrared spectra were recorded in the range 4000-400 cm^{-1} to evaluate the molecular structure.

2.4. Nuclear magnetic resonance(NMR) spectroscopy

^1H NMR and ^{13}C NMR spectra of DMA monomers were investigated on (AC-E200 Bruker) monomers with CD_3OD as the solvent and chemical shifts (δ) are given relative to tetramethylsilane as the internal standard.

2.5. SEM analyses

Surface morphology of DMA monomers was observed by SEM. The samples were coated with a thin layer of Gold (Au) by sputtering (450X, Emitechk, England) and then the morphology of them were observed on a scanning electron microscope (JSM-7001F, Jeol, Japan) that operated at the acceleration voltage of 15 kV.

2.6. DSC analysis

The DSC spectra of DMA was obtained on Perkin-Elmer DSC Model 7. Measurements were performed over the temperature range of 25-500℃ at the heating rate of 5℃/min in hermetically sealed aluminium pans.

2.7. TGA analysis

Thermal stability of the DMA was examined, from 25 to 500℃ heated at 5℃/min in nitrogen gas flushed at 200ml/min. The samples were subjected to thermo gravimetric analysis (TGA) to determine the decomposition temperatures.

3. Results and discussion

3.1. FT-IR spectra

Figure.3 shows the FTIR spectrum of the DMA. The characterization adsorption peaks at 3070cm^{-1}

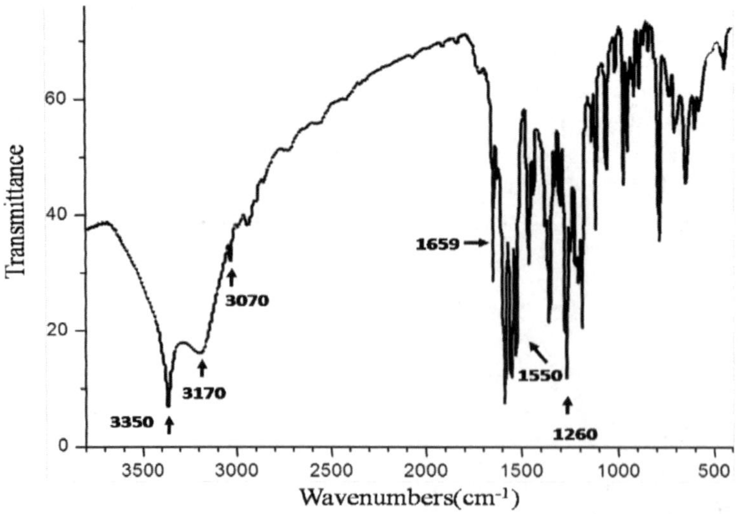

Figure 3.FTIR of DMA

and 3170~ 3350cm^{-1} (were assigned to the Chung amide N-H stretching). The characterization adsorption peak at 1659, 1550 and 1260cm^{-1} were owing to acuate peak amide I C=O stretching, double peakamide II, and C–N stretching for amide III). This is an evidence of the grafting from N atom positions of the monomer.

3.2. NMR spectra

The ^{13}C NMR and ^{1}H NMR spectra of DMA were showed in Figure4 and Figure5. A total of 12 carbon signals were found in the spectra (Figure 2). ^{13}C NMR: δ171.2 (C9), δ146.1 (C10), δ144.3-116.3 (C1-C6), δ132.0(C-11), δ42.6(C-8), δ35.7(C-7), δ18.7(C-12).

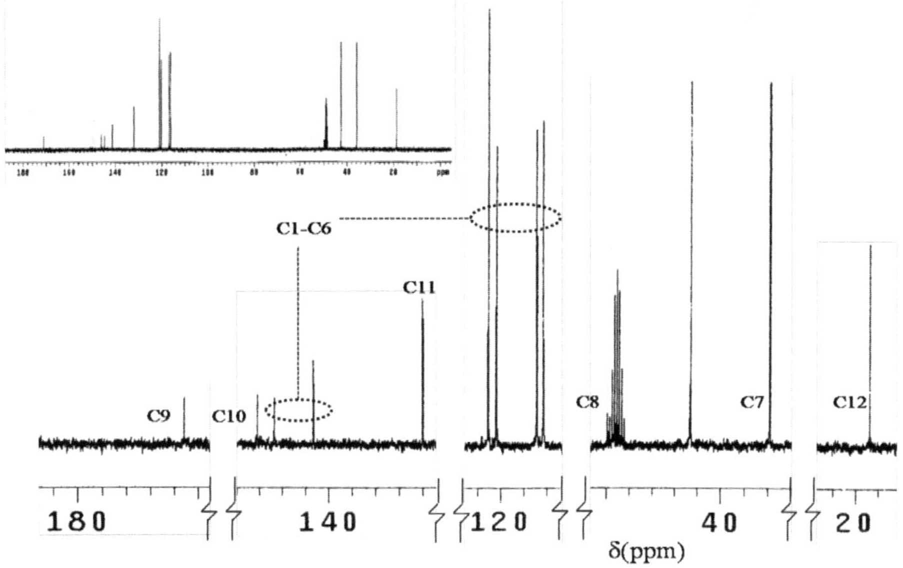

Figure 4. ^{13}C NMR of DMA

^{1}H NMRδ:1.89(3H,s,-C(=O)-C(-C$\boldsymbol{H}_3$)=CH_2),2.65(t,2H,$C_6H_3(OH)_2$-C$\boldsymbol{H}_2$-CH_2(NH)-C(=O)-),3.37(m,2H,$C_6H_3(OH)_2$-CH_2-C$\boldsymbol{H}_2$(NH)-C(=O)-),5.31(s,1H,-C(=O)-C(-CH_3)=C$\boldsymbol{H}$H),5.63(s,1H,-C(=O)-C(-CH_3)=CH$\boldsymbol{H}$),6.52(d,1H,$C_6H_2\boldsymbol{H}(OH)_2$-),6.66(m,2H,$C_6\boldsymbol{H}_2H(OH)_2$-).

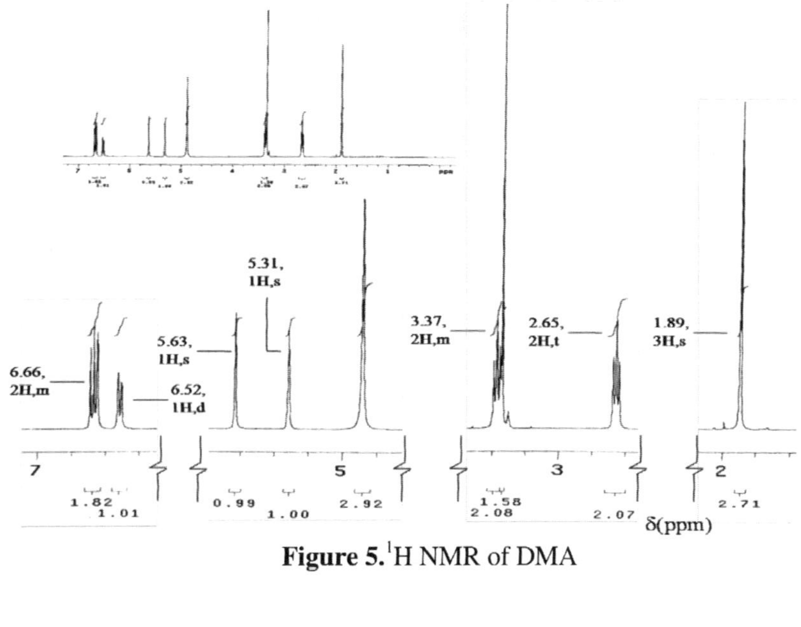

Figure 5.^{1}H NMR of DMA

3.3. SEM micrographs of DMA

The morphologies of DMA are illustrated in Figure6 and Figure7. The particles of DMA have irregular shapes. The particle sizes of DMA were ranged approximately 10 μm or more. A few big plate-like crystals due to methacrylate anhydride are clearly distinguishable from the DMA particles [15], which display an average size of 1-2 μm. SEM analyses showed the surface morphologies and sizes of DMA were altered by grafting reaction.

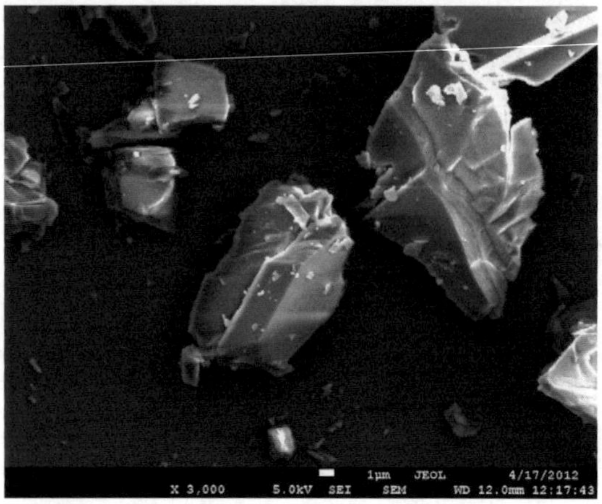

Figure 6.SEM micrograph of the DMA

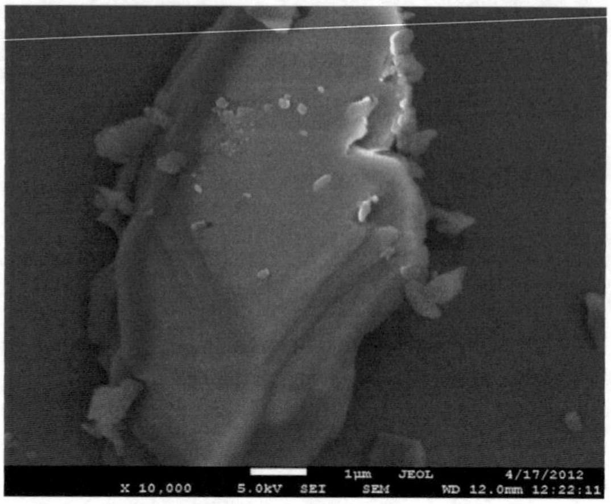

Figure 7.SEM micrograph of the DMA

3.4. Thermal analysis

The DSC and TG curve of DMA are shown in Figure 8. The obtained DSC curve showed the marked endothermal peaks at 120-160 ℃. The maximum of the endothermal peaks for DMA was at 143 ℃, which was its melting point of it. The DSC curves of DOPA did not show a marked endothermal peak like DMA in any temperature range. [16] In general, the DMA powder showed a two-step weight loss as confirmed by DSC; one with a steeper change in the weight loss curve in the temperature range from 248℃ to about 343℃, corresponding to the part of C-N bond and thermal decomposition of methacrylate anhydride. Subsequently, a comparatively slow weight loss appeared about 343-400℃ due mostly to the elimination of the main component. The melting point, DSC and TG curves of DMA were different from those of DOPA, which was due to grafting of methacryloyl on DOPA.

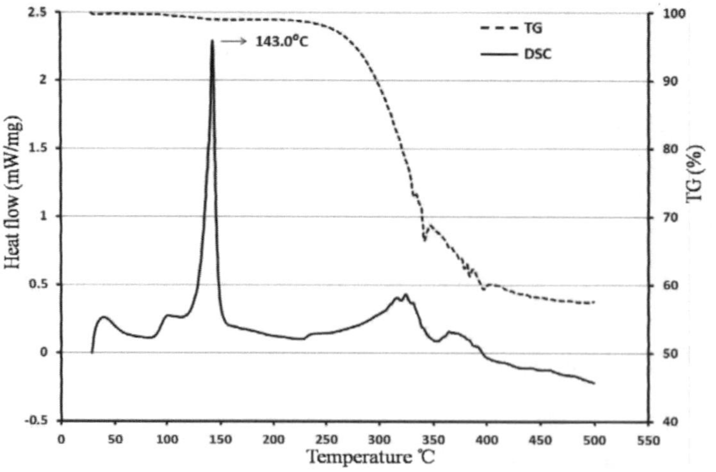

Figure 8.DSC and TG curves for DMA

4. Discussion

L-3,4-Dihydroxyphenylalanine(DOPA) is an unusual amino acid found in mussel adhesive proteins, which is believed to lend adhesive characteristics to these proteins [17]. The hydroxyl on DOPA is believed to be the function group for high adhesives of mussel secretions. Unfortunately, most previous efforts incorporating DOPA into polymer have been hypothesized to reduce the adhesive properties of DOPA due to the hydroxyl group of the catechol structure was easy to be oxidized. In this study, methacrylate anhydride was grafted on DOPA molecule to avoid its oxidization.

It was found that carbon-oxygen double bond absorption peaks appeared in FT-IR and NMR spectra of DMA, which indicateed that the new monomer, N-methacryloyl 3,4-dihydroxyl-L-phenylamine, was synthesized successfully. It was also found that peaks of hydroxyl from the FT-IR and NMR spectra, which inferred that the new monomer had the hydroxyl functional groups. The structure of the DMA, which had catechol structure, endowed its potential application as the medical adhesives .DMA was brown powder (ethyl acetate-hexane), insoluble in water, ethanol and ethyl acetate, slightly soluble in chloroform, whereas it was dissolved in DMSO and carbinol. The melting point of DMA is 143℃.

From the SEM micrographs of DMA, it was revealed that morphologies and sizes of DMA particles were altered by grafting reaction. This morphology change probably was the reason for the change of dopamine graft polymer adhesion. The obtained DSC curve of DMA show a marked endothermal peak in the temperature range 120-160℃ whereas DSC curve of DOPA did not show any endothermal peak like DMA in any temperature range. It was concluded that due to the grafting of methacrylic anhydride DMA had better crystallinity than DOPA.

5. Conclusions

It is feasible to synthesized N-methacryloyl 3,4-dihydroxyl-L-phenylamine, which still had catechol structure and hydroxyl functional groups, which might be endowed DMA the high adhesion. The present results confirmed the difference between DOPA and DMA in surface morphologies, melting point, DSC and TG spectra.

6. Acknowledgement

The authors acknowledge the financial support from the "Central University Basic Research Fund for young teachers in the Hundred Talents Program" of Southwest Jiaotong University (SWJTU2011BR026EM).

References

[1] Duarte A P, Coelho J F and Cidade M T 2012 *J.* Surgical adhesives: Systematic review of the main types and development forecast .*Progress in Polymer Science*.**37** 1031-1050.

[2] Brubaker, Carrie E, Messersmith and Phillip B.2012 *J.* The present and future of biologically inspired adhesive interfaces and materials. *Langmuir.* **28** 2000-2005

[3] Dong S H and Herbert Waite J 2012 *J.* Three intrinsically unstructured mussel adhesive proteins, mfp-1, mfp-2, and mfp-3: Analysis by circular dichroism.*Protein Science.***21** 1689-1695

[4] Yu M E, Hwang, and Deming J Y 1999 *J.* Role of L-3,4-dihydroxyphenylalanine in mussel adhesive proteins. *American Chemical Society.* **121** 5825-5826

[5] Yu M E and Deming T 1998 *J.* Synthetic polypeptide mimics of marine adhesive. *Macromolecules.***31** 4739-4745

[6] Burzio L A and Waite J H 2000 *J.* Cross-linking in adhesive quinoproteins: studies with model decapeptides. *Biochemistry.***39** 11147-11153

[7] Lee H Scherer, N F Messersmith and Proc P B 2006 *J.* A reversible wet/dry adhesive inspired by mussels and geckos. *Natl.Acad. Sci. U.S.A.***103** 12999

[8] Zhao H and Waite J H 2006 *J.* Linking adhesive and structural proteins in the attachment plaque of Mytilus californianus.*Biol. Chem.***28** 3456

[9] Haeshin L,Dellatore, Shara M Miller and William M 2007 *J.* Mussel-inspired surface chemistry

for multifunctional coatings.*Science.***318** 426–430
[10] Jonathan J Wilker 2011 *J.* Redox and adhesion on the rocks. *Nature Chemical Biology.***7** 579-580
[11] Miaoer Yu and Timothy J Deming 1998 *J.*Synthetic polypeptide mimics of marine adhesives.*Macromolecules.***31** 4739-4745
[12] Brubaker Carrie E, Hermann Kissler, Wang Ling Jia and Kaufman Dixon B 2010 *J.* Biological performance of mussel-inspired adhesive in extrahepatic islet transplantation. *Biomaterials.***31** 420-427
[13] Dalsin J L,Lee P B and Messersmith P B 2002 *J.*Synthesis and gelation of DOPA-modified poly(ethylene glycol)hydrogel.*Biomacromolecules.***3** 1038
[14] Jochen Schnurrer and Claus Michael Lehr 1996 *J.*Mucoadhesive properties of the mussel adhesive protein.*Pharmaceutics.* **141** 251-256
[15] Ye Qian, Zhou Feng and Liu Wei min 2011 *J.* Surface Characteristics of a Self-Polymerized Dopamine Coating Deposited on Hydrophobic Polymer Films. *Langmuir.* **27** 14180–14187
[16] Simonovic B,Vucelic V,Hadzi Pavlovic A and Stepien K 1990 *J.*Thermogravimetry and differential scanning calorimetry of natural and synthetic melanins.*Thermal Analysis.***36** 2475-2482
[17] Messersmith, Phillip B. 2010 *J.* Holding On by a Hard-Shell Thread. *Science.***328** 180-181

Functionally Graded Materials using Plasma Spray with Nano Structured Ceramic

E L Sioh [1,2], A I Y Tok [2]

[1]ST Kinetics, 249 Jalan Boon Lay, Singapore 619523
[2]School of Materials Science & Engineering, 50 Nanyang Avenue, Nanyang Technological University, Singapore 639798

E-mail: siohel@gmail.com, MIYTok@ntu.edu.sg

Abstract. In this paper, nano structured FGM was fabricated using DC plasma spray technique. Nano structured and micro structured powder were used as the feeding powder with steel substrate. The spray parameters was optimized and characterisation of nano-ceramic FGM and micro-ceramic FGM were done using bending test and micro-hardness test. Experimental results have shown that the nano-structured FGM exhibit 20% improvement flexure strength and 10% in hardness. A comparison was made between sintered micro ceramic tile and nano ceramic FGM using simple drop test method.

1. Introduction

Functionally-graded materials (FGM) are composites comprising of different materials with a continuous transition from one material to the other, in this case from ceramic to metal. It is believed to allow the full utilisation of the best properties of ceramics and metal, whilst also giving good impact resistance from the hard ceramic frontal layer, and thermal shock resistance due to the smooth thermal coefficient transition from ceramic to metal.

Direct Current Air Plasma spray is a technology used in many industrial applications to apply ceramic and metallic coatings, such as making hard-wearing surfaces, bio-implants, thermal-barrier and chemical-resistant coatings. It is a process which uses a plasma flame to heat up and melt metallic, ceramic or composite powders and then projects the liquefied powders onto the surface to be coated using gas pressure. It is a relatively fast process, being able to produce large surface areas covered with ceramic coatings in one single process, as compared to other methods such as hot- and cold-pressing and then sintering of ceramic tiles.

In this paper, nano-Al_2O_3 and FeNiAl powders were used to produce FGM coatings on steel substrates using the DC air plasma spray technique.

2. The Approach

The approach was to investigate the physical properties required of nano- Al_2O_3 / FeNiAl FGM coated on steel substrate using the DC plasma spray technique, by comparing the difference in properties between functionally-graded coatings using micron-sized Al_2O_3 powder, and nano-sized Al_2O_3 powder, together with FeNiAl metal-composite powder. Requirements such as hardness, fracture strength, toughness, and density were characterised.

3. Specimen Preparation of Nano Structured FGM

3.1. Raw material

The nano-alumina powder used is a commercial product. It is a granulated agglomerated alumina powder, with an average particle spherical diameter d_{50} of 70nm. It contains 99.5% alpha-phase Al_2O_3, and 0.5% MgO.

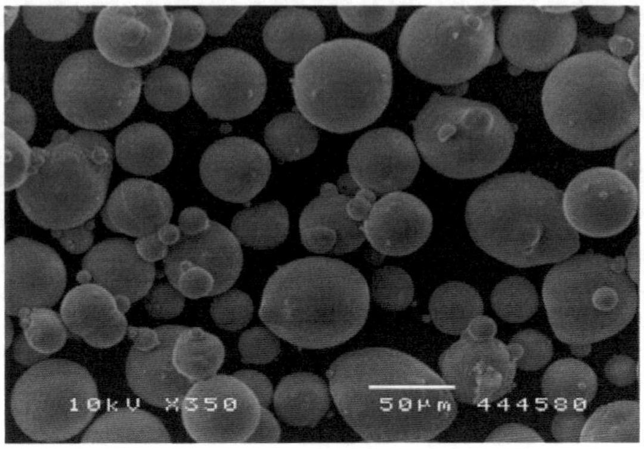

Figure 3.1. SEM image of nano-alumina powder.

Figure 3.2. SEM image of alumina powder.

The micron-alumina powder used is a commercial product from Praxair, with the average particle size of 45 micrometers. It is also alpha-phase Al_2O_3, but is made from 100% pure Al_2O_3, as compared to the nano-alumina powder used in this project.

The commercial micron metal composite powder was used as a FeNiAl powder. The average particle size is around 106 micrometers, with the 3 types of metal components clad together. This powder will undergo an exothermic reaction during spraying, and thus would be able to form a strong metallic bond with the base metal and other materials used in the thermal coating.

3.2. Process

In the specimen preparation, mild steel substrate of 0.3% carbon was used. The substrate was degreased, grit blasted and cleaned by an air gun before spraying. Iron composite powder and alumina powder (Al_2O_3) were used to fabricate the graded layers using DC plasma spray technique. These powders were weighted and premixed according to percentage by weight, followed by mechanical milling for 1.5 hours. The premixed powder was then fed into the powder feeder for plasma spraying. The coating was sprayed prior to preheating of the substrate. All steel plates were weighed and measured before being coated. The machine being used in this report is Taiwan Plasma Corp's PCS-40 Air Plasma Spray system, with a maximum power of 40kW.

3.3. Characterization of the Specimen

The graded material was characterized in terms of microhardness, porosity and micrograph examination using SEM. The Vickers microhardness of the coatings was determined by using the Matsuzawa MXT70 Microhardness Tester, with a load of 1 kgf and dwell time of 15 seconds. The microstructures and cross-sectional porosity of the coatings were examined using the Scanning Electron Microscope (Jeol 5310 and 6360). The Mirage MD-200S electronic densimeter was used to measure the bulk density of the FGM samples. The 3 points bending test was performed to measure the flexural stress, strain, flex modulus, fracture load and strain of the FGM coatings using Instron 5567. The drop impact test was performed to test the impact resistance of the nano-alumina FGM system as compared to a commercially-bought pressed and sintered alumna tile.

4. Results

4.1. Functionally Graded Material Configuration

Both the micron and nano alumina FGM specimens were of the same configuration in terms of mixture ratio with iron composite powder as shown in Figure 4.1. The spraying parameters were different due to the difference in densities of the three types of raw materials used. The powder feed rate of the powder feeding system has to be matched to the relative densities of the different powder mixtures used in the FGM coating. The heavier powder has to be fed at a slower rate to reduce clogging at the feed nozzle, while lighter powder mix can to be fed at a higher rate to ensure a more homogeneous coating.

100% Al_2O_3
25% FeNiAl / 75% Al_2O_3
50% FeNiAl / 50% Al_2O_3
75% FeNiAl / 25% Al_2O_3
Substrate

Figure 4.1. FGM specimen configuration

4.2. Micrograph of FGM Specimens

Both the morphology of the micron and nano alumina FGM specimens were observed using SEM as shown in figure 4.2, 4.3, 4.4, and 4.5. The physical properties of the both specimens are shown in table 4.1

Figure 4.2 SEM images of the cross-section of Micro-powder alumina FGM coatings

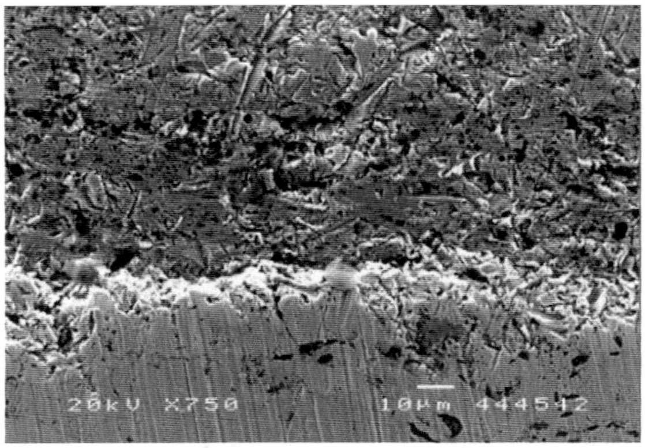

Figure 4.3 SEM images of the interface between substrate and coating

Journal of Physics: Conference Series **419** (2013) 012048 doi:10.1088/1742-6596/419/1/012048

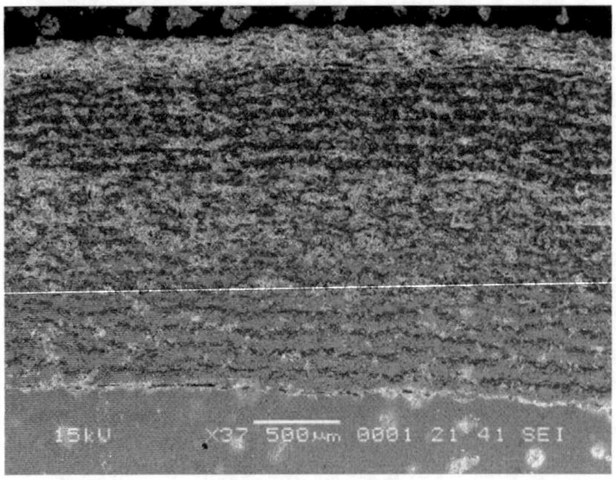

Figure 4.4. SEM images of the cross-section of nano-alumina FGM coatings

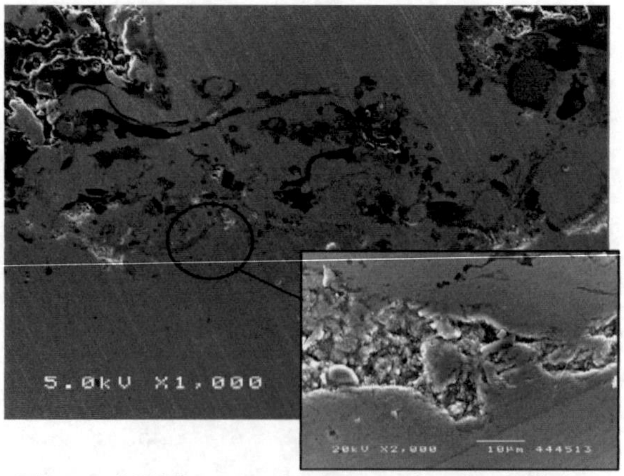

Figure 4.5 Micrograph of nano-alumina FGM, showing nano-zones containing nano-particles within.

Table 4.1. Physical properties of micro powder FGM and nano powder FGM material system.

Physical Properties	Micro powder FGM specimen	Nano powder FGM specimen
Coating thickness (mm)	2.42	2.19
Coating bulk density (g/cm3)	3.595	3.977
Approximated Porosity (%)	5	5
Vickers hardness (Hv)	1026	1531

Based on the micrograph in figure 4.2 and 4.3, micron-sized ceramic FGM had exhibited a low porosity and high hardness value. However, between the substrate and first coating interface, micro-cracks were observed. These micro cracks would combine with the tensile stresses generated by the thermal expansion, and lead to coating delamination.

In nano-sized ceramics FGM (refer to figure 4.4 & 4.5), the micrograph had shown second-phase nano-size particles dispersed within the matrix grains and the presence of the nano-zones. The Vickers micro hardness increased by 49.2% with a corresponding increased in the bulk density of 10.6% while maintaining the same percentage of porosity.

4.3. Vickers Hardness test

Vickers microhardness tests were conducted to find the FGM coating hardness of the nano alumina FGM and micron alumina FGM at different graded layers as shown in figure 4.6 and 4.7. The microhardness values of both nano and micron FGMs appear to be similar, although the hardness of the top layer of nano FGM had achieved a higher value of 1531Hv. This is coherent with findings by other researchers, who found that the hardness of nano-structured ceramic thermal coatings were similar, or slightly higher, than conventional micron-structured ceramic thermal coatings. [1]

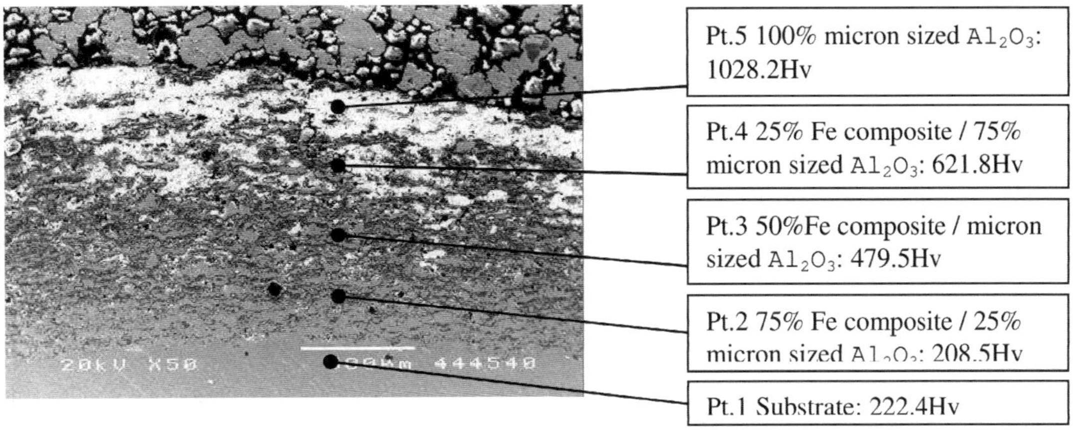

Figure 4.6 SEM images and corresponding Vickers microhardness that make up the layers of micron - alumina FGM.

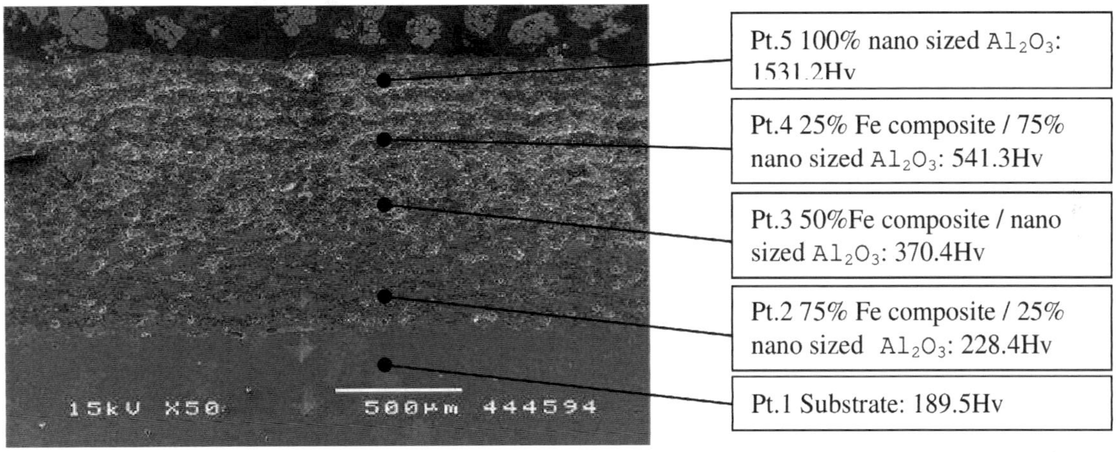

Figure 4.7 SEM images and corresponding Vickers microhardness of the layers that make up the nano-alumina FGM.

Vickers micro-indentations were also made on the polished ceramic top layers of both micron-alumina and nano-alumina FGMs. Cracks can be seen extending out from the corners of the indentation on the micron-FGM (Figure. 4.8a), whereas little cracking was observed around the indentation on the nano-alumina FGM, as shown in Figure 4.8b.

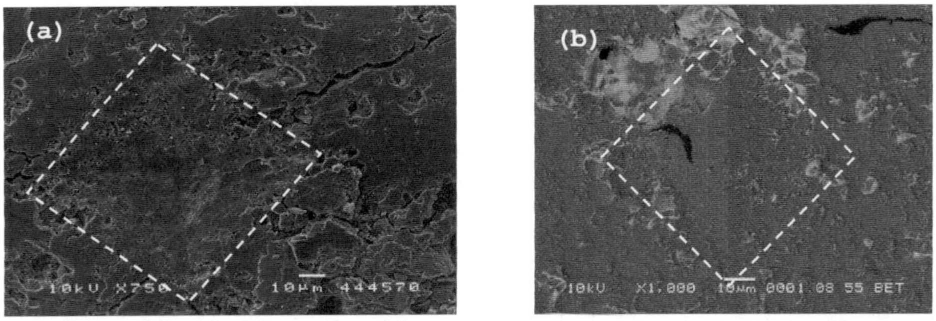

Figure 4.8 Vickers indentations on: (a) micron-alumina FGM; (b) nano-alumina FGM.

4.4. Three points bending test

From the Instron three-point bending tests conducted on the FGM samples, it was observed that the nano-alumina powder FGM, exhibited maximum flexural fracture stress before coating failure by cross-sectional shear. The flexure strength was improved by about 20%, from an average of 450MPa in micron-alumina FGM to about 600MPa in nano-alumina FGM. The maximum flexural fracture strain that the nano-FGMs can take before failure is also around 25% higher, from 0.8% in micron-alumina FGM to about 1% in nano-alumina.

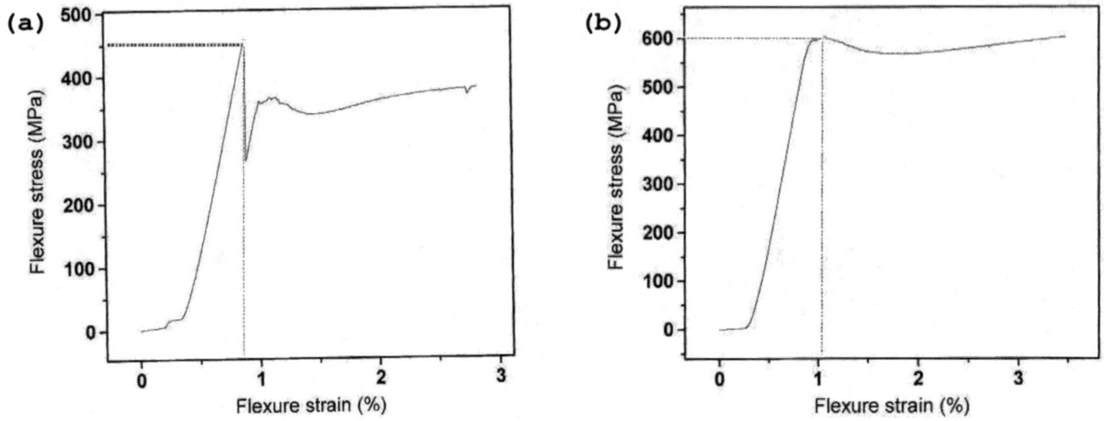

Figure 4.9 Three points bending test on: (a) micron-alumina FGM; (b) nano-alumina FGM.

From the profile of the stress-strain graphs as shown in figure 4.9, it can be observed that the nano-alumina FGMs fail less abruptly at the point of coating fracture as compared to the micron-alumina FGM, which exhibited a sharp peak at the point of brittle fracture. This abrupt initial point of failure was due to the brittle 100% micron-alumina top layer shattering under the applied load. After this layer has failed, the remaining layers underneath it then recover to take up the applied load, although not to the same stress levels as initially.

The nano-alumina FGMs on the other hand, do not exhibit this initial brittle behaviour under loading from the three-point bending test. This is due to the presence of semi-molten nano-zones that is able to arrest any interfacial and inter- splat micro-cracks [1] thereby increases the inter-splat bonding strengths of the coating. This could imply that nano-alumina FGMs are less brittle as compared to micron-alumina FGMs, thus explaining the higher fracture strains shown in the three-point bending test.

4.5. Morphology of the FGM fracture surface after bending test

After the Instron three-point bending tests was conducted, SEM images were taken on the fracture surface to study the material respond to fracture. The fracture surface of the micro-alumina FGM was found to contain a large percentage of planar cleavage planes, with the columnar sub-micron grains clearly visible within the ceramic splats. Whereas the fracture surface of the nano-alumina FGM was found to contain more step-like features, which are due to the nano-cracks within the dispersed nano-particles and the molten alumina matrix of the shells of agglomerated nano-alumina particles [2, 3]. Semi-molten agglomerates containing nano-particles within the nano-zones can also be seen in figure 4.10b. [1]

Journal of Physics: Conference Series **419** (2013) 012048 doi:10.1088/1742-6596/419/1/012048

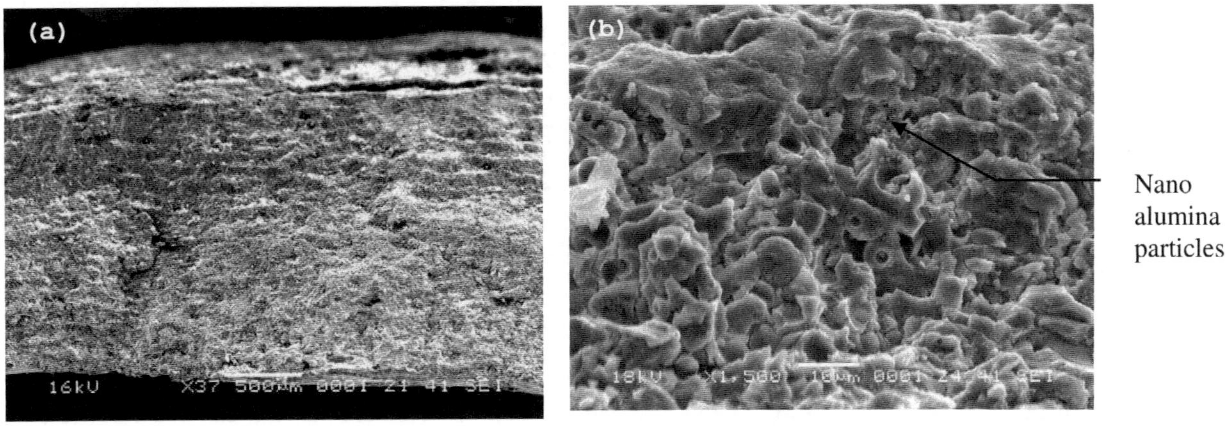

Figure 4.10 SEM image of: (a) the cross-sectional fracture surface of nano-alumina FGM specimen; (b) transgranular fracture of nano-alumina splats, also showing the large intergranular surfaces between molten alumina and semi-molten nano-alumina particles.

4.6. Drop test

This test was performed to test the impact resistance of the nano-alumina FGM system as compared to a commercially-bought pressed and sintered alumna tile.

The test involves dropping a cylinder weighing a total of 3kg, with a sharp stainless steel tip at the bottom end, with tip hardness of 458Hv / 4.5GPa, through a PVC pipe, onto the impact surface from a height of 3m. The set-up is shown in Figure 4.11. The indenter was modelled with a sharp profile.

Figure 4.11 Equipment set-up of drop impact test

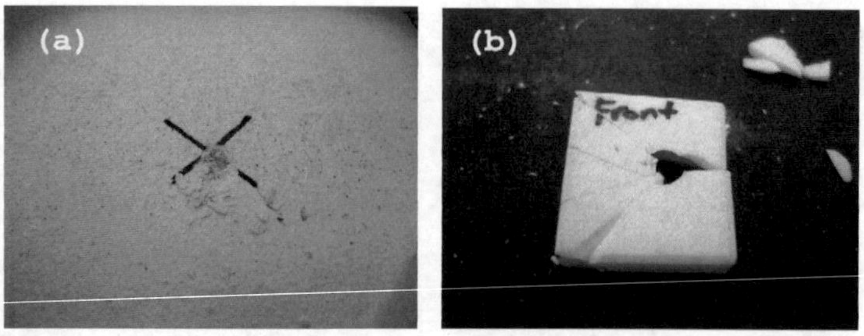

Figure 4.12 Nano-alumina FGM impact surface: (a) after 3 repeated impacts. (b) Alumina tile after 4 drop impact tests

With an impact energy of 88.3J, little visible marking was observed on the nano-alumina FGM surface after the first two impacts. The coating only broke through on the third impact, with the nano-alumina top layer chipped off, exposing the gradient layers underneath but the steel substrate. Only the point of impact was defeated by the repeated impacts, while the rest of the coating was intact. This is shown in figure 4.12a.

Similarly, the alumina tile showed little visible marking on the impact surface during the first three impacts, nor was any visible cracks observed. However, on the fourth impact, the alumina tile was found to have been shattered into many pieces as shown in figure 4.12b.

5. Discussion

From the micrograph of the nano FGM, the presence of dispersed nano-particles (nano zone) within the molten alumina matrix makes up the morphology of the splats. Together with the residual stresses around the nano particles, it act to increase the process zone area around the crack tip by initiating nano-cracks. The larger process zone requires higher fracture energy, thus increasing fracture toughness of the nano-ceramic material. Changes in the fracture mode was observed, from intergranular of micron-particle alumina to transgranular fracture of nano-alumina ceramics. Nano-zones work as crack arrestors, increasing fracture toughness as it makes it much more difficult for a crack to propagate, expanding energy by spreading the energy out over a large interface between nano-particles and splats. [1] Thus, no cracking was observed around the indentation on the nano-FGM, apart from the large holes and cracks which appeared to be due to inherent porosity and defects, also observed on the micron-FGM indentation.

The higher hardness of nano-ceramics is due to the dense packing of semi- and un-melted nano-particles, held together by molten ceramic which had infiltrated into the nano-zones packed with un-melted nano-particles. The higher inter-splat bond strength in the nano-alumina FGM also results in a higher hardness.

The presence of these nano-zones was also found to improve the adhesive bond strength of The differences in reactions to the drop impact test between the nano-alumina FGM and the alumina tile also shows the nano-structured FGM is able to better withstand tensile stress waves due to impact loading as compared to conventional pressed and sintered micron-structured alumina tiles.

6. Conclusion

Nano-alumina FGM had shown increased in fracture strength and micro-hardness with improved interface bond strength between the substrate and the first coating. The drop impact test has shown that the nano- FGM system is able to withstand multiple impact hits by a sharp point, and is able to localise impact damage and resist shattering.

Journal of Physics: Conference Series **419** (2013) 012048 doi:10.1088/1742-6596/419/1/012048

7. References

[1] Lima R S and Marple B R 2006 *Thermal Spray Coatings Engineered from Nanostructured Ceramic Agglomerated Powders for Structural, Thermal Barrier and Biomedical Applications: A Review* (Submitted September 15, 2006; in revised from December 7, 2006)

[2] Bradt R C, Munz D, Sakai M and White K W 2005 *Fracture Mechanics of Ceramics*, Springer (2005)

[3] Pavitra Bansal, Padture N P and Vasiliev A 2003 *Improved interfacial mechanical properties of Al_2O_3-13wt%TiO_2 plasma sprayed coatings derived from nanocrystalline powders*, Acta Materialia **51** (2003) 2959-2970

Analysis of mechanical properties of nanocrystalline Al+α-Al_2O_3 composites using molecular dynamics simulation

H Mei[1], L S Liu[1,2,4], X Lai[1] and P C Zhai[1,2]

[1]Department of Engineering Structure and Mechanics, Wuhan University of Technology, Wuhan 430070, P. R. China
[2]State Key Laboratory of Materials Synthesis and Processing, Wuhan University of Technology, Wuhan 430070, P. R. China

E-mail:liulish@whut.edu.cn

Abstract. In this article, mechanical properties of nanocrystalline Al+α-Al_2O_3 composites are investigated using molecular dynamics simulations. The configurations of matrix and volume fraction of α-Al_2O_3 may affect the mechanical properties of the particle reinforced metal-matrix composites and are taken into account. The potentials for the Al+α-Al_2O_3 system developed by Xin Lai et al. are adopted to depict the interactions between Al and α-Al_2O_3. Monocrystal Al and Bicrystal Al based α-Al_2O_3 particle reinforced nanocomposites are modelled respectively. Results show that: (1) volume fraction of the particles has no explicit effects on the elastic modulus and ultimate strength in both monocrystal Al and bicrystal Al based matrix nanocomposites, (2) disappearance of valley in the stress-strain curve of bicrystal Al results from existence of dislocation in matrix of various orientations.

1. Introduction

Over past decades, particle reinforced metal-matrix composites (PRMMCs) have became an important class of materials for their high specific modulus, high specific strength and good wear resistance [1]. The characteristics of mechanical properties of composites results from strengthening effect caused by ceramic particle embedding. To dominate mechanical properties of composites for application, investigation of strengthening effect on composites is necessary.

In PRMMCs, strengthening effects of ceramic particles mainly include following aspects [2]: (1) constrained plastic flow (dislocation) in the matrix, (2) load transferred to particles, (3) thermal history, (4) material texture, (5) grain size, and (6) volume fraction. The aspects (1), (2), and (3) construct the primitive mechanisms for strengthening PRMMCs directly. The last three aspects are relatively easier than the first three to be investigated and can provide specific parameters for further studying the first three aspects. In addition, they are convenient to realize in experiment and simulation.

Actually, to investigate strengthening effects of composites caused by the last three aspects, large amount of experimental and numerical simulations have been carried out [2-9]. However, results of experiments are mostly affected by more than three factors. Therefore, the relationship between single aspect and mechanical properties of material cannot be inferred. In simulation by FEM, although

[4] Corresponding author, Tel.: +86 27 87651820; Fax: +86 27 87860863
E-mail: liulish@whut.edu.cn (L S Liu)

influence factor is contracted to texture of particles, but the results are unreasonable because of ignorance of affect of the interface between particle and matrix.

Nowadays, a simulation method named molecular dynamics (MD) which is based on atomistic system can remedy it by capturing the effect of interface accurately. Vikas Tomar et al. [10] investigate the impacts of volume fractions and grain sizes on mechanical properties of PRMMCs by MD. However, original GBs in their models are destroyed by particles. Three factors that consist of the interface between particle and matrix, GBs between metal and volume fraction of particles are considered.

To investigate strengthening effect accurately, we focus on the way to contract the number of factors from three to two by removing variable on GBs in metal. The present study is devoted to explore the impact of matrix texture configuration and volume fraction effect on composites properties by MD without any operation on GBs. Particle of α-Al_2O_3 reinforced matrix of Al composites, which are abundant in nature, are employed. The material properties of composites with alteration of matrix configuration and various volume fractions with the same particle are performed.

2. Computational Methods

2.1. Potential

In MD, the potential functions, which describe the interaction between atoms, are determinant factors for the reasonable results. The potentials, which are adopted in this article, are cited from previous work by the authors [11]. It consists of three parts: a. pure Al adopts EAM potential which is proposed by Mishin [12]; b. Buckingham potential proposed by Matsui for pure α-Al_2O_3 [13]; c. Interfacial interactions between Al and α-Al_2O_3 are given by the following formulas, which are divided into two parts: Al-Al interactions(2.1) and Al-O interactions(2.2).

$$E_{Al,Al}(r) = 2\gamma f_0 \beta_{Al} b_{Al} exp(\frac{2\beta_{Al}\alpha_{Al} - r}{2b_{Al}}) \quad (2.1)$$

$$E_{Al,O}(r) = \gamma f_0 \beta_0 (b_{Al} + b_O) exp(\frac{a_{Al} + a_O - r}{b_{Al} + b_O}) +$$
$$\gamma \beta_0 D_{Al,O}[exp(-2\alpha_{Al,O}(r - r_{Al,O})) - 2exp(-\alpha_{Al,O}(r - r_{Al,O}))] \quad (2.2)$$

In which, $\gamma = 1.65$, $\beta_0 = 0.4$ and $\beta_{Al} = 1$, while some relevant potential parameters can get from table 1.

Table 1. Potential parameters

	$f_0(b_i + b_j)$ $(kcalmol^{-1})$	$a_i + a_j$ $(0.1nm)$	$1/(b_i + b_j)$ $(10/nm)$	D_{ij} $(kcalmol^{-1})$	α_{ij}	r_{ij} $(0.1nm)$
Al-O	0.2903	3.0493	3.4445	21.6893	2.0	1.8
O-O	0.3604	3.7766	2.7775	0	-	-

2.2. Modeling

As illustrated in Figure 1, two groups of models are employed for the MD simulations. A few particles that are depicted by yellow balls with the same radius (23.736Å) and the same orientation are embedded in monocrystal Al in uniform simulation box (129.6 Å×129.6 Å×129.6Å), as shown in Figure 1(a). We mark them 1-1 for one particle, 1-2 two particles, 1-3 four particles and 1-4 eight particles, volume fraction of α-Al_2O_3 in them are 2.57%, 5.14%, 10.28% and 20.56% respectively. The other parts of the box excluding the particle(s) are filled by monocrystal Al with the orientation of [100]Al//X-axis, and [001]Al//Z-axis. To demonstrate the details of the orientation relationship between matrix and particles, sketch of model 1-1 in Figure 1(a) are shown in Figure 2. The mid cross-section in Figure 1(a) is shown in Figure 2(b), in which Al and α-Al_2O_3 have an orientation relationship of [100] Al //[210] α-Al_2O_3 and {020} Al //{100} α-Al_2O_3.

Models of bicrystal Al based composite are constructed as displayed in Figure 1(b). Different from monocrystal models, the same box as Figure 1(a) is divided into eight sections, in which, Al in part two, three, five and eight have the same orientation with first four models, while remain parts being rotated about the Y axis compared to monocrystal models that leads orientation of X-axis varied from [100] Al to $[10\bar{1}]$Al.

Four particles which are the same as particles in Figure 1(a) are embedded into part three, two, five and eight one by one, we mark them 2-1, 2-2, 2-3 and 2-4 respectively, volume fraction of α-Al_2O_3 in them are 2.57%, 5.14%, 7.71% and 10.28% respectively.

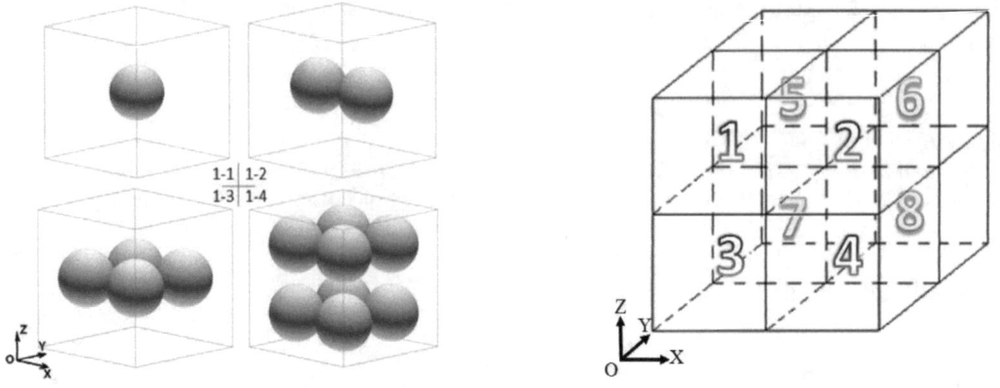

Figure 1. Two group models which are adopted for (a) monocrystal Al based and (b) bicrystal based composites respectively.

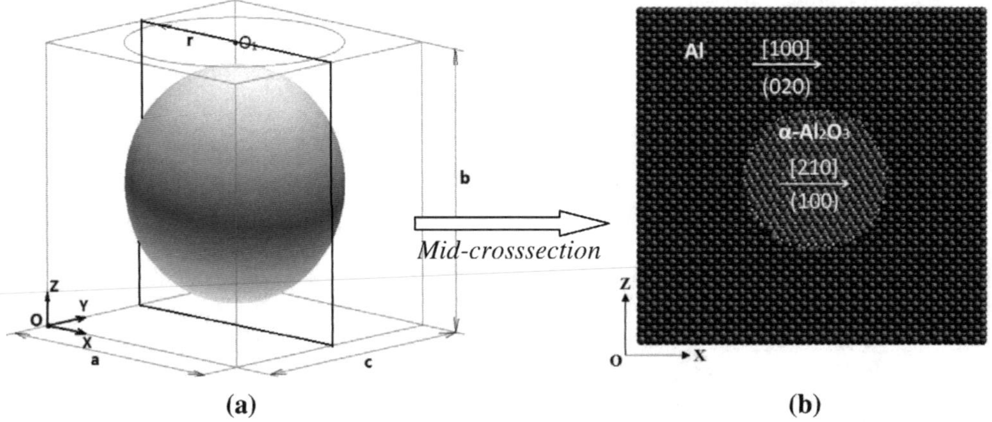

Figure 2. (a) Sketch of model 1-1 in Figure 1(a), the mid cross-section which is highlighted by black line in (a) is re-drawn at (b).

2.3. Simulation Methods

The simulations in this paper are carried out on conditions of periodic boundary conditions (PBCs) in three dimensions at room temperature (300K). A time step is set to 0.0005ps and pressure is 0Pa. To get a reasonable model configuration, equilibration is performed 30000 time steps with isothermal-isobaric (NPT) ensemble. After model configuration gets a steady condition, tension is applied to models along X-axis. In addition, for convenience and saving of calculation cost, the strain rate is set to 109/s.

The central symmetry parameter (CSP) [14] proposed by Kelchner et al. is adopted to identify the dislocation nucleation and visualize dislocations, and usually, atom is identified as defective while value of CSP is higher than six.

3. Results and Discussion

3.1. Relaxation

To search nanocrystalline composites after equilibration, Figure 3 shows atom configuration of monocrystal Al (1-1) and bicrystal Al (2-1) based composites contain one particle with atom coloured by CSP approach. The α-Al_2O_3 in composites is found to have amorphous structural order after relaxation, which is impossible to identify defect atoms by CSP. Defect analysis will focus only on the Al phase.

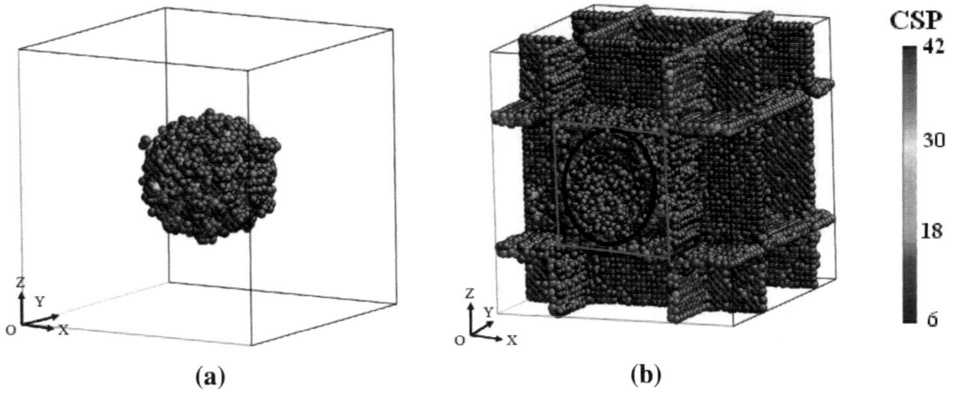

Figure 3. Atom configurations after equilibration: models (a) 1-1 and (b) 2-1.

To investigate the distribution of interior dislocation in Figure 3(b), the origin of coordinates is adjusted. The area, which red square enclosed, is the mid cross-section of part three while black line is the outline of mid cross-section of particle shell. According to CSP, the dislocation is only distributed at interface between matrix and particle in model 1-1, whereas the dislocation in model 2-1 exists at both interface and GBs. The particle and interface at GBs keep steady as original location that makes the models reliable.

3.2. Monocrystal Al+α-Al_2O_3

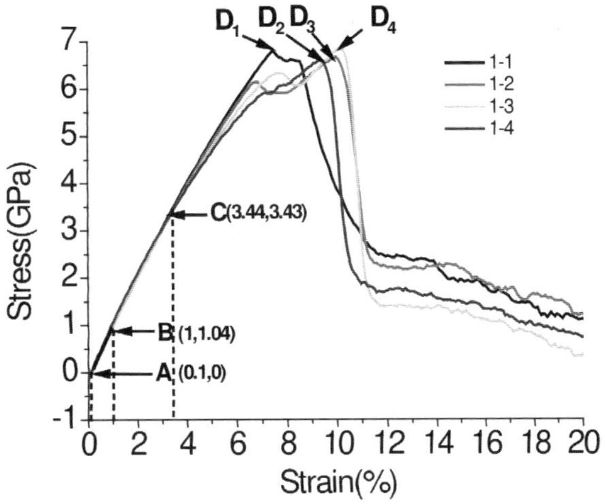

Figure 4. Stress-strain curves for monocrystal Al based composites with various volume fractions of .57%, 5.14%, 10.28% an 20.56% respectively for models 1-1, 1-2, 1-3 and 1-4.

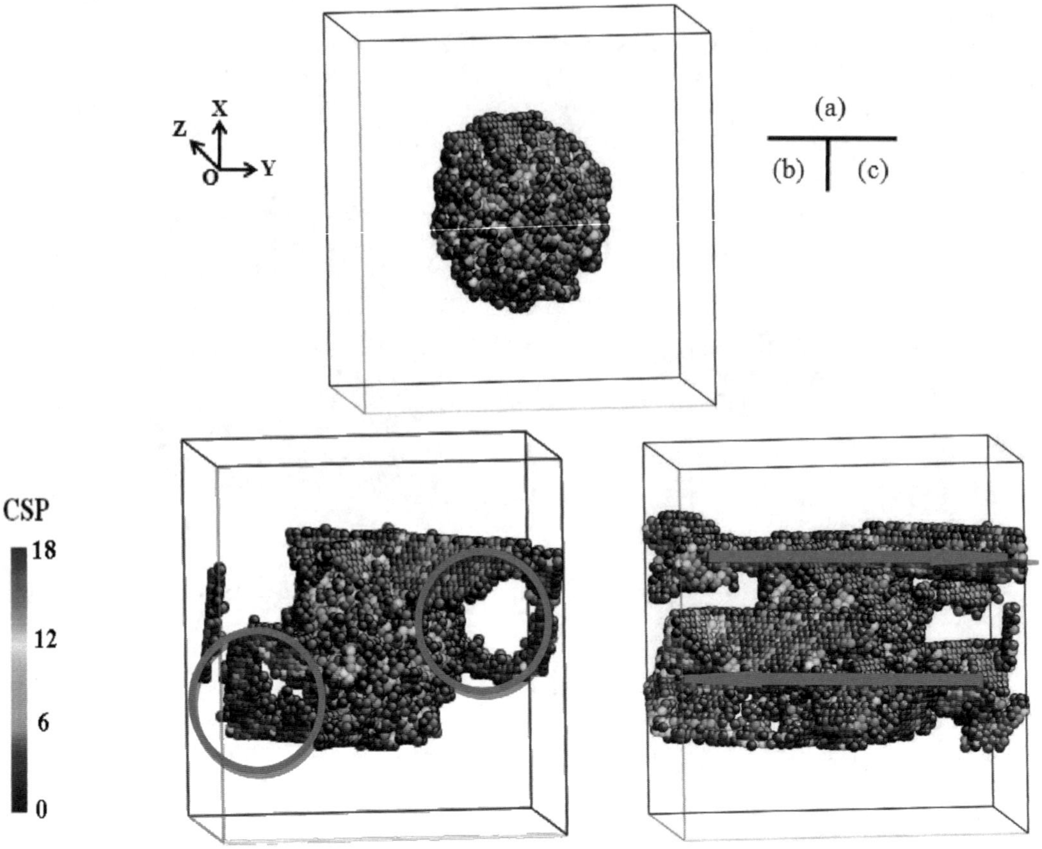

Figure 5. Atom configurations of model 1-1 at specific strains: (a) 3.44%; (b) 7.5%; (c) 7.96%.

In Figure 4, stress-strain curves of tension along X-axis about monocrystal Al based composites are illustrated together. To investigate the relationship between stress-strain curves and material structure, atom configurations of model 1-1, which are coloured by CSP, are illustrated at three specific strains in Figure 5.

In the article, the stress in the simulation is calculated over current volume. Therefore, it corresponds to true stress. The strain is calculated based on the variation in the current volume of periodic simulation cell, and therefore corresponds true strain. Due to non-linear nature and insufficient equilibration, the elastic modulus necessitates a compromise between having enough data for a reliable fit and staying within the linear region. It is found that a reasonable compromise is to use strain of 0.1%~1% for all models. Results show that, all of materials have the same elastic modulus (116GPa) which are calculated from the slope of line AB (0.1% strain to 1% strain). The range of strain is different from the traditional strain, which is defined as 0%~0.2%. This perhaps results in a slightly misestimate of elastic modulus.

Strain hardening starts at strain of 3.44% (Point C) in Figure 4, from that on, more particles lead less stress at the same strain level, which is different from substitute of particles in macro-compositing, here particles bring composites more misfit dislocation (Figure 5(a)). Therefore, model 1-4 has the smallest slope in strain hardening stage. With the dislocation piling, the curve goes to the first peak. Then, with the dislocation loops emitting (Figure 5(b)), composites start to soften, while tendency of strengthening is covered by softening effect. The structure goes to another relative steady condition once dislocation loops encounter with former dislocation and structure (Figure 5(c)). At this stage, three in four models have explicit valley in curves, which results from emission and pile of dislocation. Model 1-4, which contain particles of volume fraction of 20.56%, have no valley because the dislocation is blocked by particle and large amount of dislocation.

In simulation, with particle embedding, the interface between particle and matrix is formed. Misfit dislocation caused by interface affects the material properties of composites greatly. Materials in experiment and manufacture contain large number of defect. Such nature decreases ultimate to a small level. Matrix is so weak that the embedded particles reinforce the matrix properties obviously. Compared with results using Finite Element Method (FEM) by Y.W. Yan et al. [4], the ideal interface of particle and matrix is perfect. They identify the ideal results ignoring misfit dislocation, which closely affects material properties. Compared with simulation by molecular dynamics (MD), Vikas Tomar et al. investigate the tensile deformation of PRMMCs [10]. They replace the original nanocrystalline with reinforced material. The interface of reinforced material coincides with original GBs. The weakening effect is confused by misfit dislocation caused by GBs. The results from experiments, simulations by FEM and MD previously contain many extra factors. The relationship between interface and material properties is buried in many factors.

The authors have calculated tensile strength of pure monocrystal Al (9.03GPa) in orientation of [100] using the potential which is used here. In this article, we find that embedded particle lessen it to 6.5GPa~6.8GPa, and not changed with particle embedding and volume of particles. Accordingly, existence of original dislocation is a decisional factor inducing Al [100] to get a steady strength (6.6GPa) on average where dislocation distributes. Combination of monocrystal α-Al_2O_3 and his interface have the strength of 6.6GPa and elastic modulus of 116GPa.

3.3. Bicrystal Al+α-Al_2O_3

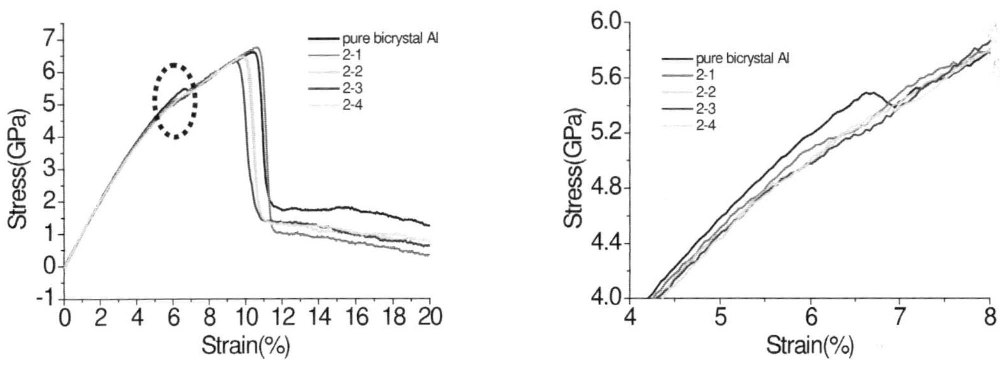

(a) (b)

Figure 6. Stress-strain curves for bicrystal Al based composites with various volume fractions, for convenience to investigate, we prepare (a) whole curves and (b) locally zoomed out curves of (a).

Figure 6 shows stress-strain curves of bicrystal Al based composites with various volume fractions and pure bicrystal. Figure 6(b) shows the locally magnified curve of Figure 6(a) by black dotted line. Figure 7 shows atom configuration of pure bicrystal Al and model 1-1 at specific strain. The orientation of figures is the same as Figure 3(b).

Results of pure bicrystal Al have the strength of 6.6GPa and elastic modulus of 107GPa which are both close to the properties of combination of monocrystal α-Al_2O_3 and his interface. With particle embedding, elastic modulus and ultimate tension strength of composites are supposed to will not change greatly. In addition, it is verified by stress-strain curves.

Although results show that the same elastic modulus and strength which is a starting and ending at curves are almost the same in models, the processes of strain hardening are different, e.g. pure bicrystal Al. In pure bicrystal Al base composite, there is a small valley in a stress-strain curve which results from large amount of original dislocations caused by GBs in structure (Figure 7). The dislocation starts to emit at place, which is enclosed by red line in Figure 7. When dislocation reaches a steady condition (Figure 7), the curve gets to valley. The particles affect the GBs around (red line in

Figure 7). Once particles are embedded to matrix in bicrystal, the valley disappears, which results from import of dislocation in interface of particles.

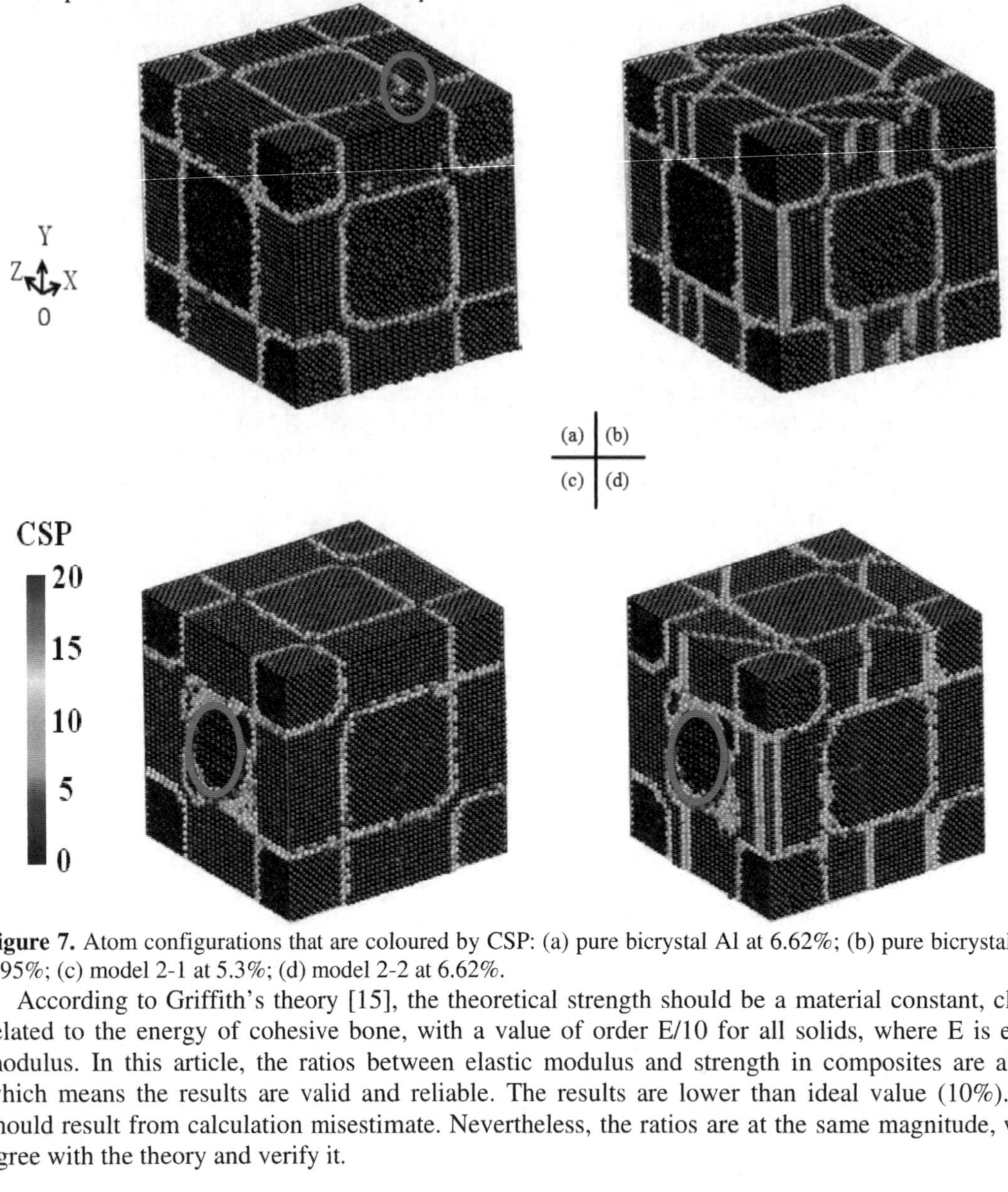

Figure 7. Atom configurations that are coloured by CSP: (a) pure bicrystal Al at 6.62%; (b) pure bicrystal Al at 6.95%; (c) model 2-1 at 5.3%; (d) model 2-2 at 6.62%.

According to Griffith's theory [15], the theoretical strength should be a material constant, closely related to the energy of cohesive bone, with a value of order E/10 for all solids, where E is elastic modulus. In this article, the ratios between elastic modulus and strength in composites are all 6% which means the results are valid and reliable. The results are lower than ideal value (10%). That should result from calculation misestimate. Nevertheless, the ratios are at the same magnitude, which agree with the theory and verify it.

4. Conclusions

The present study is devoted to investigate the impact of matrix texture configuration and volume fraction on mechanical properties of PRMMCs by MD without any operation on GBs.

After equilibration, nanocomposites keep steady as original with the interface between metal and particle homogenized. Good stability on equilibration verifies the reliability of models. Volume fraction of particles has no explicit effect on elastic modulus and tensile strength of monocrystal Al[100] and bicrystal Al based composites since embedded particles are employed with interface dislocation. Strength of all the models has a value of 0.06E, which is slightly lower than E/10 in Griffith's theory which may result from calculated misestimate. Existence of original dislocation is a decisional factor inducing pure monocrystal Al[100] to get a steady strength about 6.6GPa on average and elastic modulus of 107GPa. Dislocation emits from the interface between particle and matrix for monocrystal Al based composites while start from GBs for bicrystal Al based composites. Disappearance of valley in the stress-strain curve of bicrystal Al results from import of dislocation in interface of particles.

The observations reported here provide important data and insights for further realization of strengthening effect.

Acknowledgement

This work is supported by the Fundamental Research Funds for the Central Universities, National Nature Science Foundation of China (No. 10832008, and A3 Foresight Program-50821140308).

References

[1] T W Clyne and P J Withers 1993 *An introduction to Metal Matrix Composites* (New York: Elsevier Science)

[2] Tariq A Khraishi, Lincan Yan and Yu Lin Shen 2004 Dynamic simulations of the interaction between dislocations and dilute particle concentrations in metal-matrix composites (MMCs) *International Journal of Plasticity* **20** 1039-1057

[3] Mahmut D Mat and Kermal Aldas 2006 Experimental and numerical investigation of effect of particle size on particle distribution in particulate metal matrix composites *Applied Mathematics and Computation* **177** 300-307

[4] Y W Yan, L Geng and A B Li 2007 Experimental and numerical studies of the effect of particle size on the deformation behavior of the metal matrix composites *Materials Science and Engineering A* **448** 315-325

[5] S Groh, B Devincre, LP Kubin, A Roos, F Feyel and J-L Chaboche 2005 Size effects in metal matrix composites *Materials Science and Engineering A* **400-401** 279-282

[6] M Song and B Y Huang 2008 Effects of particle size on the facture toughness of SiC_p/Al alloy metal matrix composites *Materials Science and Engineering A* **488** 601-607

[7] Keiichiro Tohgo, Yu Itoh and Yoshinobu Shimamura 2010 A constitutive model of particulate-reinforced composites taking account of particle size effects and damage evolution *Composites: Part A* **41**313-321

[8] M Reihanian, E Bagherpour and M H Payder 2009 A model for volume fraction and particle size selection in tri-modal metal matrix composites *Materials Science and Engineering A* **513-514** 172-175

[9] T J A, Doel and P Bowen 1996 Tensile properties of particulate-reinforced metal matrix composites *Composites Part A* **27A** 655-665

[10] Vikas Tomar and Min Zhou 2007 Analyses of tensile deformation of nanocrystalline α-Al_2O_3+*fcc*-Al composites using molecular dynamics simulations *Journal of the Mechanics and Physics of Solids* **55** 1053-1085

[11] Xin L, Lisheng L, Hai M et al. 2010 A simple potential for the interaction of Al/alpha-Al_2O_3 Interface *In The 3rd International Congress on Ceramics(Osaka, Japan)*

[12] Mishin Y, Farkas D, Mehl M J et al. 1999 Interatomic potentials for monoatomic metals from experimental data and ab initio calculations *Physical Review B* **59(5)** 3393-3407

[13] Matsui M 1994 A transferable interatomic potential model for crystals and melts in the system CaO-MgO-Al_2O_3-SiO_2 *Mineralogical Magazine* **58A** 571-572

[14] C L Kelchner, S J Plimpton and J C Hamilton 1998 Dislocation nucleation and defect structure during surface indentation *Physical Review B* **58(17)** 11085-11088

[15] Brian Lawn 1993 *Fracture of brittle solids* (Cambridge: Cambridge University Press)

Ceramic/Metal Composites with Positive Temperature Dependence of Thermal Conductivity

Jianhui Li[1)], Qi Yu[1)], Wei Sun[1)], Rui Zhang[1,3)], Daisuke Ichigozaki[2)] ,Ke Wang[1)], Jing-Feng Li[1)]*

[1)]State Key Laboratory of New Ceramics and Fine Processing, Department of Materials Science and Engineering, Tsinghua University, Beijing, 100084, China

[2)]Magnetic Material & Surface Modification Dept., Metallic & Inorganic Material Engineering Division, Toyota Motor Corporation, Toyota-cho, Toyota-city, Aichi Prefecture 471-8571, Japan

[3)]Beijing Key Lab of New Energy Materials and Technology, School of Materials Science and Engineering, University of Science and Technology Beijing, Beijing, 100083, China

E-mail: jingfeng@mail.tsinghua.edu.cn

Abstract. Most materials show decreasing thermal conductivity with increasing temperature, but an opposite temperature dependence of thermal conductivity is required for some industrial applications. The present work was conducted with a motivation to develop composite materials with a positive temperature dependence of thermal conductivity. ZrO_2/ stainless steel powders (304L) composite, with 3% stearic acid, was prepared by normal sintering under the protecting of Ar after mixing by mechanical ball milling technique. With the 304L content increasing from 10% to 20%, the thermal conductivity values increased. For all samples, the thermal conductivity in the temperature range of room temperature to 700 ℃ decreased with temperature below 300 ℃, and then began to increase. The increasing thermal conductivity of the composites (within the high temperature range was attributed to the difference of the thermal conductivity and thermal expansion coefficient between ZrO_2 ceramic and 304L stainless steel powders. Two simple models were also used to estimate the thermal conductivity of the composites, which were in good agreement with the experiment results.

1. Introduction

Thermal conductivity is an important transport property of materials, and most materials show decreasing thermal conductivity with increasing temperature. Materials are used in different fields according to the characteristic of thermal conductivity, and the famous-known thermal barrier coating (TBC) materials are representative cases. TBC are used to protect and insulate hot-section metal components, can result in significant temperature reduction at the metal surface, thereby the durability and efficiency of the metal component could be improved [1-3]. The most commonly used TBC is Y_2O_3-stabilized ZrO_2 ceramic [4-6]. However, there is a growing demand for higher fuel efficiency and longer durability, providing a great motivation for developing a new material with thermal conductivity controlled to show positive temperature dependence. That is, at low temperature, the new material has lower thermal conductivity, acting as heat insulating material. While at high temperature, it has higher thermal conductivity and the heat will be diffused more quickly.

In this study, a Ceramic/Metal composite was developed, using the difference of the thermal conductivity and thermal expansion coefficient between ceramics and metals. Voids exist in the interior of the composite at low temperature as shown in Figure 1(a). The heat energy could not shed easily at low temperature because of the lower thermal conductivity of ceramic and the existence of voids. At higher temperature, the voids will be filled up due to the high thermal expansion coefficient of metals, as shown in Figure 1 (b). Accompanied by the disappearing of voids gradually, the thermal conductivity of the composite becomes increased with temperature, which could be also attributed to the high thermal conductivity of metals. This design concept will be then confirmed by the following calculate models and experiment two ways.

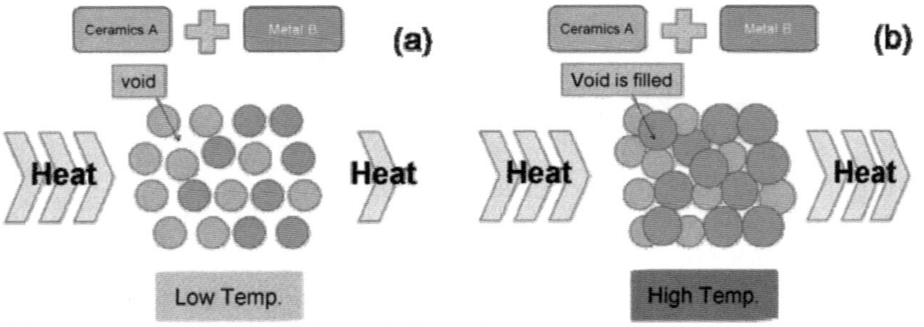

Figure 1 Diagrammatic sketch of Ceramic/Metal composite: (a) at low temperature; (b) at high temperature

2. Experiment process

The commercial elemental micro-powders of ZrO_2 (3mol. % Y_2O_3) and stainless steel powders (304L) with near-spherical were used as starting materials. Firstly, the 304L powders with 3% stearic acid were mixed in a stainless steel jar with stainless steel balls at 300 rpm for 24 h in a planetary ball mill, and the process was under the protection of Ar (99.9%) atmosphere to prevent the powders from oxidation. Next, ZrO_2 and the 304L powders with stearic acid mixed by hand according to different ratio, and the resultant powders were compacted into disk-shaped sample with ø15 mm in diameter and ~2 mm in thickness, followed by isostatic cool pressing at 200 MPa for 3 min. Then, the samples were sintering at 1100 ℃ for 4 h under the Ar atmosphere. Finally, the samples ZrO_2+x% 304L (x=0, 10, 15, 20) were prepared.

The phase structures of all samples were examined by X-ray diffraction (XRD) using Cu Kα radiation (D/max-2500). The thermal conductivity (κ) was calculated using the equation $\kappa=\lambda C_p d$, where λ is the thermal diffusivity, C_p is the heat capacity, and d is bulk density of the sample. The thermal diffusivity was measured by a laser flash technique (TC-9000, Ulvac-Riko, Japan) in Ar atmosphere in the temperature range from room temperature to 800 ℃, the heat capacity was measured using differential scanning calorimety (DSC-60) and the bulk density was obtained by the Archimedes method.

3. Results and discussion

3.1. Thermal conductivity: model analysis

Here, two simple models were used to estimate the thermal conductivity of composite we designed at low and high temperatures, respectively. In the models, the dispersion is assumed as spherical and the volume fraction of pores is not considered.

At low temperatures, the composite is composed of continuous matrix ceramic phase and the discontinuous dispersed metal phase, and the voids exist. Therefore, the effective thermal conductivity of the composite K_{eff} depends on not only the volume fraction of the dispersed phase V_d but also the dispersion size a, and the thermal barrier resistance of the voids h_c is also considered, here $h_c=10^7 W.m^{-2}.K^{-1}$ was used [7]. According to the Equation (1) [8]:

$$\frac{K_{eff}}{K_m}=\frac{[2(\frac{K_d}{K_m}-\frac{K_d}{ah_c}-1)V_d+\frac{K_d}{K_m}+\frac{2K_d}{ah_c}+2]}{[(1-\frac{K_d}{K_m}+\frac{K_d}{ah_c})V_d+\frac{K_d}{K_m}+\frac{2K_d}{ah_c}+2]} \tag{1}$$

At higher temperatures, the voids are filled up because of the thermal expansion. And K_{eff} at higher temperatures is just a function of V_d. it can be expressed as Equation (2) [9]:

$$K_{eff}=\frac{K_mV_m(dT/dx)_m+K_dV_d(dT/dx)_d}{V_m(dT/dx)_m+V_d(dT/dx)_d} \tag{2}$$

For spherical dispersion:

$$\frac{(dT/dx)_d}{(dT/dx)_m}=\frac{3K_m}{K_d+K_m} \tag{3}$$

Substituting Equation (3) into Equation (2) gives:

$$K_{eff}=\frac{K_mK_dV_m+2K_m^2V_m+3K_mK_dV_d}{V_mK_d+2V_mK_m+3V_dK_m} \tag{4}$$

Where, $K_{eff}/K_m/K_d$: the effective thermal conductivity of the composite/ the matrix / the dispersion; V_m/V_d: volume fraction of the matrix /the dispersion; a: the radium of the spherical dispersion;

Table 1 Calculated results of thermal conductivities of the composite with different V_d and a

V_d	a (μm)	$K_{low\ temperature}$	$K_{high\ temperature}$
	2	4.7	
20%	1	4.2	5.6
	0.5	3.7	
	2	5.4	
30%	1	4.6	7.1
	0.5	3.8	

The thermal conductivities of the composite with different V_d and a at low and high temperatures are calculated using function (1) and (4), respectively. From the calculating results shown in Table 1, It is found that the Ceramic/Metal composite we designed has low thermal conductivity at low temperatures and higher thermal conductivity at high temperatures as we expect, indicating our design concept is feasible in theoretically. At the same time, the calculated results also indicate that the thermal conductivity of the composite increases with the volume fracture of dispersion V_d.

3.2. Thermal conductivity: experimental data

The corresponding experiment was also performed to investigate the temperature dependence of the composite. Figure 2 shows the X-ray diffraction patterns of samples ZrO_2+x% 304L (x=0, 10, 15, 20) with 3% stearic acid. The patterns show that the main phase of ZrO_2 sintered at 1100 ℃ for 4 h is t-ZrO_2, and little m-ZrO_2 phase is also found. For the samples with 304L, 304L is also found with the intensive peaks of t-ZrO_2 and m-ZrO_2. There is no other phase is detected, indicating that no reaction occurs between ZrO_2 and 304L powders, and the stearic acid is volatilized during sintering process. Volatility of the stearic acid forms the voids in the composites.

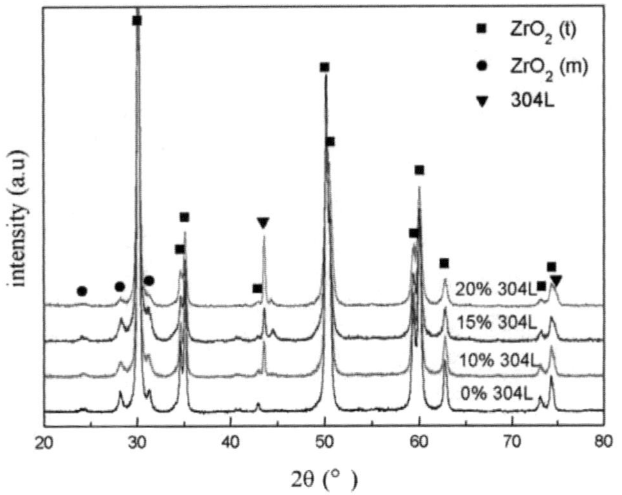

Figure 2 XRD patterns of samples ZrO_2+x% 304L with 3% stearic acid (x=0, 10, 15, 20,)

Temperature dependence of the thermal diffusivity for all samples is shown in Figure 3(a). The thermal diffusivity of all samples with 304L decreases with temperature and gets the minimum value below 500 ℃, then begins to increase. Meanwhile, with the increasing of 304L content, the values increase, and the corresponding temperature decreases from 500 ℃ to 400 ℃. The measured density values of the samples with 10%, 15% and 20% 304L are 3.501, 3.634 and 3.802 g.cm^{-3}, respectively and theoretical density value of t-ZrO_2 is 5.9 g.cm^{-3}. The density of the composites increases with 304L content increasing.

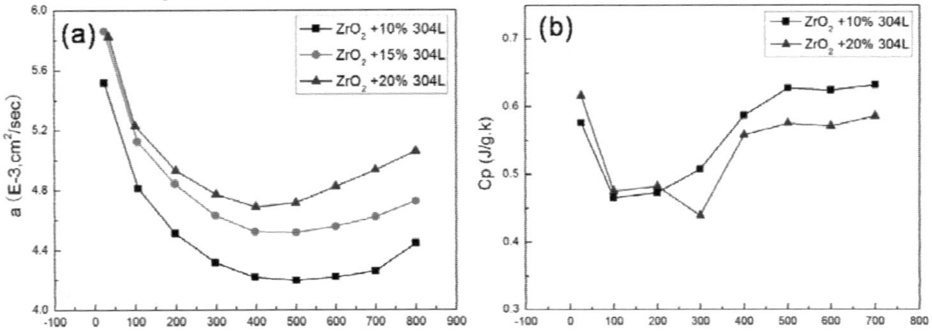

Figure 3 Temperature dependence of (a) thermal diffusivity for ZrO_2+x% 304L (x=10, 15, 20) and (b) specific heat for ZrO_2+x% 304L (x=10, 20) with 3% stearic acid

In order to calculate the thermal conductivity of the Ceramic/Metal composite, and investigate the effect of different content of 304L on the thermal conductivity of the composite, the specific heat of ZrO_2+x% 304L (x=10, 20) are measured. With temperature increasing, the specific heat of the two samples both decreases firstly, and then increases as shown in Figure 3(b). And ZrO_2+10% 304L has relatively high specific heat when the temperature above 200℃.

Temperature dependence of thermal conductivity for ZrO_2+x% 304L (x=10, 20) with 3% stearic acid is shown in Figure 4, together with the model analysis results. For comparison, the thermal conductivity of ZrO_2 from reference [10] is also listed. The thermal conductivity of ZrO_2 is almost independence of temperature, and with temperature increasing, the thermal conductivity of the two composites both decreases below 300 ℃, and then increases. And ZrO_2+20% 304L possess a relatively higher thermal conductivity. So it can be concluded that the thermal conductivity of the composite obtained according to our design concept increases with temperature and volume fraction of 304L at higher temperature, which is also consistent well with the model analysis results.

The thermal conductivity of the composite increased with temperature at high temperatures is attributed to the difference of the thermal conductivity and thermal expansion coefficient between ZrO_2 ceramic and 304L stainless steel powders. At low temperatures, the voids originating from the volatilizing of stearic acid in the composite can bring about the lower thermal conductivity of the composite. At higher temperatures, interior structure of the composite will be changed. The voids are filled up and the composite becomes more density, so the thermal conductivity also becomes higher.

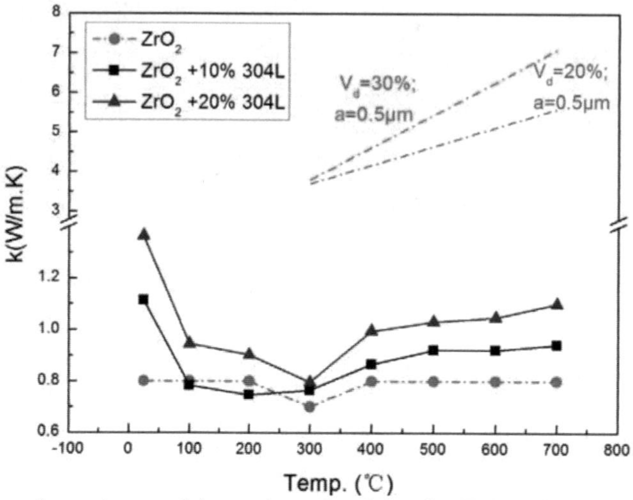

Figure 4 Temperature dependence of thermal conductivity for ZrO_2+x% 304L (x=10, 20) with 3% stearic acid and ZrO_2 [10]

4. Conclusion

A composite with positive temperature dependence was designed, according to the difference of thermal conductivity and thermal expansion coefficient between ceramic and metal. ZrO_2/304L composite had been prepared by mechanical mixing and normal sintering, and simple models were also used to estimate the thermal conductivity of composite. Excitingly, both calculated models used here and the experiment results supported our design concept. The thermal conductivity of the composites increased with temperature at higher temperatures, as well as the volume fraction of 304L powders.

References

[1] Afrasiabi A, Saremi M and Kobayashi A 2008 *Mate Sci Eng A*. **478** 264
[2] Chen H, Zhou X M and Ding C X 2003 *J Eur Ceram Soc*. **23** 1449
[3] Wu J, Padture P N and Gell M 2004 *Scripta Mater*. **50** 1315
[4] Matsumoto M, Yamaguchi N and Matsubara H 2004 *Scripta Mater*. **50** 867
[5] Dobbins T A, Knight R, Mayo M J 2003 *Journal of thermal apray technology*. **12** 214
[6] Deng Z Y, Ferreira J M F, Tanaka Y and Isoda Y 2007 *Acta Mater*. **55** 3663
[7] Smith D S, Fayette S, Grandjean S, Martin C, Telle R and Tonnessen T 2002 *J Am Ceram Soc*. **86** 105

[8] Hasselman D P H and Johnson L F 1987 *J. Comp. Mat.* **21** 508
[9] Hamilton O L and Crosser O K 1962 *Industrial & Engineering Chemistry Fundamentals*. **1** 187
[10] Miranzo P and Osendi M I 2008 *Scripta Mater.* **58** 973

Thermal Stress Analysis of W/Cu Functionally Graded Materials by Using Finite Element Method

Zhenxiao Yang [1, 2,*], Min Liu [2], Chunming Deng [2], Xiaofeng Zhang[1,2], Changguang Deng [2]

[1]School of Materials Science and Engineering, South China University of Technology, Guangzhou, 510640, China

[2]Guangzhou Research Institute of Nonferrous Metals,Guangzhou 510651, China

E-mail: yangzhenxiao@163.com

Abstract: Copper alloys with tungsten coating shows an excellent plasma irradiation resistance, however, the difference of coefficient thermal expansion between W and Cu makes it really a difficult job to prepare over 1 mm W coating with high adhesive strength. Functionally graded material (FGM) seems to be an effective method to improve the adhesive strength of thick W coating. This paper focused on the finite element simulation on thermal stress for W/Cu FGM with different graded layers, composition and thicknesses. In addition, the variance of stresses for functionally graded coatings with the steady state heat flux were simulated by finite element analysis (ANSYS Workbench). The results showed that the W/Cu FGM was effectively beneficial for the stress relief of W coating. Meanwhile, the maximum von mises stress decreased approximately by 52.8 % compared to monolithic W plasma facing material. And the four-layer FGM with a compositional exponent of 2 was optimum for 1.5 mm W coating.

1 Introduction

Tungsten is regarded as promising plasma facing material (PFM) for the nuclear fusion. Tungsten not only has the highest melting point among metals and low sputtering yields, but also owns low tritium retention and low erosion rate under plasma loading [1-2]. Copper and its alloys are being developed as heat sink materials due to their prominent thermal conductivity [3-4]. Low pressure plasma spraying (LPPS) offers the ability to coat large area, complex shapes coating and in-situ repair of the damaged coating [5]. But it is a challenge to deposit tungsten coating on CuCrZr for the larger mismatch of physical properties, which will cause thermal stress concentration on the W/Cu interface and lead to failure modes such as cracking or W coatings spalling when W/Cu exposing to severe thermal loading. The studies show that the introduction of W/Cu functionally graded materials (FGMs) can effectively alleviate the stress concentration on the W/Cu interface, release residual stress and improve the adhesive properties [6-8].

In this paper, the thermal stresses of W/Cu FGMs with different graded layers, different compositional exponent and different thicknesses, as well as the changes of stresses under the steady heat flux were simulated by finite element ANSYS Workbench code.

2 Establishment of finite element model

* Corresponding author.Tel.: +86-20-37238263; Fax: +86-20-37238531.
Email: yangzhenxiao@163.com.

2.1 Geometric model and finite element model

Figure 1(b) shows the geometric model of the W/Cu plasma facing component (PFC) and the design scheme of functionally graded coatings. The size of CuCrZr substrate is 30×40×25mm, a straight tube with diameter of 10mm, an average temperature of 30℃ and a convection exchange coefficient of 5 MW/(m^2·K) was assumed on substrate. Meanwhile, the steady state heat flux of 8 MW/m^2 was loaded on the top surface of W/Cu, the thickness of W coating was set for 1.5mm and W/Cu graded layers were designed for 0.2~5.0 mm, the number of layers ranged from 1~10. The compositional distribution was using the exponential function: $f_{Cu} = (Z / t_{FGM})_p$, where the f_{Cu} is the volume content of Cu, p is a compositional exponent. Different p represents different composition distribution discipline. When p<1, which means whole graded layers owns more volume content of Cu; when p> 1, the volume content of W is more; p = 1, the volume content of W and Cu shows a linear distribution.

The calculation was performed on ANSYS Workbench station, hexahedral eight-node elements (SOLID5) were selected as the thermal-structure model. The coating and substrate were assumed to be isotropic for simplicity in this study. The analytical model was perfect elastic body without plastic deformation. Thermal radiation and phase transformation were not considered in the calculation. The modeling consisted of two steps. Firstly, a thermal model was used to determine the temperature through the specimen during heat loading. Secondly, the resulting thermal histories were transferred to a mechanical model to compute the thermal stresses both coating and substrate. A fine mesh is introduced to model including substrate, graded layer and W coating, seen in Figure 1(a).

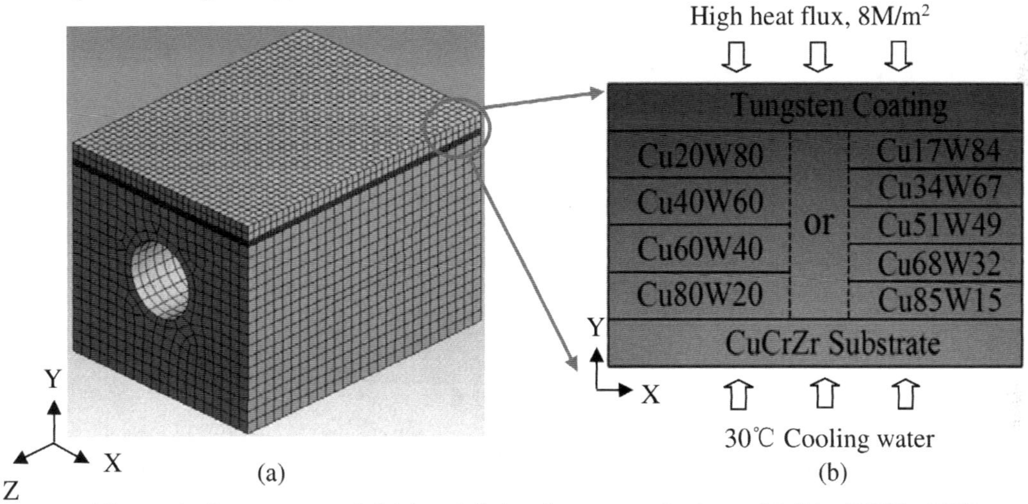

Figure 1. Geometry model (a) and finite element analysis model (b) of W/Cu PFC

2.2 Physical properties model

Table 1 lists the materials properties used in the simulation. Physical properties for different graded layers were calculated by the rules of mixture called Kern, which can be assumed from the following equations [9].

$$C = \frac{v_{Cu} c_{Cu} + v_W c_W \dfrac{3c_{Cu}}{2c_{Cu} + c_W}}{v_{Cu} + v_W \dfrac{3c_{Cu}}{2c_{Cu} + c_W}} \quad (1)$$

$$\alpha = v_{Cu}\alpha_{Cu} + v_W\alpha_W + \frac{4G_{Cu}}{K}\frac{(K-K_W)(\alpha_{Cu}-\alpha_W)v_W}{4G_{Cu}+3K_W} \quad (2)$$

Where v, v_{Cu}, v_W is volume fraction respectively; c, c_{Cu}, c_W is thermal conductivity; α, α_{Cu}, α_W is coefficient of thermal expansion and K, K_{Cu}, K_W is bulk modulus of graded layer, Cu and W. Young's modulus E, Posison's ratios γ were calculated by average field theory [10], shown as follow.

$$K = K_{Cu} + \frac{v_W}{\frac{1}{K_W - K_{Cu}} + \frac{3v_{Cu}}{3K_{Cu} + 4G_{Cu}}} \quad (3) \qquad G = G_{Cu} + \frac{v_W}{\frac{1}{G_W - G_{Cu}} + \frac{6v_{Cu}(K_{Cu} + 2G_{Cu})}{5G_{Cu}(3K_{Cu} + 4G_{Cu})}} \quad (4)$$

$$\gamma = \frac{3K - 2G}{6K + 2G} \quad (5) \qquad E = 2G(1 + g) \quad (6)$$

Where G, G_{Cu}, G_W is the shear modulus of graded layer, W and Cu, respectively. Specific heat λ, Density ρ, were determined by (7) and (8).

$$\rho = \rho_{Cu} v_{Cu} + \rho_W v_W \quad (7) \qquad \lambda \approx \lambda_{Cu} v_{Cu} + \lambda_W v_W \quad (8)$$

Where ρ, ρ_{Cu}, ρ_W is the density. λ, λ_{Cu}, λ_W is specific heat of graded layer, W and Cu, respectively.

Table 1. Physical properties of W, Cu, and different W/Cu graded layers

Name	Density (g/cm^3)	Thermal conductivity ($W \cdot m^{-1} \cdot K^{-1}$)	Specific heat ($J \cdot kg^{-1} \cdot K^{-1}$)	Coefficient of thermal expansion ($10^{-6} K^{-1}$)	Posison's ratios	Young's modulus (GPa)
CuCrZr	8.96	397	386	17	0.34	128
W	19.32	174	138	4.5	0.28	410

3 Results and discussion

3.1 Effect of number of graded layers on thermal stress

Figure 2 shows the influence of layer number on the von mises stress, surface temperature with the compositional exponent of Cu volume content is 1. It can be seen that the maximum surface temperature changed slightly with the increase of number of graded layers, nearly as a straight line, which caused by reduction of the overall thermal conductivity and modest graded layers. Further, the von mises stress decreased to 756 MPa at the joint of five-layer system, which reduced by 22.7 % in contrast to single-layer system, while its temperature increased with only 6 ℃. Additionally, the von mises stress changed within a wavy motion after the graded layers over five. The reason for this behavior that the coefficient of thermal expansion between each graded

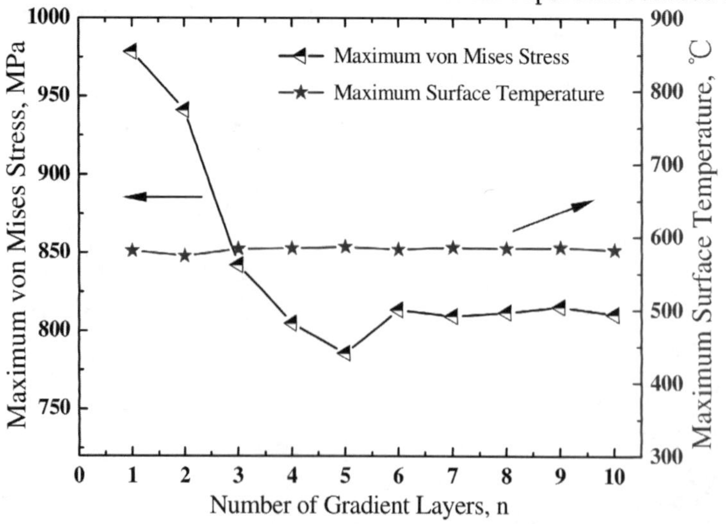

Figure 2. Influence of the layer number on the maximum von mises stress, maximum surface temperature with the compositional exponent of Cu volume content is 1.

layer reduced, which is conducive to reduce thermal stress. Moreover, the stress caused by temperature increased with increase of the surface temperature, which caused the von mises stress showing such trends.

3.2 Effect of compositional exponent on thermal stress

The thermal stress evolution is strongly influenced by compositional exponent of Cu volume content (p), shown in Figure 3. It can be observed that the surface temperature gradually increased with increase of the value of p, which is related to reduced Cu content of graded layers and increased W-content. With the increase of p-value, the von mises stress firstly decreased sharply. And then no remarkable change was observed when it increased to 10, reflecting physical properties of the changed graded layers. This shows that a reduction depends on the composition of the graded interface and significantly enhanced by the use of a partial graded with a high W-content. For five-layer system, when 0< p <1.6, the von mises stress reduced evidently, but the effect of p-value to stress was little when 1.6< p <10. The increase of surface temperature was slow with the increase of p-value.

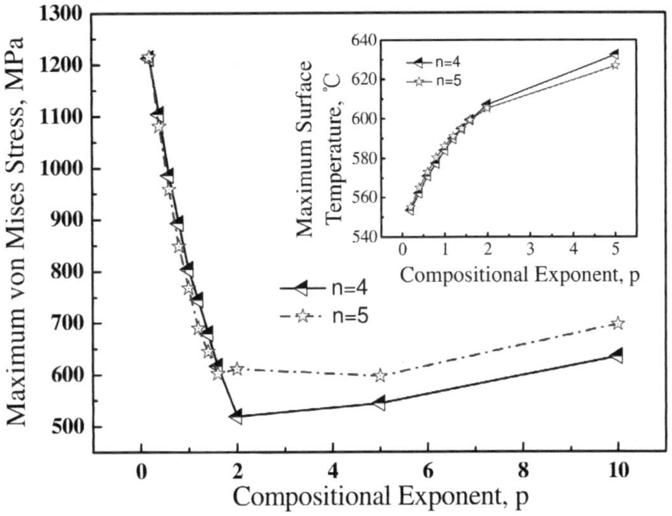

Figure 3. Influence of compositional exponent of Cu volume content on the von mises stress and the maximum surface temperature.

3.3 Effect of graded thickness on thermal stress

To optimize the thickness of W/Cu graded layers, the compositional exponent p-value of 2 and four-layer system was used. The surface temperature, von mises stress versus graded layer thickness is shown in Figure 4. The surface temperature gradually increased along with the increment of graded layer thickness. In contrast to monolithic tungsten coating without FGM model, the temperature of 5.0 mm thickness system raised nearly 170℃. However, the von mises stress rapidly reduced to 474.39 MPa from 1006.2 MPa at 5.0 mm. The reason is that with the increase of graded layer thickness, which offered a wider adjustment space for the coefficient of thermal expansion and the elastic modulus of each graded layer, and thereby achieved a more uniform distribution of the internal stress. Therefore, in order to achieve an optimum stress relaxation, the thickness can be selected larger appropriately, but surface temperature rose. Further analysis, just for the purpose of stress reduction, the thicker graded layers is a better choice. But from another perspective, with the graded layer thickness increasing, the tungsten surface will be farther away from the location of the cooling water tube, the surface temperature will rapidly rise. When the graded layer thickness reaches a certain value, the temperature may reach the melting point of Cu and then lead to Cu melting, which is bound to reduce the ability of whole layers to conduct heat. Meanwhile, it will bring the newborn holes to composite material, causing W/Cu plasma facing component to fail.

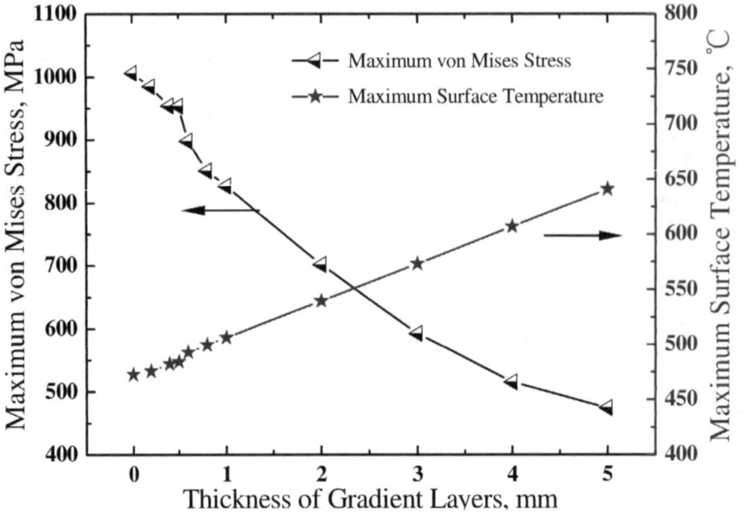

Figure 4. Influence of graded thickness on the maximum von mises stress, maximum surface temperature of four-layer system with the compositional exponent of Cu volume content is 2.

3.4 Simulation results of 4.0 mm W/Cu graded layer model

Figure 5 is the temperature and the thermal stress distribution of 4.0 mm W/Cu graded layer model when compositonal exponent p is 2. As shown in Figure 5(b), an obvious deformation and stress concentration phenomenon occured at the edges of the coating. The reason for the edge effect is that stress singularities occur at the outer surface, which characterized by element sizes. So the comparison of stresses was done at the symmetry axis (Y-axis), starting from the top of the cooling water tube to W outer surface. The von mises stress distribution of designed path (1→2) along the Y-axis is shown in Figure 5(d), and the maximum stress reduced by 48.8% lower than that of monolithic tungsten coating without FGM model. The stress at the interface between the graded layer and the copper substrate is almost the same, but with the increase of the graded layer thickness, the stress at the interface between the graded layer and W coating reduced significantly. Although the stress relaxation is very obvious, the maximum surface temperature increased by 134.26℃.

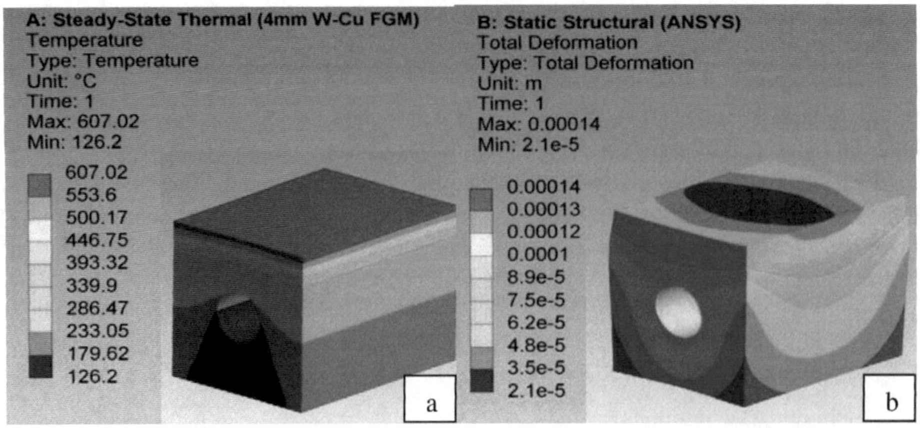

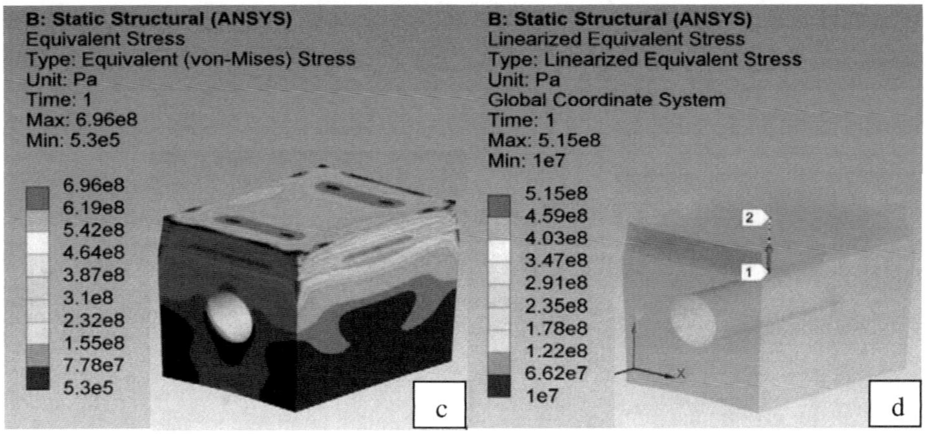

Figure 5. The total temperature distribion (a); total deformation (b); equivalent stress (c); linearized equivalent stress through Y axis (d) of W/Cu PFC with 4.0 mm graded layer model under the heat flux of 8 MW/m^2.

4 Conclusions

The introduction of W/Cu FGM effectively alleviates the stress concentration. Four-layer model with compositional exponent of 2 was optimum for 1.5 mm W coating, which induced a 23 % reduction of the maximum stress compared to the monolithic W as PFM. Meanwhile, the von mises stress was decreased with the increase in the number graded layers firstly, and then changed within a wavy motion after the graded layers beyond five. Besides, with the increase of compositional exponent, the von mises stress firstly decrease sharply, no remarkable change was observed though it increased to 10. Results also showed that the von mises stress decreased with increase of the FGM thickness.

References

[1] Montanari R, Riccardi B, Volterri R, Bertamini L 2002 *Material Letters*, *52 100*

[2] Zhangjian Zhou, Shuxiang Song, Weizhi Yao 2010 *Fusion Engineer and Design*, **10-12** 1720

[3] Kaufmann M, Neu R 2007 *Fusion Engineer and Design*. **82** 521

[4] Roedig M, Kuehnlein W, Linke J 2002 *Fusion Engineering and Design*, **61-62** 135

[5] Matejlcek J, Koza Y, Weinzettl V 2005 *Fusion Engineering and Design*, **75-79** 395

[6] Davis J W, Slattery K T, Driemeyer D E 1996 *Journal of Nuclear Materials*, **233-237** 604

[7] Chong F L, Chen J L, Li J G 2007 *Journal of Nuclear Materials*, **363-365** 1201

[8] Liu B B, Xie J X 2005 *Chinese Journal of Rare Metals*, **29** 757

[9] Kerner E H 1956 *Proceedings of the Physical Society*, B **69** 808

[10] Weng G J 1984 *International Journal of Engineering Science*, **22** 845

Processing and structure of a Nitrogen Alloyed Oxide Dispersion Strengthened Austenitic Stainless Steel by mechanical alloying

Yingli Xu, Zhangjian Zhou*
School of Materials Science and Engineering, University of Science and Technology Beijing, Beijing 100083, China

* E-mail: zhouzhj@mater.ustb.edu.cn

Abstract. Pure metallic powders of Fe, Cr, Ni, W, Ti and nano-Y_2O_3 powders were mechanical alloyed by high energy mechanical milling under N_2 atmosphere to develop a nitrogen contained oxide dispersion strengthened austenitic steel powders. The compositions of the mixed powders are Fe-18Cr-8Ni-2W-1Ti-0.35Y_2O_3. The effects of milling time on nitrogen absorption were investigated. The nitrogen contents in the as-milled powders reached 0.31 wt% after milling for 60 hours and nearly 0.8 wt% after milling for 150 hours. The as-milled powders were then annealed under vacuum at 1173K and 1373 K to promote the formation of the resultant equilibrium phases. Fully austenitic structure was achieved after annealing at 1173K.

1. Introduction

Austenitic stainless steel is a kind of important structural materials for nuclear reactors due to their excellent overall mechanical properties, good corrosion and oxidation resistance compared with ferritic-martensitic steel. The disadvantages of austenitic steel are the low ultimate tensile strength, the inferior creep strength at high temperature and the irradiation induced swelling.

High nitrogen austenitic steels are promising due to their superior mechanical and corrosion properties. An austenitic stainless steel is considered as high-nitrogen steel if it contains nitrogen higher than 0.4 wt % [1]. The nitrogen element in the austenitic steel is benefit for stabilization of austenite phase, improving tensile strengths and corrosion resistance [2,3].

Oxide dispersion strengthened (ODS) ferritic steels fabricated through powder metallurgical route have been studied extensively in the field of nuclear energy during the past two decades. It is demonstrated that ODS ferritic steels are the only materials presenting jointly two important properties (a) high dimensional stability under irradiation, i.e. high resistance to swelling and irradiation creep, and (b) a potential high temperature strength resulting from the homogeneous dispersion of nanometric oxide particles [4]. Oxide dispersion particles are much more stable than carbide at high temperature. These dispersed oxide particles in steel matrix could prevent the movement of dislocations effectively, and thus improve its creep resistance and performances at high temperature. It can be expected that ODS austenitic steels will also show the above mentioned advantages. Especially, if combining the advantages of high-nitrogen austenitic steel with those of ODS austenitic steel, this new modified austenitic steel will show promising for further improvement. However, most of the

published literature on Fe-based ODS alloys concerns ferritic or ferritic/martensitic alloys. There has very little attentions on ODS austenitic steels [5, 6] and almost no research works on nitrongen-alloyed ODS austenitic steel.
In this work, pure powders of Fe, Cr, Ni, W, Ti, and nano-Y_2O_3 powders were processed by high energy mechanical milling under N_2 atmosphere to develop a nitrogen contained ODS austenitic steel.

2. Experimental procedure

The starting materials are pure metallic powders of Fe, Cr, Ni, W, Ti and nano-Y_2O_3 with a purity of 99.9 %. The particle size of all metallic powders are less than 74 μm, while the average particle size of nano-Y_2O_3 is 30 nm. The composition of the mixed powders are designed as Fe-18Cr-8Ni-2W-1Ti-0.35Y_2O_3, with weight ratio. The mechanical alloying (MA) process of the above mentioned powders was performed at room temperature under a high purity (99.99 %) nitrogen atmosphere by using a planetary high-energy ball mill. The nitrogen pressure during MA processing was controlled at one atmosphere by using a pressure appliance. The ball-to-powder weight ratio was 10:1. The rotating speed is fixed at 380 rpm. The milling time are 30, 60, 90, 120 and 150 hours respectively. The nitrogen contents in as-milled powders at different milling time were measured by CHN/O/S elemental analyzer (CE-440, produced by EXETER ANALYTICAL, INC.).
Heat treatment of the mechanically alloyed powders at 1173 K and 1373 K respectively for 5 hours was carried out after sealing the powder in quartz tubes under vacuum (10-4 Torr). X-ray diffraction (Type: D/MAX-RB) was used for phase identification for both as-milled powders and annealed samples. The micro-structure and the composition distribution of as-milled powders were examined by using Scanning Electron Microscopy (SEM) with Energy Dispersive X-ray Spectroscopy (EDS). The size distribution of the as-milled powders was measured by laser diffraction method.

3. Results and discussions

3.1 Effects of milling time on nitrogen absorption

Figure 1 shows the nitrogen content (wt %) in the powder mixtures as a function of milling time. It should be noticed that the curve exhibits three stages during processing, which correspond to three absorption rate related to the structure and morphology of the powder particles. During the first 30 hours of milling, the nitrogen content shows a very slow absorbability. Then, in the second stage, nitrogen absorbs rather quickly with the content increasing up to 0.55 wt % after 90 hours of milling. In the third stage, the nitrogen content continues to increase with a slightly slower slope than the second stage, with almost 0.8 wt % after milling for 150 hours.
During the milling process, the ball-to-ball, ball-to-powder, ball-to-wall collided and grinded strongly and frequently, which resulted in the application of compression and shear forces to the powders and induced the rising of the local temperature. The compressed and sheared powders were flatted, fractured and welded during MA processing, thus a large amounts of defects and dislocations in the crystal structures were expected due to the severe plastic deformation. The repeated deformation and the collisions induced local temperatures rising prompt the dyanmic energy of the nitrogen atoms which cause the diffusion of nitrogen atoms along the defects and dislocations more easily. The kinetics of nitrogen infusion in the powder mixture should be controlled by the total surface area exposed to the nitrogen gas environment [7]. Figure 2 shows the size and geometry evolution of the powder grains with milling time of 30, 60, 90 and 120 hours. Figure 3 shows the apparent particle size of powders after milling at different time. The size of milled powders decreased obviously when milling time was less than 90 hours, but the size increased after milling for 120 hourrs. From figure 2 and figure 3, it can be seen that at the first stage of mechanical alloying (milling time less than 30 hours), the powders were deformed but not fractured obviously; the powder sizes were still rather large, which indicates the specific surface area of milling powders is still rather small. This could be an explanation for the very low nitrogen absorption at first 30 hours of milling. When further

increasing the milling time to 90 hours, the accumulated energy in the as-milled powders reached at a very high point, which results in the strong fractures of milling powders and increasing the specific surface area significantly. Thus the nitrogen absorption rate increases obviously during this stage. It is interested to find that when increasing the milling time to 120 hours, the powders became coarse severely. Actually, most of the powders adhered to the vial's wall. This explained why the nitrogen absorption rate will be decreased at this stage, as shown in figure 1.

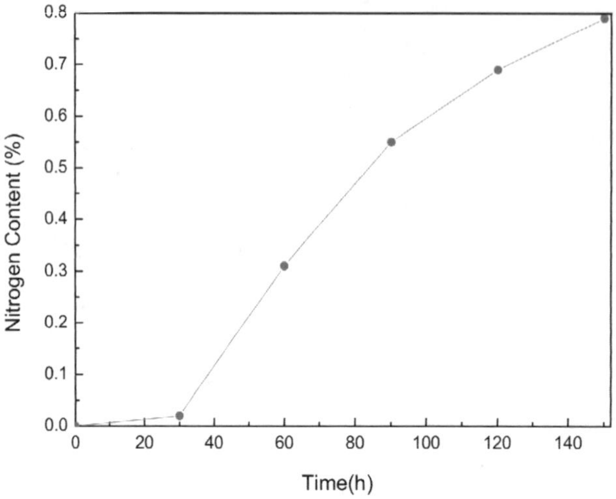

Figure 1. Nitrogen content as a function of milling time

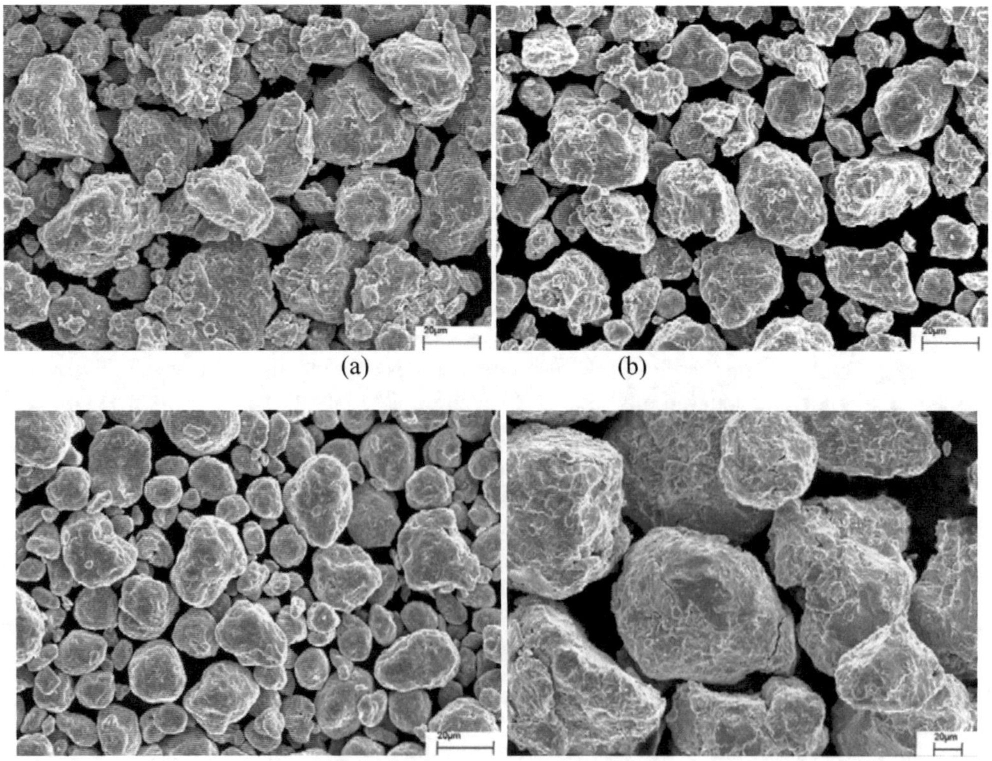

(a) (b) (c) (d)

Figure2. SEM micrographs of the powders milled for (a) 30h, (b) 60h, (c) 90h, and (d) 120h

Figure 4 shows the XRD patterns of powders milled for 30 h, 60 h, 90 h and 120 h respectively. It should be noticed that the peaks corresponding to the initial mixed powders tend to broaden or disappear with the prolongation of milling time. The reasons for this phenomenon should be (1) high density of defects and dislocations were generated during milling, (2) the infinite solid solution was formed between Cr and Fe because of the subtle atomic radius of these two metal elements, (3) the size of as-milled powders become smaller due to the fractures during milling, (4) the lattice instant and plane distance were enlarged as the alloying elements (including Y_2O_3 and Ti) diffused into the Fe matrix [8-11].

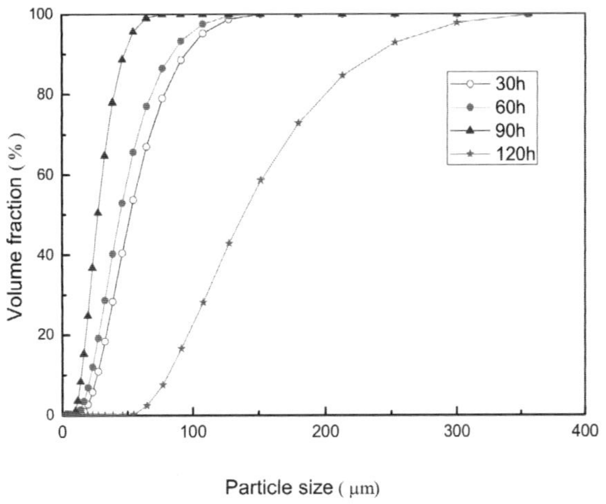

Figure 3. The powder size evolution of as-milled powders as a function of milling time

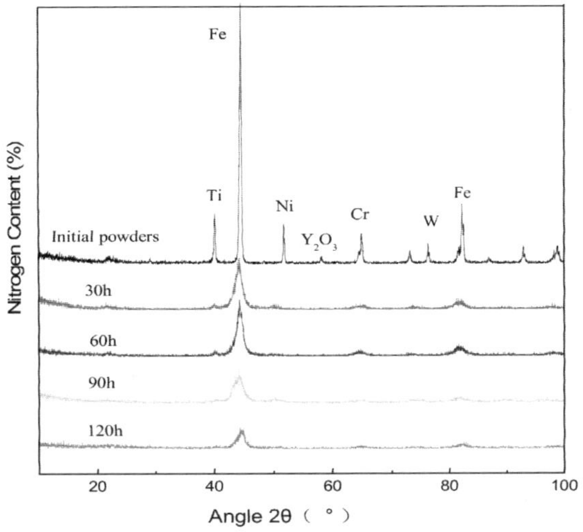

Figure 4. XRD patterns of initial and as-milled powders (30h, 60h, 90h, and 120h)

3.2 Annealed structure

As-milled powders with different milling time were encapsulated in quartz capsule under vacuum circumstance and annealed at 1173K and 1373K for 5 hours to promote the formation of massive equilibrium austenitic phase. The phases of annealed powders were analysed qualitatively by XRD, as shown in figure 5. The equilibrium phase of as-milled powders after annealing were austenite and no other phases appeared (figure 5 (a)).

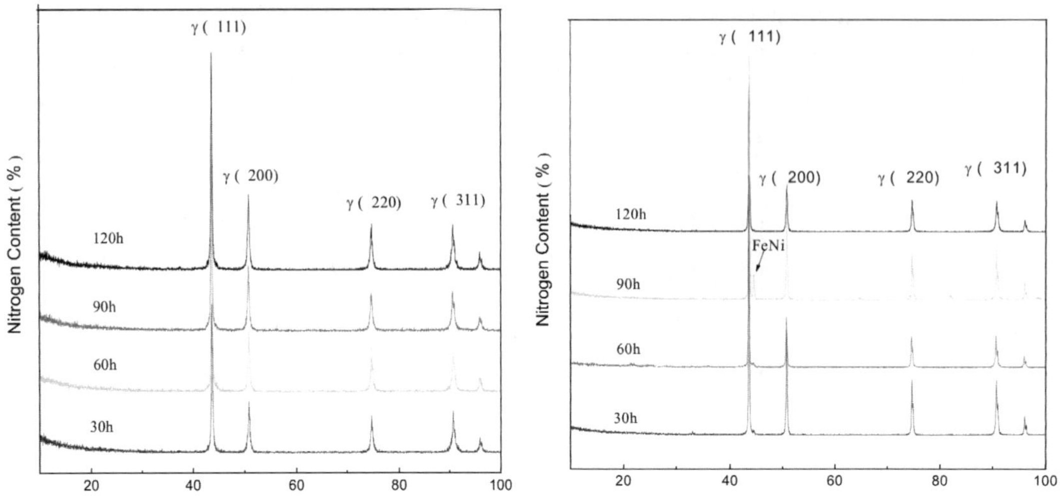

(a) After annealing at 1173K (b) After annealing at 1373K

Figure 5. XRD of N_2 alloyed ODS austentic steel powders after annealing at 1173K (a) and 1373K (b)

The results of annealing indicated that fully austenitic structure could be obtained by mechanical alloying. The outcome of desired structure could be explained by isothermal section of the equilibrium for Fe-Cr-Ni system. At both annealing temperature, no nitride and carbide were found. However, at temperature of 1373 K, FeNi solid solution phase appeared. But only for powders milling for 90 hours, this phase appeared apparently.

4. Conclusion

The nitrogen alloyed ODS authentic steel powders with composition of Fe-18Cr-8Ni-2W-1Ti-0.35Y_2O_3–xN were prepared by mechanical alloying successfully. The alloying compositions of Y_2O_3 and Ti diffused into the Fe matrix after milling for 30 hours.

The amount of nitrogen introdouruced into the structure increased with the milling time. The absorption of nitrogen content during milling time showed three apparent stages. At the first stage of 30 hours of milling, the amount of nitrogen increased very slowly, and then saw a much quicker growth in the following time of milling (30 to 90 hours). From 90 hours to 120 hours of milling, the absorption rate slowed down a little.

Fully austenite was achieved after annealing the as-milled powders at 1173K. FeNi solid solution formed when annealing at 1373K.

References

[1] J. W. Simmons. *Materials Science and Engineering A* 207 (1996) 159-169.

[2] A.C.Filemonowicz, D.Clemens, W.J.Quadakkers. *J. of Materials Processing and Technology* 53 (1995) 93-99.
[3] Z. Z. Yuan, Q.X.Dai, X.N.Cheng, C.H.Zhang. *J. of Jiangsu University (Natural Science)* 23 (2002) 72-77.
[4] C.Suryanarayana, E.Ivanov, V.V.Boldyev. *Materials Science and Engineering A* 304(2001)151-158.
[5] M.P. Phaniraj, Dong-Ik Kim, Jae-Hyeok Shim, Young Whan Cho. *Acta Materialia* 57 (2009) 1856 - 1864.
[6] T. Kim, C. Baei, D. Kim, J. Jang, S. Kim, C. Lee, D. Hahn. *Nuclear Engineering and Technology* 40 (2008) 305-310
[7] M. Mendez, H. Mancha, M. M. Ciseros, G. Mendozam, J. I. Escalante, H. F. Lopez. *Metallurgical and Materials Transactions* 33A (2001) 3273-3278.
[8] C. Suryanarayana. *Progress in Materials Science* 46 (2001) 1-184.
[9] J. C. Rawers, D. Govier, R. Doan. *Materials Science and Engineering A* 220 (1996) 162-167.
[10] I. F. Vasconcelos, S. D. Figueiredo. *Nanostructured Materials* 111 (1999) 935-946.
[11] G. A. Nematollahi, E. M. Rad, A. R. Aghaee. *Materials Science and Engineering A* 492 (2008) 455-459.

Dielectric properties of the compositionally graded $BaZr_xTi_{2-x}O_5$ thin films prepared by sol-gel method

YiPing Gong, DongYun Guo*, ChuanBin Wang, Qiang Shen, LianMeng Zhang

State Key Laboratory of Advanced Technology for Materials Synthesis and Processing, and School of Materials Science and Engineering, Wuhan University of Technology, Wuhan 430070, China
E-mail: guodongyun@gmail.com

Abstract. The downgraded and upgraded $BaZr_xTi_{2-x}O_5$ (x=0, 0.01, 0.02, 0.03 and 0.04) films normal to $Pt/Ti/SiO_2/Si$ substrates were prepared by sol-gel method. The microstructure and dielectric properties of the compositionally graded $BaZr_xTi_{2-x}O_5$ films were investigated. The single-phase downgraded and upgraded films were obtained as the films were annealed at 900 °C for 30 min. The downgraded $BaZr_xTi_{2-x}O_5$ film had dense surface, while there were pores in the surface of the upgraded $BaZr_xTi_{2-x}O_5$ film. The thicknesses of both upgraded and downgraded $BaZr_xTi_{2-x}O_5$ films were about 500 nm. The values of permittivity (ε_r) for the downgraded and upgraded $BaZr_xTi_{2-x}O_5$ films were 84 and 100, respectively, at 1 MHz.

1. Introduction

Since Akashi et al [1,2] reported that the $BaTi_2O_5$ single crystal showed ferroelectric properties and high relative permittivity (ε_r) only along the *b*-axis with high Curie temperature (T_C = 750 K), the $BaTi_2O_5$ ferroelectrics, as a new lead-free ferroelectrics, has been widely investigated [3-6]. However, the remnant polarization (P_r) of $BaTi_2O_5$ ceramics and single crystal was very small. Yue et al [7] improved the ε_r value of $BaTi_2O_5$ ceramics using Zr substitution. Due to the demand of device miniaturization, the Zr-substituted $BaTi_2O_5$ should be applied as the film form. To the best of our knowledge, there was no literature reported on the preparation of Zr-substituted $BaTi_2O_5$ thin film.

Recently, much attention has been paid to compositionally graded ferroelectric films due to many intriguing properties, such as the enhancement of dielectric and ferroelectric properties [8-11]. In this study, the compositionally graded $BaZr_xTi_{2-x}O_5$ (x=0, 0.01, 0.02, 0.03 and 0.04) thin films normal to $Pt/Ti/SiO_2/Si$ substrates were prepared by sol-gel method, and then their microstructure and dielectric properties were investigated.

2. Experiment

The compositionally graded $BaZr_xTi_{2-x}O_5$ (x=0, 0.01, 0.02, 0.03 and 0.04) films were prepared on $Pt/Ti/SiO_2/Si$ substrates by sol–gel method. The details of preparation of $BaZr_xTi_{2-x}O_5$ precursors were introduced elsewhere [12]. The $BaZr_xTi_{2-x}O_5$ films were spin-coated on $Pt/Ti/SiO_2/Si$ substrates in sequence at 3000 r/min for 25 s to achieve the compositionally graded $BaZr_xTi_{2-x}O_5$ films. And then the as-coated films were heated at 350 ℃ for 15 min. Figure 1 shows the cross-sectional schematics of compositionally graded $BaZr_xTi_{2-x}O_5$ thin films. From the substrate to the film surface, the ''upgraded'' (Fig. 1a) and ''downgraded'' (Fig. 1b) preparations were described with increasing and decreasing Zr content, respectively. Finally, the films were annealed at 900 °C for 30 min in the air ambient.

The crystalline structure of samples was analyzed by an X-ray diffractometer (XRD, Rigaku

Ultima III) with Cu K*a* radiation. The morphologies were observed by a field emission scanning electron microscope (FESEM, JSM-6700F). To measure the dielectric properties, the top electrode with the area of 2.0×10^{-3} cm^2 was fabricated by sputtering Au target. Dielectric properties were measured using a precision impedance analyzer (Agilent 4294 A) at room temperature.

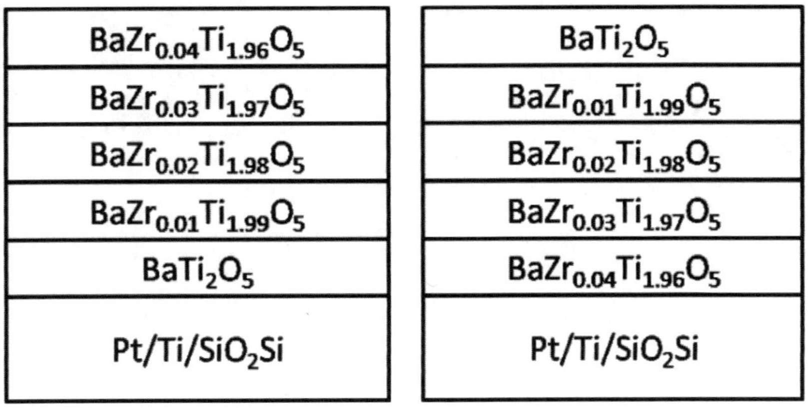

(a) (b)

Figure1 The cross-sectional schematics of the compositionally graded $BaZr_xTi_{2-x}O_5$ films: (a) Upgraded and (b) Downgraded

3. Result and discussion

Figure 2 shows the XRD results of the graded $BaZr_xTi_{2-x}O_5$ films. The single-phase upgraded and downgraded $BaZr_xTi_{2-x}O_5$ films were obtained. The microstructures of the graded $BaZr_xTi_{2-x}O_5$ films are shown in Fig. 3. The downgraded $BaZr_xTi_{2-x}O_5$ film had dense surface, while there were pores in the surface of the upgraded $BaZr_xTi_{2-x}O_5$ film. The thicknesses of both upgraded and downgraded $BaZr_xTi_{2-x}O_5$ films were about 500 nm, according to the cross-sectional morphologies.

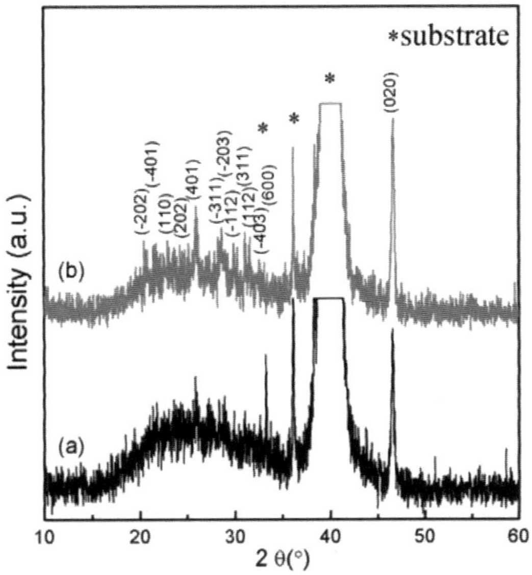

Figure 2 XRD patterns of the compositionally graded $BaZr_xTi_{2-x}O_5$ films: (a) Downgraded, (b) Upgraded

Figure 4 depicts the frequency dependence of ε_r for the downgraded and upgraded $BaZr_xTi_{2-x}O_5$ films. With increasing the frequency, the ε_r decreased at low-frequency region and kept at the almost constant values at high-frequency region. The ε_r values of the downgraded and upgraded $BaZr_xTi_{2-x}O_5$ films were 84 and 100, respectively, at 1 MHz.

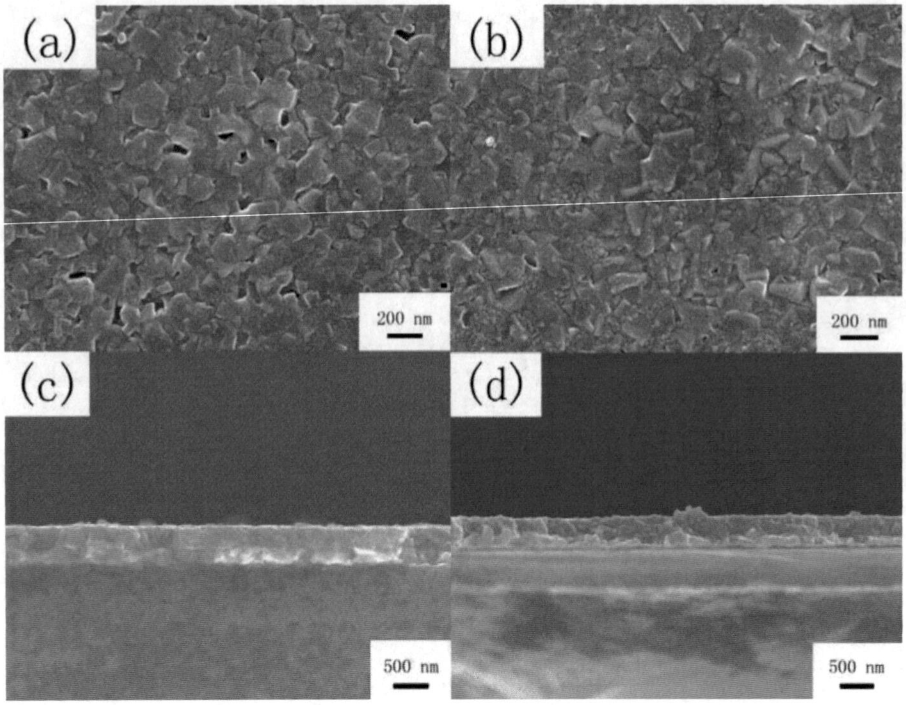

Figure 3 Surface and cross-sectional images of the compositionally graded $BaZr_xTi_{2-x}O_5$ films: Upgraded (a, c) and downgraded (b, d)

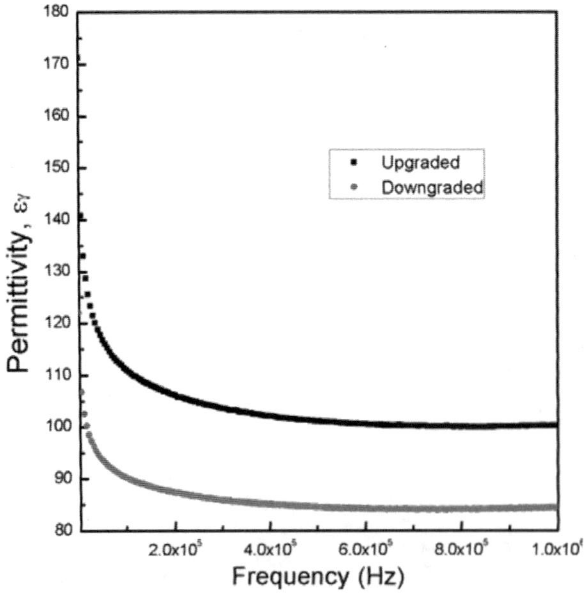

Figure 4 Permittivity (ε_r) as a function of frequency for the graded $BaZr_xTi_{2-x}O_5$ films

4. Conclusion

The downgraded and upgraded $BaZr_xTi_{2-x}O_5$ films were prepared on $Pt/Ti/SiO_2/Si$ substrate by sol-gel method. The single-phase downgraded and upgraded films were obtained as the films were annealed at 900 °C for 30 min. The downgraded $BaZr_xTi_{2-x}O_5$ film had dense surface, while there were pores in the surface of the upgraded $BaZr_xTi_{2-x}O_5$ film. The thicknesses of both upgraded and downgraded $BaZr_xTi_{2-x}O_5$ films were about 500 nm. The ε_r values of the downgraded and upgraded $BaZr_xTi_{2-x}O_5$ films were 84 and 100, respectively, at 1 MHz.

Acknowledgement

This work was supported by the International Science and Technology Cooperation Program of China (Grant No. 2009DFB50470). This work was also supported in part by International Science and Technology Cooperation Project of Hubei Province (Grant No. 2010BFA017), New Century Excellent Talents in University of China (Grant No. NCET-10-0662) and the 111 Project of China (Grant No. B13035).

5. Reference

[1] T. Akashi, H. Iwata, T. Goto 2003 Mater. Trans. 44(4): 802-804

[2] T. Akashi, H. Iwata, T. Goto 2003 Mater. Trans. 44(8): 1644-1646

[3] D. Guo, A. Ito, R. Tu, T. Goto 2012 Key Eng. Mater. 508: 185-188

[4] J.J. Wang, C.B. Wang, Q. Shen, L.M. Zhang 2012 J. Alloy. Compd. 512(1): 140-143

[5] A. Ito, D.Y. Guo, R. Tu, T. Goto 2012 J. Eur. Ceram. Soc. 32(10): 2459-2467

[6] D.Y. Guo, A. Ito, R. Tu, T. Goto 2012 Appl. Surf. Sci. 259: 178-185.

[7] X.Y. Yue, R. Tu, T. Goto 2008 Mat. Trans 49(1): 120-124

[8] D. Maurya, N. Wongdamnern, R. Yimnirun, S. Priya 2010 J. Appl. Phys. 108: 124111

[9] J. Sigman, P. Clem, C. Nordquist, J. Richardson, J. Dawley 2007 J. Appl. Phys. 102: 054106

[10] J. Mantese, N. Schubring, A. Micheli, A. Catalan, M. Mohammed, R. Naik, G. Auner 1997 Appl. Phys. Lett. 71: 2047-2049

[11] C. Liu, Y. Gong, D. Guo, C. Wang, Q. Shen, L. Zhang 2012 J. Mater. Sci.: Mater. Electron. 23: 1711-1714

[12] Y.P. Gong, C.Y. Liu, D.Y. Guo, C.B. Wang, Q. Shen, L.M. Zhang 2012 Mat. Lett 73: 72-74

Numerical investigation on anti-penetration behavior of ceramic/metal target under ballistic impact

H Mei[1], Y C Wang[1], X Liu[1], D F Cao[2], L S Liu[2,4]

[1]Department of Engineering Structure and Mechanics, Wuhan University of Technology, Wuhan 430070, P. R. China
[2]State Key Laboratory of Materials Synthesis and Processing, Wuhan University of Technology, Wuhan 430070, P. R. China

E-mail:Liulish@whut.edu.cn

Abstract. In the paper, we used the LS-DYNA FE code to simulate the bullet penetration against the target plate with different ceramic-steel ratio of thickness. The main stages of the bullet penetration and damage contours of the target were studied by analyzing the residual velocity-time curves. We also studied energy absorption of the ceramic/metal target. Considering curves of residual velocity-time, we reckon the process of penetration contains four stages. Ceramic performed good resistance before the formation of damage cone of ceramic. But after the damage cone formed, the anti-penetration behavior kept declining. When the bullet started to penetrate the layer of metal, the anti-penetration behavior of target rose slightly. Compared with thickness ratio of 0.4 and 0.6, ceramic with 0.2 absorbed more energy and works longer. Of several different thicknesses, layers of ceramic and steel were studied. Steel per cm absorbed more energy than ceramic per cm.

1. Introduction

The extremely complex problem of a projectile impacting a ceramic backed by a metal plate has already been studied in the past decades. Penetration of composite armors by a high-speed projectile has been a subject of considerable research interest. Wilkins [1] and Florence [2] were among the early investigators of the subject of penetration mechanics of composite armors. Florence [2] developed an analytical model for the two-component armor. The model assumes a ceramic hard facing and a ductile backing plate impacted by a rigid projectile. It predicts the protection ballistic limit velocity which is a measurement of the resistance of protective armor to projectile penetration. Woodward [3] later proposed a one-dimensional model for the penetration of a projectile into ceramic armor using a lumped mass approach, which takes into account the erosion of both projectile and target in a simple way. Reijer [4] gave a more complicated model which allows projectile erosion, mushrooming, and different deformation modes for the backing plates. Ben-Dor et al. [5] proposed a slightly modified Florence's model by introducing a coefficient in order to increase the accuracy of the

[4] Corresponding author, Tel.: +86 27 87651820; Fax: +86 27 87860863
E-mail: liulish@whut.edu.cn (L S Liu)

predictions.

Armor design based on the developed impact analysis models has received increasing attention during the past three decades. As Florence's model demonstrated satisfactory agreement with experimental data [6-8], it has been widely used as a guide for the optimum design of two-component armors. Based on that model, Hetherington [9] developed an equation to obtain the optimum thickness ratio of the front and backing plates, which provides the best protection for a given aerial density. Wang and Lu [10] proposed a design criterion for calculating the optimum thickness ratio in a way that provides for optimum performance of two-component ceramic armor under a given total armor thickness. Ben-Dor et al. [5] presented an optimum design solution for the two-component armor in terms of dimensionless variables, whereby all the characteristics of the projectile and the armor are expressed as a function of two.

In this study, we used the LS-DYNA FE code to simulate the bullet penetration against the target plate with different ceramic-steel ratio of thickness. The main stages of the bullet penetration and damage contours of the target were studied by analyzing the residual velocity-time curves. We also studied energy absorption of the ceramic/metal target. In this paper, the ceramic AlN, the metal target Al and the bullet steel are selected for this investigation.

2. Numerical model

The finite element model and material parameters in this numerical analysis are describes in this section.

2.1. Finite element model

In the LS-DYNA FE code, the 2D, axi-symmetric, Lagrangian, dynamic-explicit and non-linear analysis has been utilized. The sketch for the projectile, ceramic and metal layers are shown in Figure 1. A 2D axi-symmetric quadratic shell element with four nodes with y axes of symmetric has been used. The projectile and ceramic brick elements size is 0.5mm×0.5mm and the every layer of metal material is 0.5mm×0.2mm element size. The FE model considers realistic boundary conditions of the system during perforation. A zero values are imposed to the displacement in the z-direction and rotations in x and y direction for all the elements, also the edges of ceramic and metal layers is fixed. The initial impact velocity of projectile is considered as initial condition.

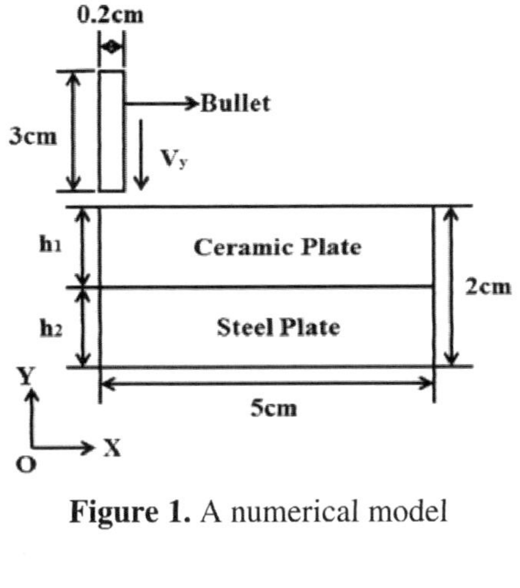

Figure 1. A numerical model

2.2. Material parameters

The brittle failure model proposed by the Johnson-Holmquist is well suited to the numerical constitutive modeling of brittle failure in ceramic material. The ceramic material is subjected to loading conditions that include high pressures, high strain rates and large deformations.

Table1. JH2 parameters of ceramic (AlN)

Density	Elastic Constants		Damage Constants		Equation of State			
ρ (kg/m³)	G(GPa)	K(GPa)	D1	D2	K1(GPa)	K2(GP)	K3(GPa)	β
3226	127	201	0.02	1.85	201	260	0	1.0

Strength Constants

HEL (GPa)	σ_{HEL} (GPa)	P_{HEL} (GPa)	T	T*	A	B	C	N	M	σ*fmax
9.0	6.0	5.0	0.32	0.064	0.85	0.31	0.013	0.29	0.21	N/A

The constitutive model is based upon a GRMNEISEN equation, which evaluates the current state of pressure as a function of the volumetric change. The intact and fractured ceramic material strengths are evaluated as nonlinear functions of normalized pressure (P*), tensile strength (T*) and the normalized total incremental strain rate. The AlN material parameters are listed in table 1.

The Johnson-Cook material behaviour equation, which considers the effects of work hardening, strain rate and temperature in mechanical behaviour of projectile, has been used for metal targets (table 2). To characterize the plastic material behaviour at high pressures, typical for highly dynamic processes, the relation among the hydrostatic pressure, the local density (or specific volume), and local specific energy has been used. This relation is known as equation of state (EOS). The most commonly used reference curve to establish the Mie-Gruneisen EOS for solid materials is the shock Hugoniot and is defined in the LS-DYNA code (table 3).

Table 2. JC parameters of steel

Density ρ(g/cm³)		Shear modulus G(100GPa)				D5	C2/P
7.80		0.77				0	0
A	B	N	C	M	TM	TR	EPSO
0.00792	0.0051	0.26	0.014	1.03	1793	294	1.0E-6
CP	PC	SPALL	IT	D1	D2	D3	D4
4.77E-6	-9	3.0	0	3.0	0	0	0

Table 3. EOS_GRMNEISEN parameters of steel

C	S1	S2	S3	GAMAO	A	EO	VO
0.4569	1.49	0	0	2.17	0.46	0	1.0

The *MAT_PLASTIC_KINEMATIC, which is suited to model isotropic and kinematic hardening plasticity with the option including rate effects, is used for projectile (listed in table 4).

Table 4. Material properties of bullets

Density ρ(g/cm³)	Elastic modulus E(GPa)	Poisson ratio ν	Yield stress σ_S(GPa)	Tangent modulus σ_t(GPa)	Beta β
7.65	200	0.29	1.6	80	1.0

3. Result and discussion

3.1. Bullet penetration and damage contour

To study the main stages in bullet penetration, we consider several different kinds of models and find a suitable model which contains integer stages of penetration and considered finally. Figure 2 shows residual velocity-time curve in the process of penetration when the steel layer sets as 0.4cm and the ceramic layer 1.6cm. The curve shows that the process of bullet penetrating contains four stages. Stage one corresponds to line OA in Figure 3, and maintains $1\mu s$. Stage one is similar with ditching in theory of penetration; Stage two (line AB) is the start of form of cone, in which damage cone is constructed and the target deformation resist penetration. Stage three (line BC) is the process of bullet penetrating cone and target. Stage four (line CD) is the process of bullet penetrating over ceramic layer and move on to deformed metal layer up to the end. Stage one is transient and ignored for raising anti-penetration behavior of ceramic and steel, while stage two is significant in armor design.

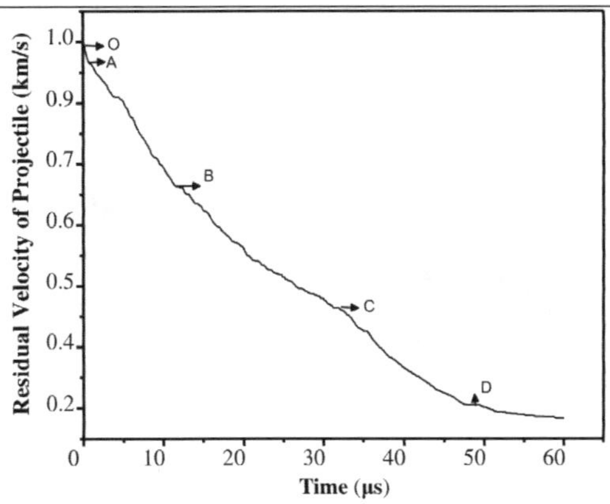

Figure 2. Residual velocity-time curve in the process of penetration when the steel layer sets as 0.4cm and the ceramic layer 1.6cm

The main stages can also be explored in damage contours of target. In figure 3, we choose the same model as figure 2 to explore the main stages by simulate the damage contour of target. Figure 3(a) corresponds to line OA in figure 2 and Figure 3(b) line AB. Figure 3(c) corresponds to the start of line BC, while figure 3(d) the end of line BC. Figure 3(e) corresponds to CD. When damage cone forms well (Figure 5(c)), the bullet is far from the conjunction place of ceramic layer and steel layer. But since the acceleration is declining, ceramic works at the stage of OA and AB in figure 2. The stage of ditching is limited in time. Hence the increment of ceramic will not work at contact time between ceramic and bullet. It will work to raise thickness of ceramic in improving the anti-penetration behaviour. But the increment of ceramic is will be a part of damage cone.

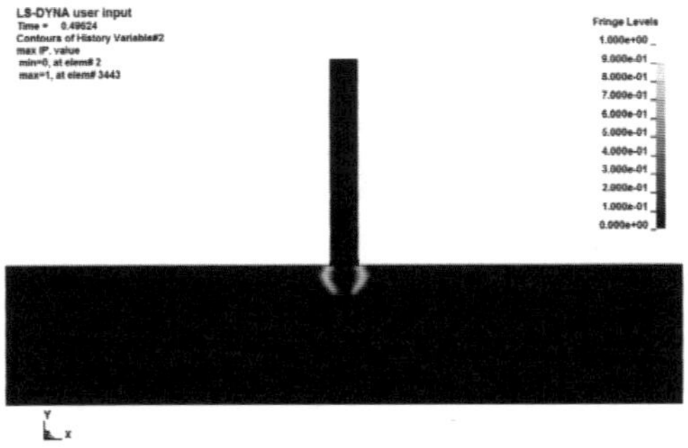

(a)

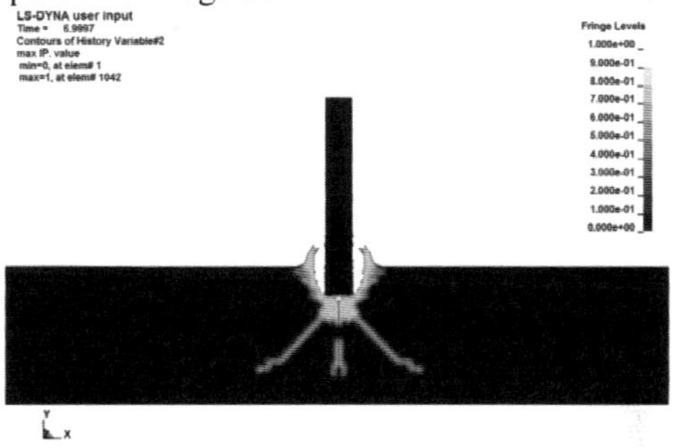

(b)

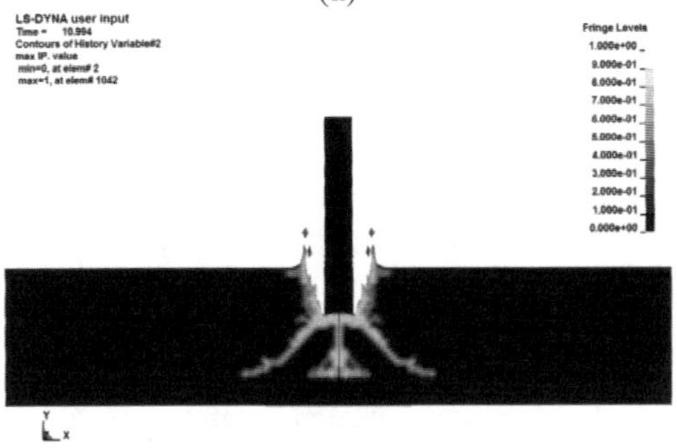

(c)

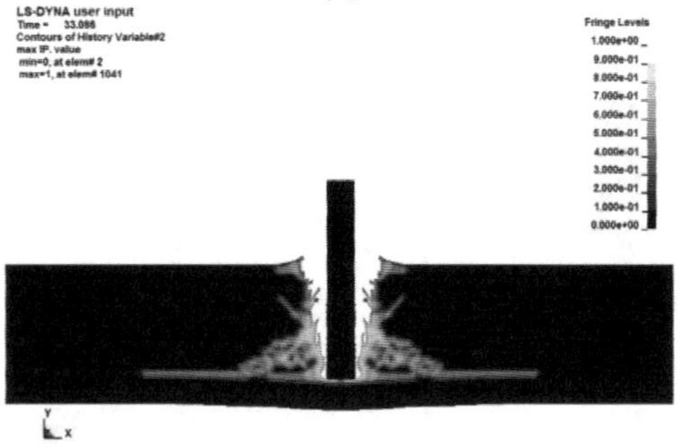

(d)

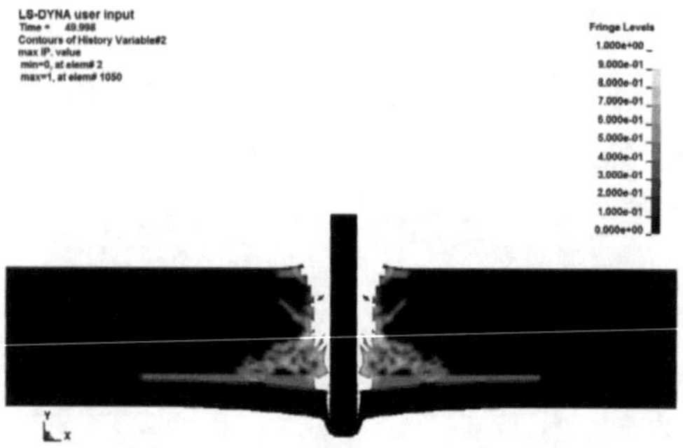

(e)

Figure 3. Damage of target which has a 0.4cm steel layer and a 1.6cm ceramic layer in five moments: (a) 0.4μs, (b) 7μs, (c) 11μs, (d) 33μs, (e) 50μs

3.2. Energy absorption of target with different thickness ratios

Energy absorption of target is studied in the constraint of total layers with a 2cm thickness. The layer of ceramic and metal varies. In figure 4, curves of energy absorption-time in ceramic slab thickness ratios set 0.2, 0.4 and 0.6 separately when the speed of bullets set as 1000 m/s. Thickness ratio is a ratio between the thickness of a ceramic layer and a metal layer. Three curves in figure 4 are almost the same in variation. At first, the energy is linearly increasing with time until it reaches stable. Model with ratio of 0.2 has highest absorption energy at last and has longest effective time in all of them.

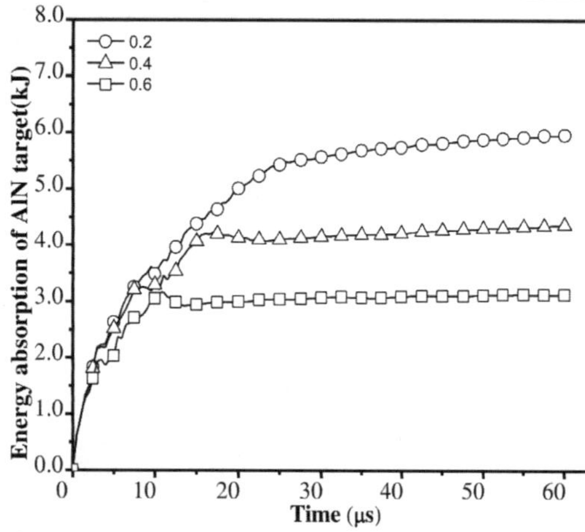

Figure 4. Curves of energy absorption-time when thickness ratios set 0.2, 0.4 and 0.6 separately.

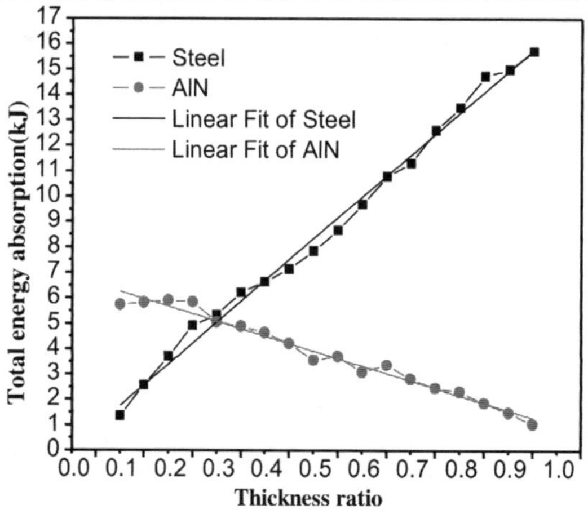

Figure 5. Relation of a layer of ceramic and steel in different thickness ratio.

Figure 5 shows the compare among models with various thickness ratios in total energy absorption with a linear fit. It shows that with the increase of thickness ratio, the total energy absorption of AlN goes low, while that of steel target increases. According to the slope of fitted curves, the layer of ceramic in total energy absorption is 0.25kJ/cm, while the lay of metal in that is 1kJ/cm. In few words, when the whole thickness is fixed, steel per cm performs better in anti-penetration behavior than ceramic per cm.

4. Summary

In the article, penetrations of bullet and damage features are studied using simulation software LS-DYNA. We also investigate anti-penetration of composite armor when thickness is constant and ratios of thickness vary. Conclusions are as follow:

Considering curves of residual velocity-time, we reckon the process of penetration contains four stages. Ceramic performs good resistance before the formation of damage cone of ceramic. But after the damage cone has formed, the anti-penetration behavior keeps declining. When the bullet starts to penetrate the layer of metal, the anti-penetration behavior of target rises slightly.

Compared with thickness ratio of 0.4 and 0.6, ceramic with 0.2 absorbs more energy and works longer.

With several different thicknesses, layers of ceramic and steel have been studied. Steel per cm absorbs more energy than ceramic per cm.

Acknowledgements
The work was supported by Fundamental Research Funds for the Central Universities.

Reference

[1] Wilkins ML 1978 Mechanics of penetration and perforation *Int J EngSci* **16** 793–807
[2] Florence AL 1969 *Interaction of projectiles and composite armour Part II* Standford Research Institute Menlo Park California AMMRC-CR-69-15 August
[3] Woodward RL 1989 *A basis for modelling ceramic composite armour defeat* Material research laboratory DSTO Ascot Vale Victoria,Australia MRL-RR-3-89
[4] Den Reijer PC 1991 *Impact on ceramic faced armour* Ph.D. thesis Delft Univ Tech Delft The Netherlands
[5] Ben-Dor G, Dubinsky A, Elperin T and Frage N 2000 Optimisation of two-component ceramic armor for a given impact velocity *Theor Appl Fract Mech* **33** 185–190
[6] Prior AM 1988 July *The ballistic impact of small calibre ammunition on ceramic composite armour* Ph.D. thesis RMCS Shrivenham
[7] Gagne MP 1989 *The penetration mechanics of small arms projectiles in ceramic-faced vehicle armours* 16 MVT Course RMCS Shrivenham UK
[8] Rajagopalan BP 1989 *The experimental validation of an analytical model for use in composite armour design* 17 MVT Course RMCS Shrivenham UK
[9] Hetherington JG 1992 The optimization of two-component composite armors *Int J Impact Eng***12** 409–414
[10] Wang B and Lu G 1996 On the optimization of two-component plates against ballistic impact *J Mater Process Tech* **57** 141–145

Journal of Physics: Conference Series **419** (2013) 012055 doi:10.1088/1742-6596/419/1/012055

Stress Wave attenuation in SiC_{3D}/Al Composite

YUAN Chunyuan[1,2], WANG Yangwei[1,2], LI Guoju[1,2], Zhang Xu[1,2], Gao Jubin[1,2]

[1] School of Materials Science and Engineering, Beijing Institute of Technology, 100081, Beijing, China,
[2] National Key Laboratory of Science and Technology on Materials under Shock and Impact, Beijing, 100081, China

*Corresponding author e-mail: wangyangwei@bit.edu.cn

Abstract. SiC_{3D}/Al composite is a kind of special composite with interpenetrating network microstructure. The attenuation properties of stress wave propagation along the SiC_{3D}/Al composite are studied by a Split Hopkinson Pressure Bar system & FEM simulations, and the attenuation mechanism is discussed in this paper. Results show that the attenuation rate of the stress wave in the composite is up to 1.73MPa • mm^{-1}. The reduction of the amplitude of waves is caused by that plenty of interfaces between SiC and Al within the composite acting with stress waves. When the incident plane wave reaches the SiC_{3D}/Al interface, reflection wave and transmission wave propagates in different directions along the irregular interface between SiC phase and aluminium phase due to the impedance mismatch of them, which leads to the divergence of stress wave. At the same time, some stress micro-focuses occurs in the aluminium phase for the complex wave superimposition, and some plastic deformation may take place within such micro-regions, which results in the consumption of stress wave energy. In conclusion, the stress wave attenuation is derived from divergence and consumption of stress wave.

1. Introduction

SiC_{3D}/Al composite is a kind of special composite with interpenetrating network microstructure. For well integration of the properties of SiC and aluminum, it has better impact resistance performance in comparison with high hardness SiC and good plasticity of aluminum, exhibiting very attracting application potential in the field of high-speed brake pads.

The better impact resistance performance of SiC_{3D}/Al composite depends closely on the stress wave attenuation in the composite under impact load [1, 2].Previous studies have shown that mechanical properties of material and its microscopic structure affect the stress wave propagation in it, and the divergence of stress wave or absorption of stress wave energy will occur when stress wave spreads through the material. There are a large number of experiments about stress wave propagation in the rock and results demonstrate that stress wave attenuates when it spreads through the rock [3~5]. Nail et al [6] found that when the planar longitudinal stress wave encounters an interface between the grain and the grain boundary, reflection and transmission of the incident stress wave take place because of the impedance mismatch at the interface, clearly this leads to attenuation of stress wave intensity. And Lagoudas et al [7] found that NiTi SMA can transform or absorb the stress wave energy in the form of phase transformation and/or detwinning.

There are a great many interfaces between SiC and Al in the microstructure of SiC_{3D}/Al composite, but up to now research about the stress wave propagation in such material is relative rare. In this paper, interaction between the stress wave propagation and microstructure of SiC_{3D}/Al composite was studied, and the mechanism of stress wave attenuation was discussed, which laid a solid foundation for the further application of SiC_{3D}/Al composite on impact resistance.

2. Experiment and Results

Stress wave propagation is studied by Split Hopkinson Pressure Bar system, and the data is recorded by the strain gauges [3~9]. The diameter of the SHPB system is 14.5mm and the experimental apparatus is shown in Fig. 1. In order to avoid superposition of the incident pulse and the reflected wave at the interface between specimen and the transmission bar, the distance from the last strain gauge to the transmission rod should be enough for stress wave reflection. And because the length of incident pulse is about twice of the striker bar length, for reducing the total length of the sample, 37mm short striker bar is used in the experiment and three short composite bars with 70 mm are connected one by one along the axis of wave conducted bar, and the final length of specimen is 210mm. Three strain gauges were placed at the position of 20 mm, 50 mm and 80mm from the impact end of the specimen respectively, and strains was real-time monitored at different positions during the wave propagation. The intensity of incident wave is low enough in order to make the specimen deformed in elastic stage during the testing.

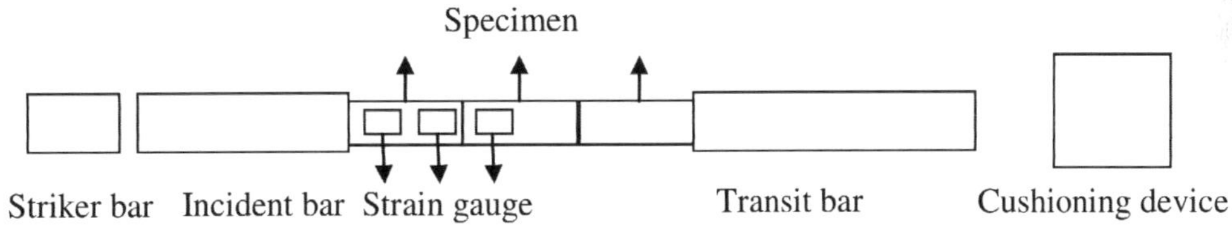

Fig. 1. The apparatus of testing the stress wave propagation

The curves of strain signals at different locations along SiC_{3D}/Al composite are shown in Fig. 2. The stress wave amplitudes in SiC_{3D}/Al composite decreased gradually with the position of strain gauges away from the impact point. And the attenuation rate of the stress wave in the composite is up to 1.73MPa • mm^{-1}, which is twice than the rate of the stress wave attenuation in the SiC_p/Al composite, it's only 0.81MPa • mm^{-1} [10]. The area of the curves enclosed is the momentum, and the attenuation rate of the momentum in SiC_{3D}/Al composite is 0.02MP • ms • mm^{-1}.

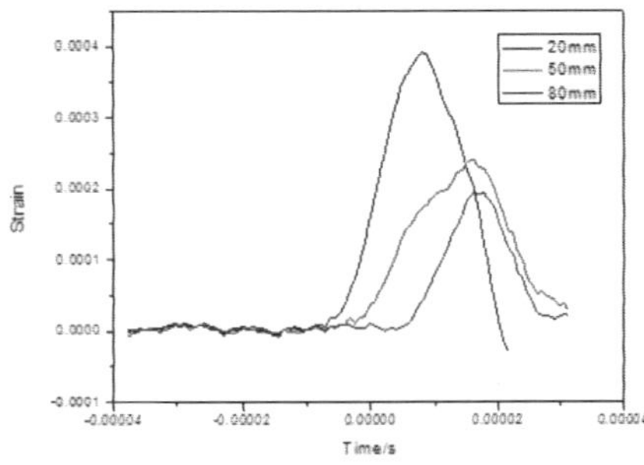

Fig. 2. Strain-time curves in SiC_{3D}/Al composite at different location

3. Numerical studies and Results

The physical experiments about stress wave propagation show that the stress wave attenuates when it passes the SiC_{3D}/Al composite. Obviously this phenomenon relates to the microstructure of the composite greatly. However, the interaction between the stress wave and complex micro interfaces has not been formulated clearly. So in this section, the numerical simulation, based on microstructure technology, is used to study the stress wave propagation to discuss the mechanism of stress wave attenuation in the SiC_{3D}/Al composite.

Fig.3 is the diagrammatic sketch of the model and loading pulse. To avoid the influence of reflection wave at the end of the model, the total solving time is set as 0.02μs, where T=L/C, L is length of the model, C is the wave velocity of SiC_{3D}/Al composite.

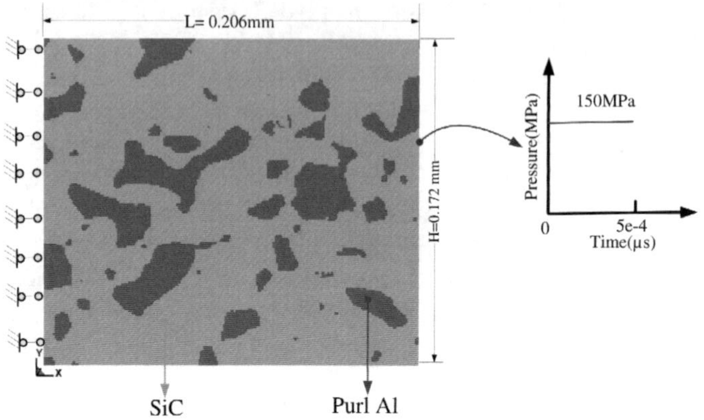

Fig. 3. Finite element model and the loading method

Plastic-Kinematic model was used for Aluminium phase of the SiC_{3D}/Al composites and Material properties used in the model are shown in Table 1.

Table 1. Purl Al Plastic-Kinematic Model

Density (kg/m^3)	Modulus of elasticity(GPa)	Poisson's ratio	Yield strength (MPa)	Bulk modulus (MPa)	Hardening coefficient
2.7×10^3	70	0.36	70	700	1

And JH2 model was used for the SiC_{3D} Ceramic phase. Material properties for SiC_{3D} Ceramic phase are shown in Table 2.

Table 2. SiC JH2 Model

A	B	C	N	M	Beta	HEL (GPa)	HEL Pressure (GPa)
0.96	0.35	0	0.65	1.0	1.0	14.567	5.9
D1	D2	K1 (GPa)	K2 (GPa)	K3 (GPa)	Tensile Strength(GPa)		Normalized Fracture Strength
0.48	0.48	204.785	0	0	0.37		0.8

Three elements were picked out along the loading direction corresponding to the positions of the strain gauges in experiment. The numbers of the elements are 12210, 12244 and 12289 respectively and there are aluminium phase between them as shown in Fig. 4.

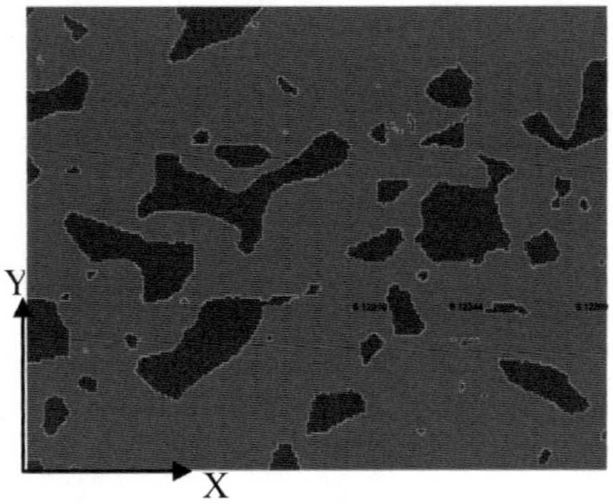

Fig. 4. The positions of the selected elements in the model

Effective stress of three selected elements is shown in Fig. 5. It can be seen that, with the same Fringe Levels, effective stress of the first element is higher than the second and third element which corresponding to the middle and lower levels in Fringe Levels respectively. It means that the stress wave attenuated gradually from the first element to third one. At the same time, stress-time curves in the three elements are shown in Fig.6, as can be seen that the amplitudes of stress wave decreased significantly too.

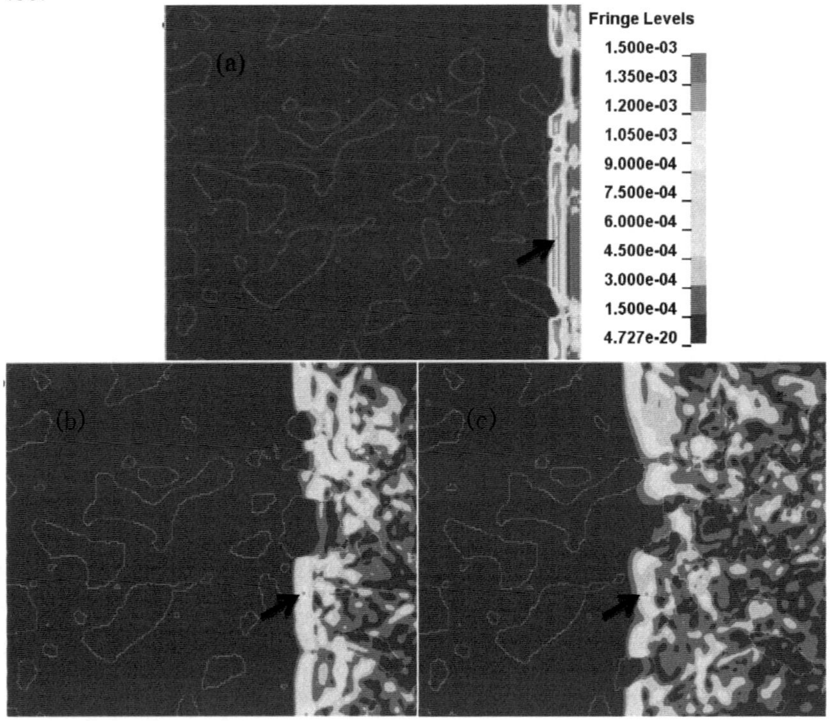

Fig. 5. The effective stress cloud charts of the three moments: (a) the stress wave reaches to position of the first element(b) the stress wave reaches to position of the second element(c) the stress wave reaches to position of the third element

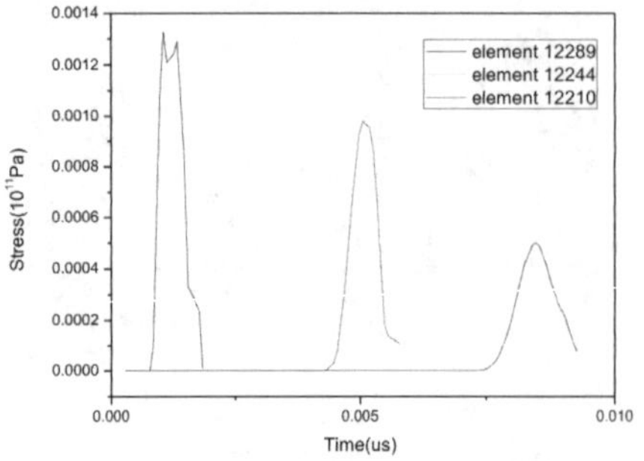

Fig. 6. Stress-time curves in the three elements

The reason for the stress wave attenuation in SiC_{3D}/Al composite is that, for the impedance mismatch between SiC phase and aluminum phase, stress wave reflects and transmits in different directions along the irregular interface, which leads to the divergence of stress wave. As shown in Fig. 7(a), with the same Fringe Levels, effective stress on the front edge of stress wave pulse is higher than the other parts and when the stress wave propagates through the interface effective stress decreased gradually. It means that the stress wave intensity reduced when the stress wave passed through the interfaces of SiC and Al.

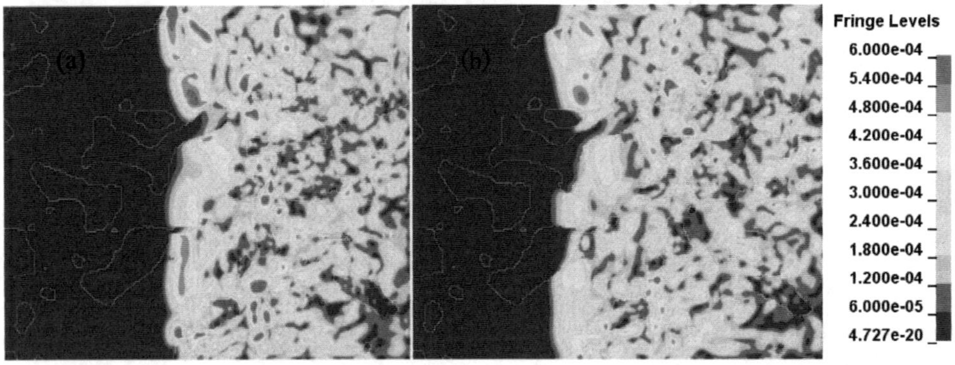

Fig. 7. Effective stress cloud charts at different moments (a) at 0.0104μs, (b) at 0.0114μs

When stress wave propagates in SiC_{3D}/Al composite, some stress concentration regions can be found in the aluminum phase which resulted from the complex wave superimposition. And some plastic deformation was seen within such regions, which resulted in more consumption of stress wave energy and enhanced the stress wave attenuation rate. To be clear, effective stress of aluminium phase under the Fringe Levels in which the maximum stress is about 70MPa was shown in Fig. 8. It can be seen that stress of aluminium phase has reached yield strength in the red regions. But such area is too small, which is less than one percent of the overall aluminum phase, so the plastic deformation just consumed some stress wave energy, not be enough for macroscopic damage in SiC_{3D}/Al composite.

In summary, when stress wave passes through SiC_{3D}/Al composite and encounters an internal interface, for the impedance mismatch between SiC phase and aluminum phase, reflection and transmission of the stress wave take place along the irregular interface in different directions, which leads to the divergence of stress wave. At the same time, plastic deformation can be found in some aluminum phase, part of the stress wave energy was absorbed during this process. So the farther the stress wave propagates, the lower the stress wave amplitude is.

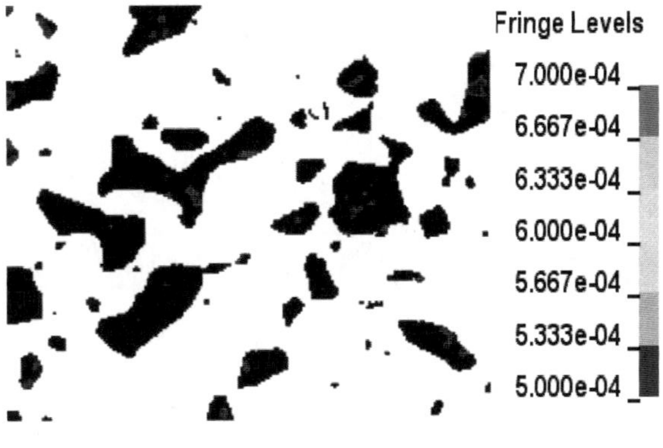

Fig. 8. Effective stress cloud charts of aluminium phase

4. Conclusions

From above discussion we can draw the conclusion that stress wave attenuation in SiC_{3D}/Al comes from the divergence and consumption of stress wave by the irregular internal micro-SiC_{3D}/Al interfaces and slight plastic deformation of aluminium phase within the stress concentration region.

References

[1]Ming Cheng, Clint Hedge, Jean-Philippe Dionne, etal. Ballistic resistance enhancement by adjusting stress wave propagation paths. Internal Journal of Solids and Structures, 2004, 12(4): 2-23.
[2]Claudio Pecorari. Attenuation and dispersion of Rayleigh waves propagation on a cracked surface: an effective field approach . Journal of Materials Processing Technology, 2000, 38: 754-760.
[3]Yang Ju, Les Sudak, Heping Xie. Study on stress wave propagation in fractured rocks with fractal joint surfaces. Iternational Journal of Solids and Structure, 2007, 44: 4256-4271.
[4]J.C.Li, G.W.Ma. Experimental study of stress wave propagation across a filled rock joint. International Journal of Rock Mechanics and Mining Sciences, 2009, 46: 471-478.
[5]M.M. Dehghan Banadaki, B. Mohanty. Numerical simulation of stress wave induced fractures in rock. International Journal of Impact Engineering, 2012, 40: 16-25.
[6]N. K. Naik, Rahul Goel, Mandar D. Kulkarni. Stress wave attenuation in ceramic plates. Juurnal of Applied Physics, 2008, 103: 1-6.
[7]D.C.Lagoudas etal. Dynamic loading of polycrystalline shape memory alloy rods. Mechanics of Materials, 2003, 35:689-716.
[8]Yangwei Wang,Fuchi Wang,Xiaodong Yu,etal. Effect of interlayer on stress wave propagation in CMC/RHA multi-layered Structure. Composites Science and Technology, 2010, 70: 1669-1673.
[9]A. Tasdemirci, I. W. Hall, B.A. Gama. Stress wave Propagation Effects in Two and Three-layered Composite Materials. Journal of Composite Materials, 2004,38: 995-1009.
[10]Xiaopeng Kang. Study on the rule of propagation for stress wave in ceramic-metal composites [the dissertation of the degree master in materials science and engineering]. Beijing: Beijing Institute of Technology, 2005.

The Photocatalytic Activity of Nano-TiO_2 (anatase) Gradient Coating on Tourmaline Beads

Huang Fengping, Fan Yingge and Sun Jingjing

Key Laboratory of Auxiliary Chemistry & Technology for Chemical Industry, Ministry of Education. Shaanxi University of Science & Technology, Xi'an Shaanxi, 710021, PR China

e-mail: fanyg0622@163.com

Abstract. Gradient nano-TiO_2(anatase) /tourmaline beads were prepared by sol impregnated method, and the phase composition and microstructure of gradient coating were characterized by XRD and SEM, and the photocatalytic activity were evaluated by degradation of methyl orange(MO), compared with that of pure nano-TiO_2 (anatase), tourmaline beads. The result indicated that the MO degradation ratio by tourmaline beads is 5.8%; that by pure nano-TiO_2 (anatase) is 38.2%; while the MO degradation ratio approached 70.4% by gradient nano-TiO_2 (anatase) /tourmaline beads, much better than the other two types of catalysts. When the gradient nano-TiO_2 (anatase) /tourmaline beads are illuminated by UV irradiation, the electron–hole pairs are generated and the photogenerated carrier will be adsorbed tightly on the anode of tourmaline, owing to its strong electrostatic field, which suppresses the recombination of the electron–hole pairs, and enhances photocatalytic efficiency.

1. Introduction

With the rapid development of the global economy, the global environment is becoming increasingly serious, all kinds of poisonous and harmful material seriously pollutes people's living environment, and has attracted extensive attention all over the world. Tourmaline mineral material has natural electrical polarity, thermoelectricity, piezoelectricity, radiation wavelength 4-14 μ m infrared performance, release negative ions performance[1]. The use of natural environmental mineral tourmaline material management air or water heavy metal ions pollution has begun to research reports [2].TiO_2 by researchers' attention is due to its higher oxidation activity, good chemical stability, non-toxic to the human body, low cos, pollution-free and wide range of applications[3]. However, Titania can only operate under UV light irradiation because of its wide bandgap (Eg=3.2ev, $\lambda g < 384$ nm) [4], and the fast recombination of the photogenerated electron-hole pairs restrict its technological utility for limited applications [5] The method by doping and modified has been proven to be an efficient route to improve the photo-catalytic activity of TiO_2[6-10].

In this paper, Gradient nano-TiO_2 (anatase) /tourmaline beads prepared enhance titanium dioxide light response range. And the phase composition and microstructure of gradient coating were characterized by XRD and SEM, and the photocatalytic activity were evaluated by degradation of methyl orange.

2. Experimental procedure

At room temperature, 15ml of Ti $(OC_4H_9)_4$ and diethanol amine was mixed with 60 ml of anhydrous ethanol (solution A). Solution B consisted of 0.9mL deionized water, 12mL anhydrous ethanol in the required stoichiometry. Then solution B was added dropwise into solution A under vigorous stirring at room temperature for 0.5 h until the transparent sol was obtained. The above sol was allowed to stand for 20 h to form a viscous sol. The change in viscosity with time was monitored using Brookfield viscometer.

The tourmaline beads were degreased, cleaned thoroughly and dried in an oven at 70℃ before deposited on it. Afterwards, the tourmaline beads were impregnated into the viscous sol. The viscous TiO_2 sol adhered to the tourmaline beads was dried in open atmosphere at room temperature. A very thin and uniform film of TiO_2 was further dried in an oven at 70℃ for 3 h. Following this method the more coatings were prepared. Gradient nano-TiO_2 (anatase) /tourmaline beads were prepared.

The crystalline phase of the samples was analyzed by XRD (Shimadzu D/max – 2200 /PC). The particle morphology and microstructure of the samples were performed on SEM (Hitachi S4800).

The photocatalytic activity of each simple was evaluated by the degradation of MO in TiO_2 aqueous solution under UV irradiation. The different Gradient nano-TiO_2 (anatase)/tourmaline beads photo-catalyst were loaded into the reactor column, respectively. The solution MO (50 ml, 1×10^{-5} $mol \cdot L^{-1}$) was poured into the reactor column and kept 30 min in dark place. After that the reactor was irradiated under UV irradiation for different time intervals. Then the sample was separated and analyzed by UV-visible spectrophotometer (Lambda 950, PerkinElmer). The absorbance at 464 nm for methyl orange (absorption peak methyl orange) was measured. Photocatalytic activity performance will be indicated by the apparent degradation rate of MO, as shown in Formula (1).

$$\text{Degradation rate} = A_0 - A / A_0 \times 100\% \quad (1)$$

3. Results and discussion

3.1. Structural evaluation of the gradient coatings

The crystalline structure of the gradient coatings was evaluated by XRD. The TiO_2 calcined at different temperatures were shown in Fig.1. The peak at 25.3° corresponded to characteristic peak of anatase (101) crystal plane. The width of (101) plane diffraction peak of anatase became narrower with the calcination temperature increasing. The diffraction peak of anatase phase has appeared at 380℃, It could be shown that the gel have been crystallized into the anatase structure. At 400℃ anatase and rutile coexist to form a mixed crystal, indicating that the anatase phase have begun to transform to rutile. The diffraction peaks of rutile phase were significantly enhanced with the annealing temperature increasing. According to our knowledge, Titanium (IV) oxide exhibits three polymorphs viz., anatase, brookite and rutile. Out of the three polymorphs, the anatase phase draws wide attention because of its activity as a photocatalyst [11]. Therefore, the optimum calcining temperature must be 400℃, and the obtained product has a pure anatase phase and good crystallinity.

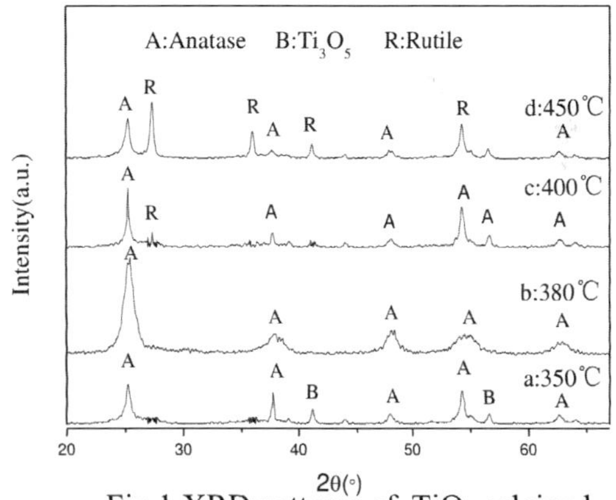

Fig.1 XRD patterns of TiO_2 calcined at different temperatures

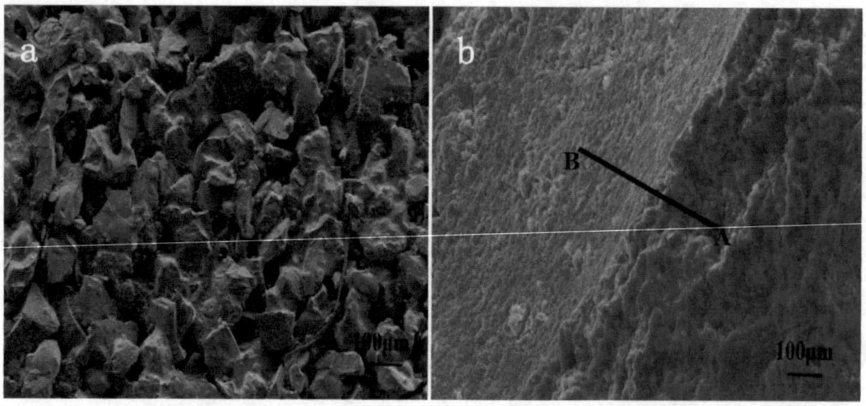

Fig. 2 displayed the SEM images of tourmaline beads (Fig.2a) and TiO_2 gradient coatings (Fig.2b) on tourmaline beads

Morphological characterization of TiO_2 gradient coatings were carried out by SEM, as shown in Fig.2. It could be seen that the gradient coatings of TiO_2 is uniform distribution and highly adherent on tourmaline beads. It could enlarge specific surface area of the catalyst films to improve its photocatalytic activity. From A to B line scan shows that titanium elements have infiltrated into the tourmaline carrier (Fig.3).

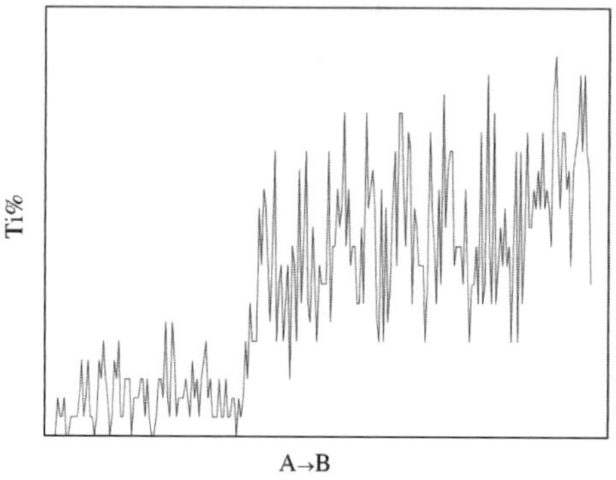

Fig.3 A→B line scan

3.2. Photocatalytic activity of Nano-TiO_2 (anatase) Gradient Coating on Tourmaline Beads

Fig.4 shows the photocatalytic degradation efficiency of methyl orange (MO). The result indicated that pure tourmaline beads have a weak effect on degradation of methyl Orange, and the MO degradation ratio is 5.8%. Under UV excitation the anatase TiO2 powders prepared by sol-gel method, have certain catalytic effect of MO. The rate of degradation of MO is 38.2% after reaction 4h. While the MO degradation ratio approached 70.4% by gradient nano-TiO_2 (anatase) /tourmaline beads, much better than the other two types of catalysts. This is because the surface electric field of tourmaline can inhibit the re-composite of TiO_2 photo-electron and hole, and the photo-electron were firmly adsorbed on the anode of tourmaline, while the photo-hole stayed the surface of TiO_2 will combine with surface hydroxyl of gradient material to form hydroxyl free radicals, and improve the photocatalytic reaction efficiency[12].

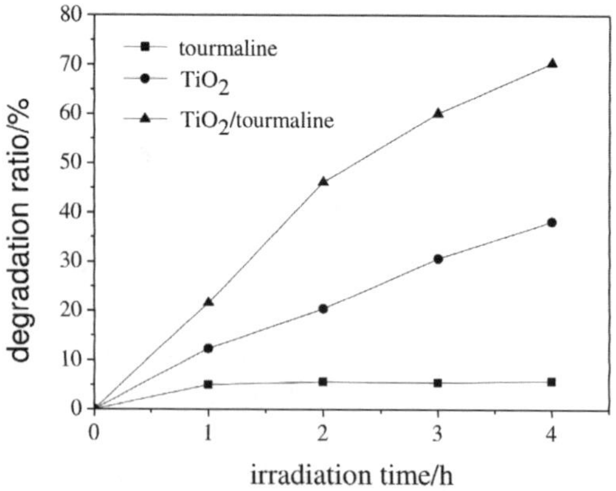

Fig.4 the photocatalytic efficiency of TiO_2/tourm aline gradient material 、TiO_2、 tourmaline

The photocatalytic activities of TiO_2 gradient material at different coating times were shown in Fig.5. With the coating time increase, the photocatalytic efficiency of gradient material can be improved. But when the coating times exceed a certain value, the catalytic effect is decreased. Photocatalytic activity is mainly related to the crystal structure of TiO_2. When the coating time is less, nanoparticle film load on tourmaline surface is not uniform, even a small amount of blank area is not covered, TiO_2 anatase structure is not perfect and has a big defect, which make the composite risk of photo-electron and hole increased, and activity of gradient material lower. When the load increases the number, the TiO_2 nano-granular layer on tourmaline surface also thicken, the activity of gradient composite catalyst rise. Load for 3 times, tourmaline surface of TiO_2 anatase crystalline phase structure has been more perfect, and then increase the amount of TiO_2 crystalline structure of the TiO_2 film has little effect. Under this load, TiO_2 nanoparticles completely cover the surface of the carrier, and even a small amount of stuffing. Photocatalytic reaction generally occurs on catalyst surface, the migration of electron-hole has a certain depth. So in actual reaction, real motivated to produce photo generated carrier only outer certain thickness of the catalyst, the inner layer of TiO_2 role is minimal. Preparation of composite photocatalyst, if the load is too many times, the coating layer is too thick, the outer layer of TiO_2 particles farther away from the tourmaline surface distance. By tourmaline surface field strength formula($Er = (2/3)\ Eo\ (a/r)^3$) known, the electric field around tourmaline abate rapidly with the increase distance between outer layer of TiO_2 and tourmaline surface, and the interaction The interaction between the two weakened, tourmaline not achieve the role of the electron-trapping agent, the activity is decreased. Therefore optimums coating time was three times.

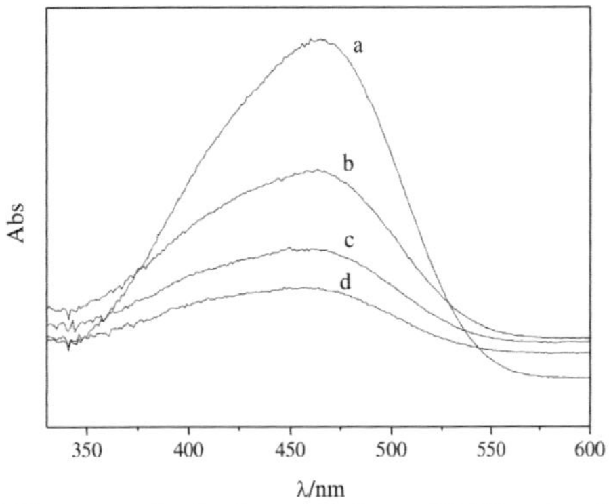

Fig.5 Photocatalytic degradation of MO using TiO_2 gradient material at one time (b), three times (d), five times(c) and (a) initial concentration of methyl orange.

4. Conclusions

We reported the preparation of gradient nano-TiO_2 (anatase) /tourmaline beads by sol impregnated method. The photocatalytic activity was determined by degradation of MO under UV light. Gradient materials have significant effect on enhancing the photocatalytic efficiency than pure nano-TiO_2 (anatase), tourmaline beads. In this work, it was found that the MO degradation ratio approached 70.4% by gradient nano-TiO_2 (anatase) /tourmaline beads. The surface electric field of tourmaline can inhibit the re-composite of TiO_2 photo-electron and hole, and the photo-electron were firmly adsorbed on the anode of tourmaline, while the photo-hole stayed the surface of TiO_2 will combine with surface hydroxyl of gradient material to form hydroxyl free radicals, and improve the photocatalytic reaction efficiency.

Acknowledgements
The work was supported by the Doctoral Research Fouds of Shaanxi University of Science and Technology (BJ0903).

References

[1] Dong F-A, He D-L and Yuan Ch-L 2005 Journal of Functional Materials **10** 1485

[2] Zhang X-J, Wu R-H and Tang Y-H 2004 China Non-metallic Mining Industry Herald **3** 39
[3] Ji H-W, Ma W-H and Huang Y-P 2003 Chinese Science Bulletin **21** 2199
[4] Xue W, Zhang G and Xu X 2011 Chem. Eng. J **167** 397
[5] Ch S-Zh, Zhang P-Y and Zhu W-P 2004 Progress In Chemistry **4** 613
[6] Yuan Ch-L and Dong F-Q 2006 Acta Minalogica Sinica **2** 233
[7] Ren M, Zhang Y-J, Liu S-W, Ma X and Xiu Zh-L 2006 Journal of Ji Nan University(Science and Technology) **4** 15
[8] Peng F, Huang L and Chen Sh-H 2006 Modern Chemical Industry **2** 18
[9] Zhang X-H, Wu R-H and Ma H-W 2006 Non-Metallic Mines **3** 325
[10] Wu H-J, Wu M, Xie M-S , Liu H, Yang M and Sun F-X 2000 Chinese Journal of Catalysis **5** 399
[11] L. Jong, H.N. Wooseok and K. Misook 2003 Appl. Catal. A Gen **244** 49
[12] Feng Y-W, Liang J-Sh, Liang G-Ch, Ding Y and Wang J 2003 Journal of Synthetic Crystals **6** 569

Research on self-propagating high temperature synthesis prepared ZrC-ZrB_2 composite ceramic

Cheng Yong[1], Su Xunjia[1], Hou Genliang[1], Xing YaKun[2]
The Second Artillery Engineering College, Xi'an 710025, China;
Chong Qing Communication Constitution)

E-mail: wenzibinbin2009@126.com

Abstract: ZrC-ZrB_2 composite ceramic material is prepared by self-propagating high temperature synthesis, using Zr powders, CrO_2 powders and Al powders as raw materials. Samples are studied by XRD and SEM, the results show that: ZrC-ZrB_2 composite ceramic is attained after self-propagating high-temperature reaction, with Zr+ B_4C as the main reactive system, and which is added respectively different content (CrO_3 + Al) system. The study finds that the ceramic composite products are mainly composed of ZrC and ZrB_2 phase, and other subphase. Compared to the main reactive system composite ceramic, composite ceramic grains grow up obviously, after introduction of the highly exothermic system (CrO_3 + Al) in the main reactive system, and with the gradual increase of the content (CrO_3 + Al).
Key words: SHS; ZrC-ZrB_2; composite ceramic.

Self-propagating high-temperature synthesis (SHS), also known as combustion synthesis, is a way to use reactant own adiabatic reaction heat released by situ synthesis of new materials. It has the outstanding advantages with simple process, fast synthesis, less energy consumption [1-3]. In recent years, the self-propagating high-temperature synthesis technology has been widely applied in material preparation with high temperature, refractory and wear resistant, such as intermetallics, ceramic and composite materials [4-6]. At present, most researches focus on carbide, boride such as Ti, W, Si, Mo, but the research on ZrC and ZrB_2 ceramic is relatively less.

ZrC possess a very high melting point (3530℃), a good ultra-temperature performance, therefore, ZrC can be used as high temperature structural material which has been applied in the aerospace field. Because of the high melting point of ZrB_2 (as high as3050℃), especially its outstanding resistance to high temperature oxidation, it is often used for the resistance of ablative materials in the environment of high temperature oxidation [7]. Carbide has the highest melting point, but its fracture toughness is very low, and it is easy to be oxidized. Boride has a good comprehensive performance, its thermal shock resistance in high temperature, low fracture toughness and creep resistance are very good [8]. In this report, Zr powders and B_4C powders are used as raw materials, ZrC-ZrB_2 composite ceramic with high melting point, excellent ablation resistance and good thermal shock performance is prepared by self-propagating high temperature synthesis.

1、 The experimental process

The test put to use B_4C powders with particle size of 3.5 μm, purity of greater than 98%, Zr powders with particle size less than 43μm, purity more than 99%, and chemical pure Al powders and CrO_3 powders with particle size of between 43~ 61μm. Zr and B_4C as the main reactive system, ingredients are respectively added to the mass fraction of the highly exothermic system 0, 12.3%, 15.5%, 19.3%, (CrO_3+Al) of which is for reaction to provide heat, and its combustion synthesis is ZrC-ZrB_2 composite ceramic. Thereinto, (Zr+B_4C) and (CrO_3+Al) system is prepared at a molar ratio, see below.

$$3Zr + B_4C \rightarrow Zr\,C + 2\,ZrB_2 \quad (1)$$
$$CrO_3 + 2Al \rightarrow Al_2O_3 + Cr + 1094\ \text{kJ} \quad (2)$$

Ceramic phase which based on Rigaku D/Max 2250 PC type X-ray diffraction analysis use Cu target, operating voltage 40 kV, operating current of 200 ma , and scanning speed of 10° / min. Observation of ceramic fracture is carried out by the characterized Sirion - 200 type field emission scanning electron microscopy (FESEM) microscopic .

2、 Results and discussion

Figure1 is the XRD patterns of the composite ceramic, due to the main reactive system (Zr+ B_4C) and the added different (CrO_3+Al) content. Figure1 (a) is the main reactive system composite ceramic XRD spectrum, and it is known that the composite ceramic is mainly composed of ZrC and ZrB_2. Figure1 (b, c, d) is the added different (CrO_3 + Al) content of the composite ceramic XRD spectrum, and it is found that after the added different (CrO_3 + Al) content, the composite ceramic is mainly composed of ZrC、 ZrB_2、 α-Al_2O_3 and $(Zr,Cr)_3B_2$. The composite ceramic compared to the main reactive system doesn't change, still ZrC and ZrB_2,but introduction of the new-product phase α-Al_2O_3 and $(Zr,Cr)_3B_2$. α-Al_2O_3 is Al restore CrO_3 to generate its product .

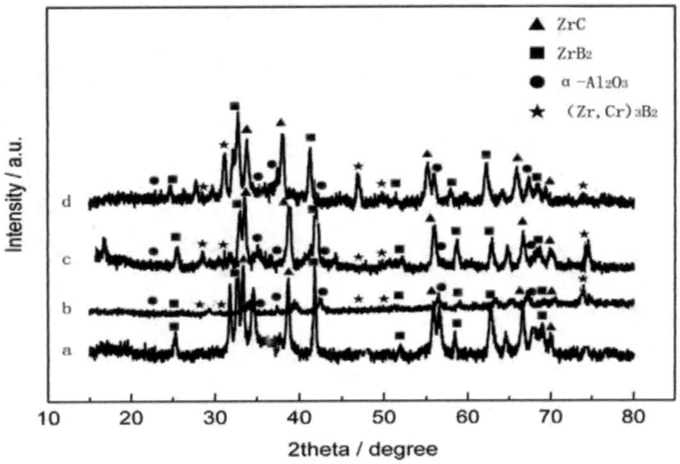

Fig.1 XRD patterns of composite ceramic with different（CrO3+Al）contents

Figure 2 shows the fracture microstructure of the combustion synthesis product. The gray tissue is the reaction synthesis of ZrC-ZrB_2 matrix, the white particles are Al_2O_3 particles. The black holes appear in Figure 2, is caused by the loosen raw powders and the volatilized trace impurities as well as the volatized gases of the raw powders in the combustion process. In the raw reaction process, if imposed pressure from the external (i.e. the SHS / Qip process), the relatively dense ZrC- ZrB_2 composites can attain . (Figure 2 (a) - (d)) can be seen, when excluding (CrO_3 + Al), ZrC-ZrB_2 grain is refinement, and shows variety of irregular sphericities. When adding a certain amount (CrO_3 + Al), ZrC-ZrB_2 particles are significantly large, while the holes aperture accordingly also becomes small. The trend of ZrC- ZrB_2 particles growing up is strengthened, when (CrO_3 + Al) content increase to 19.3% from 12.3%, which is closely related to the combustion synthesis reaction temperature within the range of the content. The formation and growth of a new phase is not only with the thermodynamic factors, but also with kinetic factors. Due to the grain growth is exponential relationship with the combustion temperature [9-10], the (CrO_3 + Al) is a highly exothermic system, increase of (CrO_3 + Al) content raises the combustion temperature of the reaction, and thus with the (CrO_3 + Al) content increases, it gradually raises the combustion temperature of the reaction, and promotes the growth of ZrC-ZrB_2, so that the grains appear more obviously. While grain growth, due to the high-calorie, cooling time accordingly becomes long, the gas generated by the reaction can be fully released, so that the holes aperture corresponding becomes smaller.

(Figure 2 (a)-(d)) can be seen, as (CrO_3+Al) content gradually increase, the composite ceramic grain is increasingly large, the holes aperture becomes small, while corresponding introduction of the more Al_2O_3 and Cr_2C impurity phases. Al_2O_3 possess the melting point (2303℃),which is lower than the ZrB_2（3050℃）,and thus the more impurity phases content will greatly affect the performance of the ZrC-ZrB_2 composite ceramic, so (CrO_3 + Al) synthesizes ZrC-ZrB_2 composite ceramic in high exothermic system at the same time try to reduce the impurity phase.

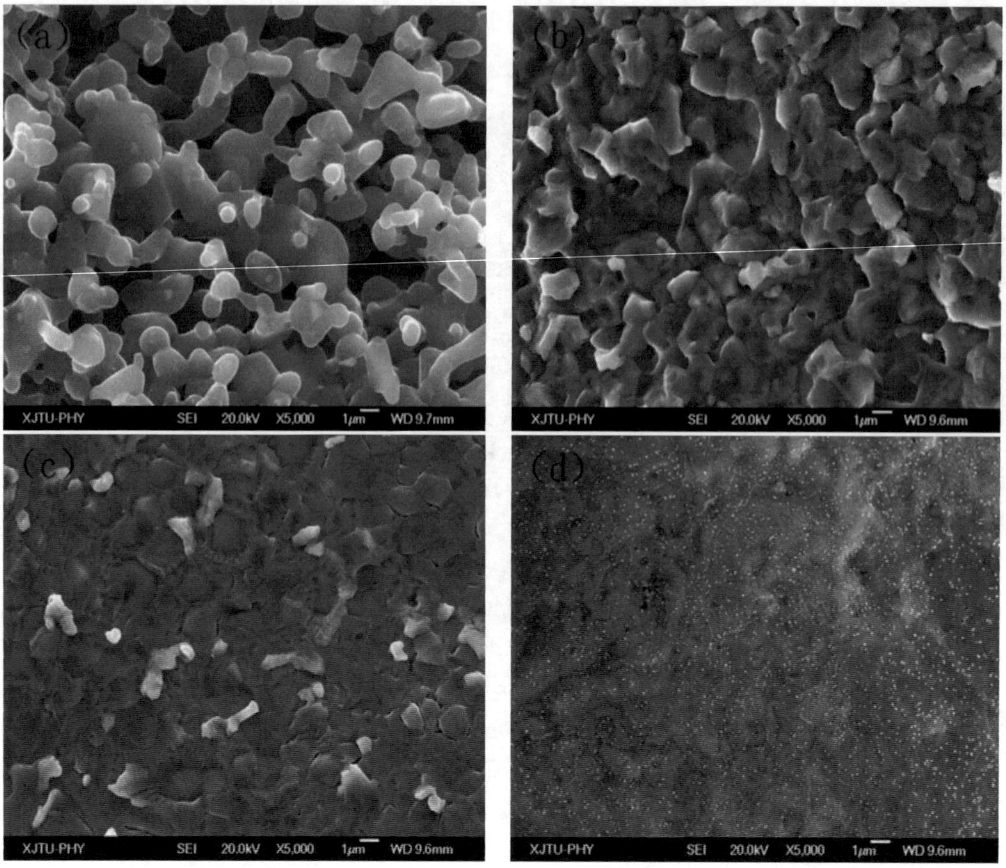

Fig.2 SEM of fracture of the self-reactive spray formed preforms with different(CrO_3+Al) content

3、 Conclusion

- Zr and B_4C powders as raw materials, according to Zr/B_4C for 3/1 (mole ratio) ingredients, it is known that ZrC-ZrB_2 composite ceramic with a high performance is attained.
- When excluding (CrO_3 + Al), ZrC-ZrB_2 grain is refinement, and it shows variety of irregular sphericities. When adding a certain amount (CrO_3 + Al), particles are significantly large, while the holes aperture accordingly also becomes small.
- As (CrO_3+Al) content gradually increase, the corresponding introduction of the more impurity phases, (CrO_3 + Al) synthesize ZrC-ZrB_2 composite ceramic in high exothermic system at the same time try to reduce the impurity phase.

References:

[1] Khanra A K, Godkhindi M M, Pathak L C. [J]. Materials Science and Engineering A,

2007:281-287.

[2] Shen Ping, Zou Binglin.[J]. Material Science and Engineering A, 2007: 300-309

[3] Lu Dajun, Quan Wanglin, An Zhanjun.[J].Journal of Functional Materials, 2006, 37(3): 456-458.

[4] Mossino P. [J]. Ceramics International, 2004, 30(3): 311-332.

[5] Yeh C L, Yeh C C. [J]. Journal of Alloys and Compounds, 2005, 388(2): 241-249.

[6] Nassaj E T, Mirhosseini S H. [J]. Journal of Materials Processing Technology, 2003, 142(2): 422-426.

[7] Jiang Danyu. [J]. Rare Metals Letters, 2006, 28(2): 22-26.

[8] Wang Yanbing. [D]. The Second Artillery Engineering College, 2007: 1-3.

[9] Wu Shengqin . [J].Journal of Materials Science and Engineering, 2004: 45

[10] Lu Dajun. [J].Journal of Functional Materials, 2006:32-33.

AUTHOR INDEX

Amigo, R C R 11
Boussaa, D 33
Cai, H 268
Cao, D F 239, 247, 259, 276, 328
Cao, M 121, 205
Chen, F 27, 61, 183
Chen, J L 222
Chen, P 51
Chen, Y 121, 158, 205
Cheng, Y 345
Cheng, Z P 75
Deng, C 312, 312
Duan, Y 188
Fan, Q 110, 132, 268
Fan, Y 340
Feng, S 254
Galaviz-Pérez, J A 61
Gao, J 334
Gao, R 110
García, J R V 61
Gong, Y P 324
Guo, D Y 324
Guo, L C 82
Guo, Q 170
He, J 188
He, Y 102
Hebert, R 19
Hou, G 345
Hu, H 165, 211
Huang, F 340
Huang, X 150
Ichigozaki, D 306
Itakura, Y 229
Jin, Y 175
Kawasaki, A 15
Kim, J-H 19
Kirihara, S 229
Kwon, H 15
Lai, X 259, 298
Lee, G-G 15
Lei, Z T 239, 247, 276
Leparoux, M 15
Li, B 86
Li, C 188
Li, F 27
Li, G 132, 334
Li, J 1, 92, 170, 306
Li, J-F 306
Li, M 97, 117
Li, N 183
Li, S 37, 165, 211
Li, Y 121, 145, 158, 205
Liang, S 188
Liang, X 46, 233
Liao, L 165
Liu, J 268
Liu, L S 239, 247, 259, 276, 298, 328
Liu, M 312
Liu, Q 145, 188
Liu, S 97, 188
Liu, W 233
Liu, X 239, 247, 276, 328
Liu, Y 51, 282
Liu, Z 158
Luo, G 46, 51, 92, 97, 117, 128, 197
Ma, Z 86, 170
Matsuura, K 218
Mei, H 239, 247, 259, 276, 298, 328
Mei, Z 75
Min, J 233
Nie, Y 170
Ohno, M 218
Park, H 65
Ping, S D 75
Qu, S 282
Ran, A 233
Rosin, A 65
Sato, H 7
Shen, Q 1, 27, 46, 51, 61, 92, 97, 117, 128, 145, 170, 183, 197, 254, 324
Shen, W 138
Shi, G P 254
Shibuya, M 7
Shukla, A 19
Silva, E C N 11
Sioh, E L 289
Su, X 345
Sun, D 175
Sun, H 37
Sun, J 340
Sun, W 55, 306
Sun, X 233
Tao, Z J 75
Tasaki, S 229
Tok, A I Y 289
Vatanabe, S L 11
Wang, C 183
Wang, C B 324
Wang, D 55, 102
Wang, F 86, 110, 132, 268
Wang, H 188
Wang, K 306
Wang, L 55, 233
Wang, M 37, 165, 211
Wang, Y 86, 117, 128, 132, 145, 175, 334
Wang, Y C 328
Wang, Z 82, 254, 259
Watanabe, Y 7
Wen, L 175
Willert-Porada, M 65

AUTHOR INDEX

Wright, J T 19
Wu, K 65
Wu, Z 132
Xing, Y 345
Xiong, X 282
Xiong, Y 46, 92
Xu, Y 318
Yan, R 222
Yang, A 197
Yang, J 121, 158, 205
Yang, S Y 276
Yang, Z 312
Yao, J 102
Ye, B 233
Ye, J 197
Yu, Q 306
Yuan, C 334
Yuan, H 1, 46, 92
Zhai, P C 259, 298
Zhang, G 211
Zhang, J 117, 128, 233
Zhang, L 1, 19, 27, 37, 46, 51, 61, 82, 92, 97, 117, 128, 150, 170, 183, 197, 211, 254, 324
Zhang, P 138
Zhang, Q 138
Zhang, R 306
Zhang, X 132, 218, 312, 334
Zhao, X 188
Zhao, Z 150
Zhou, Y 138
Zhou, Z 37, 165, 211, 318
Zhou, Z J 222
Zhu, D H 222
Zou, L 37, 211